BIOMECHANICAL ANALYSIS OF THE MUSCULOSKELETAL STRUCTURE FOR MEDICINE AND SPORTS

BIOMECHANICAL ANALYSIS OF THE MUSCULOSKELETAL STRUCTURE FOR MEDICINE AND SPORTS

A. Seireg
Mechanical Engineering Department
University of Wisconsin–Madison

Mechanical Engineering Department
University of Florida, Gainesville

R. Arvikar
Bell Laboratories, AT&T

● HEMISPHERE PUBLISHING CORPORATION
A member of the Taylor & Francis Group

New York   Washington   Philadelphia   London

BIOMECHANICAL ANALYSIS OF THE MUSCULOSKELETAL STRUCTURE FOR MEDICINE AND SPORTS

Cover design by Debra Eubanks Riffe.

1 2 3 4 5 6 7 8 9 0   E B E B   8 9 8 7 6 5 4 3 2 1 0 9

Library of Congress Cataloging in Publication Data

Seireg, Ali.
    Biomechanical analysis of the musculoskeletal structure
for medicine and sports by A. Seireg and R. Arvikar.
      p. cm.
    Includes bibliographies and index.

    1. Human mechanics. 2. Kinesiology. 3. Musculoskeletal system—
Physiology. I. Arvikar, R. II. Title.
    [DNLM: 1. Biomechanics. 2. Musculoskeletal System—anatomy &
histology.   3. Musculoskeletal System—physiology. 4. Sports
Medicine. WE 101 S461b]
QP303. S4 1989
612'.76—dc19
DNLM/DLC
ISBN 0-89116-423-5

88-24739
CIP

To Aristotle, Avicenna, and Leonardo da Vinci
whose perceptive thoughts inspired
the undertaking of this study

Contents

Preface

Interest in understanding the mechanics of the human structure has intrigued investigators since ancient times. The first known treatment of the subject is in a treatise by Aristotle (384–322 B.C.), "Parts of Animals, Movement of Animals, and Progression of Animals." The detailed analysis of muscle action as illustrated in the celebrated notebooks of Leonardo da Vinci (1452–1519) laid the foundation for the comprehensive modeling of the different muscles which contribute to the support of human posture and movement. The modern era for mathematical studies of human biomechanics was initiated by Braune and Fischer at the end of the Nineteenth Century.

During the last two decades, there has been a rapidly increasing interest in musculoskeletal biomechanics as applied to medicine and sports. This book presents detailed information on the human musculoskeletal model that was developed at the University of Wisconsin–Madison. The model represents the first known attempt at a unified comprehensive tool for the analysis of muscle and joint forces in the entire musculoskeletal system. The approach followed in this work is to model the human structure as a collection of skeletal bodies articulated through the joints and connected together by various muscles and ligaments and kept in equilibrium at all times by the tensile forces in the muscles. The lines of muscular actions are defined for any posture or configuration of movement by utilizing coordinate transformations to identify the points of origin and insertion with respect to a set of reference axes. Equilibrium equations for three-dimensional force and moment balance can be obtained for each of the skeletal segments. These equations include joint tensile reactions and joint unbalanced moments which represent the loads to be carried by the ligaments if an unbalance exists. The problem is statically indeterminate since the number of unknowns far exceeds the number of equations and is accordingly treated as a linear program optimizing a linear objective function. Numerous forms of the objective function which controls the muscle load sharing were tried. It appears from the quasi-static investigations of the electromyographic activities of the muscles that a criterion which minimizes a weighted sum of all muscle forces and the forces on the ligaments gives muscle action patterns which correlate well with the electromyographic activity patterns. In all the considered situations, the generated solutions showed no reliance on any of the ligaments to carry load with the exception of the crucial ligament of the knee joint and the ligament maintaining the arch of the foot. It is interesting to note that the total sum of muscle forces may be considered as an indication of the physiological work necessary to maintain the quasi-static equilibrium. It can be represented by the total number of muscle fibers in contraction without regard to the number of fibers available for action in any of the muscles. The total sum of muscle forces can be considered the equivalent of the total sum of muscle fibers necessary for equilibrium.

The complexity of the musculoskeletal models described in the book can render its use difficult and prone to human modeling and programming errors. The incorporation of interactive and graphical display techniques into an automated analysis system facilitates the reformation of existing models so they can be used to investigate many different conditions and complex activities. Such systems have been developed by Williams and Seireg (1977, 1979) and described in the Appendixes. This interactive procedure addresses all of the modeling and analysis phases, and simplifies many of the tedious and time-consuming tasks for the user. It provides a convenient means to modify the models and automatically formulates the system equations for any selected posture or activity. It also allows for the modeling of the entire human structure or a small segment of it. The system removes the burden of writing the equilibrium equations which can be a formidable task especially when different geometric changes are made in the anthropometric dimensions or points of muscle attachment. The graphics capability of the system can be invaluable for visually checking the adequacy of the constructed models and would allow users to perform analysis without extensive training in engineering analysis methods.

The authors are aware of a considerable number of excellent investigations on the muscle load sharing in different body segments, especially the lower extremity. Many of them are documented in numerous archival papers and chapters of books. The approaches used are diverse and consider linear, nonlinear, two-dimensional, three-dimensional, static and dynamic models for the system and the muscle behavior. We have not attempted to review these studies in this book or make comparative evaluations of the results obtained by the different approaches. This would have been a formidable task if we wanted to incorporate all the studies with meaningful evaluations. We therefore chose to concentrate on our own approach and left it to the many students and researchers in this field to perform their own evaluation on the advantages and disadvantages of the different modeling methods.

We hope that this book contributes some new data and information to the continuing quest for understanding of the forces in the musculoskeletal structure. Some of the studies reported here were supported by the National Science Foundation. The authors would also like to acknowledge the help of Mrs. Mary Stampfli and Mrs. Lorraine Loken in the preparation of the manuscript.

A. Seireg
R. Arvikar

CHAPTER I

Introduction

"Nature is written in
mathematical symbols."

Galileo
1564-1642

HISTORICAL DEVELOPMENT

The interest in the mechanics of human and animal posture and motion is probably as old as the human race. Techniques of bearing loads, pulling, pushing, kicking, hammering, throwing and other activities of daily life are likely to have been developed without conscious control by instinctive reflexes, adaptive learning and acquired skills through training.

It is not unreasonable to postulate that evolutionary experimentation took place to determine the proper posture for work and recreation. Experiences on the best ways were naturally passed on from one generation to another and utilized in everyday activities to maximize the human ability to perform work with minimum ill effects or injury to the musculoskeletal structure.

One of the earliest known records of organized analytical treatments of the subject can be found in a Treatise by Aristotle[1] (384-322 B.C.) on "Parts of Animals, Movement of Animals and Progression of Animals". After detailed descriptions of the parts of animals he concludes:

"We have now spoken severally of all animals: we have described their parts and stated the reason why each is present in them. Now that this is concluded, the next thing is to describe the various ways in which animals are generated."

He then proceeds to analyze the cause of movement and how it is achieved:

"For all animals move and are moved with some object, and so this, namely their object, is the limit of their movement. Now we see that things which move the animal are intellect, imagination, purpose, wish and appetite. Now all these can be referred to mind and desire."

".....For example when you conceive that a person ought to walk and you yourself are a person you immediately walk."

> ".....""Now all animals clearly both possess an innate spirit
> and exercise their strength in virtue of it."

And finally through remarkable insight he explains the complex control of
motion by an analogy to a well governed state:

> ".....We have now stated what is the part by the movement of
> which the soul creates movement and for what reason. The
> constitution of an animal must be regarded as resembling that
> of a well-governed city-state. For when order is once
> established in a city there is no need of a special ruler
> with arbitrary powers to be present at every activity, but
> each individual performs his own task as he is ordered, and
> one act succeeds another because of custom. And in the
> animals the same process goes on because of nature and
> because each part of them, since they are so constituted, is
> naturally suited to perform its own function, so that there
> is no need of soul in each part, but since it is situated in
> a central origin of authority over the body, the other parts
> live by their structural attachment to it and perform their
> own functions in the course of nature."

Acknowledged as the undisputed master of reason long after his death, his
teachings dominated the field until Leonardo da Vinci's time (1452-
519). Da Vinci's works[3-5] on the mechanics of the human body and his
detailed anatomical sketches represent the true birth of anatomy as a
discipline and mechanics as the science governing human and animal motion
(Fig. 1). Mechanics as an applied mathematical science was held in the
highest esteem by Leonardo and he used it to explain many biological
phenomena.

Biomechanics was undoubtedly the subject of his most detailed
attention. This is an area where he as an inspired mechanician can grasp
and analyze with considerable detail. His absolute faith that movement
in animals follows mechanical laws is evident in his writing "Why nature
cannot give the power of movement to animals without mechanical
instruments as is shown by me in this book on the works of movement which
nature has created in animals. And for this reason, I have drawn up the
rules of the four powers of nature without which nothing through her can
give local movements to these animals."

The four powers presented by Leonardo are movement, weight, force
and percussion. This concept of mechanics is in part Aristotelian and in
part of Leonardo's own making. He used it well to rationalize mechanical
body movement. Even with his handicaps as a fifteenth century
mechanician, he was determined, to understand the mechanism of animal and
human movement and the forces producing them: **"After the demonstration
of all parts of the limbs of men and of the other animals, you will
represent the proper method of action of these limbs that is in rising
after laying down, in moving, running and jumping in various attitudes,
in lifting and carrying heavy weights, in throwing things to a distance
and in swimming, and in every act you will show which limbs and which
muscles are the causes of the said actions and especially in the play of
the arms."**

His interest in the muscles lay primarily in their action. His
treatment was similar to that of Avicenna[2] (980-1037) though with greater
detail and with the addition of excellent illustration. The pattern of

Postures and movements

Figure 1-1. Leonardo daVinci's sketches for human posture and muscle action.

muscle play in body movement interested him as artist, anatomist and mechanician.

> ".....You will make the rule and the measurement of each muscle and you will give the reason of all their function and the manner in which they use them."

> ".....Every muscle uses its power along the line of its length."

> ".....The muscles always begin and end in the bones that touch one another and they never begin and end in the same bone for it would not be able to move anything unless this was itself in a state of rarity or density."

> ".....In all parts where man has to work with greater effort, nature has made the muscles and tendons of greater thickness and breadth."

> ".....It is the function of the muscles to pull and not to push except in the cases of the genital member and the tongue."

> ".....No muscle uses its power in contracting but always in drawing to itself the parts conjoined by it."

He developed muscle dynamometers to study its force. He also realized that muscles work in pairs:

> ".....In fact, there may be found as many muscles as there are movements in the lips and as many more to counteract these movements; and these I intend to describe and figure in full, proving the movements by my mathematical principles."

Although Leonardo wrote extensively on body mechanics, the man generally credited to be the father of modern biomechanics is Giovanni Alfonso Borelli[6] (1608-1679). His book "De Motu Animalium" shows a generally qualitative graphical solution to many problems in skeleton-joint-muscle mechanics. He treated the bones as mechanical levers actuated by the muscles according to mathematical principles (Fig. 2). His statical work was accurate but with pre-Newtonian mechanics he could do nothing with problems of body dynamics. For almost two centuries after him no one else elaborated on or extended his promising approach to body dynamics.

The significant developments on the subject during the eighteenth century were described by Barthez (1798), including the work of Haller, Camper and Blumenback[7]. He also criticized Borelli's treatment of such motions as walking, jumping, swimming and flying.

Many of 19th century biomechanicians were more concerned with the anatomico-physiological rather than the anatomico-mechanical aspects of posture and motion. The Weber brothers[8], R. Fick[14], Vom Meyer, Marey, [9-11] Braune[13] (1830-1892) and Fischer[12] (1861-1917) and Du Bois Reymond[15] are among the prominent contributors in this period. The fundamental laws of motion formulated by Sir Isaac Newton (1642-1727) were available during that period when the Weber brothers; Ernst (1795-1878), Wilhelm (1804-1891), and Eduard (1806-1871) investigated the motion of the center

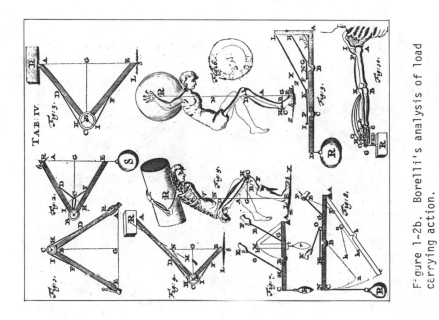

Figure 1-2b. Borelli's analysis of load carrying action.

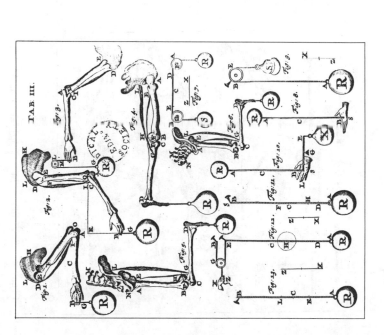

Figure 1-2a. Borelli's analysis of muscle forces in lifting weights.

5

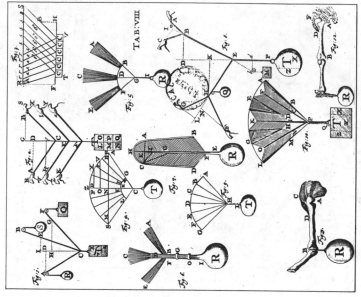

Figure 1-2d. Borelli's evaluation of load sharing by muscles.

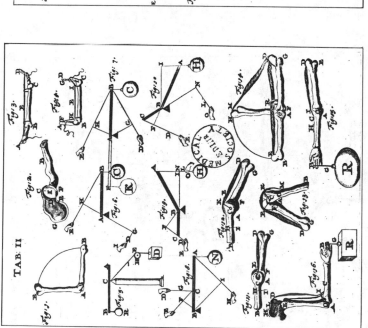

Figure 1-2c. Borelli's application of the law of levers.

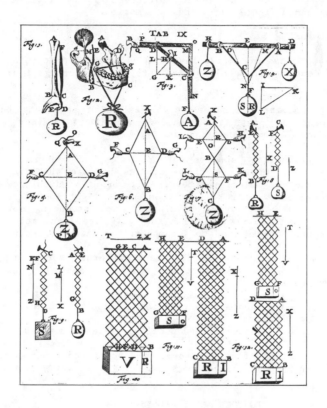

Figure 1-2e. Borelli's model for muscle action.

gravity of the body and the mechanics of muscular action on a scientific basis. Also during that period some of the early investigations of human gait were conducted by Braune and Fischer who made experimental determinations of the location of the center of gravity of the body and its parts but a thorough analysis of gait was sparked off by the development of serial photography. The actual sequence of motions in animals and humans was revealed for the first time when Eadweard Muybridge[16] (1955) obtained moving pictures of the trotting horse and later of the human figure and animals in motion.

MODERN DEVELOPMENT

The major areas of activities in investigating posture and movement during this century can be classified into six main categories:

1. Anthropometric studies which deal with pertinent body dimensions, mass distribution, center of gravity and moment of inertia of body parts.

2. Kinesiological studies of posture and patterns of movement. Most of these studies utilize photographic techniques, accelerometers, and angular displacement transducers. More recently automatic light spot tracing techniques are under development for producing motion patterns from targets on a subject. The information is directly transmitted to a computer for data processing.

3. Dynamical analysis of different patterns of motion and attempts at mathematical modeling and trajectory synthesis.

4. Electro-myographic investigation of muscular activities.

5. Physiological studies and modeling skeletal muscles.

6. Calculations of muscle forces and joint reactions during different postures and motions.

THE CONTROL OF HUMAN POSTURE AND LOCOMOTION

Although interest in understanding the mechanisms which operate in the control of posture and locomotion has a long and colorful history, there is yet no quantitative means for predicting the patterns of human motion in work or play. Observations such as "......the whole of this ensemble acts in unison with a single and complete rhythm, fusing the whole enormous complexity into clear and harmonious simplicity"* may be true but provide no clues to the understanding of the mechanisms involved. Continued interest in quantitative evaluation of the postural patterns and of the forces in the different muscles, bone and joints of the human body during its various activities is evident in the extensive list of investigations covering many fields of medicine, physical

---

*Quoted from Bernstein et. al., "Investigation on Biodynamics of loco-motion," Vols. 1,2, Moscow 1935, 1940 by Contini.

anthropology and sports.

The skeletal system, connected together through ligaments and muscles, provides the vital structural support for the human body. With the aid of muscular actions the human body can perform a plethora of coordinated limb movements through the numerous articulating joints. In performing such movements the joints are inescapably subjected to forces, the severity of which varies with the type of action performed. In recent years the mechanical aspects of the human body have been attracting increasing attention of investigators who have attempted to study the function and architecture of the musculoskeletal system. These include investigations of the forces produced in the joints, bones, ligaments and muscle actions necessary to maintain the structural equilibrium of the body.

Walking is one of the most common of all human activities and it is no surprise that it has been one of the first to be investigated. Simple though as it may appear, it is controlled by complicated and meticulous coordination between various elements. Interest in the study of locomotion was sparked off by the early work of the Weber brothers[8] (1836) who claimed that during the swing phase of walking, muscular control was not necessary and the motion of the leg occurred much like a simple pendulum. Further contributions to study of human gait were made by Marey et al.[9-11] (1873,1887,1895) in France, Braune and Fischer[13] (1889) in Germany, and Bernstein[17] (1935) in Russia. Elftman[18-21] (1934, 1939) studied the distribution of pressure in the human foot, the function of arms in walking, the rotation of the body and the functions of muscles in locomotion. Using basograh recordings, the changing pressures on various parts of the foot during walking have been analyzed extensively by Schwartz et al.[23] (1964). Saunders et al.[30] (1953) recognizing the concept that fundamentally locomotion is the translation of the center of gravity through space along a pathway requiring least expenditure of energy have outlined six major determinants as being essential for qualitative analysis of gait namely: pelvic rotation, pelvic tilt, knee and hip flexion, knee and ankle interactions and lateral pelvic displacement.

During recent times, extensive studies on walking in both normal and abnormal subjects have been conducted at the University of California (1947).[28] Murray et al. (1964) investigated the displacements associated with locomotion for normal men spanning a wide range of age and height. Similar kinesiological studies are available for investigating the pattern of motion in many of the common normal activities and in sports (see Refs. 27,77,78). Studies have also been conducted for the analysis of the motion of children and disabled persons (Refs. 24-26 for example).

Human body motions and supportive forces during a wide range of activities has been the subject of many investigations utilizing specially designed monitoring platforms and associated instrumentation[29-32].

Mathematical models with different degrees of sophistication have been developed to describe the experimental findings.

For example, in a study by Murray, Seireg and Scholz [29] the magnitude and orientation of the vertical supportive force were measured with a force platform (Fig. 3) during the following activities of a normal male: descending to and ascending from squatting and seated

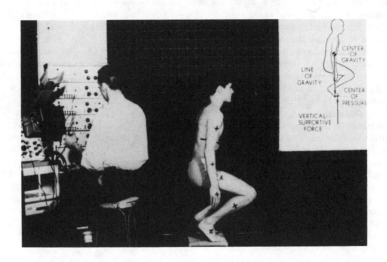

Figure 1-3. Analysis of center of mass and center of support.

postures, and jumping. Simultaneous photographic records were made of the displacements of the mass centers of body segments. This combination of methods has provided a means to 1) compare vertical forces calculated from the photographic records with the force-platform measurements, 2) differentiate between changes in the applied force and changes in the position of the center of gravity of the body, and 3) differentiate between the excursions of the line of gravity and the action line of the vertical supportive force (center of pressure). The vertical force fluctuated above and below body weight during all test activities. The calculated force patterns approximated the measured patterns. Distinctly different pathways were seen for the center of pressure and the line of gravity with the former fluctuating and the latter moving smoothly. The interaction between the two suggests a fundamental servomechanism operable in the control of human posture and motion. Bresler and Frankel[35] (1950) using the force plate developed by Cunningham and Brown[36] (1952) to measure the ground-to-foot forces and simultaneously recording the positions of the leg in space obtained curves showing the variation with time of the three force components and three moment components transmitted at the ankle, knee and hip joints during walking on a level surface but did not estimate the values of muscle forces and joint reactions.

The supportive forces and moments resulting from human activities in an underwater environment were investigated by Seireg, Baz and Patel[32]. A platform capable of monitoring all components of supporting forces and moments has been used in a water tank (Fig. 4a). A mathematical model has been developed based on the experimental results and can be utilized for the analysis of human body dynamics underwater. The study was extended to develop a simplified mathematical model for analysis of crawl swimming[33] (Fig. 4b).

A mathematical programming method for evaluating the optimal trajectories and controls for systems of coupled rigid bodies which can be used for the analysis of human and animal locomotion has been developed by Townsend and Seireg[34].

Trajectories and controls for bipedal locomotion are synthesized for optimum stability and energy expenditure. The model represents rigid body idealization of human locomotion. The effect of Model Complexity and gait criteria on the synthesis of bipedal locomotion has also been investigated[34].

Chaffin[37] (1969) treated the human body as a series of seven solid links articulated at the ankles, knees, hips, shoulders, elbows and wrists to develop a computerized model for certain gross body actions. The model was specifically designed to investigate body movements that occur during the lifting and carrying of materials. However, the several constraints that are employed in the development of the model limit its practical applicability.

Studies of athletic events such as that by Lascari[31] (1970) showed that what is considered beautiful may not be most efficient.

Numerous other mathematical models can be found in the literature for analysis of human body dynamics using rigid body idealization with active controls at the different joints. Examples of such studies are those of Smith and Kane[38] (1968), Kane and Scher[39] (1970), Huston and Passerello[40] (1971), Frank and Vukobratovic[41] (1969), etc.

Figure 1-4a. Underwater platform in tank.

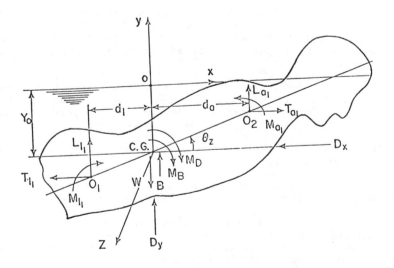

Figure 1-4b. Forces and moments on the swimmer's body.

The mathematical analysis of the musculoskeletal system with a view towards calculating the forces in the muscles called upon to maintain a particular posture is an important consideration in biomechanical studies. The action of muscles in the static and dynamic activity of upper limb has been studied by Morecki et al.[42] (1966) who measured the activity of a muscle group during the process of flexion and extension of the forearm. They used a special device to synchronously register the static and dynamic, mechanical and electrical parameters of the upper limb for various positions, velocity and external forces to determine the coefficient of participation of muscles acting in a given moment and to establish the equality of external mechanical moment to the sum of all muscle moments.

Troup and Chapman[43] (1969) applied static bending moment to the trunk in the sagittal plane via arms stretched horizontally forward with the subjects standing or seated with pelvis fixed. They calculated the mean lumbar extensor moments for males in the range of 500 kg-cm and extensor muscle force at between 500-600 kg.

McLeish and Charnley[44] (1970) attempted to determine the adduction forces in the one-legged stance using force measurements and disposition of the skeleton of subjects standing on a forceplate in front of a radiograph screen. This enabled them to locate the center of gravity for different postures. By estimating the load sharing by the adductor muscles, the force in the muscles and joints were determined. They found that the ratio of joint force to body weight ranges from 1.8 to 2.7 as the pelvis changed from elevated to sagging relative to the supporting leg. However their analysis did not take into account the three-dimensional configuration of the skeleton and muscle system.

Merchant[45] (1965) positioned a dried male pelvis articulated with the lower 2 lumbar vertebrae, and femur to represent a man standing on one extremity. He attached chains and strain gage load cells to represent the hip adductor muscles (i.e., gluteus minimus, gluteue medius, tensor fasciae latae and the iliotibial tract). The model was loaded at the approximate center of gravity of the body with different weights and different hip positions to determine the counter force on the load cells and thereby approximate the force exerted by the muscles. He obtained the total abductor muscle force for a 150 lb man to be 154 lb and concluded that abductor muscle force is least with the pelvis in abduction and most with the pelvis in adduction. Rotation of the femur internally or externally from its anatomical or neutral position necessitates increased adductor muscle force to maintain a level pelvis.

To study the biomechanical considerations of lumbricalis behavior in the human figure, Thomas et al.[46] (1968) developed mathematical expressions to include the passive and active tensions in the muscles controlling a hypothetical finger with a metacarpophalangeal joint and only one interphalangeal joint. They utilized average anatomical data about tendon locations relative to the joint centers of rotation at each joint along with expressions for the stiffness of each of the four muscles presumed to be controlling these joints so that the angle at each joint is defined in terms of the active and passive muscle tensions.

Clinical fractures in the lower extremities placed emphasis of muscle forces and joint reactions. Direct determination of joint force has been performed by Rydell[47-48] (1965, 1966) who fitted two of his patients, one male and one female, with a hip-joint prosthesis carrying

13

strain gages. The greatest value of joint force recorded was 4.33 times the body weight which occurred in the female subject when running. For level walking, the greatest force was 3.3 times the body weight, Pauwels[49] (1935) estimated the total hip joint force to be three times body weight for a person standing on one leg. He also states that under the dynamic conditions of walking, the force may increase to 4.5 times the body weight during the stance phase.

Other analyses of hip forces include those of Inman[50] (1947); Blount[51] (1956); Strange[52] (1963); McLeish and Charnley[44] (1970); Williams and Svenson[53] (1968).

Recently, the reaction forces at the hip joint during walking have been analyzed by Paul[54-56] (1966, 1967, 1971) and at the knee joint by Morrison[57-58] (1969, 1970). Both Paul and Morrison measured photographically the three dimensional configuration of the leg segments during a walking cycle in which the ground-to-foot actions were measured by a force plate dynamometer. Using the technique of Bresler and Frankel (loc. cit.) the resultant forces and moments transmitted between segments were calculated. The moments are transmitted by tensions in muscles or ligaments. Allotting these forces to the respective muscles at the joints (shown to be active by electromyographic signals which were also recorded) and making various simplifying assumptions to facilitate solution of the statically indeterminate system of equilibrium equations, the joint reactions were calculated. Maximum loads in the range of 2.3-5.8 times the body weight at the hip joint are reported for a series of male and female subjects of varying structure. Morrison reports a value between 2 and 4 times the body weight at the knee joint.

Many electromyographic studies have been conducted to investigate the manner by which the skeletal structure maintains its posture under the action of muscle forces. One such investigation is that of Houtz and Walsh[59] (1959) who reported an electromyographic analysis of the function of the muscles acting on the ankle during weight-bearing with special reference to the triceps surface. Surface electrodes are used on 10 normal young adults to monitor action potentials from the tibialis anterior, extensor digitorum longus, tibialis posterior, peroneus longus and brevis, medial and lateral gastrocnemius, soleus, flexor digitorum longus. The activities performed are relaxed standing, standing on an inclined and on a declined plane, shifting the body weight from the fore part of the foot to the heel, standing on one extremity, standing on the toes of both feet and on the fore part of one foot, relaxed walking, and stair climbing. The findings of Houtz and Walsh regarding indirect knee stabilizing function of the ankle plantar flexors in walking on the level are supported by Sutherland[60] (1966).

To observe the role of muscles in the arch support of foot, Basmajain and Stecko[61] (1963) used simultaneous electromyography of six muscles (tibialis anterior and posterior, peroneus longus, flexor hallucis longus, adductor hallucis, flexor digitorum brevis) in the leg and foot in 20 subjects to reveal that only heavy loading elicits muscle activity. They concluded that the first line of defense of the arches is ligamentous; the muscles form a dynamic reserve, called upon reflexly by excessive loads, including the take-off phase in walking. These findings are supported by Mann and Inman[62] (1964) who investigated into the phasic activity of intrinsic muscles of the foot with the help of an electromyographic study of six intrinsic muscles of the foot.

A direct electromyography of the Psoas major muscles was conducted by Keagy, Brumlik and Bergan[63] (1966). Linge[64] (1961) investigated with the aid of an 8-channel electromyograph and a moving belt the behaviour of the quadriceps muscle during walking at various speeds. The complex kinesiology of the hand has been the subject of research of many authors. (For example, Long and Brown[65] (1964), Close and Kidd[66] (1969) and Forrest and Basmajian[67] (1965).

There are numerous studies of the electromyographic signals from all important muscles during the walking cycle. Comprehensive data can be found, for example, in reports from the University of California[28] (1947). Such studies provide information about the pattern of muscle participation but do not quantify the muscle forces.

Kinesiological electromyography has been increasingly used in recent years to investigate the anatomical functions of the various muscles of the human body connecting the different segments of the skeletal structure. An extensive compilation of some of the significant EMG work can be found in the book "Muscles Alive" by Basmajian[68]. Evaluations of energy transfer, useful work done during human activity and efficiency are also available in the literature. Examples of the work in this area include that done at the University of California (Ralston[69] 1958, Ralston and Lukin[70] 1969, Saunders et al.[30] 1953, Lukin et al.[71] 1967), and at Milan (Cavagne et al.[72], 1963, Margaria[73] 1968). Texts such as those of Morton and Fuller[74] (1952) and Hill[75] (1965) have exhaustive discussions on the biophysical aspects of muscle motion, energy and efficiency.

The extent and scope of recent investigations in Musculoskeletal Biomechanics is to large to review in this historical introduction. The references cited in this chapter are the ones which were relevent to the development of the material presented in this book and are by no means a complete bibliography on the subject. Recent literature is rich with many excellent books and publications covering the wide spectrum of musculoskeletal biomechanics.

CONCEPTUAL FRAMEWORK FOR MOTION CONTROL

It can be seen from the introduction that inspite of the diversity, scope and depth of the investigations cited, the basis question of the nature of the mechanism controlling posture and motion remains to be answered.

In this section a conceptual framework for motion control is postulated in order to shed some light on the different possible levels and criteria for the control.

The different analyses, measurements and observations would suggest that in human motion as in most other operations there appears to be three distinct levels of control:

STRATEGIC CONTROL: It is expected that at this level the goal for the action is set and the task to be accomplished is defined. A decision is made on the feasibility of the action and consequently sets into motion the mechanisms to execute it. A possible criterion for the decision may be the level of confidence in accomplishing the task. It is primarily a subjective judgment based on past experiences and favorable evidence and expectations.

TACTICAL CONTROL: At this level it can be postulated that all the possible alternative patterns are considered and the most suitable postural configurations or action patterns are selected. A possible criteria for decision may include the minimization of the effort necessary to perform the act, the maximization of stability, and the aesthetics of the posture or movement. Such a criteria is partially objective and partially subjective and consequently the selected pattern of movement would vary according to the personal preference and capability of each individual. Such patterns may be approximated but not fully predicted solely on mathematical basis.

TECHNICAL CONTROL: Once a decision on the postural configuration or the motion trajectory has been made, the final level of control is to ensure that the motor locomotor system optimally executes the task, that is, to optimally attain and maintain the stability of the posture or movement. This implies a decision on the load sharing between the different muscles in executing the task and maintaining the equilibrium of the selected posture. As in all technical decisions the criterion on this level is expected to be primarily objective and the muscular load sharing can therefore be expected to follow a well established totally objective order.

In summary, it is postulated that there are three levels of controls for human posture of locomotion:

Strategic: or goal setting; the results of which is a decision to start the task. Such decision would be based on primarily subjective criteria.

Tactical: for the selection of the optimum pattern or trajectory; this selection is based partly on objective and partly on subjective criteria.

Technical: for the selection of the muscles necessary to execute the task and the load sharing between these muscles.

This selection is expected to be based on an objective and well established mathematically definable criterion. It is this level of control which appears to define the extent of muscular participation and is the main theme of this work. On all the previous levels of controls, the main constraint is that the motions resulting from all forces applications and the forces resulting from all motions must be related by the Newtonian Laws of motion. A brief review of these laws is given in the following chapter.

It should be, however, emphasized that unlike a truly mechanical control system which may possess the three levels of control mentioned above, the human locomotor control system is unique in that it is equipped with an additional level of control, namely, the adaptive feedback control. In routine or repetitive skeletal activities decision implementation can be initiated by a simple trigger signal, without any active assistance at the other two levels. If the environment or the existing conditions change and in cases of emergency situations, such implementation on the technical level alone cannot be tolerated. In such case decisions on the strategic and tactical levels are needed and an overriding mechanism is essential in handling these changes of environment or the emergency situations. Thus adaptive feedback mechanisms, based on instinct and learning are back into play to render such actions or "technical control decision" routine, if repeated in the future.

REFERENCES

1. Aristotle, "Parts of Animals," with an English translation by A.L. Pech. "Movement of Animals" and "Progression of Animals", with an English translation by E.S. Foster. Harvard Univ. Press, Cambridge, Mass. and Wm. Heinemann, Ltd., London, 1955.

2. Gruner, O.C., A Treatise on the Canon of Medicine of Avicenna. London, 1930.

3. Leonardo da Vinci, "On the Human Body", Charles D. O'Malley and J.B. and C.M. Saunders, Henry Schuman, N.Y., 1952.

4. MacCurdy, E., The Notebooks of Leonardo da Vinci, New York, George Braziller, 1958.

5. Seireg, A. Leonardo Da Vinci——The Biomechanician, (Eds. D. Bootzin and H.C. Muffley) Plenum Press, N.Y., 1969.

6. Govanni Alfonso Borelli, De Motu Animalium, Opus Postuhumum Romae, Extypographia A. Bernabo, 1680-1681.

7. Barthez, P.J., Nonvelle Mechanique des Mouvements de R'Homme et des Animaux, Carassonne, P. Polere, 1798.

8. Weber, E. and Weber, W., Mechanik der menschlichen Gehwerkzeuge, Göttingen, Germany, 1836.

9. Marey, E.J., "Da La Locomotion terrestre chez les pipedes et les quadrupedes. Journal de l'Anat. et de la Physiol., 9:42-80, 1873.

10. Marey, E.J. and Demeny, G., "Etudes experimentales de la locomotion humaine", Comptes Rendus Acad. des Sciences, 105:544-552, 1887.

11. Marey, E.J., Movement, D. Appleton Co., N.Y., 1895.

12. Fischer, O., "Der Grang des Menschen", Abhandl. d. Cl. d. k. Sächs. Gesellsch. Wissensch., Vol. 21-28, 1898-1904.

13. Braune, W. and Fischer, O., "Über den Schwerpunkt des menschlichen Körpers mit Ruecksicht auf die Aurs. des deutchen Infant", Abhandl. der Koenigl. Sächs. Gesellsch. Wissensch, Vol. 15, 1872.

14. Fick, R., "Handbuch der Anatomie und Mechanik der Gelenke", 1-3 Teil. Fischer, Jeno. 1911.

15. DuBois-Reymond, R., "Die Grenzen der Unterstützungsfläche beim stehen", Arch. Anat. Physiol., 562-564, 1900.

16. Muybridge, E., The Human Figure in Motion, Dover Publishers, Inc., N.Y., 1955.

17. Bernstein, N.A. et al., "Biodynamics of Locomotion", VIEM Moscow, USSR, Vol. 1, 1935.

18. Elftman, H., "A Cinematic Study of the Distribution of Pressure in the Human Foot", Anat. Rec., 59:481, 1934.

19. Elftman, H., "The Functions of the Arms in Walking", Human Biol., 2:529, 1939.

20. Elftman, H., "The Rotation of the Body in Walking", Arbeits-physiologie, 10:477, 1939.

21. Elftman, H., "The Function of the Muscles in Locomotion", Am. J. Physiol., 125:357, 1939.

22. Murray, M.P., Drought, A.B. and Kory, R.C., "Walking Patterns of Normal Men", J. Bon Jt. Surg., 46A:335, 1964.

23. Schwartz, R.D., Heath, A.L., Morgan, D.W. and Towns, R.C., "A Quantitative Analysis of Recorded Variables in the Walking Patterns of Normal Adults", J.B. Jt. Surg., 46A:324, 1964.

24. Murray, M.P., Gore, D.R. and Clarkson, B.H., "Walking Patterns of Patients with Unilateral Hip Pain due to Osteo-arthritis and Avascular Necrosis', J. Bone Jt. Surg., 53A:259, 1971.

25. Carlsöo , S., "Kinetic Analysis of the Gait in Patients with Hemiparesis and in Patients with Intermittent Claudication", Scand. J. Rehab. Med., 6:166, 1974.

26. Richards, C. and Knutsson, E., "Evaluation of Abnormal Gait Patterns by Intermittent-light Photography and Electromyography", Scand. J. Rehab. Med., 3:61, 1974.

27. Frankel, V.H. and Yang, Y., "Recent Advances in the Biomechanics of Sports Injuries", Anat. Rec., 1974.

28. University of California, "Fundamental Studies of Human Locomotion", A report to the National Research Council, Committee on Artificial Limbs, Berkeley, 1947.

29. Murray, M.P., Seireg, A. and Scholz, R.C., "Center of Gravity, Center of Pressure, and Supportive Forces during Human Activities", J. Appld. Physiol., 23:831, 1967.

30. Saunders, J.B., Inman, V.T. and Eberhardt, H.D., "The Major Determinants in Normal and Pathological Gait", J. Bone Jt. Surg., 35A:543.

31. Lascari, A.T., "The Felge Handstand——A Comparative Kinetic Analysis of a Gymnastic's Skill", PhD Thesis, U. of Wis., 1970.

32. Seireg, A., Baz, A. and Patel, D., "Supportive Forces on the Human Body During Underwater Activities", J. Biomechanics, Vol. 4, pp. 23-30, Pergamon Press, 1971.

33. Seireg, A. and Baz, A., "A Mathematical Model for Swimming Mechanics, Proceedings First Int. Symposium on Biomechanics in Swimming, L. Lewillie and J.P. Clarys, eds., Brussels, 1970.

34. Townsend, M. and Seireg, A., 1. "The Synthesis of Bipedal Locomotion", J. Biomechanics, Vol. 5, pp. 71-83, 1972; 2. "Effect of Model Complexity and Gait Criteria on the Synthesis of Bipedal Locomotion", IEEE Trans. on Biomedical Engr., Nov. 1973, Vol. BME-20, No. 6.

35. Bresler, B. and Frankel, J.P., The Forces and Moments in the Leg during Level Walking", Trans. ASME, 72:27-36, 1950.

36. Cunningham, D.M. and Brown, G.W., "The Devices for Measuring the Forces Acting on the Human Body during Walking", Proc. Soc. Exp. Stress Anal. IX, 2:75, 1952.

37. Chaffin, D.B., "A Computerized Biomechanical Model, Development Of and Use in Studying Gross Body Actions", J. Biomechanics 2(4), 429-441, 1969.

38. Smith, P.G. and Kane, R.R., "On the Dynamics of the Human Body in Free Fall", J. Appl. Mech., 35, 167-168, 1968.

39. Kane, T.R. and Scher, M.P., "Human Self-rotation by Means of Limb Movements", J. Biomechanics, 3, 39-40, 1970.

40. Huston, R.L. and Passerello, C.E., "On the Dynamics of a Human Body Model", J. Biomechanics, 4, 369, 1971.

41. Frank, A.A. and Vukobratovic, M., "On the Synthesis of Biped Locomotion Machines", 8th Int. Conf. Medical and Biological Engng., Evanston, Illinois, 1969.

42. Morecki, A. et al., "The Participation of Muscles in the Static and Dynamic Activity of Upper Limb", Archiwun Budowy Maszyn, 13, (3), 329-355, (In Polish), 1966.

43. Troup, J.D.G. and Chapman, A.E., "The Strength of the Flexor and Extensor Muscles of the Trunk", J. Biomechanics, 2 (1), 49-62, 1969.

44. McLeish, R.D. and Charnley, J., "Adduction Forces in the One-Legged Stance", J. Biomechanics 3 (2), 191-209, 1970.

45. Merchant, A.C., Hip Abductor Muscle Force, J. Bone Jt. Surg. 47A (3), 1965.

46. Thomas, D.H., Long, C. and Landsmeer, J.M.E., "Biomechanical Considerations of Lumbricalis Behavior in the Human Finger", J. Biomechanics, 1 (2) 107-115, 1968.

47. Rydell, N., Forces in the Hip Joint (ii); Intravital Studies, In Biomechanics and Related Bio-Engineering Topics. (Edited by Kenedi, R.M.). Pergamon Press, Oxford.

48. Rydell, N., "Forces Acting on the Femoral Head Prosthesis", Acta. Orthop. Scand. Suppl., 88, 1966.

49. Pauwels, F., Der Schenkelhalsbruch ein Mechanisches Problem, Stuttgart, Fesd. Enke., 1935.

50. Inman, R.T., "Functional Aspects of the Abductor Muscles of the Hip", J. Bone Jt. Surg., 39(3), 607, 1947.

51. Blount, W., "Don't Throw Away the Cane", J. Bone Jt. Surg., 38A, 695, 1956.

52. Strange, F.G.S.C., The Hip, Heinemann, London, 1963.

53. Williams, J.F. and Svensson, N.L., "A Force Analysis of the Hip Joint", Biomed. Engng., 3 (8) 365-370.

54. Paul, J.P., "Bio-engineering Studies of the Forces Transmitted by Joints: (II). Engineering Analysis", In <u>Biomechanics and Related Bio-Engineering Topics</u> (Edited by R.M. Kenedi). Pergamon Press, Oxford. (Proceedings of a Symposium held in Glasgow, Sept. 1964.)

55. Paul, J.P., "Forces Transmitted by Joints in the Human Body", Proc. Inst. Mech. Engrs. 181 (3J), 8. Presented at the Inst. of Mech. Eng. Symp. in Lubrication and Wear in Living and Artificial Human Joints. Paper 8, London, April 1967.

56. Paul, J.P., "Load Actions on the Human Femur in Walking and Some Resultant Stresses", Exp. Mech., 3, 121, 1971.

57. Morrison, B.B., "Bioengineering Analysis of Force Actions Transmitted by the Knee Joint", Bio-med. Engng, 4, 164, 1969.

58. Morrison, J.B., "The Mechanics of the Knee Joint in Relation to Normal Walking", J. Biomechanics, 3, 51, 1970.

59. Houtz, S.J. and Walsh, E.P., "Electromyographic Analysis of the Function of the Muscles Acting on the Ankle During Weightbearing with Special Reference to the Triceps Surface", J. Bone Jt. Surg., 51A (8), 1959.

60. Sutherland, D.H., "An Electromyographic Study of the Plantar Flexors of the Ankle in Normal Walking on the Level", J. Bone Jt. Surg., 48-A, (1), 66, 1966

61. Basmajian, J.V. and Stecko, G., "The Role of Muscles in the Arch Support of the Foot", J. Bone Jt. Surg., 45A, (6), 1963.

62. Mann, R. and Inman, V.T., "Phasic Activity of Intrinsic Muscles of the Foot", J. Bone Jt. Surg., 40A, 1964.

63. Keagy, R.D., Brumlik, J. and Bergan, J.J., "Direct Electromyography of the Psoas Major Muscle in Man", J. Bone Jt. Surg., 48A, 1966.

64. Linge, B.V., "Behavior of Quadriceps Muscle during Walking", Proceedings of Netherlands Orthop. Soc., November 1961.

65. Long, C. and Brown, M.E., "Electromyographic Kinesiology of the Hand: Muscles Moving the Long Finger", J. Bone Jt. Surg., 46A (8), 1964.

66. Close, J.R. and Kidd, CC., "The Functions of the Muscle of the Thumb, the Index and Long Fingers", J. Bone Jt. Surg., 51-A (8), 1969.

67. Forrest, W.J. and Basmajian, J.V., "Functions of Human Thenar and Hypothenar Muscles", J. Bone Jt. Surg., 47-A (8), 1965.

68. Basmajian, J.V.: <u>Muscles Alive</u>, Williams and Wilkins Co., Baltimore, 1974.

69. Ralston, H.J., "Energy-speed Relation and Optimal Speed during Level Walking", Int. Z. Angew. Physiol., 17, 277-282, (Formerly

Arbeitsphysiologie.), 1958.

70. Ralston, H.J. and Lukin, L., "Energy Levels of Human Body Segments during Level Walking", Ergonomics, 12, 39-46, 1969.

71. Lukin, L., Polissar, M.J. and Ralston, H.J., "Methods for Studying Energy Costs and Energy Flow during Human Locomotion", Human Factors, 9, 603-608, 1967.

72. Cavagna, G.H., Saibene, F.P. and Margaria, R., "External Work in Walking", J. Appl. Physiol., 18, 1-9, 1963.

73. Margaria, R., "Positive and Negative Work Performance and Their Efficiencies in Human Locomotion", Int. Z. Angew Physiol., 25, 339-351, 1968.

74. Morton, D.J. and Fuller, D.O., "Human Locomotion and Body Form, p. 285, Williams and Wilkins, Baltimore, 1952.

75. Hill, A.V., Trails and Trials in Physiology, p. 374, Williams and Wilkins, Baltimore, 1965.

76. Bellman, R. and Tomović, R., "A Systems Approach to Muscle Control", Mathematical Biosciences, 8, 265-277, 1970.

77. Northrip, J.W., Logan, G.A. and McKinney, W.C. Introduction to Biomechanic Analysis of Sport. W.C. Brown, Co., Dubuque, Iowa, 1974.

78. Rasch, P.J. and Burke, R.K., Kinesiology and Applied Anatomy, Lea and Febiger, Philadelphia, 1971.

CHAPTER II

Newtonian Mechanics

"In this sense rational mechanics will be the science of motions resulting from any forces whatsoever, and the forces required to produce any motions accurately prepared and demonstrated."

Isaac Newton
1642-1727

DEFINITION

Mechanics is the science dealing with force and motion. Its objective is to describe all motions occurring in nature in the simplest manner and to relate them to the forces producing them on the basis of the simplest possible laws. This section gives a brief review of some of the laws governing the equilibrium or motion of passive as well as living elements and systems. These laws were first proposed and demonstrated by Sir Isaac Newton (1642-1727) and are consequently referred to as Newton's Laws of Motion.

MASS, CENTER OF MASS AND MOMENT OF INERTIA

Mass is a measure of the quantity of matter contained within a body and is related to the space it occupies by the expression:

$$\text{mass} = \text{density} \times \text{volume} \qquad [2\text{-}1]$$
$$[\text{mass}: \text{ lb or Kg.}]$$
$$[\text{density}: \text{ lb/in.}^3 \text{ or Kg/m}^3]$$
$$[\text{volume}: \text{ in.}^3 \text{ or m}^3]$$

If the body volume can therefore be found, its mass can be obtained once its density is known. For objects of known geometrical shapes, this is a particularly easy task. The volume can be conveniently obtained by simple algebraic formulae based on its dimension if the body is of a regular, geometrically defined shape. If the body shape is irregular then its volume can be approximated by dividing it into finite, discrete elements of known geometric shapes whose volumes can be individually computed and then summing these up to yield the total body volume.

The mass of a body represents a measure of the resistance of the body to motion in a straight line. This resistance is measured in terms of 'force', which is defined as the product of mass and acceleration.

The weight of the body is the gravitational pull on its mass. Thus, if the body mass is 'm', its weight is:

$$w = mg \qquad [2\text{-}2]$$

23

where w = st. of the body (lbf or N.)
      m = mass (lbm or Kgm)
      g = gravitational acceleration (32.2 ft/sec^2, 0.98 cm/sec^2)

Since the total weight is a force, it is a vector quantity and by notation directed downwards to represent the pull of the earth. Being a force, it must have a point of application commonly referred to as the center of mass or center of gravity (CG). It is the point through which the resultant of weights of all the particles (essentially any body may be considered to be made of particles) passes through the body. Since all the body particles (or masses) are subjected to the same gravitational acceleration it implies that the center of mass and the center of gravity are one and the same, and will be used interchangeably.

As described above, for bodies with known geometrical shapes one can readily identify the location of center of gravity by simple algebraic formulae. Since segments of the human body are not of uniform volume or mass one may evaluate the location of the CG through experimental procedures.

For a collection of rigid bodies with different masses and therefore, different locations of centers of gravity, the resultant CG can be evaluated as follows. If, for example, there are n bodies of weights $w_1, w_2, \ldots, w_n$ etc. and their locations of centers of gravity in cartesian coordinates with respect to some known reference axes are given by (x_1, y_2, z_2), $(x_1, y_2, z_2), \ldots, (x_n, y_n, z_n)$ etc., as shown in Fig. 2-1, then the center of gravity of this system is given by $(\bar{x}, \bar{y}, \bar{z})$ where,

$$\bar{x} = \frac{w_1 x_1 + w_2 x_2 + \ldots + w_n x_n}{w_1 + w_2 + \ldots + w_n} = \frac{\Sigma w_i x_i}{\Sigma w_i} \qquad i = 1 \rightarrow n \quad [2-3]$$

$$\bar{y} = \frac{w_1 y_1 + w_2 y_2 + \ldots + w_n y_n}{w_1 + w_2 + \ldots + w_n} = \frac{\Sigma w_i y_i}{\Sigma w_i} \qquad i = 1 \rightarrow n \quad [2-4]$$

$$\bar{z} = \frac{w_1 z_1 + w_2 z_2 + \ldots + w_n z_n}{w_1 + w_2 + \ldots + w_n} = \frac{\Sigma w_i z_i}{\Sigma w_i} \qquad i = 1 \rightarrow n \quad [2-5]$$

Note that in selection of the positive directions for the three reference axes, the right hand rule is applied as illustrated in Fig. 2-1. It can be seen that the thumb points to the positive z-direction if the other four fingers are folded in the direction from x to y.

The Moment of Inertia of a body about an axis is a measure of the resistance to rotation about that axis and is related to the distribution of the body mass with respect to the axis. Referring to Fig. 2-2, the moment of inertia of a mass m about an axis x-x is calculated as:

$$I_{xx} = m \cdot d^2 \qquad\qquad [2-6]$$

where d is the perpendicular distance between the center of mass and the axis.

It can be seen that the same mass can have considerably different values for the moment of inertia depending on its distance from the

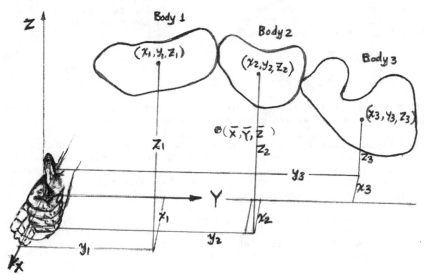

FIG. 2-1   C.G. of a collection of rigid bodies.

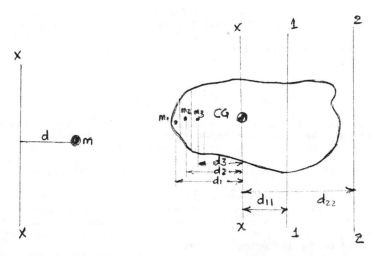

FIG. 2-2   Moment of inertia of
a mass m about an
axis $x\text{-}x$.

FIG. 2-3   Moment of inertia of a
rigid body and the
parallel axis theorem.

axis. Therefore, whenever describing the moment of inertia of a mass, the axis of reference should always be stated.

The expression [2-6] above for the moment of inertia of a single mass, can be easily extended to that for a body. The latter being simply a collection of several masses or particles, the resultant moment of inertia about a desired axis is the sum of the moments of inertia of the individual masses about the same axis. Referring to Fig. 2-3, if the axis x-x passes through the CG of the body then the moment of inertia of the body about this centroidal axis is

$$I_{CG} = m_1 d_1^2 + m_2 d_2^2 + \ldots + m_n d_n^2$$
$$= \Sigma \, m_i \cdot d_i^2 \qquad i = 1 \to n$$

[2-7]

The expressions for the moment of inertia of some geometric solids about the three centroidal axes are given in Table 2-1.

Note that since the moment of inertia is a scalar quantity, the summation in expression [2-7] is scalar. In other words, irrespective of the position of the mass with respect to the axis x-x, whether positive or negative, the contributions of the individual masses to the total moment of inertia are always additive.

It is frequently required to calculate the moment of inertia of a body about an axis parallel to, but other than the centroidal axis. Then, if the moment of inertia about the centroidal axis is known, the moment of inertia about the desired parallel axis, say 1-1 (Fig. 2-3) is obtained by:

$$I_{11} = I_{CG} + M \cdot d_{11}^2$$

[2-8]

where m is the mass of the body and d is the distance between the centroidal axis x-x and parallel axis 1-1. The desired axis may pass through the body or even lie outside but it should be parallel to the centroidal axis. Thus, for example, the moment of inertia about the axis 2-2 lying outside the body but parallel to x-x is given by:

$$I_{22} = I_{CG} + m \cdot d_{22}^2$$

[2-9]

Note that the parallel-axis principle also applies if the moment of inertia of the body were known about a noncentroidal axis and it was required to obtain the moment of inertia of the body about a parallel axis through the CG. This will be demonstrated in Chapter III for the experimental determination of moments of inertia of human body limbs.

Further details on evaluation of moments of inertia of bodies of known geometric shapes and some numerical examples are given in Appendix B. Methods for evaluation of masses of human body segments, their centers of masses and moments of inertia and currently available data are described in the following chapter.

FREE BODY OR SYSTEM ISOLATION

In order to study and analyze a portion of the universe, the portion selected is imagined to be detached from its surroundings. The isolated body can be a planet, a rocket, the crank shaft of an engine or a part of a living system. The effects exerted by the surroundings should be

TABLE 2-1
Mass Moment of Inertia of Some Common Solids

Solid	Figure	Moment of inertia
Slender rod		$I_{xx} = I_{yy} = 1/12 \ m\ell^2$ $I_{zz} = 0$ $I_{xy} = I_{xz} = I_{yz} = 0$
Sphere		$m = 4/3 \ \pi R^3 \rho$ $I_{xx} = I_{yy} = I_{zz} = 2/5mR^2$ $I_{xy} = I_{xz} = I_{yz} = 0$ For thin spherical shell, use $I = 2/3 \ mR^2$ where $m = 4\rho\pi^-R^2t$
Hemisphere		$I_{xx} = I_{yy} = 0.259 \ mR^2$ $I_{zz} = 2/5 \ mR^2$ $I_{xy} = I_{xz} = I_{yz} = 0$
Cylinder (thin disk)		$I_{xx} = I_{yy} = 1/12 \ m(3R^2+h^2)$ $I_{zz} = 1/2 \ mR^2$ $I_{xy} = I_{xz} = I_{yz} = 0$ $m = \rho\pi R^2 h$ For thin disk or circular plate use the given I with $h = 0$

Table 2-1 (cont.)

Solid	Figure	Moment of inertia
Semicylinder		$I_{xx} = 0.0699\ mR^2 + m/12\ h^2$ $I_{yy} = 0.320\ mR^2$ $I_{zz} = 1/12\ m\ (3R^2+h^2)$ $I_{xy} = I_{xz} = I_{yz} = 0$
Hollow right cylinder (Thin washer)		$M = \pi R^2 \ell \rho$ $m = \pi r^2 \ell \rho$ $I_{xx} = (MR^2-mr^2)/2$ $I_{yy} = I_{zz} = [3(MR^2-mr^2) + \ell^2(M-m)]/12$ $I_{xy} = I_{yz} = I_{xz} = 0$ For thin washer use the given I values with $\ell = 0$
Cylindrical shell (Thin ring)		$m = 2\pi R + \ell \rho$ $I_{xx} = mR^2$ $I_{yy} = I_{zz} = m/12(6R^2+\ell^2)$ $I_{xy} = I_{yz} = I_{xz} = 0$ For thin ring use the given I values with $\ell = 0$
Rectangular parallel pipe (Thin or rectangular plate)		$m = m/12\ \rho ab\ell$ $I_{xx} = m/12\ (a^2+b^2)$ $I_{yy} = m/12\ (b^2+\ell^2)$ $I_{zz} = m/12\ (a^2+\ell^2)$ $I_{xy} = I_{xz} = I_{yz} = 0$ For thin or rectangular plate use the given I values with $\ell = 0$

Table 2-1 (cont.)

Solid	Figure	Moment of inertia
Ellipsoid		$m = 4\rho\pi abc/3$ $I_{xx} = m(b^2+c^2)/5$ $I_{yy} = m(a^2+c^2)/5$ $I_{zz} = m(a^2+b^2)/5$ $I_{xy} = I_{xz} = I_{yz} = 0$ For ellipsoid of revolution use given I values with b=c
Torus		$m = 2\rho\pi^2 r^2 R$ $I_{xx} = m(4R^2+3r^2)/4$ $I_{yy} = I_{zz} = m(4R^2+5r^2)/8$ $I_{xy} = I_{xz} = I_{yz} = 0$
Right circular cone		$m = 1/3 \; \pi R^2 h\rho$ $I_{xx} = I_{yy} = 3m/80(4R^2+h^2)$ $I_{zz} = 3/10 \; mR^2$ $I_{xy} = I_{yz} = I_{xz} = 0$ For thin bevelled disk use given I with h = 0

considered as external actions on that portion when treated as a free body. The system should then be idealized in order to make a solution possible through the laws of mechanics. This idealization is usually based on certain assumptions which enable the translation of the physical system into a mathematical model.

Mechanical systems can be idealized into one or a combination of the following categories:

1. A PARTICLE:

A particle is an object whose mass and not its size is of significance in the analysis of its motion. A particle can only move in a translatory fashion. This motion is represented by the change in position of the center of mass and since no rotation is expected the distribution of mass is of no consequence.

In the general sense, a particle has three degrees of freedom. The number of degrees of freedom that a system possesses is the number of independent coordinates which must be specified in order to define its motion in space. In the case of a particle these coordinates are conveniently represented by the displacements x, y, z in the direction of three fixed orthogonal axes X, Y, Z defining an inertial reference. The sequence of the three axes should follow the right hand rule as shown in the figure. The thumb will point to the direction of the positive Z axis when the other four fingers close in the X-Y plane moving from the positive X direction to the positive Y direction.

The body shown in Fig. 2-4, for example, is dynamically treated as a particle since the line of action of all the forces (F_1, F_2, F_3) acting on it pass by its center of mass. The body will therefore translate without rotation. The motion of all points in the body is the same as that of its center of mass and the body dimensions will have no effect on this motion.

2. A RIGID BODY:

A rigid body is an object whose dimensions and mass distribution are of primary concern in the analysis of its equilibrium and dynamic behavior. All points in a rigid body are also assumed to remain at fixed distances from each other during its motion and consequently the coordinates of the points of load application remain unchanged with respect to its center of mass.

Any displacement of a rigid body may be reduced to a succession of translation and rotation. Such body has, in general, six degrees of freedom. Three of them can be conveniently considered to represent the general translation (x,y,z) of the center of mass and the other three to represent the rotation ($\theta_x, \theta_y, \theta_z$) of the three reference axes as shown in Fig. 2-5.

A DEFORMABLE BODY OR A CONTINUOUS MEDIUM

In the previous definition of rigid bodies, the assumption was made that all points in the bodies remain at constant distances from each other in spite of the external forcing effects. This assumption is acceptable when the motion of the body as a whole is the principal interest. However, the assumption is an idealization for most situations since it is well known that all bodies deform to a certain extent when external effects are applied to them. The relative displacements between the different particles result for the load distribution within the body and are mutually dependent. A continuum may therefore have infinite degrees of freedom since it is composed of infinite particles which can move relative to each other.

NEWTON'S LAW

The laws of mechanics used to analyze the motion of the previous systems are based on Newton's Laws of motion which are stated in the following:

1. A body continues in the state of rest or of uniform motion in a straight line unless it is compelled to change that state by forces impressed upon it.

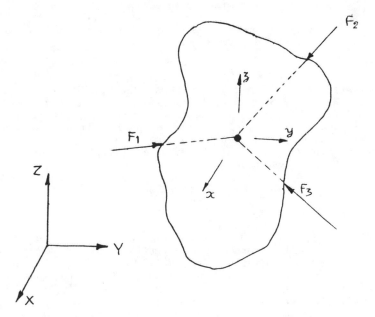

FIGURE 2-4 Rigid body as a particle.

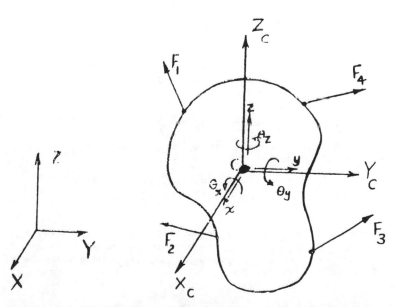

FIGURE 2-5 Six degrees of freedom of a rigid body in space

31

2. The change of motion is proportional to the motive force impressed and is made in the direction of the straight line in which the force is impressed.

3. To every action there is always a reaction equal in magnitude and opposite in direction.

KINEMATICS

Kinematics is the geometry of motion. The kinematics of a particle deals with the motion of a point in space while that of rigid bodies and continuous media takes into consideration changes in position as well as orientation. The position of a point in space is known when its three coordinates with respect to a fixed system of axes are given. Its motion is completely described if its coordinates (x,y,z) are known at all instants of time. These coordinates may be given as:

$$x = f_1(t)$$
$$y = f_2(t)$$
$$z = f_3(t)$$

where f_1, f_2, f_3 are continuous functions of time t since it is not possible in an actual motion that a moving point disappear in one position to reappear after a very small interval of time in a new position at a finite distance from the old. These functions are also assumed to have definite derivatives for every value of time. The first and second derivatives give the components of the velocity (V_x, V_y, V_z) and acceleration (a_x, a_y, a_z) in the direction of the reference axes. This is expressed analytically as

$$V_x = \frac{dx}{dt} \quad , \quad a_x = \frac{d^2x}{dt^2}$$

$$V_y = \frac{dy}{dt} \quad , \quad a_y = \frac{d^2y}{dt^2}$$

$$V_z = \frac{dz}{dt} \quad , \quad a_z = \frac{d^2z}{dt^2}$$

The distance S between any two points (x_1,y_1,z_1) and (x_2,y_2,z_2) in space is shown in Fig. 2-6 to be:

$$S = \sqrt{(x_2-x_1)^2 + (y_2-y_1)^2 + (z_2-z_1)^2} \qquad [2\text{-}10]$$

The direction of the vector S connecting (x_1,y_1,z_1) to (x_2,y_2,z_2) is defined by its orientation with respect to the reference axes. This is generally given as cosine the angles $(\theta_{sx}, \theta_{sy}, \theta_{sz})$ made with the X, Y, Z axes respectively. Accordingly,

$$\ell = \cos \theta_{sx} = \frac{x_2 - x_1}{S} \qquad [2\text{-}11]$$

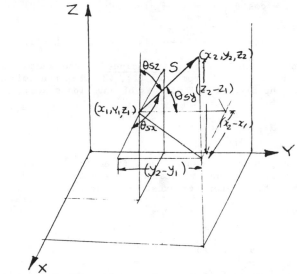

FIGURE 2-6   Direction cosines of a vector in space.

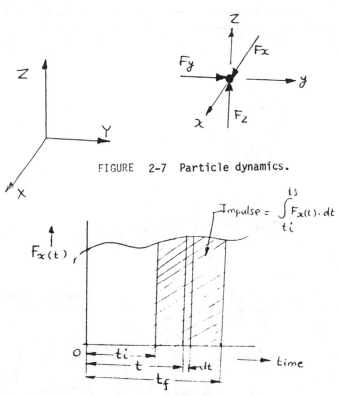

FIGURE   2-7   Particle dynamics.

FIGURE 2-8   Definition of impulse as the area
under the force-time curve.

$$m = \cos\,\theta_{sy} = \frac{y_2 - y_1}{S} \qquad\qquad [2-12]$$

$$n = \cos\,\theta_{sz} = \frac{z_2 - z_1}{S} \qquad\qquad [2-13]$$

where ℓ, m, n are known as the <u>directional cosines</u> of the vector under consideration. It can be concluded from Eqs. (1), (2) that the following relation exists between the directional cosines of any line space:

$$\ell^2 + m^2 + n^2 = 1 \qquad\qquad [2-14]$$

The rotation of a rigid body with respect to a fixed frame of reference can be defined by the change in the orientation of any fixed line in the body with respect to the three reference axes. This orientation is given at any instant by:

$$\theta_x = f_4(t)$$

$$\theta_y = f_5(t)$$

$$\theta_z = f_6(t)$$

where θ_x, θ_y, θ_z are the angles in radians between any body-fixed line and the three axes x, y, z respectively. The functions f_4, f_5, f_6 are also continuous and have definite derivatives. The components of the angular velocity (w_x, w_y, w_z) and the angular acceleration ($\alpha_x, \alpha_y, \alpha_z$) are therefore

$$w_x = \frac{d\theta_x}{dt} \quad , \quad \alpha_x = \frac{d^2\theta_x}{dt^2}$$

$$w_y = \frac{d\theta_y}{dt} \quad , \quad \alpha_y = \frac{d^2\theta_y}{dt^2}$$

$$w_z = \frac{d\theta_z}{dt} \quad , \quad \alpha_z = \frac{d^2\theta_z}{dt^2}$$

PARTICLE DYNAMICS
EQUATIONS OF MOTION

Since a particle is in effect a mass point, its motion in a fixed frame of reference (Fig. 2-7) is completely described by direct application of Newton's Second law:

$$m\,\frac{d^2x}{dt^2} = m\,\frac{dV_x}{dt} = F_x \qquad\qquad [2-15]$$

$$m\,\frac{d^2y}{dt^2} = m\,\frac{dV_y}{dt} = F_y \qquad\qquad [2-16]$$

$$m \frac{d^2 z}{dt^2} = m \frac{dV_z}{dt} = F_z \qquad [2-17]$$

where   m   = mass of the particle
         x,y,z   = components of the displacement
         F_x, F_y, F_z = the forces acting on the particle

WORK AND ENERGY

The previous equations can be used to derive the work and energy equations. Equation (4a) for example can be rewritten as

$$m \frac{dV_x}{dx} \frac{dx}{dt} = F_x$$

or
$$m V_x \, dV_x = F_x \, dx$$

and by integrating this equation between an initial position i and a final position f, we get

$$\frac{1}{2} m (V_x)_f^2 = \frac{1}{2} m (V_x)_i^2 = F_x [x_f - x_i] \qquad [2-18]$$

Similarly,

$$\frac{1}{2} m (V_y)_f^2 - \frac{1}{2} m (V_y)_i^2 = F_y [y_f - y_i] \qquad [2-19]$$

and

$$\frac{1}{2} m (V_z)_f^2 - \frac{1}{2} m (V_z)_i^2 = F_z [z_f - z_i] \qquad [2-20]$$

Since $1/2 \, m \, V^2$ represents the kinetic energy of the particle at any instant therefore these equations illustrate the relationship between work and energy.

> **"The work done by all forces acting on a particle**
> **in any direction equals the change in the kinetic**
> **energy of the particle in that direction."**

Notice that the work and energy are scalar quantities which make this formulation more advantageous, in many situations, than the differential equations of motion. Scalar quantities have only magnitude and no orientation and can therefore be added and subtracted as algebraic quantities.

IMPULSE AND MOMENTUM

Newton's Second Law can also be rewritten to derive the relation between impulse and momentum. Impulse is defined as the area under the force-time curve during the considered interval of time (Fig. 2-8). Momentum on the other hand is defined as the product of mass and velocity

35

at any instant in time. Both impulse and momentum are vector quantities with a magnitude and direction. Using Eq. (4a) as an example, therefore

$$m \frac{dV_x}{dt} = F_x$$

or

$$m \, dV_X = F_x \, dt$$

and by integrating between an initial condition i and a final condition f we get:

$$m(V_x)_f - m(V_x)_i = \int_i^f F_x \, dt \qquad [2\text{-}21]$$

Similarly,

$$m(V_y)_f - m(V_y)_i = \int_i^f F_y \, dt \qquad [2\text{-}22]$$

and

$$m(V_z)_f - m(V_z)_i = \int_i^f F_z \, dt \qquad [2\text{-}23]$$

The left hand side of these equations represents the change in momentum and the right hand side is the impulse in the particular direction during the time interval $(t_f - t_i)$.

RIGID BODY DYNAMICS
EQUATIONS OF MOTION

The rigid body motion (Fig. 2-5) can be described generally by the following six equations:

$$F_x = m \frac{d^2x}{dt^2} \qquad [2\text{-}24]$$

$$F_y = m \frac{d^2y}{dt^2} \qquad [2\text{-}25]$$

$$F_z = m \frac{d^2z}{dt^2} \qquad [2\text{-}26]$$

which govern the translation of the center of mass, and

$$M_x = I_{x_c x_c} \left(\frac{d^2\theta_x}{dt^2}\right) + \left(\frac{d\theta_z}{dt} \cdot \frac{d\theta_y}{dt}\right)(I_{z_c z_c} - I_{y_c y_c}) \qquad [2\text{-}27]$$

$$M_y = I_{y_c y_c} \left(\frac{d^2\theta_y}{dt^2}\right) + \left(\frac{d\theta_x}{dt} \cdot \frac{d\theta_z}{dt}\right)(I_{x_c x_c} - I_{z_c z_c}) \qquad [2\text{-}28]$$

$$M_z = I_{z_c z_c} \left(\frac{d^2\theta_z}{dt^2}\right) + \left(\frac{d\theta_y}{dt} \cdot \frac{d\theta_x}{dt}\right)(I_{y_c y_c} - I_{x_c x_c}) \qquad [2\text{-}29]$$

36

which govern the rotation of the body as a whole. In these equations:

m	= mass of the body
$I_{x_c x_c}, I_{y_c y_c}, I_{z_c z_c}$	= the principal moments of inertia* about the axes through the center of mass (X_c, Y_c, Z_c are body-fixed axes)
x,y,z	= the displacements of the center of mass in the reference frame X, Y, Z
$\theta_x, \theta_y, \theta_z$	= the angular rotation of the body about the reference axes
F_x, F_y, F_z	= the forces acting on the body in the X, Y, Z directions
M_x, M_y, M_z	= the moments of all forces acting on the body about the center of mass in the three reference directions.

The first three equations can be easily recognized as Newton's Second Law. The last three equations are known as <u>Euler's Equations</u> which represent an extension of Newton's Law to account for the rotation of the rigid body about its center of mass. Euler's equations are also applicable when the rotation is about a fixed point in space. The reader who is interested in the proof of these equations should consult any of the texts on the dynamics of rigid bodies.

WORK AND ENERGY

Since the motion of rigid bodies combines a translation of the center of mass and a rotation about it, the kinetic energy for a rigid body should include both the energy of displacement as well as the energy of rotation. It is defined by:

$$\text{Kinetic Energy} = \frac{1}{2} m V^2 + \frac{1}{2} I_{x_c x_c} \left(\frac{d\theta_x}{dt}\right)^2$$
$$+ \frac{1}{2} I_{y_c y_c} \left(\frac{d\theta_y}{dt}\right)^2$$
$$+ \frac{1}{2} I_{z_c z_c} \left(\frac{d\theta_z}{dt}\right)^2 \qquad [2\text{-}30]$$

where

$$V = \left[\left(\frac{dx}{dt}\right)^2 + \left(\frac{dy}{dt}\right)^2 + \left(\frac{dz}{dt}\right)^2\right]^{1/2}$$

V = magnitude of the velocity of the center of mass.

---

*The principal axes of inertia of a rigid body at any point are mutually perpendicular axes, one of them has the highest moment of inertia for the body at the considered point, the other has the lowest moment of inertia and the third is mutually perpendicular to the other two axes. Principal axes are easily determined by inspection if the body has axes of symmetry (an axis of symmetry is a principal axis). Otherwise, they can be determined mathematically as the axes as the given point where the products of inertia vanish.

The work and energy equation can therefore be written as:

Change in Kinetic Energy = work done by all forces and moments acting on the body during a particular motion.

IMPULSE AND MOMENTUM

A rigid body may be subjected to linear impulse due to the resultant force at the center of mass, as well as angular impulse due to the moment of all forces acting on it about the center of mass. It will accordingly undergo a change in both its linear and angular momentum. The linear momentum is obtained by the product of the mass and the velocity V of the center of mass. The angular momentum about each principal axis at the center of mass is obtained by multiplying the moment of inertia by the angular velocity about that axis. This is also true for the angular momentum about a fixed axis of rotation as well as for a principal axis at a fixed center of rotation.

SUPPORT CONDITIONS

The contact or connection between two rigid bodies can occur in a variety of ways. The transmittal of force from one body to another is influenced by the type of support conditions prevalent at the junction. For example, in the human skeletal structure the segments articulate with respect to each other through numerous joints of different types and construction. The motions that can occur at the joint and the transmittal of forces between segments is dependent on the geometry of joint. Several different types of common support conditions are illustrated in this section. In general, the nature of the force system at a particular support can be established by assessing the possibility of relative motion between the bodies. In those directions where relative translation is impeded or obstructed by the support or joint there will be a reaction force component at this support or joint in a free-body diagram of either body. Similarly in the directions about which relative rotation is impeded or obstructed at the junction there will be a moment component at the support or joint in the free body diagram of either body. Some of the common types of support and joints are briefly described in the following:

a) The Pin Joint: (Fig. 2-9)
If the reference orthogonal axes are selected as shown, it is evident that the only motion that can occur at this joint is planar rotation, that is, rotation about the x-axis in the given case. There is no relative translation along any of the three axes indicating reaction force constraints in these directions. There is also no relative rotation about either y or z axis, suggesting moment constraints about these axes. Accordingly, the support conditions at a joint of this type can be described by:

$$F_x = R_x$$
$$F_y = R_y$$
$$F_z = R_z$$
$$M_x = 0$$
$$M_y = M_y'$$
$$M_z = M_z'$$

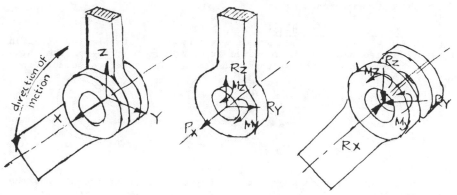

FIGURE 2-9 The pin joint.

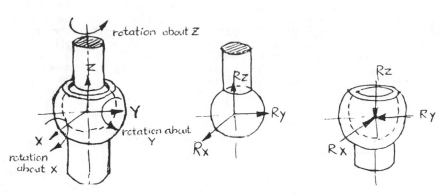

FIGURE 2-10 Ball and socket joint.

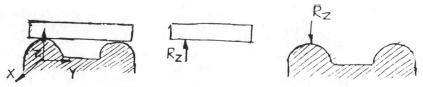

FIGURE 2-11 Simple support without friction.

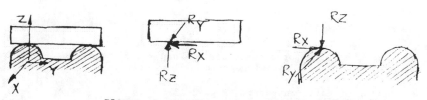

FIGURE 2-12 Simple support with friction

39

An example of such a joint occurs at the elbow, between the upper arm (humerus) and the forearm (ulna).

b) <u>Ball and Socket Joint</u>: (Fig. 2-10)
The only permissible motions at this type of joint is the rotation about the three axes, while relative translation along any of the three axes is constrained. The support conditions can therefore be described by:

$$F_x = R_x$$
$$F_y = R_y$$
$$F_z = R_z$$
$$M_x = M_y = M_z = 0$$

Examples of the ball and socket type of joints occur at the hip and the shoulder joint.

c) <u>Simple Support (Without Friction)</u>: (Fig. 2-11)
In this type of support only normal reaction force is possible at the contact. An example of this occurs at the knee joint, where the curved surfaces of the femoral condyles are supported by the tibial surface. A model of this type of joint is illustrated in Fig. 8 where it is seen that relative translation between the two bodies can occur along the x and y axes while rotation is allowed about all the three axes. Along the z axis motion is possible for separation only, and consequently the support can be described by the following:

$$F_x = 0$$
$$F_y = 0$$
$$F_z = R_z \quad (compressive)$$
$$M_x = 0$$
$$M_y = 0$$
$$M_z = 0$$

d) <u>Simple Support (With Friction)</u>: (Fig. 2-12)
The support conditions for transmission of moments remain unaltered from the preceding case (that is, $M_x = 0$, $M_y = 0$, $M_z = 0$). Similarly, in the vertical direction, the joint is still capable of transmitting a compressive force ($F_z = R_z$). For linear motions which can occur along the x and y axis however, the force constraints along these axes must not exceed the maximum allowable frictional forces resulting from the compressive force R_z. In other words, if μ is the coefficient of friction between the two surfaces then the following inequalities must hold:

$$- \mu R_z \leqslant R_x \leqslant \mu R_z$$
$$- \mu R_z \leqslant R_y \leqslant \mu R_z$$

where R_x and R_y are the force constraints along x and y axis respectively and R_z is the compressive normal reaction on the joint.

e) <u>String or Cable Support</u>: (Fig. 2-13)
As illustrated in the diagram a body supported by string or cable

allows both translation along and rotations about the three axes. The forces T provided by the cable however will always be tensile. The support conditions are therefore:

$$F_x = R_x = T \cos \theta_x = T(\ell)$$
$$F_y = R_y = T \cos \theta_y = T(m)$$
$$F_z = R_z = T \cos \theta_z = T(n)$$
$$M_x = 0$$
$$M_y = 0$$
$$M_z = 0$$

where θ_x, θ_y, θ_z are the angles between the string and the positive directions of the x, y, z axes respectively, and ℓ, m, n are their direction cosines.

EXAMPLES

The study of bodies in equilibrium deals with situations when there is no resultant force acting on the body trying to alter its state of rest or of uniform motion. In this section the analysis of forces keeping particles and rigid bodies in static equilibrium is illustrated by general examples.

STATIC EQUILIBRIUM OF PARTICLES

Let us consider a particle in space which is in equilibrium under the action of three forces F_1, F_2 and F_3 and its own weight W acting through point O (Fig. 2-14). Let us assume that the directions of the three forces F_1, F_2 and F_3 are completely known in terms of their respective direction cosines and it is required to find the magnitudes of these forces.

Three equations are necessary to define the equilibrium of this particle:

$$\Sigma F_x = 0 = F_1 \ell_1 + F_2 \ell_2 + F_3 \ell_3$$
$$\Sigma F_y = 0 = F_1 m_1 + F_2 m_2 + F_3 m_3$$
$$SF_z = 0 = F_1 n_1 + F_2 n_2 + F_3 n_3 - W$$

The above three equations contain three unknown variables: F_1, F_2 and F_3 and therefore can be solved conveniently to yield their values in terms of the known direction cosines;

$$F_1 = \frac{W(\ell_2 m_3 - \ell_3 m_2)}{\ell_1(m_2 n_3 - m_3 n_2) - \ell_2(m_1 n_3 - m_3 n_1) + \ell_3(m_1 n_2 - m_2 n_1)}$$

$$F_2 = \frac{W(m_1 \ell_3 - m_3 \ell_1)}{\ell_1(m_2 n_3 - m_3 n_2) - \ell_2(m_1 n_3 - m_3 n_1) + \ell_3(m_1 n_2 - m_2 n_1)}$$

and

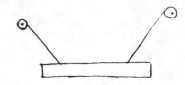

FIGURE 2-13 String or cable support.

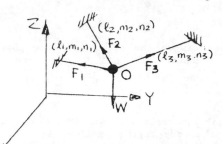

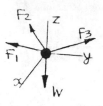

FIGURE 2-14 Equilibrium of a particle.

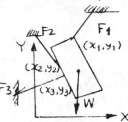

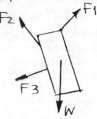

FIGURE 2-15 Planar equilibrium of a rigid body.

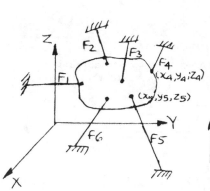

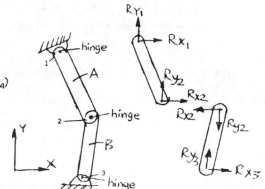

FIG. 2-16 Spatial equilibrium of a rigid
body.

FIG. 2-17 Rigid bodies in planar
motion.

42

$$F_3 = \frac{W(\ell_1 m_2 - \ell_2 m_1)}{\ell_1(m_2 n_3 - m_3 n_2) - \ell_2(m_1 n_3 - m_3 n_1) + \ell_3(m_1 n_2 - m_2 n_1)}$$

It should be re-emphasized that there are three equations to solve for 3 unknowns and therefore the system is statistically determinate. Therefore, so long as the number of equations equals the number of unknown variables contained in them one can obtain an unique solution. This could be further demonstrated by restating the above problem. Suppose the values of forces F_2 and F_3 and their directions are known and it is required to find F_1 and its direction. Since the direction of F_1 is fully established if two of its direction cosines are known*.

The three equilibrium equations formulated above still hold and now contain the three desired unknown variables F_1, ℓ_1, m_1, which can be obtained from the algebraic rules for solution of simultaneous equations as:

$$F_1 = \sqrt{(F_2 \ell_2 + F_3 \ell_3)^2 + (F_2 m_2 + f_3 m_3)^2 + (F_2 n_2 + F_2 n_3 - W)^2}$$

$$\ell_1 = - \frac{(F_2 \ell_2 + F_3 \ell_3)}{\sqrt{(F_2 \ell_2 + F_3 \ell_3)^2 + (F_2 m_2 + F_3 m_3)^2 + (F_2 m_2 + F_2 n_3)^2}}$$

and

$$m_1 = - \frac{(F_2 m_2 + F_3 m_3)}{\sqrt{(F_2 \ell_2 + F_3 \ell_3)^2 + (F_2 m_2 + F_3 m_3)^2 + (F_2 n_2 + F_2 n_3)^2}}$$

EQUILIBRIUM OF A RIGID BODY

The equilibrium of a rigid body is illustrated by two cases — the rigid body (a) in planar and (b) in spatial conditions. The equilibrium of planar motion of a rigid body can be described by three equations—two for force balance along two of the three orthogonal axes and one for moment balance about the third axis. In Fig. (2-15) a rigid body constrained to motion in the x,y plane is acted upon by three forces F_1, F_2, F_3 whose directions and points of application in this plane are completely known. The three equilibrium equations are written as:

$$\Sigma F_x = 0 = F_1 \ell_1 + F_2 \ell_2 + F_3 \ell_3 + f_x$$

$$\Sigma F_y = 0 = F_2 m_1 + F_2 m_2 + F_3 m_2 + f_y$$

$$\Sigma M_z = 0 = F_1(m_1 X_1 - \ell_1 Y_1) + F_2(m_2 X_2 - \ell_2 Y_2)$$
$$+ F_3(m_3 X_3 - \ell_3 Y_3) + (f_y X - f_x Y)$$

where (ℓ_1, m_1, n_1), (ℓ_2, m_2, n_2), (ℓ_3, m_3, n_3) are directional cosines of F_1, F_2, F_3 respectively.

---

* the problem amounts to determining F_1, ℓ_1, and m_1 given F_2, F_3 and (ℓ_2, m_2, n_2) and (ℓ_3, m_3, n_3).

$(X_1,Y_1),(X_2,Y_2),(X_3,Y_3)$ = coordinates of points of application of the forces F_1, F_2, F_3 respectively on the body.

$(f_x,f_y),(X,Y)$ = external forces acting on the body along x,y axes respectively at a point whose coordinates are X,Y.

Note that it is required in this case to determine F_1,F_2,F_3 from the three above equations when all other pertinent information is known. This being a statically determinate problem, it can be readily solved for the three unknowns. This is left as an exercise for the reader. In case the body was acted upon by more than three forces, no determinate solution could have been obtained.

The spatial equilibrium of a rigid body is described by six equations. Since a maximum of six equations will be available at all times for such a case, a determinate solution can be obtained only if there are six unknown variables. This general case of a rigid body in space is illustrated in Fig. (2-16) where six unknown forces F_1,F_2,F_3,F_4,F_5 and F_6 act as shown. The directions of these forces and their points of application on the body and the external forces on the body (such as body weight, inertia forces and moments etc.) and their directions and locations are known.

The six equations of equilibrium are written as:

$$\Sigma F_X = 0 = \sum_{\ell=1}^{i=6} F_i \ell_i + f_x$$

$$\Sigma F_Y = 0 = \sum_{i=1}^{i=6} F_i m_i + f_y$$

$$\Sigma F_Z = 0 = \sum_{i=1}^{i=6} F_i n_i + f_z$$

$$\Sigma M_X = 0 = \sum_{i=1}^{i=6} F_i (n_i \cdot Y_i - m_i \cdot Z_i) + (F_z \cdot Y - F_y \cdot Z) + M'_x$$

$$\Sigma M_Y = 0 = \sum_{i=1}^{i=6} F_i (\ell_i \cdot Z_i - n_i \cdot X_i) + (F_z \cdot Z - F_z \cdot X) + M'_y$$

$$\Sigma M_Z = 0 = \sum_{i=1}^{i=6} F_i (m_i \cdot X_i - \ell_i \cdot Y_i) + (F_y \cdot X - F_x \cdot Y) + M'_z$$

where ℓ_i, m_i, n_i = directions of cosines along X,Y,Z respectively of the force F_i

X_i, Y_i, Z_i = coordinates of point of application of the forces F_i on the body

$(f_x, f_y, f_z), (X, Y, Z)$ = external forces on the body (such as body weights, inertia forces, etc.) and coordinates of its point of application

M_x', M_y', M_z' = external moments on the body (such as those from inertia, or externally applied torques).

The above six equations contain a total of six unknowns and can be readily solved for the latter.

TWO RIGID BODIES IN PLANAR MOTION

A linkage of two rigid bodies capable of motion in the plane of the paper is illustrated in Fig. (2-17). Body A is connected at one of its ends to a support hinge and at its other end to body B. The latter is connected to a second support hinge at its other extremity. As described in Fig. (2-9) each of the two hinges for this case provide a planar reaction which can be resolved into two components along x and y directions respectively. At the pin connection between A and B similarly, there will be a planar reaction force transmitted from one body to the other which can also be resolved into its own two x and y components as shown in Fig. (2-17). Note that in the free body analysis, these reaction components between two adjoining joints whether between A and B or at the hinges will be equal in magnitude but opposite in direction (Newton's Third Law) from one segment to another.

Because of planar motion, each body will be described by three equilibrium equations each, resulting in a total of six equations. We therefore have a system of six equations to solve for the six unknowns and consequently a unique solution can be obtained.

Consider next the situation if an additional member is added to the structure as shown in Fig. (2-15). In this case we have a system of three bodies in planar motion and hence a total of 9 equations. However addition of the new member, increases the total number of unknowns to 10 as two more planar hinge joints with two reaction components (along X, Y) at each, will now have to be considered. Since there are more unknowns than we have equations to solve for them the structure is statically indeterminate.

APPLICATION OF NEWTONIAN MECHANICS TO THE HUMAN BODY

The principles of mechanics outlined in this chapter are the basic laws for the investigation of the musculoskeletal structure. The structure is considered as a system of rigid bodies connected together by different types of joints and acted upon by the tensile forces produced by a large number of muscles and ligaments. Since the system is in equilibrium at all times during normal static postures or motion occurring in everyday life, the forces developed by the muscles acting on the skeletal elements and the reactions at the different joints must maintain each individual segment of the structure in static equilibrium. It can be expected that for almost all instances, the analysis of a particular segment or a linkage of segments of the musculoskeletal structure invariably yields a statically indeterminate problem and therefore special analytical techniques are necessary to obtain a feasible solution. This is discussed in detail in the chapters to follow.

In any of these analyses it is necessary to know the geometrical and inertial characteristics of the human structure. The following chapter deals with this aspect of the problem.

CHAPTER III

Anthropometric Data
of the Human Structure

"After the demonstration of all the parts of the
limbs of men and of other animals, you
will represent the proper method of action of
these limbs."

Leonardo da Vinci
1452-1519

INTRODUCTION

Having noted the similitude between the mechanics of passive systems
and living beings, we will discuss in this chapter, certain physical
parameters of body segments that play a major role in the performance of
dynamic activities and in maintenance of static postures. We will also
discuss the evaluation of such properties from a historical vantage
point, the accepted norms in biomechanical analyses at present time and
techniques of their determination.

Any biomechanical analysis to investigate the influence of different
forces and moments acting on the body or an individual member of the body
skeletal mechanism in performing any normal activity, will be seriously
lacking without adequate knowledge of body segment masses (weights),
locations of centers of gravity and mass moments of inertia. The need
for such information arises in almost all areas of biomechanics, such as
in the design of seats and work environment of aircraft personnel; for
manufacturing of dummies in investigation of motor vehicle crashes; in
physical therapy for rehabilitation of the handicapped; in physical
education for understanding of human motion and for betterment of
athletic performances; and in assessment of body maneuvers of astronauts
in space.

The purpose of this chapter is to briefly review the methods, both
experimental and empirical in determination of segment masses, locations
of centers of masses, and mass moments of inertia and recapitulate the
important results. We shall also illustrate how these results can be
utilized in the analysis of human motions and postural equilibrium.

ANTHROPOMETRIC INVESTIGATIONS OF CADAVERS
A HISTORICAL PERSPECTIVE ON HUMAN ANTHROPOMETRIC MEASUREMENTS

Interest in human anthropometry can be traced back to the ancient
eras. In those times, body measurements were based on established rules
and canons, in which the length of a particular body segment was commonly
used as a standard unit of measure. Around 3000 B.C. the canon in
existence used the distance between the floor (sole) and the ankle joint

47

as the standard module. The height of the body was then set as 21.25 times this module. This was later replaced by a new canon in which the length of the middle finger was used as the module. On this basis the body height was adopted as 19 units (Fig. 3-1a). In the first century B.C., a Roman architect named Vitruvius designed the "square of the ancients", in which the body height was set equal to the distance between the extremities of the middle fingers with the arms outstretched (Fig. 3-1b). This square was later adopted by Leonardo da Vinci who also modified the square by drawing a circle (Fig. 3-1c) around the human figure. In recent times, a decimal canon was constructed by Kollmann by dividing the body height into ten equal divisions which in turn were divided into ten equal parts. According to this, the respective lengths of the head, the whole arm and the leg are illustrated in Fig. 3-1c.

WEIGHTS AND CG"S OF BODY SEGMENTS

Volumetric measurements of dismembered body segments have been accomplished mostly by the process of immersion. The amount of resulting displaced liquid (usually water) is used as a measure of the volume of the segment. This simple procedure is conveniently adaptable to measurements of segments in living subjects as well.

The mass of a cadaver segment can be obtained through direct measurements. The location of center of gravity of a segment is obtained by balancing the segment on a balance-plate to yield the position of the point along the length of the segment (Fig. 3-1). By repeating this process with different surfaces of the segment uppermost on the balance-plate and drawing transverse lines perpendicular to the segment, the true anatomical location of center of mass can be identified.

Leonardo Da Vinci was probably the first to develop norms for the measurements of man and experimental procedures for evaluating the center of gravity for living objects. Although concepts of measurements and the importance of body weight distribution were discussed by Leonardo, no quantitative data were reported by him.

The first known published data is that of Harless[3] (1860) who dissected two young adult cadavers and measured the length, the weight, the density and the location of the center of gravity for various body segments. His results for the weights of segments and their locations with respect to the apex of the head are shown in Tables 3-I and 3-II respectively. To demonstrate the distribution of the body weight over its various segments, Harless created a link model wherein dimensions of the links represent segment lengths and the volumes of the spheres placed at corresponding centers of gravity represent the segment weights (see Fig. 3-2). Since the specimens employed by Harless were decapitated criminals who might have lost significant amount of blood his results for segment weight distribution may not be accurate.

The extensive works of Braune and Fischer[4,5] (1889,1892) and Fischer[6] (1906) who refined the earlier procedures, have come to be regarded as classic and proved to be generally accepted sources of data on human body segment characteristics. Their results for body segment masses and locations of centers of mass based on three male cadavers whose height and weight were representative of that of an average German infantryman are shown in Table (3-III). A later study by them based on dissection of two cadavers of slightly lower average weight compared to

48

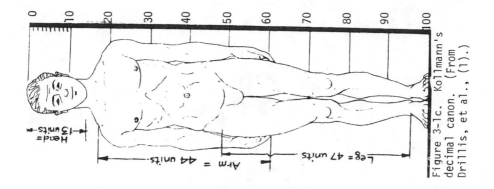

Figure 3-1c. Kollmann's decimal canon. (From Drillis, et al., (1).)

Head = 13 units

Arm = 44 units

Leg = 47 units

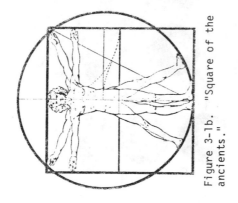

Figure 3-1b. "Square of the ancients."

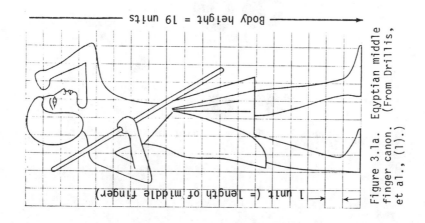

Body height = 19 units

1 unit (= length of middle finger)

Figure 3.1a. Egyptian middle finger canon. (From Drillis, et al., (1).)

49

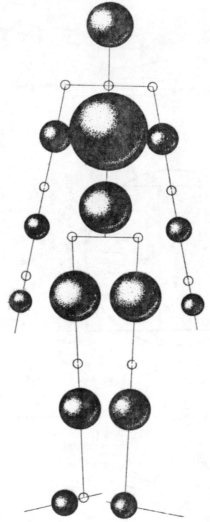

Figure 3-2. Body mass distribution.
(After E. Harten).

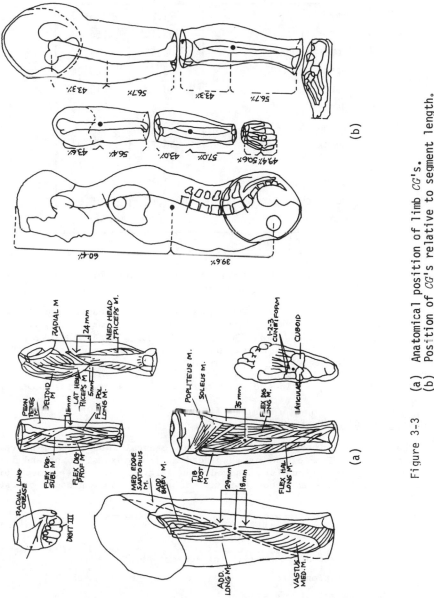

Figure 3-3   (a)   Anatomical position of limb CG's.
               (b)   Position of CG's relative to segment length.
                     (After Dempster (1955a).)

TABLE I.   Relative weights of body segments.*

| Segment | Body wt. = 1000.00 | |
	Cadaver 1	Cadaver
Head	71.20	75.11
Upper trunk	360.40	356.43
Lower trunk	102.43	97.56
Upper arm	32.35	29.04
Forearm	18.13	15.94
Hand	8.44	7.69
Thigh	112.00	118.00
Shank	43.77	45.04
Foot	18.28	19.74
Both upper extremeties	117.86	105.34
Both lower extremeties	348.11	365.56
Whole body	1000.00	1000.00

*After Harless (1860).

that of the three dissected earlier yielded slightly different results for body weight distribution as shown in Table 3-IV.

With a view of analyze limb movements of a seated operator a thorough study of the body segment characteristics was conducted by Dempster[7] (1955) who dissected eight frozen cadavers aged 52 to 83 years old and measured the masses, locations of centers of mass and specific gravities of body segments. The average values for the segment mass distribution and the location of centers of mass are depicted in Table 3-V. The anatomical positions of the centers of mass are illustrated in Fig. 2-3.

MOMENTS OF INERTIA OF SEGMENTS

The moment of inertia of a segment is generally determined by the pendulum method. The segment is suspended at its proximal joint and its frequency of oscillation is measured (see Fig. 3-4). The moment of inertia of such a system about the axis of rotation is expressed by the relation

$$I_0 = \frac{Wd}{4\pi^2 f^2}$$

where I_0 = moment of inertia of the segment about the axis of rotation

W = weight of the segment

52

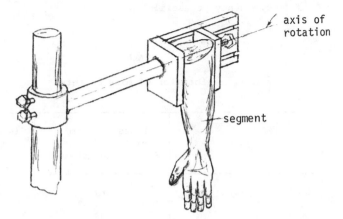

Figure 3-4. Determination of segment moment of inertia of oscillation.

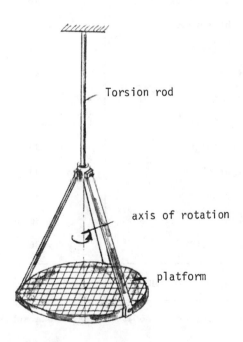

Figure 3-5. Torsional pendulum for determination of moment of inertia of segments.

d = distance of the segment CG from the point of   suspension

f = frequency of oscillation

TABLE II

Location of mass centers.*

Segment	Distance from proximal joint Segment length=1.00	Distance from apex of head sole=1000
Head (from apex)	.361	43.530
Upper trunk	.497	276.585
Lower trunk	.518	413.00
Right upper arm	.427	235.245
Left upper arm	.432	235.245
Right forearm	.417	404.290
Left forearm	.402	402.805
Right hand	.361	608.230
Left hand	.357	605.245
Right thigh	.430	571.850
Left thigh	.509	568.875
Right shank	.493	841.680
Left shank	.494	841.680
Right foot (from heel)	.436	974.955
Left foot	.436	974.955
Mean for all segments	.432	

*After Harless (1860). (Data on one subject.)

If the segment mass and its CG position are known beforehand, I_o can be calculated using the above expression. The moment of inertia of the

TABLE III. Body mass distribution and center of mass.*

Segment	Segment mass (% of total body weight)	Center of mass from proximal joint (% of segment length)
Upper arm	3.3	47.0
Forearm	2.1	42.1
Forearm & hand	2.9	47.2[1]
Hand	0.85	
Upper leg	10.75	43.9
Lower leg	4.8	41.95
Foot	1.7	43.4[2]
Lower leg & foot	6.5	51.9[3]
Head, neck & trunk minus the limbs	53.0	
Head	6.95	
Torso	46.3	
Entire body	100.0	

*After Braune & Fischer (1889). Average values for three cadavers.
[1]Segment length is from arbital axis to lower edge of flexed fingers.
[2]Segment length is from front to rear of foot.
[3]Segment length is from knee joint to sole of foot.

TABLE IV. Body mass distribution and center of mass.*

Segment	Segment mass (% of total body weight)	Center of mass from proximal joint (% of segment length)
Upper arm	2.93	45.9
Forearm & hand	2.54	46.0
Upper leg	11.23	43.4
Lower leg	4.53	42.4
Foot	1.88	41.7
Lower leg & foot	6.42	53.4
Head & trunk	51.31	58.1
Entire body	100.00	

*After Braune and Fischer (1892). Average values for two cadavers.

segment about a parallel axis through the center of gravity (I_{CG}) can then be computed from:

$$I_{CG} = I_0 - \frac{W}{g} d^2$$

An alternate method is based on the torsional pendulum concept. The torsional pendulum consists of a torsion rod attached to a platform carrying the segment (see Fig. 3-5). If for example the location of the center of mass of the segment is known, the segment can be so positioned on the platform as to have the axis of torsional oscillations pass through the CG. In this way, the moment of inertia of the segment about its CG can be obtained by the following procedure: Respective time periods of torsional oscillations of the platform are measured with the platform empty (T_0), with the platform carrying an object of known moment of inertia (T_A) and with the platform carrying the segment whose moment of inertia is to be determined (T_B). The segment moment of inertia is then calculated from the following expression:

$$I_B = I_A(T_B^2 - T_0^2)/(T_A^2 - T_0^2)$$

where I_A = moment of inertia of the object
       I_B = moment of inertia of the segment about the axis of rotation

The moments of inertia of body segments using dissected cadavers as reported by Braune and Fischer[5] (1892) and Fischer[6] (1906) are shown in Table 3-VI. The radii of gyration ($r = \sqrt{I/m}$) expressed as percent of segment lengths are also shown in the same table. The average results for moments of inertia about the center of gravity of body segments as obtained by Dempster[7] (1955) are shown in Table VII. Since he did not report the average value for the eight cadavers dissected, data for each individual cadaver is presented. Recent works in this field are those of Hodgson et al.[2] (1972) and Becker[2] (1972) who have measured the masses and principal moments of inertia of cadaver heads. According to the former, the average weight of 13 cadavers was found to be 10.0 lb while the average value of moments of inertia about the principal axis perpendicular to the sagittal plane was 0.20 lb-sec^2-in.

ANTHROPOMETRIC INVESTIGATIONS ON LIVING SUBJECTS
WEIGHTS AND CG'S OF BODY SEGMENTS

Different techniques have been reported by several authors for determining body segment characteristics in living subjects. Understandably, the biological aspects of human body often prove to be an insurmountable source of error during direct determination of such properties of living subjects. For example the respiration process associated with inhaling or exhaling air into the lungs may cause a change in body weight and therefore in its position of center of gravity. Similarly the blood circulation is affected by the body postures assumed and the type of activity performed. Thus migration of blood to and from the different segments of the body may have some effect on mass distribution. Moreover, during the course of an experiment substantial shifting of tissue and of internal organs may also occur depending on the posture, thus seriously altering the body segment characteristics. Inevitably, these physiological and anatomical body

57

TABLE V. Body mass distribution and CG locations.*

Segment		Percent of total weight	Distance from center of gravity to reference dimension stated as %
Trunk minus limbs		56.5	60.4% to vertex; 39.6% to hip axis
Trunk minus shoulders		46.9	64.3% to vertex; 35.7% to hip axis
Both shoulders		10.3	
Head and neck		7.9	43.3% to vertex; 56.7% to seventh cervical vertebra
Thorax		11.0	62.7% to first thoracic centrium[1], 37.3% to twelfth thoracic centrium
Abdomen plus pelvis		26.4	59.9% to first lumbar centrium; 40.1% to hip axis
Entire upper extremity	L R	4.8 4.9	51.2% to gleno-humeral axis; 48.8% to ulnar styloid
Upper arm	L R	2.6 2.7	43.6% to gleno-humeral axis; 56.4% to elbow axis
Forearm and hand	L R	2.1 2.2	67.7% to elbow axis; 32.3% to ulnar styloid
Forearm	L R	1.5 1.6	43.0% to elbow axis 57.0% to wrist axis
Hand	L R	0.6 0.6	50.6% to wrist axis; 49.4% to knuckle III (in position of rest)
Entire lower extremity	L R	15.7 15.7	43.7% to hip axis; 50.6% to medial malleolus
Thigh	L R	9.7 9.6	43.3% to hip axis; 56.7% to knee axis
Leg and foot	L R	6.0 5.9	43.4% to knee axis, 56.6% to medial malleolus
Leg	L R	4.5 4.5	43.3% to knee axis; 56.7% to ankle axis
Foot	L R	1.4 1.4	24.9% of foot link dimension to ankle axis (oblique); 43.8% of foot link dimension to heel (oblique); 59.4% of foot link dimension to toe II (oblique).[2]

*Dempster (1955a). Average values of eight cadavers.

[1] Center of vertebral body.

[2] Alternately, a ratio of 42.9:57.1 along the heel to toe distance establishes point above which the CG lies on a line between ankle axis and ball of foot.

58

functions associated with normal human life are bound to influence the measurements on a living subject; to the extent of accuracy of information desired from such experiments, one may, with a certain degree of innovation fabricate setups to ameliorate the shortcomings.

A technique for determining the limb weight is to record the shift in the total body CG resulting from movement of the limb. For example, in the procedure adopted by Drillis[1] (1959), the subject is made to lie on a board supported on two ends--a fixed end and on a sensitive weighing scale (see Fig. 3-6). If the CG location of the arm is known (for example, from the results of experiments on cadavers as described above) its mass can be determined by simple analysis in the following manner:

$$m = \frac{(S-S_o)D}{d}$$

where
S_o = reaction force with the arm held horizontal along side the body
S = reaction force with the arm held vertical
D = distance between the two board supports
d = distance of segment mass center from proximal joint

Segment volumes can be determined by direct immersion or by photogrammetry. In the former process, the overflow caused by submerging a segment into a full tank of water is measured and indicates the volume of the segment. Illustrative sketches for finding volumes of the limbs (adopted by Dempster[7] (1955)) are shown in Fig. 3-7. A variant of this technique is described by Drillis et al.[1] (1964) wherein the volume of a segment is measured in finite increments along the length of the segment. The total volume of the segment then is equal to the sum of the incremental volumes; and by taking the moments of these incremental volumes about the desired axis of rotation one may locate the center of volume of the segment.

The photogrammetry method consists of photographing the segment with the resulting picture treated as an aerial photograph of terrain upon which contour levels are applied. The volumes of portions between successive contour levels are found using a polar planimeter on the photograph and totaled up to yield the total volume.

In the procedure followed by Dempster[7] (1955) a pantograph-planimeter setup was used to develop body-contour plots in living subjects. By measuring the areas of body cross section at various heights the body-contour plot is obtained, the area under this plot being the body volume. Volumetric measurements may be used to supplement the information on body mass and center of mass obtained by other means, and serve as a check for the accuracy of these methods.

Meeh[9] (1884) measured body segment volumes of 10 living subjects (8 males and 2 females) 12 to 56 years old and presented the results in terms of one thousandth parts of the total body volume. His results for the body volume distribution, based on a segment subdivision shown in Fig. 3-8, are tabulated in Table 3-VIII. He also reports the whole body specific gravity during quiet respiration to vary between 0.946 and 1.071. Spivak[1] (1915), from his volumetric measurements on 15 males estimated the specific gravity of total body to be between 0.916 to

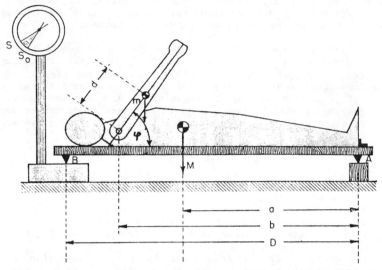

Figure 3-6. Determination of weight of arm. (From Drillis (1959).)

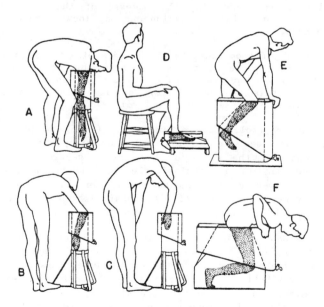

Figure 3-7. Determination of segment volumes by the method of immersion. (From Dempster (1955).)

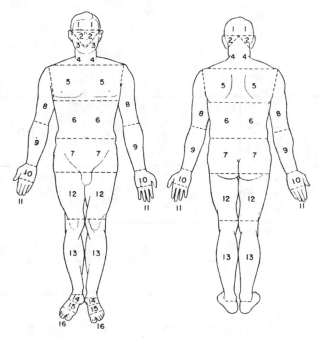

Figure 3-8. Body segments for volume measurement. (After Meeh (1895).)

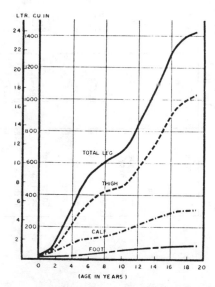

Figure 3-9. Mean leg volume change with age. (After Zook (1930).)

61

TABLE VI. Moments of inertia of body segments.*

Segment		Cadaver 1		Cadaver 2			
		Axis perpendicular to segment longitudinal axis		Axis perpendicular to segment longitudinal axis		Axis parallel to segment longitudinal axis	
		Radius of gyration (% of segment length)	Moment of inertia about the CG ($gmcm^2 \times 10^6$)	Radius of gyration (% of segment length)	Moment of inertia about the CG ($gmcm^2 \times 10^6$)	Radius of gyration (cm)	Moment of inertia about the CG ($gmcm^2 \times 10^6$)
Head & trunk		0.23	9.17	0.29	10.56		
Trunk				0.29	5.57		
Head				0.43	0.18		
Entire upper extremity	R	0.29	1.04	0.31	0.80		
	L	0.28	1.11	0.30	0.76		
Arm	R	0.27	0.10	0.31	0.08	2.8	0.010
	L	0.27	0.12	0.29	0.08	2.7	0.009
Forearm & hand	R	0.26	0.14	0.29	0.12	2.8	0.008
	L	0.28	0.15	0.32	0.15	2.7	0.009
Entire lower extremity	R	0.30	6.78	0.32	4.94		
	L	0.30	7.26	0.32	4.80		
Thigh	R	0.26	0.70	0.31	0.59	4.6	0.100
	L	0.27	0.81	0.31	0.63	4.6	0.100
Shank	R	0.25	0.25	0.24	0.17	3.1	0.020
	L	0.26	0.26	0.26	0.18	3.0	0.018
Lower leg & foot	R	0.32	0.83	0.33	0.62		
	L	0.32	0.83	0.35	0.64		
Foot	R	0.30	0.04	0.30	0.03	6.2	0.035
	L	0.31	0.04	0.30	0.03	6.2	0.035

*From Braune and Fischer (1892) and Fischer (1906).

TABLE VII. Moments of inertia of body parts about their centers of gravity (in gmcm²x10⁶).*

Cadaver Number	Body Weight (Kg)	Entire Upper Extremity		Arm		Forearm and hand		Forearm		Hand		Entire Lower Extremity		Thigh	
		Right	Left	Right	Left	Right	Left	Right	Left	Right	Left	Right	Left	Right	Left
14815	51.364	0.90	1.10	0.190	0.122	0.220	0.187	0.058	0.059	0.007	0.005	4.60	4.90	0.21	0.26
15059	58.409	.79	.78	.130	.118	.137	---	.041	.051	.005	.004	6.70	5.70	.70	.64
15062	58.409	.99	.96	.102	.115	.180	.155	.055	.043	.004	.005	5.67	6.05	.76	.82
15095	49.886	.58	.79	.062	.079	.128	.128	.039	.035	.003	.005	5.20	4.60	.69	.65
15097	72.500	1.10	.98	.220	.222	.298	.287	.072	.055	.011	.009	9.20	9.10	1.27	1.14
15168	71.364	1.40	1.42	.145	.191	.197	.188	.061	.072	.003	.002	8.60	10.00	3.12	3.29
15250	60.455	1.57	1.35	.166	.155	.232	.218	.068	.074	.004	.003	9.50	9.80	1.44	1.22
15251	55.909	0.91	1.02	.122	.112	.152	.146	.054	.050	.003	.003	5.20	5.60	0.61	0.61

Cadaver Number	Body Weight (Kg)	Leg and Foot		Leg		Foot		Trunk Minus Limbs	Trunk Minus Shoulders	Shoulders		Head and Neck	Thorax	Abdomino-pelvic Region
		Right	Left	Right	Left	Right	Left			Right	Left			
14815	51.364	0.81	0.82	0.307	0.321	0.018	0.021	---	---	---	---	---	---	---
15059	58.409	.85	.73	.340	.330	.025	.025	15.5	14.0	0.378	0.355	0.22	0.45	---
15062	58.409	.86	.55	.298	.308	.028	.029	23.7	15.9	.324	.500	---	---	---
15095	49.886	.75	.73	.275	.260	.013	.026	13.4	13.0	.421	.417	---	---	---
15097	72.500	1.65	1.56	.620	.650	.040	.037	22.1	21.0	.700	.800	.31	1.19	3.24
15168	71.364	1.64	1.66	.620	.560	.038	.043	24.3	23.1	.800	.520	.23	2.18	9.70
15250	60.455	1.40	1.29	.620	.560	.035	.035	14.9	16.4	.425	.425	.32	0.96	2.44
15251	55.909	0.96	0.94	.360	.340	.032	.033	14.9	14.0	.420	.480	.39	0.99	1.96

*From Dempster (1955).

63

TABLE VIII. Distribution of body volume.*

Segment	Males (8 Subjects)		Females (2 Subjects)	
Cranimm (1)* & upper jaw (2)	71.64		57.67	
Lower jaw (3) & neck (4)	38.32		29.83	
Head & neck (1+2+3+4)		109.96		87.50
Chest (5)	186.10		137.76	
Abdomen (6)	137.47		144.68	
Pelvis (7)	182.95		215.83	
Whole trunk (5+6+7)		506.52		498.27
Upper arm (8)	28.04		27.56	
Forearm (9)	14.90		13.51	
Palm & thumb (10)	5.20		3.72	
The four fingers (11)	1.95		2.07	
The whole hand (10+11)	7.15		5.79	
Both upper extremities		100.19		93.73
Thigh (12)	81.63		100.42	
Shank (13)	43.56		46.51	
Base of foot (14)⎫ Middle foot (15)⎬	13.77		10.92	
The five toes (16)	2.70		2.40	
The whole foot (14+15+16)	16.47		13.32	
Both lower extremities		283.33		320.50
Total body		1,000.00		1,000.00

*After Meeh (1895). The segment numbers are as indicated in Fig. 5.
1.049. Unlike passive materials, whose properties are invariant with time, living objects undergo property change with age. One instance of such transformation is reported by Zook[10] (1930) who studied the relationship between body segment volume and age in a sample of young people 5 to 19 years old. As an example, his results for mean leg volume change with age are shown in Fig. 3-9.

Since the distribution of body segment composites (muscle, fat, bone, etc.) is not uniform over the segment length, its specific gravity may not be constant along the linear dimensions of a segment. However, according to Salzgeber[11] (1949), a group of workers led by Bernstein in the 1930's in Russian found evidence to the contrary. They observed the mass and volume centers of the extremities to be almost coincident and therefore the density along the segment appeared to be fairly uniform. In general however, the center of mass for the whole body does not coincide with the center of volume, because of the trunk which normally is less dense compared to the remainder of the body. Bernstein and coworkers[12] (1936) carried out comprehensive studies of the body segment characteristics as a prelude to their investigations into biodynamics of human motion and a summary of their results for body weight distribution based on 76 males and 76 females, 12 to 75 years is shown in Table 3-IX, along side those reported by Braune and Fischer[4] on three male cadavers (1889). The results from Dempster[7] (1955) for distribution of volume according to body physique are shown in Table 3-X. His results for mean body segment density based on eight cadavers are shown in Table 3-XI. The mean density values from Table 3-IX are used to estimate weights of body segments based on knowledge of segment volumes (Table 3-X) and shown

TABLE IX. Body mass distribution.

Segment	Bernstein (1935)			Braune and Fischer (1889)
	76 males	76 females	Average	
Upper arms	5.31	5.20	5.26	6.72
Forearms	3.64	3.64	3.64	4.56
Hands	1.41	1.10	1.26	1.68
Thighs	24.43	25.78	25.11	23.16
Shanks	9.31	9.68	9.49	10.54
Feet	2.92	2.58	2.75	3.66

TABLE X. Average volumetric distribution of body limbs according to build.*

Segment	Rotund (percent)	Muscular (percent)	Thin (percent)	Median (percent)
Entire upper extremity	5.28	5.60	5.20	5.65
Arm	3.32	3.35	2.99	3.46
Forearm plus hand	1.96	2.24	2.22	2.16
Forearm	1.52	1.70	1.63	1.61
Hand	0.42	0.53	0.58	0.54
Entire lower extremity	20.27	18.49	19.08	19.55
Thigh	14.78	12.85	12.90	13.65
Leg plus foot	5.52	5.61	6.27	5.97
Leg	4.50	4.35	4.81	4.65
Foot	1.10	1.30	1.46	1.25

*After Dempster (1955).

TABLE XI. Average density of body segments.*

Body Segment		Density
Trunk minus limbs		1.03
Trunk minus shoulders		1.03
Shoulders		1.04
Head and neck		1.11
Thorax		0.92
Abdomino-pelvic		1.01
Entire lower extremity	R	1.06
	L	1.06
Thigh	R	1.05
	L	1.05
Leg and foot	R	1.08
	L	1.09
Leg	R	1.09
	L	1.09
Foot	R	1.09
	L	1.10
Entire upper extremity	R	1.11
	L	1.10
Arm	R	1.07
	L	1.07
Forearm and hand	R	1.11
	L	1.12
Forearm	R	1.13
	L	1.12
Hand	R	1.17
	L	1.14

*From Dempster (1955). Based on experiments on 8 cadavers.

in Table 3-XII. It is apparent that there are significant differences between weight distribution in cadavers and in living subjects notably, the segment mass for upper and lower extremities are a much larger proportion of the total mass for the latter category. This may be due to the large anthropometric variations between the cadaver sample and the living sample. The cadavers were 52 to 83 years old while the living subjects were a select group of male university students. A recent study for determining body segment characteristics has been conducted by Drillis and Contini[13] (1966) on twenty living subjects, who also provide an excellent summary of important findings of previous researchers.

MOMENTS OF INERTIA OF SEGMENTS

The moment of inertia of a segment about its proximal joint in a living subject can be directly evaluated by the "quick-release" method. The method consists of causing the test limb to rotate about its proximal joint under the action of a known torque (see Fig. 3-10). From the laws

TABLE XII. Body mass distribution according to body build.*

Segment	Living Subjects*				Average of 8 cadavers	
	Rotund (percent)	Muscular (percent)	Thin (percent)	Median (percent)	Left Side	Right Side
Entire upper extremity	5.53	5.87	5.45	5.92	4.8	4.9
Arm	3.38	3.42	3.05	3.53	2.6	2.7
Forearm plus hand	2.08	2.39	2.38	2.29	2.1	2.2
Forearm	1.62	1.82	1.74	1.72	1.5	1.6
Hand	0.46	0.59	0.64	0.60	0.6	0.6
Entire lower extremity	20.47	18.69	19.29	19.74	15.7	15.7
Thigh	14.79	12.86	12.92	13.65	9.7	9.6
Leg plus foot	5.68	5.78	6.46	6.14	6.0	5.9
Leg	4.67	4.52	5.00	4.83	4.5	4.5
Foot	1.15	1.36	1.53	1.31	1.4	1.4
Trunk minus limbs	47.99	50.89	50.52	48.67	56.5	56.5

*Computed from results of Dempster (1955). Entries for volume in Table X are multiplied by corresponding density from Table XI to obtain the mass distribution for living subjects.

TABLE XIII-A. Average body mass distribution according to different authors.*

BODY PART	Author			
	Harless (1860) (Average of 2 cadavers)	Braune & Fischer (1889) (Average of 3 cadavers)	Bernstein (1936) (Average of 76 living males)	Dempster (1955) (Average of 8 cadavers)
Arms	6.14	6.72	5.31	5.30
Forearms	3.41	4.56	3.64	3.1
Hands	1.61	1.68	1.41	1.2
Thighs	23.00	23.16	24.43	19.30
Shanks	8.88	10.54	9.31	9.00
Feet	3.80	3.66	2.92	2.80
Head, neck and trunk minus limbs	53.16	49.68	52.98	56.50

*Expressed in % of total body mass.

TABLE XIII-B. Location of segment center of mass according to different authors.*

BODY PART	Author			
	Harless (1860) (Average of 2 cadavers)	Braune & Fischer (1889) (Average of 3 cadavers)	Bernstein (1936) (Average of 76 living males)	Dempster (1955) (Average of 8 cadavers)
Arms	48.5	47.0	46.6	43.6
Forearm	44.0	42.0	41.2	43.0
Thigh	46.7	44.0	38.6	43.3
Shank	36.0	42.0	41.3	43.3

*Expressed in % of segment length, from proximal joint.

69

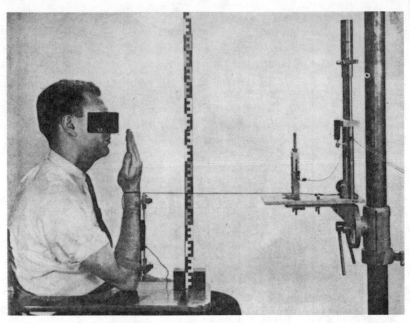

Figure 3-10. The quick-release method of determining the moment of inertia of body segments in a living subject. (From Drillis, et al., (11).)

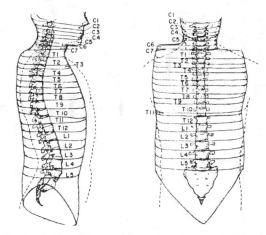

Figure 3-11. Approximate location of trunk dissections. The dashed outline represents the pre-death configuration. (From Lin, et al., (1971).)

of dynamics, the rotary acceleration induced in the segment is related to the applied torque by the expression,

$$T = I\alpha$$

where T = applied torque
   I = moment of inertia of the body about the axis of rotation
   α = angular acceleration of the body

If the torque is computed as a product of force F applied on the limb and the distance 'd' of the force from the proximal joint, then $I = F.d./\alpha$. I can therefore be conveniently evaluated if F, d and α are measured. In the experimental setup, the limb is held so that its proximal joint does not move and a force is exerted on the distal end by means of a cord pulling a spring. The limb is set in rotary motion by sharply cutting the cord and the resulting acceleration is measured.

Several other methods based on the pendulum concept have also been tried. In the compound pendulum method, the segment is allowed to oscillate about its proximal joint. Time periods of oscillation are recorded for three cases: one, with the body segment alone; two, with a known weight attached to the segment at a known distance from the joint; and three, with another weight attached to the segment at the same point. Measurement of these three time periods, the weights added and their locations from the axis of oscillation yields the values for the effective point of suspension, the center of mass and the moment of inertia on solving the three equations of motion. A variant of this method involves making a plaster casting of the segment and determination of its moment of inertia by obtaining its time period of oscillation. The center of mass of the casting is then determined by a balance-plate, and the measured period corrected on the basis of relative weights to represent the desired parameter (center of mass or moment of inertia) of the actual segment if its weight and that of the casting are known. The torsional pendulum may also be used for determining moments of inertia of body segments and of the total body. Details of these experimental setups and the ones described above can be found in Drillis et al.[1] (1963,1964).

Fenn[14] (1938) has reported several measurements obtained by the "quick-release" method and compared them with calculated values based on Braune and Fischer's results. His results shown in Table 3-XIV indicate close agreements with those of Braune and Fischer.

Also shown in Table 3-XIV are the results of Bresler and Frankel[15] (1950) who measured the moments of inertia of the lower leg and foot.

Weinbach[17] (1938) has obtained moments of inertia of the body or body segments by direct measurement of volume increment along the segment length. By using front and side view photographs contour map of the product of incremental volume and $(length)^2$ is drawn; its area being the moment of inertia about segment end. His assumptions that cross-sections of body segments are elliptical and that the density of the body is uniform throughout (=1.00 gm/cm^3) raise doubts as to the validity of the results. Bartholomew (1952) measured the time periods of oscillation of the leg and foot in five living subjects and in one freshly amputated leg. His measurements for the static moment about the joint center (product of weight and distance of center of gravity from the joint) and

71

TABLE XIV. Moments of inertia about proximal end.

Fenn (1938)[1]

Subject	Measured	Calculated[3]
M.N.	5.35	5.74
W.B.L.	3.77	3.79
A.P.	4.44	4.80

Bresler and Frankel (1950)[2]

Subject	Measured	Calculated[3]
1	0.28	0.274
2	0.20	0.265
3	0.24	0.258

[1] Results from Fenn (1938) are for lower leg (units: $gmcm^2 \times 10^6$)

[2] Results from Bresler and Frankel (1950) are for lower leg with shoe (units: $slug-ft^2$)

[3] Calculated from data about segment mass, location of CG and radius of gyration from Braune and Fischer (1889) and Fischer (1906)

the moment of inertia are compared in Table 3-XV against corresponding values calculated from Braune and Fischer's data. His results for static moments seem to agree well with the calculated values but the results for moments of inertia match poorly. Duggar[18] (1963) obtained the moments of inertia of the arm for 10 subjects by the pendulum method and found them to average only 74 percent of the calculated values based on Braune and Fischer's[5] (1892) radius of gyration data. He concludes that the use of the pendulum method in a living subject may not give reliable results because of the damping effect produced by the muscles and joint ligaments. More recently Bouisset and Pertuzon[49] (1968) determined the moment of inertia of the forearm and hand for several subjects using the quick release method. They found the mean value for eleven subjects to be 0.0599 kg m^2 for a range of 0.0430-0.0797 kg m^2.

SEGMENT PROPERTIES FOR THE SPINE

As outlined in the above review determination of segment parameters has centered around the torso and the limbs only. Very few studies have dealt with evaluation of inertial parameters for the vertebral column. An empirical relation suggested by Ruff[20] (1950) assume 60 percent of the body weight to act on the fifth lumbar vertebra with decrements of 2-3 percent as one moves up the lumbar region. In the thoracic region the decrement is of 3-4% starting with about 47% of body weight on the T_{12}. The only known work to date which deals with complete assessment of properties of vertebral segments of the trunk is that by Liu et al.[19] (1971). The authors sectioned one cadaver trunk (weight 152 lb) into

TABLE XV. Static moments and moments of inertia of lower leg and foot about the proximal joint.

Subject	Height (cm)	Weight (kg)	Static moment (Kgcm)		I_o (Gmcm^{2}x10^6)	
			Measured	Calculated	Measured	Calculated
A	175.3	65.8	108.5	118.8	1.19	4.18
B	185.4	68.0	115.2	139.6	2.50	5.37
C	188.0	72.6	138.2	149.9	2.47	6.12
D	172.7	86.2	128.3	151.4	2.41	5.20
E	193.0	77.1	141.0	152.1	3.39	5.71

horizontal segments each containing one vertebra and measured the mass, the location of the center of mass with respect to the center of the vertebral body, and the mass moment of inertia about each of the three principal axes for each segment. The limbs were sawed off at their proximal joints and the trunk was sectioned as shown in Fig. 3-11. Their results for mass distribution, mass eccentricity distribution corrected for tissue shift and twist of cadaver, and mass moments of inertia distribution along the various vertebral levels are shown in Figs. 3-12, 3-13, 3-14 respectively. The mass and mass moments of inertia are generally seen to increase down the vertebral column. Also, it can be seen that at each level $I_z > I_x > I_y$ (Fig. 3-14). Such information is of great assistance in constructing mechanical analogs of the human spine for simulation of the human body in impact and acceleration situations.

TOTAL—BODY ANTHROPOMETRIC INVESTIGATIONS

Numerous investigations dealing with measurement of locations and moments of inertia of living subjects in a variety of postures can be cited. As mentioned earlier, the reliability of measurement is vastly influenced by the dynamic nature of the human body. Even during erect standing for example it is well known that the body continually oscillates in the sagittal and frontal planes affecting the location of the center of gravity (Murray et al.[21] (1975), Akerblom[22] (1948), Hellebrandt et al.[21] (1937,1942), Reynolds and Lovett[24] (1909) etc.) and never attains a truly steady state Fig. 3-15. The respiratory process involving inhaling and exhaling of air into the lungs and the circulation of blood related to changes in posture may also alter the mass distribution and hence the location of the center of gravity (Mosso[25] (1884), Cotton[26] (1932) etc.).

Reynolds and Lovett[24] (1909) obtained the position of the body CG by locating the intersection of the planes containing it during standing and in the supine position respectively. Cotton[26] (1932) reports that for supine subjects the average height of the CG from the platform of rest is 10.9 cm for man and 10.1 cm for woman. Akerblom[22] (1948) observed that in the erect postures, the average CG line fell 1.0 to 5 cm in front of the ankle joints in the sagittal plane, with fluctuations from 1.5 to 5 cm about these values due to continual body sway. In the frontal plane he noted the CG line oscillations to be from 1.5 to 3.5 cm. Swearingen[27] (1953) made extensive measurements on CG positions on five subjects aged

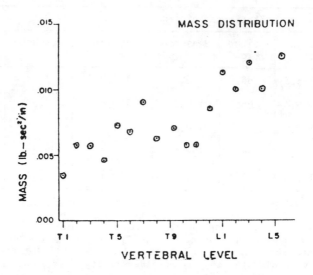

Figure 3-12. Mass distribution of cadaver trunk at various vertebral levels. (From Lin, et al., (1971).)

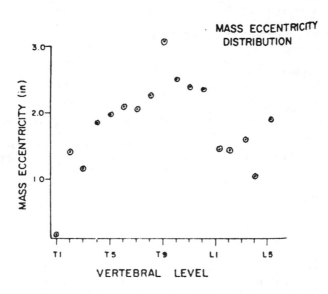

Figure 3-13. Mass eccentricity distribution with vertebral level. The mass eccentricity is the anterior distance from center of the vertebral body in sagittal plane. (After Lin, et al., (1971).)

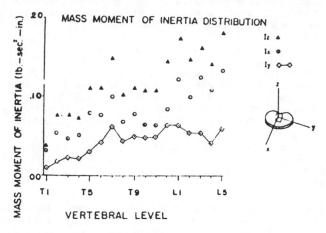

Figure 3-14. Distribtuion of the principal mass moments of inertia with vertebral level. (After Lin, et al., (1971).)

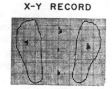

X-Y RECORD

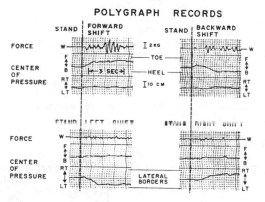

Figure 3-15. Sample X-Y plot and portion of the simultaneous polygraph recordings of a normal man during standing and weight-shifting in each of four directions. The X-Y recordings show the excursions of the center of pressure with respect to the base of support. The top channel on each of the four sample polygraph recordings shows the vertical force fluctuations above and below body weight (W); the middle and bottom channels show the simultaneous forward backward (F-B) and right-left (RT-LT) fluctuations of the center of pressure with respect to the outermost limits of the supporting base.

75

29 to 60 years and weighting 51.4 to 102 kg in 67 positions. His report indicates the CG to fall in a circle of 3.8 cm radius at a distance of 97.2 cm from the floor during erect standing.

Several studies have been conducted with a view to assess the body parameters in its working environment, especially for aircraft personnel equipped with their accounterments. For example the CG of ejection seat-pilot combination based on nine Air Force personnel has been reported by Hertzberg and Daniels[28] (1950) to fall within a 2.8 cm radius of the average location. These values seem to agree well with the results of Swearingen[27] (1953) for seated subjects. Santschi, DuBois, and Omoto[29] (1963) determined the center of gravity and moments of inertia about three axes through the center of gravity for sixty-six living subjects in eight selected body positions. Their mean values of CG locations and moments of inertia of the whole-body for the different postures are shown in Table 3-XVI.

The locus of center of gravity during dynamic activities has been the subject of interest of several authors. A group led by Eberhart[30] (1947) at the University of California have performed extensive investigations on human locomotion in normal subjects and amputees. Using a force plate they recorded the ground-to-foot supportive forces and torque and calculated the displacements of the body center of gravity. Since at any instant of walking, the resultant vertical force on the foot is equal to the body weight superimposed by the inertia force (body mass times vertical acceleration), by integration one can obtain first, the vertical velocity and with subsequent integration, the vertical displacement of the body CG. The vertical displacement of CG during level walking in a normal subject is shown in Fig. 3-16. These correspond to the total vertical force on the foot and its location (that is, the center of pressure) in various configurations assumed during the act as shown. A study of motion of center of gravity during descending to and ascending from squatting and seated postures, and jumping in a normal male was conducted by Murray et al.[31] (1967). The authors recorded photographically the displacements of centers of masses of various body segments and calculated their accelerations by graphical differentiation. Their computed and experimental results for the ground-to-foot vertical forces, the line of gravity and the center of support for one of the activities (descending to squat posture at free and at fast speed) is shown in Fig. 3-17.

COMPARISON BETWEEN PUBLISHED DATA

The results for body weight distribution according to Harless[3] (1860). Braune and Fischer[4] (1889) and Dempster[7] (1955) based on cadaver measurements and according to Bernstein[12] (1936) based on analysis of living subjects is summarized in Table 3-XIII-A. Data from the above authors are also compiled in Table 3-XIII-B for comparing their results on locations of centers of mass of the extremities.

As can be seen in Table 3-XVII for any particular body segment, there appears to be substantial differences between the results reported by these four authors. The most significant deviation is observed for the mass of the head, neck and trunk (without limbs). According to Braune and Fischer, this segment is 49.68% of body weight while according to Dempster the figure is somewhat higher, namely 56.50 percent.

76

Table XVI. Average moments of inertia and centers of gravity location.*

Position	Axis	Center of Gravity** (Cm)		Moment of Inertia (Gm Cm^{2}x10^6)	
		Mean	S.D.	Mean	S.D.
Standing	X	8.9	0.51	130.0	21.8
(arms at	Y	12.2	0.99	116.0	20.6
sides)	Z	78.8	3.68	12.8	2.5
Standing	X	8.9	0.56	172.0	29.5
(arms over	Y	12.2	0.99	155.0	28.6
head)	Z	72.7	3.38	12.6	2.1
Spread	X	8.4	0.48	171.0	30.6
Eagle	Y	12.2	0.99	129.0	24.1
	Z	72.4	4.82	41.4	8.9
Sitting	X	20.1	0.91	69.1	10.6
(elbows	Y	12.2	0.99	75.4	13.1
at 90°)	Z	67.3	2.89	37.9	6.6
Sitting	X	19.6	0.86	70.5	11.0
(forearms	Y	12.2	0.99	77.0	13.6
down)	Z	68.1	2.95	38.2	6.7
Sitting	X	18.3	0.94	44.2	6.8
(thighs	Y	12.2	0.99	43.0	6.6
elevated)	Z	58.7	1.98	29.7	5.8
Mercury	X	20.1	0.86	74.4	10.6
Position	Y	12.2	0.99	85.1	15.8
	Z	68.8	2.89	38.7	6.3
Relaxed	X	18.5	0.84	104.0	15.0
(weight-	Y	12.2	0.99	99.8	15.0
less)	Z	69.9	3.66	40.6	6.1

*From Santschi et al (1963) based on a study of 60 living subjects.

**Location of CGS are w.r.t. the back plane, anterior superior spine of the ilium, and top of the head.

The small number of subjects used in their investigations and the lack of any anthropological information about body physique of the subjects, cast serious doubts as to the applicability of these results for a large heterogeneous living population consisting of different ages, body builds, ethnic types and sex. Similarly, since body build may substantially influence locations of centers of mass, universal coefficients to locate portions of segment centers of masses in the living subjects cannot be conveniently obtained.

MATHEMATICAL MODELS FOR COMPUTING INERTIAL PARAMETERS

With the availability of improved computational tools, mathemtical models of the human body for evaluating inertial properties in living

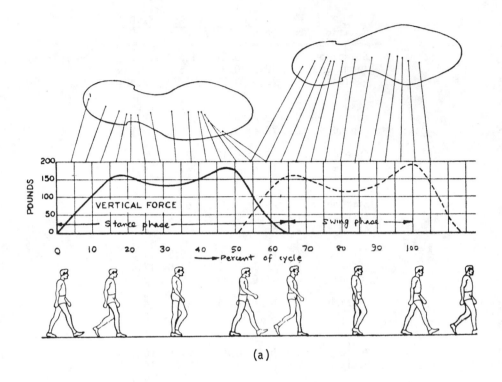

(a)

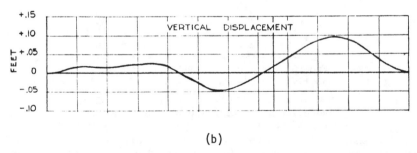

(b)

Figure 3-16. (a) Vertical force on the foot and the center of pressure.
(b) Vertical displacement of the center of gravity of a
normal subject during level walking.
(From University of California (1947).

78

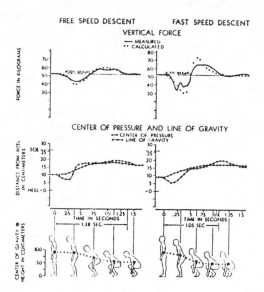

Figure 3-17. Normal male, descending to squat posture at free and at fast speeds. Upper graphs show vertical force patterns measured from platform and values calculated from photographic records of displacements of mass centers of body segments. Middle graphs show fore-aft excursions of center of pressure and line of gravity on the supporting surface. Lower graphs show vertical trajectory of the center of gravity. (From Murray, et al., (1967).)

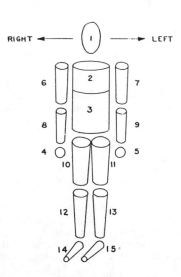

Figure 3-18. Geometrical representation of man. (From Hanavan, 1964).)

subjects are becoming more commonplace. These models treat the body segments as simple, geometrical solids of resolution of uniform density. The shapes most commonly selected are ellipsoids, cylinders, frustums of cones or spheres, and are assigned suitable dimensions to reflect true anthropometric measurements on living persons. Once the dimensions of the various solids representing different body segments are known, their inertial characteristics can be conveniently computed by algebraic expressions describing them.

The model segments are assumed to articulate through hinge joints and can be oriented to simulate normal human postures if the relevant data on segmental kinematics is known. Once a desired configuration is selected, pertinent information as to the location of center of gravity of the body and its moments of inertia about the three centroidal axes can be readily computed. In general the calculated values of desired parameters yielded by the models show fairly close agreement to the experimental results obtained for similar postures in living subjects.

Models of the human body based on geometrical simplifications have been extensively used in investigations of work space requirements of aircraft pilots (Boeing Company[33] (1969), Kroemer[34] (1972), Chaffin and Snyder[35] (1972) etc.), for assessment of body maneuvers in space orbital devices (Griffin[36] (1962), Kuluicki et al.[37] (1962) etc.) and in designing zero-gravity propulsion devices (Flexman et al.[38] (1963) etc.). A mathematical model of the human body to analyze some dynamic response characteristics of a weightless man has been described by Whitsett[39] (1963). A personalized mathematical model of the human body for evaluating inertial properties based on anthropological data and body mass distribution has been developed by Hanavan[40] (1964). His work has been widely used in recent times by several authors (see Refs. 41-48) for investigations of a variety of activities.

Since the basic concepts underlying the development of such mathematical modes are invariably common, the work by Hanavan[40] (1964) and its application in computing the center of gravity and mass moments of inertia of the whole body will be discussed in greater details in the following pages.

Hanavan's Model

The model can be used to evaluate the inertial parameters of the human body in any desired body configuration. These parameters are:

1. mass of body segments
2. centers of mass of body segments
3. mass moments of inertia of body segments
4. center of gravity of the whole body
5. principal axes of inertia of the whole body

The model assumes that the body is symmetrically built and that the inertial properties are not influenced by change in body positions or by application of external forces. The body is treated as a collection of 15 simple geometric solids articulated through hinge joints at their proximal ends, as illustrated in Fig. 3-18. A total of twenty-five body

80

TABLE XVII. List of anthropometric dimensions required to define the model.

1. Weight
2. Stature
3. Shoulder height
4. Substernesle height
5. Trochanteric height
6. Tibiale height
7. Upper arm length
8. Forearm length
9. Chest depth
10. Waist depth
11. Buttock depth
12. Chest breadth
13. Waist breadth
14. Hip breadth
15. Axillary arm circumference
16. Elbow circumference
17. Wrist circumference
18. Fist circumference
19. Thigh circumference
20. Knee circumference
21. Ankle circumference
22. Sphyrion height
23. Foot length
24. Sitting height
25. Head circumference

dimensions as listed in Table 3-XVII are necessary to define the input parameters of the model. Their values were selected from the information reported by Santschi et al.[29] (1963) in a study of the CG's and moments of inertia of sixty-six Air Force personnel in various postures.

The body mass is distributed among the various segments by using Barter's[32] (1957) regression equations (derived from the results of Braune and Fischer[4,5] (1889,1892) and Dempster[7] (1955)) as follows:

$$
\begin{aligned}
\text{Weight of head, neck and trunk} &= 0.47\ W + 12.0 \\
\text{Weight of both upper arms} &= .08\ W - 2.9 \\
\text{Weight of both forearms} &= .04\ W - 0.5 \\
\text{Weight of both hands} &= 0.01\ W + 0.7 \\
\text{Weight of both upper legs} &= 0.18\ W + 3.2 \\
\text{Weight of both lower legs} &= 0.11\ W = 1.9 \\
\text{Weight of both feet} &= .02\ W + 1.5
\end{aligned}
$$

With the exception of the head which is modelled as a right circular ellipsoid and the hands which are modelled as spheres, most of the body segments are represented as frustums of right circular cones. The upper and lower torso however are modelled as right elliptical cylinders. These solids of revolution are then assigned dimensions based on actual dimensions of living subjects. For example, the geometry of the head of the body-model (Fig. 3-19) is defined by:

major radius= R = 0.5 (stature – shoulder height)

minor radius= RR = head circumference/2π

segment length = SL = stature – shoulder height

distance of CG = η = 0.5

segment weight = SW = 0.079 * body weight

segment mass= SW/g

delta= sp. gravity = $SW/(4R(RR)^2)$

segment mass moment of inertia about centroidal Y axis

S1XX= 0.2 SM (R^2+RR^2)

segment mass moment of inertia about centroidal Y axis

S1YY = S1XX

and finally, the mass moment of inertia about Z axis

S1ZZ= 0.4 $SM(RR)^2$

The upper torso is modelled as right elliptical cylinder (Fig. 3-20) and is defined by the following parameters:

R= 0.5 chest breadth
RR= 0.25 (Chest depth + waist depth)
SL= Shoulder height – substernate height
η = 0.5
The masses of the upper and lower torso are taken proportional to their densities

V_2= upper torso volume = $\pi \cdot R \cdot RR \cdot SL$

V_3= lower torso volume

delta= $\dfrac{\text{weight head, neck \& trunk} - \text{weight of head}}{V_2 + \dfrac{1.01}{0.92} V_3}$

SW= delta V_2
SM = SW/g
S1XX= SM$(3R^2+SL^2)$
S1YY= SM$(3RR^2+SL^2)$
S1ZZ= SM(R^2+RR^2)

For the lower torso which is also represented by a right elliptical cylinder, the dimensions are given by:

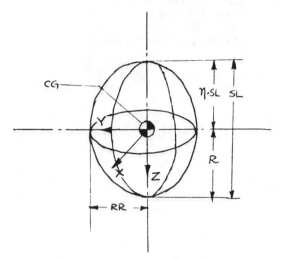

Figure 3-19.   Model of the head.

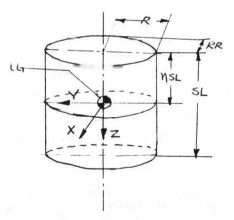

Figure 3-20.   Model of the upper or lower torso.

83

R   = 0.5 hip breadth
RR = 0.25 (waist depth + buttock depth)
SL = Sitting height − (Stature − substernal height)
η = 0.5
V_3 = lower torso volume = π RR R SL
SW = weight of head, neck and trunk − weight of head − weight of upper torso

The mass moments of inertia are given by expressions similar to those for the upper torso.

The hand is modelled as a sphere (Fig. 3-21) and its geometry is derived from:

$$R = \frac{\text{fist circumference}}{2\pi}$$

RR   = R
SL   = 2R
η = 0.5
SW= 0.5 (weight of both hands)
delta= $3SW/(4\pi R^3)$
S1XX= $0.4\ SM\ R^2$
S1YY   = S1XX
S1ZZ   = S1XX

The upper arm is modelled as a frustum of a right circular cone (Fig. 3-22a). Its dimensions are obtained by:

$$R = \frac{\text{Axillary arm circumference}}{2\pi}$$
$$RR = \frac{\text{Elbow circumference}}{2\pi}$$

SL= Upper arm length
SW= 0.5 (Wt. of both upper arms)

The forearm dimensions (Fig. 3-22b) are obtained by:

$$R = \frac{\text{elbow circumference}}{2\pi}$$
$$RR = \frac{\text{wrist circumference}}{2\pi}$$

SL= forearm length
SW= 0.5 (Wt. of both forearms)

The upper leg is modelled as a frustum of a right circular cone (Fig. 3-23a). Its properties are derived from

$$R = \frac{\text{thigh circumference}}{2\pi}$$
$$RR = \frac{\text{knee circumference}}{2\pi}$$

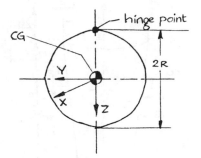

Figure 3-21.   Model of the hand.

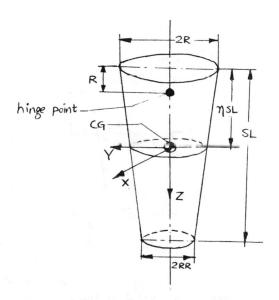

Figure 3-22a.   Model of the upper arm.

85

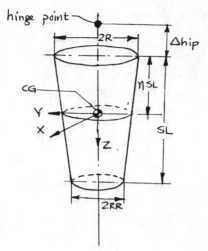

Figure 3-23a. Model of the upper leg.

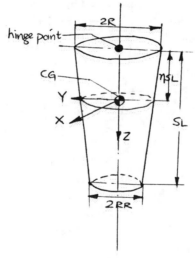

Figure 3-23b. Model of the lower leg.

SL= Stature – sitting height – tabiale height

Δ_{hip}= sitting height – (stature-trochanteric   height)

SW= 0.5 (Wt. of both upper legs)

For the lower leg (Fig. 3-23b) the geometry is expressed by

$R = \dfrac{\text{knee circumference}}{2\pi}$

$RR = \dfrac{\text{ankle circumference}}{2\pi}$

SL= Tibiale height – Sphyrion height

SW= 0.5 (Wt. of both lower legs)

The foot like the rest of segments of the extremities is modelled as a frustum of a right circular cone (Fig. 3-24); its dimensions formulated as:

$$R \;\; = 0.5 \text{ Sphyrion height}$$
$$SL \; = \text{foot length}$$
$$\eta \;\; = 0.429$$
$$SW = 0.5 \text{ (Wt. of both feet)}$$

Since the upper arm, forearm, upper leg, lower leg and foot are all frusta of right circular cones their properties are derived from a common set of formulae given as:

$$\text{delta} \;\; = \frac{3SW}{SL(R^2 + R \cdot RR + RR^2)\pi}$$

$$\mu \;\; = RR/R$$

$$\sigma \;\; = 1 + \mu + \mu^2$$

$$\eta \;\; = \frac{1 + 2\mu + 3\mu^2}{4\sigma}$$

$$AA \;\; = \frac{9}{20\pi}\frac{1 + \mu^2 + \mu^3 + \mu^4}{\sigma^2}$$

$$BB \;\; = \frac{3}{80}\frac{1 + 4\mu + 10\mu^2 + 4\mu^3 + \mu^4}{\sigma^2}$$

$$S1XX \;\; = \frac{AA(SM)^2}{(\text{delta})(SL)} + BB \cdot (SM)(SL)^2$$

$$S1YY \;\; = S1XX$$

$$S1ZZ \;\; = \frac{2(AA)(SM)^2}{(\text{delta})(SL)}$$

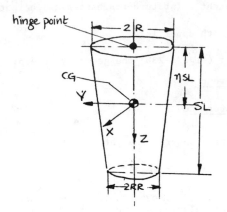

Figure 3-22b. Model of the forearm.

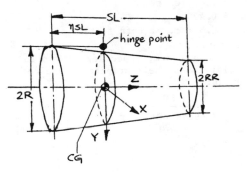

Figure 3-24. Model of the foot.

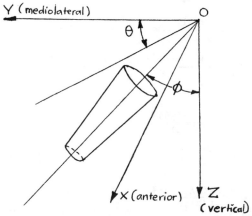

Figure 3-26. Inertial properties for a segment with known orientation.

Using the data for anthropometric dimensions of the Air Force flying population as reported by Hertzberg et al[28] (1950), five composite subjects are defined by using the 5, 25, 50, 75 and 95th percentile (See Table 3-XVIII). The inertial parameters of segments of the 50th percentile male are computed on lines described above and are shown in Table XIX. The main dimensions of the model of such a subject are illustrated in Fig. 3-25.

The model can be used to evaluate the whole body inertial properties for different types of postures if segment configurations are known. For each moving segment two enter angles are defined, θ and φ is shown in Fig. 3-26. θ is the angle between the segment and the Z axis, and θ is the angle between the plane containing the segment and the Z axis makes with Y axis. Hanavan used to model to compare the accuracy of the predicted model results with the experimental values of whole-body CG locations and mass moments of inertia in 66 living subjects as studied by Santschi et al[29] (1963). He found the calculated CG to generally fall within 5/10 of an inch of the experimental data in the X-direction and within 7/10 of an inch of the experimental data in the Z-direction. He also noted that the moments of inertia generally fell within 10% of the experimental data. His predicted results for body segment center of mass location and specific gravity are compared with the experimental results for the same from Dempster[7] (1955) in Table 3-XX show very close agreement. It appears therefore that the data offered by Hanavan (Table 3-XIX) for mass distribution and centers of masses as well as the moments of inertia of the segments may be conveniently adopted for general use in biomechanical analyses. Since these data are based on the experimental works on Braune and Fischer (loc. cit.) and Dempster (loc. cit.) the segmental properties can be readily computed to match a given subject of known dimensions.

89

TABLE XVIII. Anthropometric data of models.

Dimension	Percentile				
	5	25	50	75	95
Weight	123.5	148.7	161.9	167.6	200.8
Stature	65.2	67.5	69.1	70.7	73.1
Shoulder Height	52.8	55.0	56.7	58.0	60.2
Substernate Height	45.6	57.4	48.7	50.1	52.1
Trochanteric Height	32.6	35.0	36.1	37.3	39.0
Tibiale Height	16.6	17.4	18.0	18.7	19.6
Upper Arm Length	12.7	13.1	13.5	13.8	19.6
Forearm Length	10.2	10.6	10.9	11.2	11.6
Chest Depth	8.0	8.6	9.0	9.6	10.4
Waist Depth	6.7	7.3	7.9	8.5	9.5
Buttock Depth	7.6	8.7	8.8	9.4	10.2
Chest Breadth	10.8	11.5	12.0	12.5	13.4
Waist Breadth	9.4	10.0	10.6	10.2	12.3
Hip Breadth	12.1	12.7	13.2	13.7	14.4
Axillary Arm Circumference	11.9	11.8	12.4	13.2	14.4
Elbow Circumference	9.9	10.5	10.9	11.4	12.0
Wrist Circumference	6.3	6.6	6.8	7.1	7.5
Fist Circumference	10.7	11.2	11.6	11.9	12.4
Thigh Circumference	19.6	21.2	22.4	23.6	25.3
Knee Circumference	13.2	13.8	14.3	14.9	15.6
Ankle Circumference	8.1	8.6	8.9	9.3	9.8
Sphyrion Height	2.7	2.8	2.9	3.1	3.2
Foot Length	9.8	10.2	10.5	10.8	11.3
Sitting Height	33.8	35.1	36.0	36.8	38.0
Head Circumference	21.5	22.1	22.5	22.9	23.5

WEIGHT IN LB., DIMENSIONS IN INCHES

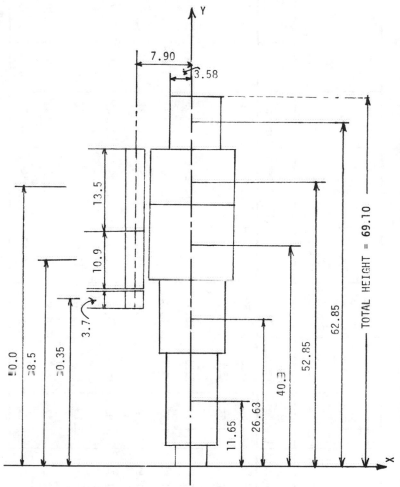

Figure 3-25. Main dimensions of the model of a 50th percentile male (weight=161.9 lbs.).

TABLE XIX. Intertial properties of a 50th percentile male (weight = 161.9 lb).

Segment	Half Length of Minor Axis R (inches)	Half Length of Major Axis RR (inches)	Segment Length SL (inches)	Relative Position of Center of Mass η (inches)	Segment Weight SW (lbs)	Segment Mass $SM-1$ (lb-ft sec^2)	Segment Density Δ (lb/in.^{3})	Moment of Inertia (Slug-in.^{2})		
								I_{xx}	I_{yy}	I_{zz}
Head	6.25	3.58	12.5	0.50	12.8	0.397	0.040	4.11	4.11	2.04
Upper Torso	6	4.25	7.9	0.50	22.5	0.72	0.036	119.0	81.0	37.9
Lower Torso	6.6	4.18	15.6	0.50	52.2	1.62	0.039	605.0	479.0	98.5
Hand	1.85	1.85	3.7	0.50	1.16	0.036	0.043	0.049	0.049	0.049
Upper Arm	1.98	1.74	13.5	0.49	5.0	0.155	0.034	2.34	2.34	0.0065
Fore-Arm	1.74	1.08	10.9	0.422	3.0	0.093	0.043	0.87	0.87	0.0022
Upper Leg	3.57	2.29	15.1	0.428	16.4	0.51	0.040	9.3	9.3	0.054
Lower Leg	2.29	1.42	15.1	0.422	7.9	0.245	0.048	4.38	4.38	0.01
Foot	1.45	0.94	10.5	0.429	2.37	0.074	0.050	0.65	0.65	0.0013

TABLE XX. Location of C.G. of body segments.*

| Body Segments | Model | | | Experimental** |
	High	Low	Average	
Head & torso	73.2	61.3	64.5	60.4
Upper arm	49.6	44.6	47.3	43.6
Forearm	45.0	39.8	42.8	43.0
Upper leg	45.3	47.0	43.7	43.3
Lower leg	47.6	39.8	41.6	43.3

*Distance from upper end in % of segment length
**Experimental results based on dissection of 8 cadavers
as reported by Dempster (1955).

| Body Segments | Model | | | Experimental** |
	High	Low	Average	
Head	1.47	0.90	1.15	1.11
Upper torso	1.00	0.72	0.84	0.92
Lower torso	1.10	0.80	0.92	1.01
Hand	1.72	1.02	1.29	1.17
Upper arm	1.22	0.79	0.97	1.07
Forearm	1.56	1.04	1.30	1.13
Upper leg	1.32	0.88	1.13	1.05
Lower leg	1.44	0.83	1.19	1.09
Foot	2.14	1.12	1.60	1.00

**Experimental results based on dissection of 8 cadavers
as reported by Dempster (1955).

REFERENCES

1. Drillis, R., Contini, R. and Bluestein, M.: Body segment parameters: a survey of measurement techniques. Artificial Limbs: 8,1,1964.

2. Damon, A., Stoudt, H.W. and McFarland, R.A.: The Human Body in Equipment Design. Harvard University Press, Cambridge, Mass., 1966.

3. Harless, E.: Die Statischen Momente der menschlichem Gliedmassen. Abh. d. Math.-Phys. Cl. d. Könige . Baver. Akad. d. Wiss., 8:69-96 and 257-294. 1860.

4. Braune, W. and Fischer, O.: The center of gravity of the human body as related to the equipment of the German infantryman (in German), Treatise of the Math.-Phys. Class of the Royal Acad. of Sc. of Saxony, 26:1889.

5. Braune, W. and Fischer, O.: Bestimmung der Tragheitsmomente des menschlichen Körpers und seiner Glieder. Abh. d. Math. Phys. Cl. d. k. Sachs. Gersell. d. Wiss. 18: 409, 1892.

6. Fischer, O.: Theoretische Grundlagen für eine Mechanik der lebenden Körper , B.G. Teubner, Berline, 1906.

7. Dempster, W.T.: Space Requirements of the Seated Operator. WADC Tech. Rept. 55-159. Wright-Patterson Air Force Base, Ohio, 1955.

8. Drillis, R.: The use of gliding cyclograms in the biomechanical analysis of movement. Human Factors, 1(2): 1959.

9. Meeh, C.: Volumn ess ungen des menschlichen Körpers under seiner einzehner Teile in der vergchiedenen Altersstufen. Ztschr. für Biologie, 13: 125,1895.

10. Zook, D.E.: The physical growth of boys. Am. J. Dis. Children, 1930.

11. Salzgeber, O.A.: A method of determination of masses and location of mass centers of stumps (In Russian) Trans. Svent. Res. Inst. Prosth. Moscow, Vol. 3, 1949.

12. Bernstein, B.A., Salzgeber, O.A., Parlenko, P.P. and Gurvich, N.A.: Determination of location of the centers of gravity and mass of the limbs of the living human body (In Russian). All-Union Inst. Exptl. Med. Moscow, 1936.

13. Drillis, R. and Contini, R.: Body segment parameters. NYU Tech. Rept. No. 1166.03, September, 1966.

14. Fenn, W.O.: The mechanics of muscular contraction in man. J. Appl. Physiol, 9: 165, 1938.

15. Bresler, B. and Frankel, J.P.: The force and movement in the leg during level walking. Trans. ASME, 72: 27, 1950.

16. Bartholomew, S.H.: Determination of knee moments during the swing phase of walking, and physical constants of the human shank. Univ. of Calif., Berkeley, 1950.

17. Weinbach, A.D.: Contour maps, center of gravity, moment of inertia, and surface area of the human body. Human Biology, 10: 356, 1938.

18. Duggar, B.C.: Continuous control performance of the human upper extremity. Unpubl. diss., Harvard School of Public Health, Boston, Mass. 1963.

19. Liu, Y.K., Laborde, M. and Van Buskirk, W.C.: Inertial properties of a segmented cadaver trunk and their implications in acceleration injuries. Aerospace Med., 42: 650, 1971.

20. Ruff, S.: Brief acceleration: less than one second. German Aviation Medicine in WW II, Vol. I., pp. 584-597, U.S. Govt. Printing Office, Washington, D.C. (1950).

21. Murray, M.P., Seireg, A. and Sepic, S.: Normal postural stability and steadiness, qualitative assessment. The Journal of Bone and Joint Surgery, Vol. 57-A, No. 4, pp. 510-516, June 1975.

22. Akerblom, B.: Standing and sitting posture. A.B. Nordiska Bokhandeln, Stockholm. 1948.

23. Hellebrandt, F.A., Genevieve, G. and Tepper, R.H.: The relation of the center of gravity to the base of support in stance. Am. J. Physiol., 119: 331, 1937. Also,
Hellebrandt, F.A. and Fries, E.C.: The eccentricity of the mean vertical projection of the center of gravity during standing. Physiotherap. Rev., 22: 186, 1942.

24. Reynolds, E. and Lovett, R.W.: A method of determining the position of the center of gravity in its relation to certain bony landmarks in the erect position. Am. J. Physiol, 24: 289, 1909.

25. Mosso, A.: Application de la balance a'l'etude de la circulation du sang chez l'homme. Arch. Italiennes de Biologie. J: 130, 1884.

26. Cotton, F.S.: Studies in center of gravity changes. Australian J. of Exptl. Biol., 10: 16 and 225, 1932.

27. Swearingen, J.J.: Determination of centers of gravity of man. Civil Aeronautics Medical Research Lab., Oklahoma City, Okla. (1953).

28. Hertzberg, H.T.E. and Daniels, G.S.: The center of gravity of a fully loaded F-86 ejection seat in the ejection position. Rept. No. MCREXD-45341-4-5, USAF Air Material Command. Wright-Patterson Air Force Base, Ohio, 1950.

29. Santschi, W.R., DuBois, J. and Omoto, C.: Moments of inertia and centers of gravity of the living human body. Tech. Documentary Rept. No. AMRL-TDR-63-36, Wright Patterson Air Force Base, Ohio, 1957.

30. University of California: Fundamental studies of human locomotion and other information relating to the design of artificial limbs. Rept. to Natl. Res. Council. Committee on Artificial Limbs, Berkeley, 1947.

31. Murray, M.P., Seireg, A. and Scholz, R.C.: Center of gravity, center of pressure, and supportive forces during human activities. J. Appl. Physiol., 23: 831, 1967.

32. Barter, J.T.: Estimation of the mass of body segments. WADC Tech. Rept. 57-260, Aerospace Medical Lab, Wright Air Dev. Cntr., Wright-Patterson Air Force Base, Ohio, 1957.

33. Boeing Company: Cockpit geometry evaluation. JANAIR Report 630104, Boeing Report No. D162-10128-1. Jan. 1969.

34. Kroener, K.H.E.: COMBIMAN: COMputerized BIomechanical MAN-model. AMRL-TR-72-16, 1972.

35. Chaffin, D.B. and Snyder, R.G.: New advances in volitional human mobility simulation. Proceed. Human Impact Response Symp. General Motors Research Lab., Warren, Mich. Oct. 1972.

36. Griffin, J.B.: Feasibility of a self-maneuvering unit for orbital maintenance workers. ASD Tech. Documentary Rept. 62-278. Wright-Patterson A.F. Base, Ohio, August 1962.

37. Kulwicki, P.V. et al.: Weightless man: Self-rotation techniques. AMRL Tech. Rept. 62-129, Aerospace Medical Research Laboratories, Wright-Patterson AF Base, Ohio, October 1962.

38. Flexman, R.E. et al.: Development and test of the below zero-G.belt. AMPL Tech. Rept. 63-23, Wright-Patterson Air Force Base, Ohio, 6570th AMRL, March 1963

39. Whitsett, C.E.: Some dynamic response characteristics of weightless man. AMRL Tech. Report 63-18, Wright-Patterson Air Force Base, Ohio, 6570th AMRL, April 1963.

40. Hanavan, E.P.: A mathematical model of the human body. Aerospace Medical Research Lab., AMRL-TR-102, October, 1964.

41. Lepley, D.A.: A mathematical model for calculating the moments of inertia of individual body segments. GM Report TR67-27; May 1967.

42. Patten, J.S. and Theiss, C.M.: Auxiliary program for generating occupant parameter and profile data. CAL Report No. VJ-2759-V-1R, Jan. 1970.

43. Bartz, J.A. and Gianotti, C.R.: A computer program to generate input data sets for crash victim simulations. Celspan Rept. No. ZQ-5167-V-1, Jan. 1973.

44. Bartz, J.A. and Gianotti, C.R.: Computer program to generate dimensional and inertial properties of the human body. ASME paper no. 73-WA/Bio-3. Winter Annual Meeting, Detroit, Mich., November 1973.

45. Bartz, J.A.: A 3-D computer simulation of a motor vehicle crash victim-Phase 1-development of the computer program. CAL Report No. VJ-2978-V-1, PB No. 204172, July 1971.

96

46. Bartz, J.A.:   Validation of a 3-D mathematical model of the crash victim.   Proc. Human Impact Response Symp., G.M. Research Lab., Warren, Mich., Oct. 1972a.

47. Bartz, J.A.:   Development and validation of a computer simulation of a crash victim in three dimensions.   SAE Transactions, SAE, New York, 1972b.

48. Danforth, J.P. and Randall, C.D.:   Modified ROS occupant dynamics simulation user manual.   Research Publ. GMR-1254, Research Lab. G.M.C., Oct. 1972.

49. Bouisset, S. and Pertuzon, E.:   Experimental determination of the moment of inertia of limb segments.   Biomechanics I. (Ist. Int. Seminar Zurich. 1976).   Karger, Basel/New York, 1968, pp. 106-109.

50. Hill, A.V.:   The dynamic constants of human muscle.   Proc. Roy. Soc. B., 128:   263, 1940.

51. Wilkie, D.R.:   The relation between force and velocity in human muscle.   J. Physiol. (Lond.), 110:   249, 1950.

CHAPTER IV

Modeling of the Musculoskeletal System for the Upper and Lower Extremities

"You will make nothing but confusion in your demonstration of the muscles and their positions, beginnings and ends, unless first you make a demonstration of the fine muscles by means of threads; and in this way you will be able to represent them one above another as nature has placed them; and so you will be able to name them according to the member that they serve, that is, the mover of the point of the big toe, and of the middle bone, or the first bone, etc.

And after you have given these details you will show at the side the exact shape and size and position of each muscle; but remember to make the threads that denote the muscles in the same positions as the central lines of each muscle and in this way these threads will show the shape of the leg and their distance in rapid movement and in repose."

Leonardo da Vinci
1452-1519

INTRODUCTION

The mechanics of all bodies in the universe, whether at rest or in motion is based on Newton's laws. It follows, therefore, that animal mechanics which deals with functioning of segments of the musculoskeletal structures of humans and animals at rest or in motion, must invariably adopt these principles.

The maintenance of static and dynamic human body postures is brought about by the well-coordinated activity of the various muscles. Under neurological commands the muscles are stimulated producing tensile forces which act on the individual body segments articulated through different types of joints to attain a stable posture.

As discussed in the Introductory Chapter a decision on the strategic level from the central nervous system sets into motion the mechanisms which execute a posture or dynamic activity. The controlling criterion for this decision is the feasibility of the act and is primarily subjective. Also with central nervous system participation, a decision on the tactical level defines the best pattern of movement to attain the desired motion or posture. The criterion used for this decision is partially objective (for example, maximization of stability and minimization of expenditure of energy) and partly subjective (reflecting the physical limitation and personal preference of each individual).

Once these decisions are made and the preferred patterns of movement or posture are established, the locomotor system is called upon to achieve the defined task. The participation of the muscles has to be compatible with the physical laws of mechanics and should follow primarily objective rules which are mathematically predictable.

This chapter gives an illustrative description of the general modeling procedure which can be utilized for the evaluation of the muscular load sharing and joint forces for different static postures and dynamic activities. The modeling technique is exemplified by the extensively investigated cases of the human upper and lower extremities. Because of the numerous investigations available in the literature which treat these segments of the musculoskeletal structure, the considered examples would also serve to elucidate the differences between the comprehensive modeling procedures and other simplified methods.

THE MUSCULOSKELETAL MODEL FOR THE UPPER EXTREMITIES

The skeletal structure of the upper extremity is shown in Fig. 4-1. It is composed of the hand, with its multi-segmented structure, complex musculature and ligamentation, connected to the forearm at the wrist joint. The forearm is composed of two bony elements: the ulna and the radius connected together by an interosseous membrane. The radius can rotate with respect to the ulna about a longitudinal axis to produce such motions as supination and pronation.

The upper arm, on the other hand, is a single bone, the humerus, and articulates at its distal end with the forearm through the humeroulnar or the elbow joint. The proximal end of the humerus is connected to the shoulder girdle at the glenohumeral joint (or shoulder joint) formed by the head of the humerus and the glenoid cavity of the scapula. Mechanically, this functions as a ball and socket joint allowing rotary movements about three axes—about a longitudinal or vertical axis during actions of medial and lateral rotation, about a transverse axis such as during arm flexion and extension, and about an anterior-posterior axis for acts of abduction and adduction of the arm.

THE ARM MUSCULATURE

The skeletal muscles spanning different segments are the tensile members of the musculoskeletal structure. They develop the pull at their points of attachments on the segments to ensure the stability of body actions at all times. One may, therefore, assume that a muscle develops tension along a straight line in space connecting its point of origin to its point of insertion. Since muscles do not originate and insert in a truly straight line fashion considerable judgement is exercised in defining the best possible approximation for their line of action in space. As will be shown later, in several instances one may need more than one line to describe the different parts or components of the same muscle, in order to simulate its capability for producing widely differing actions.

The major muscles of the arm which are included in the considered model are the following:

1.	Latissimus dorsi	(Fig. 4-2)
2.	Pectoralis major	(Fig. 4-3, 4-4)
3.	Deltoideus	(Fig. 4-5)
4.	Supraspinatus	(Fig. 4-6)
5.	Infraspinatus	(Fig. 4-6)
6.	Teres major	(Fig. 4-7)
7.	Teres minor	(Fig. 4-6)

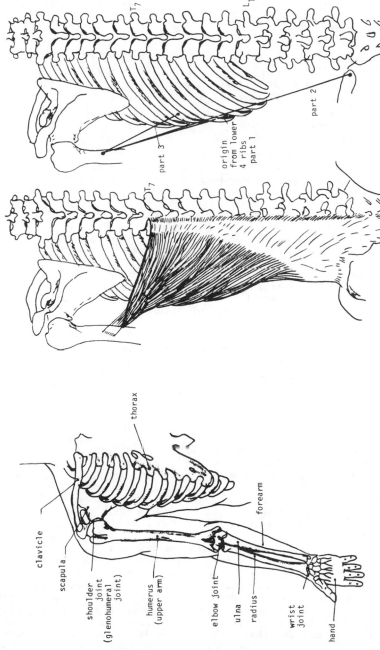

Figure 4-2. Muscle model for the latissimus dorsi.

Figure 4-1. The skeletal structure of the upper extremity.

8.	Subscapularis	(Fig. 4-6)
9.	Coracobrachialis	(Fig. 4-5)
10.	Biceps Brachii	(Fig. 4-8)
11.	Brachioradialis	(Fig. 4-9)
12.	Triceps brachii	(Fig. 4-7)
13.	Brachialis	(Fig. 4-5)
14.	Anconeus	(Fig. 4-7)
15.	Supinator	(Fig. 4-11)
16.	Pronator teres	(Fig. 4-9)
17.	Extensor carpi radialis longus	(Fig. 4-10)
18.	Extensor carpi radialis brevis	(Fig. 4-10)
19.	Extensor digitorum communis	(Fig. 4-10)
20.	Extensor carpi ulnaris	(Fig. 4-10)
21.	Flexor carpi ulnaris	(Fig. 4-9)
22.	Flexor carpi radialis	(Fig. 4-9)
23.	Flexor pollicis longus	(Fig. 4-12)
24.	Flexor digitorum sublimis	(Fig. 4-9)

Since the hand is considered to be an integral part of the forearm, muscles that originate on the forearm and insert on the hand are excluded from the above list. Similarly, since the forearm which is composed of the ulna and the radius is assumed to be one single unit, muscles attaching the ulna to the radius are also omitted from the model.

With the exception of the latissimus dorsi, the pectoralis major, the deltoideus, the biceps brachii, and the triceps brachii, which will be dealt with later in greater detail, all of the muscles in the above list can be generally modelled by one straight line connecting the point of muscle origin to its point of insertion. For example the supraspinatus muscle (Fig. 4-6), takes its origin from the supraspinous fossa above the spina of the scapula and inserts into the greater tubercle of the humerus. It is modelled as a line connecting its proper site of origin on the scapula to its corresponding point of insertion on the humerus. Other muscles, namely the infraspinatus (Fig. 4-6) the teres major (Fig. 4-7), the teres minor (Fig. 4-6), the subscapularis (Fig. 4-6) and the coracobrachialis (4-5) which have their origins on the scapula and insert on the humerus are treated likewise.

On similar lines, the brachioradialis muscle as shown in Fig. 4-9 arises from the lateral supracondylar ridge of the humerus and inserts into the alteral port of the radius above the styloid process. It is modelled as a line attaching its origin on the humerus to its insertion on the radius. Examples of some other muscles connecting the upper arm to the forearm and modelled by direct straight lines spanning the origins and insertions are the brachialis (Fig. 4-5), the anconeous (Fig. 4-7), the supinator (Fig. 4-11) and the pronator teres (Fig. 4-9).

The extensors and flexors of the hand (numbered 17 through 24 in the list of muscles), some of which insert in the hand as distal as the phalanges, are modelled somewhat differently albeit by straight lines. Since the hand is assumed to be a rigid part of the forearm, these muscles which are primarily responsible for extension and flexion of the fingers would have no functional significance in the considered model. Only the portions of these muscles which lie in the forearm in their course of distal insertions have to be considered and their portions beyond the wrist joint may be omitted. They are thus modelled as straight lines connecting their origins (usually either from the humeral epicondyles or from the ulna-radius) to appropriate points in the vicinity of the wrist joint along their course of insertions (see Figs. 4-9 through 4-12).

The following muscles due to their relatively complex anatomy require special attention:

a) Latissimus dorsi:

The latissimus dorsi (Fig. 4-2) is a broad, triangular muscle which arises from the thoracolumbar fasica attached to the spinous processes of the lower six thoracic and the lumbar vertebrae, the spine of the sacrum, posterior part of the iliac crest and from the outer surface of the lower four ribs by fleshy digitations. It inserts into the bottom of the intertubercular groove of the humerus by means of a flat tendon and functionally acts to rotate the arm medially and extend and adduct the humerus. Because of its extensive origin this muscle was simulated by means of three lines:

(i)   a straight line connecting the thorax to the humerus representing the part of the muscle arising from the lower four ribs.

(ii)   a straight line connecting the pelvis to the humerus representing the part of the muscle originating from the iliac crest and the spine of the sacrum.

(iii)   a straight line connecting the thorax to the humerus representing the part of the muscle originating on the spines of the thoracic and the lumbar vertebrae. Because of the curvature of the rib cage, it is not possible to model the muscle fibers arising from the vertebrae by a direct line connecting the latter to the humerus as the thorax is interposed between the two locations. Since the muscle will 'wrap around' the thorax during its course of humeral insertion and the thorax being considered to be 'fixed' for the purpose of the model one may select an alternate point on the thorax to serve as a substitute for the origin on the spines of the vertebrae. Consequently the muscle may be modelled as a straight line joining this point on the thorax to the insertion on the humerus. The muscle model for latissimus dorsi is shown in Fig. 4-2.

b) Pectoralis major:

The pectoralis major arises from the clavicle, from the anterior surface of the sternum and from the costal cartilages of the upper seven ribs, and inserts on the lateral margin of the intertubercular groove of the humerus. Because of its extensive origin, the muscle is modelled by three lines—two from the thorax to represent the upper and lower portions of the muscle arising from the sternum and the costal cartilages, and one from the clavicle to represent the clavicular portion of the muscle. All three lines converge to attach to the point on the humerus where the muscle inserts (see Figs. 4-3 and 4-4).

c) Deltoideus:

This muscle is modelled by three lines to simulate its anterior, lateral and posterior portions respectively.

The anterior portion arises from the lateral third of the clavicle, the lateral part from the lateral margin of the acromion of the scapula

103

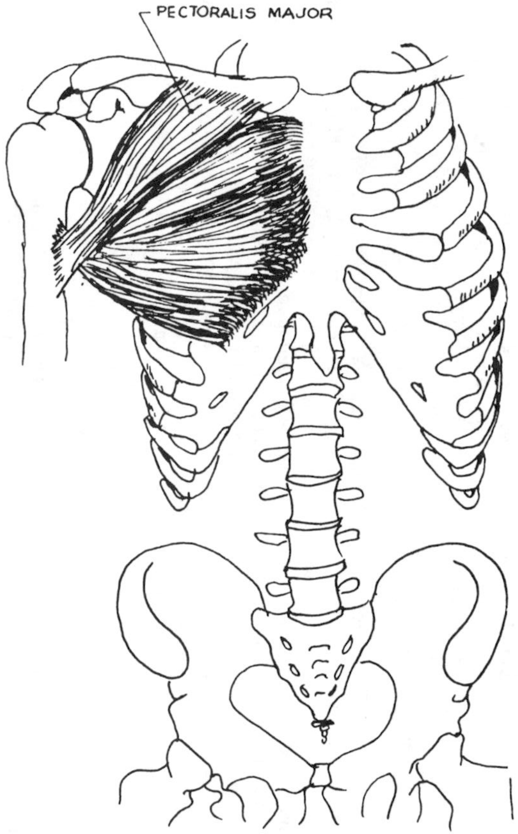

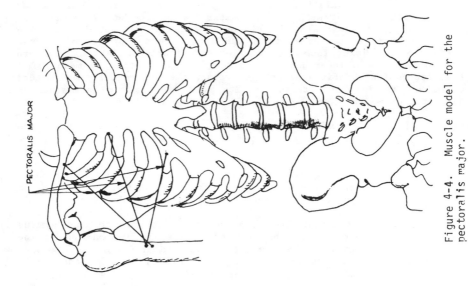

PECTORALIS MAJOR

Figure 4-4. Muscle model for the pectoralis major.

PECTORALIS MAJOR

Figure 4-3. The pectoralis major muscle.

and the posterior part from the spine of the scapula. Though all three
parts are inserted into the deltoid tuberosity of the humerus, only the
anterior and the lateral part are modelled to be so. The posterior part
curves or wraps around the humerus to insert into the deltoid tuberosity
which implies it is not possible to model this portion of the muscle as a
straight line stretching between the points of origin and insertion. An
alternate location for the insertion is therefore selected, situated
somewhat posterior to that for the anterior and the medial parts. (See
Fig. 4-5).

d) Biceps brachii:

The biceps brachii arises from two locations——the short
head from the coracoid process of the scapula and the long head from the
supraglenoid tubercle of the scapula. The long head after passing
through the shoulder joint capsule lies in the intertubercular groove of
the humerus and unites with the short head to be inserted by a tendon
into the tuberosity of the radius (Fig. 4-8). Accordingly, the two
components of the muscle are modelled in the following fashion:

(i)   The long head is assumed to wrap around the head of the
        humerus and is therefore represented by two different
        lines——a line connecting the supraglenoid tubercle on the
        scapula to a point on the head of the humerus in the
        vicinity of the proximal end of the intertubercular groove
        and a second line joining a point in the intertubercular
        groove in the vicinity of the two tubercles to the point
        of insertion on the radial tuberosity. The first line
        represents the course of the long head tendon near its
        origin and is almost horizontal, while the second line
        represents the course of the tendon along the humerus
        after passing through the space between the greater and
        the lesser tubercles, and is directed somewhat
        vertically. In a practical sense thus the long head can
        be likened to a cable wrapping around a pulley formed by
        the tubercular prominence.

(ii)   The short head is conveniently simulated by a straight
        line connecting the coracoid process of the scapula to the
        point of insertion on the radial tuberosity (Fig. 4-8).

e) Triceps brachii:

The triceps (Fig. 4-7) is located in the posterior part of
the upper arm and as the name signifies, has three heads of origin. The
long head originates from the infraglenoid tubercle of the scapula while
the medial and the lateral heads arise from the posterior surface of the
humerus below and above the radial groove respectively. The three heads
unit in a common tendon to insert into the olecranon process of the
ulna. This muscle is consequently modelled by three straight lines, one
for each of the three heads respectively joining their corresponding
points of origin to the common point of insertion on the ulna (Fig. 4-7).

The muscle models described above provided the best linear
representations for the actions of the various muscles under
consideration. As can be seen from above, if the situation so requires,
a muscle may have to be modelled by more than one line to completely
describe its function. In other instances, when a muscle wraps around a

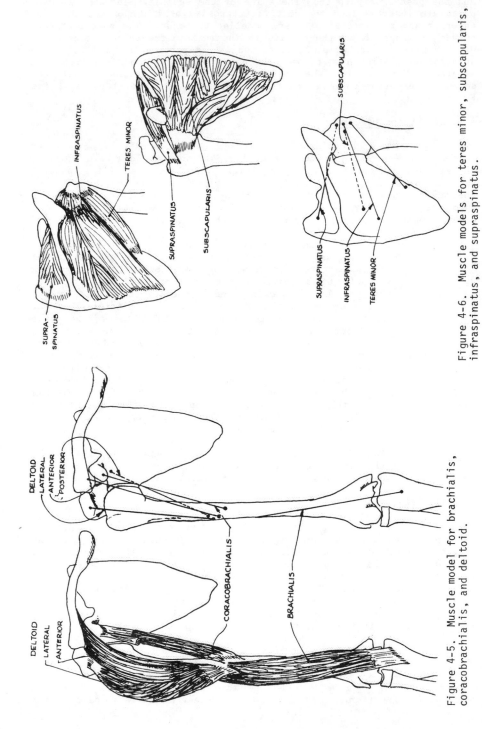

Figure 4-6. Muscle models for teres minor, subscapularis, infraspinatus, and supraspinatus.

Figure 4-5. Muscle model for brachialis, coracobrachialis, and deltoid.

106

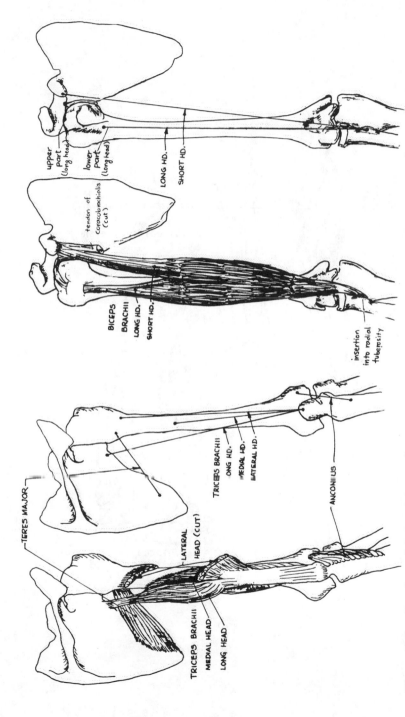

upper part (long head)
lower part (long head)
tendon of Coracobrachialis (cut)
LONG HD.
SHORT HD.
BICEPS BRACHII
LONG HD.
SHORT HD.
insertion into radial tuberosity

Figure 4-8. Muscle model for the biceps brachii.

TERES MAJOR
TRICEPS BRACHII
LONG HD.
MEDIAL HD.
LATERAL HD.
ANCONEUS
LATERAL HEAD (CUT)
TRICEPS BRACHII
MEDIAL HEAD
LONG HEAD

Figure 4-7. Muscle model for teres major, anconeus and triceps brachii.

107

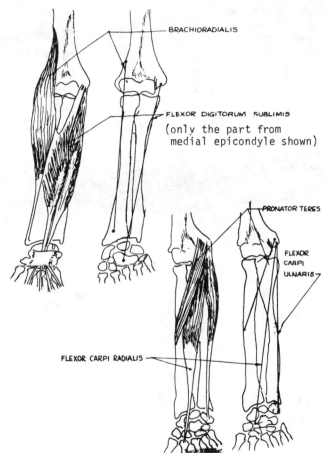

Figure 4-9. Muscle model for brachioradialis, flexor digitorum sublimis, pronator teres, flexor carpi radialis, flexor carpi ulnaris.

segment and therefore cannot be modelled by a single straight line it is simulated by two, as in the case of the long head of the biceps brachii (Fig. 4-8). For some muscles, their true anatomical origins or insertions may have to be disregarded and alternate locations along their course of origin or insertion found, such as for the latissimus dorsi (Fig. 4-2).

SELECTION OF BODY-FIXED AXES AND COORDINATES

The third major step in the modeling process involves identification of the points of muscle origins and insertions with respect to suitable "body-fixed" coordinate axes. The axes are termed 'body-fixed' because they are assumed to be fixed with the skeletal elements and translate or rotate as if they were an integral part of them. The origin of the body-fixed coordinate-axes system for a particular skeletal element is conveniently selected to coincide with the estimated center of rotation of the segment at the joint that it forms with the adjoining element. For example, for the humerus the origin of the body-fixed axes is chosen to be at the center of the head and the three axes are chosen to be parallel to the three axes of rotation of the humerus at the shoulder joint. The three axes follow the right-handed screw system described earlier (Chapter II). With the three coordinate axes thus located at the center of the head of the humerus, the coordinates of the points of interest on this body can be measured using either a skeleton or a scaled diagram. The measurements given in this chapter were made from the scaled diagram as given in the book by Braus (11). The selected body-fixed axes and the coordinates of the muscle attachment points are shown in Figs. 4-13 through 4-16 for the humerus, the radius and the ulna, the hand, the scapula and the clavicle respectively. The diagrams are drawn to scale in two views, anterior and lateral.

The muscle numbers in these figures are consistent with those assigned to the various muscles in Table I. In general the plane formed by the x,z coordinate axes is the same as the sagittal plane (anterior-posterior plane), while the y,z plane coincides with the coronal plane.

The kinematics of the forearm require special attention in view of the fact that the radius can rotate about the ulna such as in supination and pronation. The radius rotates about an oblique axis defined by joining the point of articulation between the head of the radius and the capitulum of the humerus, to the distal end of the ulna. Since movements of flexion and extension at the elbow joint can be performed with the forearm supinated, semi-prone or pronated, it is necessary to establish the position of the radius with respect to the ulna. Accordingly, both ulna and radius have their own body-fixed axes and the coordinates of muscle origins and insertions on each of these bones are established independently. The origins of these coordinate axes for the ulna and the radius are selected as shown in Fig. 4-13. For the hand, the coordinate-axis system is chosen to have its origin at the distal end of the radius at the wrist joint. Although the hand is treated as an integral part of the forearm for the purpose of evaluating the muscle force requirements in the actions to be studied, nonetheless the position of the hand with respect to the forearm needs to be established for each case so that points of interest on this segment can also be determined. In this respect the hand can move with respect to the forearm. The body-fixed axes for the hand as shown in Fig. 4-14 can be used to identify the position of the hand at any articular condition.

TABLE I. Coordinates with respect to body fixed axes.

No.	Muscle Name	Origin	x^1	y^1	z^1	Insertion	x^2	x^2	x^2
1	Latissimus dorsi	Thorax*	-10.5	-9.0	-23.0	Humerus	1.0	0.0	-5.8
2	Latissimus dorsi	Thorax	-8.0	-11.5	-28.0	Humerus	1.5	0.6	-4.0
3	Latissimus dorsi	Pelvis	-11.0	-5.5	-38.0	Humerus	1.5	0.6	-4.0
4	Pectoralis major	Thorax	4.0	1.0	7.0	Humerus	2.0	-0.9	-4.5
5	Pectoralis major	Thorax	4.0	-2.0	-19.0	Humerus	2.0	-0.9	-4.5
6	Pectoralis major	Clavicle	1.8	-5.0	0.7	Humerus	2.0	-0.9	-4.5
7	Deltoideus	Clavicle	0.8	-13.0	0.5	Humerus	1.0	-0.8	-12.0
8	Deltoideus	Scapula	0.0	-2.5	-1.5	Humerus	1.0	-1.0	-12.0
9	Deltoideus	Scapula	-6.5	6.0	-2.5	Humerus	-1.0	0.0	-12.0
10	Supraspinatus	Scapula	-6.0	7.5	-0.6	Humerus	1.9	-2.0	2.5
11	Infraspinatus	Scapula	-9.5	6.0	-8.0	Humerus	-0.8	-2.3	0.4
12	Teres major	Scapula	-11.0	6.0	-14.0	Humerus	1.4	0.8	-6.4
13	Teres minor	Scapula	-7.0	2.5	-9.5	Humerus	-0.6	-2.3	-0.5
14	Subscapularis	Scapula	-9.0	6.5	-8.5	Humerus	1.5	1.2	0.8
15	Coracobrachialis	Scapula	1.0	3.6	-2.5	Humerus	1.0	0.8	-13.2
16	Biceps Brachii (SH)	Scapula	1.0	3.6	-2.5	Radius	0.8	0.8	-5.5
17	Biceps Brachii (LH)	Scapula	-0.5	2.0	-2.0	Humerus	0.5	0.5	2.5
18	Biceps Brachii (LH)	Humerus	2.5	0.0	1.0	Radius	0.8	0.8	-5.5
19	Brachioradialis	Humerus	0.0	-1.5	-27.0	Radius	0.0	-2.0	-24.0
20	Triceps brachii	Scapula	-1.5	0.5	-6.0	Ulna	-1.5	0.0	2.5
21	Triceps brachii	Humerus	-1.0	1.2	-20.0	Ulna	-1.5	0.0	2.5
22	Triceps brachii	Humerus	-0.8	-1.0	-16.0	Ulna	-1.5	0.0	2.5
23	Brachialis	Humerus	1.0	-0.6	-20.0	Ulna	0.0	0.8	-3.0
24	Anconeus	Humerus	-1.6	-2.5	-30.0	Ulna	-1.8	-0.6	-4.0
25	Supinator	Humerus	-1.3	-2.8	-30.0	Radius	0.6	-0.6	-6.5
26	Pronator teres	Humerus	-0.5	3.5	-29.0	Radius	0.5	-1.0	-13.0
27	Extensor carpi radialis longus	Humerus	-1.0	-2.8	-28.0	Hand	-0.6	-1.3	-3.4
28	Extensor carpi radialis brevis	Humerus	-1.0	-3.0	-29.0	Hand	-0.6	-0.9	-3.6
29	Extensor digitorum communis	Humerus	-1.2	-3.0	-29.5	Hand	-0.7	0.0	-1.5
30	Extensor carpi ulnaris	Humerus	-1.4	-2.7	-30.0	Hand	-0.3	2.5	-3.0
31	Flexor carpi ulnaris	Humerus	-1.2	+3.0	-29.5	Hand	0.7	1.8	-1.2
32	Flexor carpi radialis	Humerus	-0.3	3.6	-29.5	Hand	0.8	-1.8	-2.5
33	Flexor pollicis longus	Humerus	-0.5	3.5	-29.5	Hand	0.7	-3.0	-3.0
34	Flexor digit sublimis	Humerus	-0.5	3.5	-29.5	Hand	0.7	0.0	-2.5

*True anatomical origin from lumbar vertebrae. Since the upper and lower torso are fixed, an alternate point on the thorax has been selected.

COORDINATES OF JOINTS

Shoulder joint - (with respect to axes on the scapula at the acromioclavicular joint)
        x = 0, y = 0, z = -4.0 cm
Elbow joint   - (with respect to axes on the humerus) x = 0, y = 2.0, z = -30.5
Radiohumeral articulation - (with respect to axes on the humerus) x = 0, y = -1.0,
        z = -30.5
Wrist joint   - (with respect to axes on the radius at the radiohumeral articulation)
        x = 0, y = 0, z = -26.5 cm

(Measurements in the standing erect posture with the arm hanging vertically, forearm supinated.)

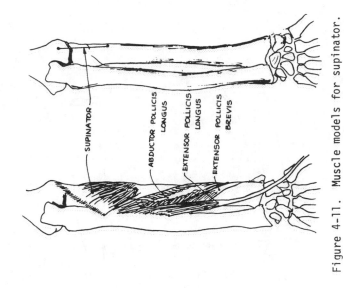

Figure 4-11. Muscle models for supinator.

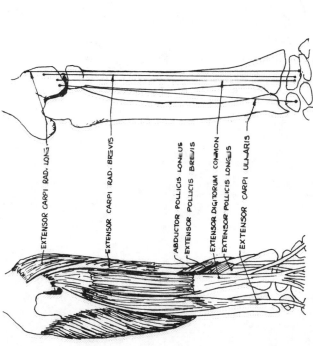

Figure 4-10. Muscle models for extensor carpi radialis longus and brevis, extensor digitorum communis, and extensor carpi ulnaris.

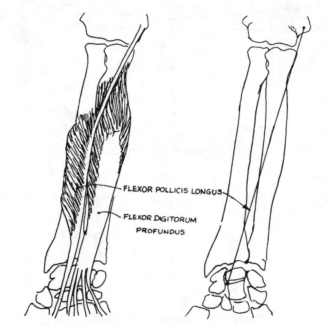

Figure 4-12. Muscle model for flexor pollicis longus.

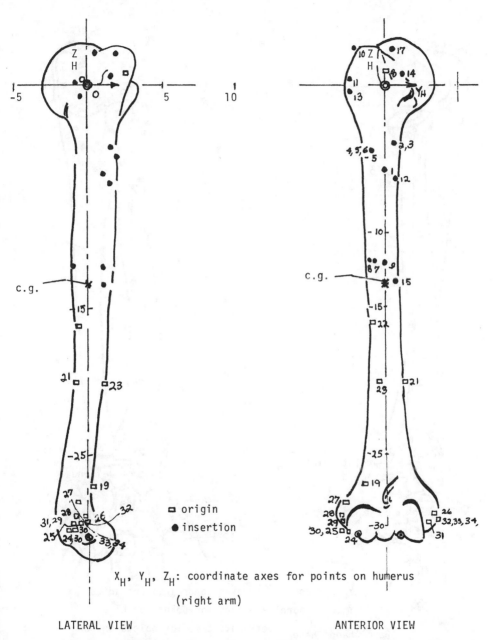

X_H, Y_H, Z_H: coordinate axes for points on humerus

(right arm)

LATERAL VIEW   ANTERIOR VIEW

Figure 4-13. Coordinates of muscle attachment points on the humerus.

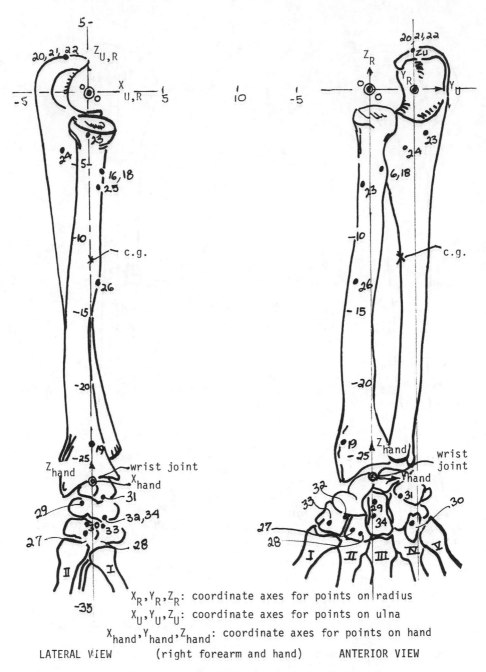

X_R, Y_R, Z_R: coordinate axes for points on radius

X_U, Y_U, Z_U: coordinate axes for points on ulna

$X_{hand}, Y_{hand}, Z_{hand}$: coordinate axes for points on hand

LATERAL VIEW   (right forearm and hand)   ANTERIOR VIEW

Figure 4-14. Coordinates of muscle attachment points on the ulna-radius.

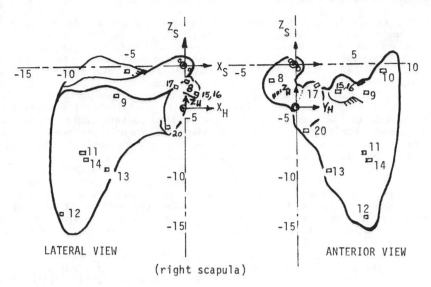

Figure 4-15. Coordinates of muscle attachment points on the scapula.

□ origin
● insertion

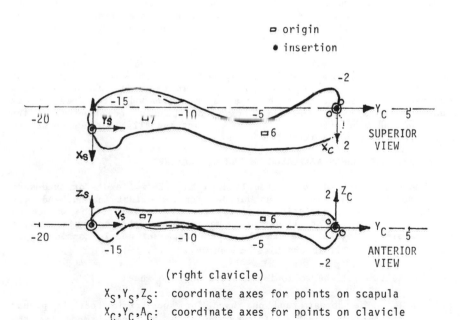

(right clavicle)

X_S, Y_S, Z_S: coordinate axes for points on scapula
X_C, Y_C, A_C: coordinate axes for points on clavicle

Figure 4-16. Coordinates of muscle attachment points on the clavicle.

SELECTION OF GROUND-FIXED AXIS

The segmental configuration, that is, the posture of the arm in any action under analysis has to be pre-established. This involves knowledge of the spatial orientations of the various segments. The thorax along with the rest of the shoulder girdle may be considered to be rigid or immobile, and hence one may select a suitable set of reference axes lying on this body which will be fixed at all times. Thus by knowing the position of the hand with respect to the forearm, the position of the forearm with respect to the upper arm and the position of the upper arm with respect to the 'fixed' thorax, the desired postural orientation can be achieved. It is possible though, that the scapula and the clavicle interposed between the thorax and the arm which complete the upper extremity linkage may also undergo motion along with the arm (in raising the arm, for instance). Since several of the upper extremity muscles have their points of origin on the scapula, on the clavicle or on the thorax and cross the shoulder joint to insert on either the upper arm or the forearm, it may be desirable to consider the motions of the three segments. If the orientation of the upper arm with respect to the scapula, of the scapula with respect to the cavicle and of the clavicle with respect to the thorax is known, then the kinematics of the upper-extremity linkage can be fully established. Since the thorax is assumed immobile, the point of articulation between the thorax and the clavicle, the sternoclavicular joint, appears to be the most favorable location for the ground-fixed axes (see Fig. 4-17). It is with respect to these axes that the coordinates of all points of interest lying on different segments will then be evaluated. This is performed through coordinate transformations, which relate the coordinates of points on a particular segment which is freely movable, with respect to its own "body-fixed" axes to the ground-fixed axes. The details of these transformations are discussed in Appendix (A). It would only suffice to mention here that the transformation requires the identification of the orientation of the segment-fixed axes and their origin with respect to the ground-fixed axes.

It can, therefore, be seen that even a relatively simple segment of the human structure has many elements with many degrees of articulation. A general model of the upper extremity would therefore require the treatment of each bone as a rigid body described by six equations representing the equilibrium condition of the bone under the action of the muscle forces, ligament support and all the reaction forces at the joints with the adjoining bones.

SIMPLIFIED ARM MODELS AVAILABLE IN THE LITERATURE

The evaluation of forces required in the elbow flexors or extensors during normal activities such as during carrying weight in the hand with the elbow flexed or resisting traction forces applied at the wrist has often been used as an illustrative example of the analysis of the musculoskeletal forces. An example of the types of models considered is that adopted by Williams and Lissner[1] shown in Fig. 4-18.

In their model the following assumptions are made:

1. The forearm and the hand form one rigid body in planar motion capable of rotation about the z-axis only at the elbow.

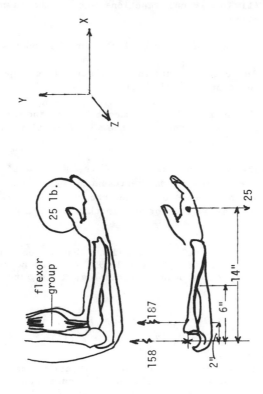

Figure 4-18(a). Calculating the flexor force required to lift a 25 lb weight in the hand. After Williams and Lissner (1).

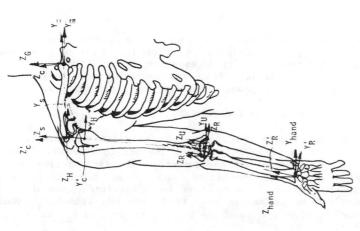

Figure 4-17. Kinematic linkage for evaluating coordinates with respect to ground-fixed axes X_G, Y_G, Z_G.

2. All forces and reactions act in the plane of motion (x-y plane).

3. The rest of the body including the upper arm are fixed.

4. The elbow joint is a planar hinge capable of resisting reactions R_x and R_y only.

5. All applied forces, muscle forces and reactions are either in the x or in the y direction to simplify the calculations.

6. The different actions of the muscles of the flexor group are combined and represented by a one single force F_{bb} which is considered as the resultant of all the muscular actions on the forearm in this configuration.

The problem is therefore reduced to that of equilibrium of a rigid body in planar motion (see pages 192-199) and can be described by three equations with only three unknown variables; R_x, R_y and F_{bb} as:

$$\Sigma F_x = 0 \quad R_x$$
$$\Sigma F_y = 0 = F_{bb} - 4 - 25 - R_y$$
$$\Sigma M_z = 0 = F_{bb}(2) - 4(6) - 25(14)$$

The magnitude of the unknown variables can consequently be determined by simultaneous solution of the equations. The first equation shows that $R_x = 0$. The value of F_{bb} can be calculated from the last equation as:

$$F_{bb} = \frac{24 + 350}{2} = 187 \text{ lbs}$$

Substituting this value in the second equation we can solve for R_y as:

$$R_y = 187 - 4 - 25$$
$$= 158 \text{ lbs}$$

The representation of all muscle actions by a single force F_{bb} is necessary in this type of analysis in order to obtain a determinate solution. Note that there were three equations of equilibrium and three variables were to be evaluated. Since the number of equations equalled the number of variables, the problem was statically determinate.

THE STATICAL INDETERMINANCY OF THE PROBLEM

Let us now investigate what happens if an additional muscle force F_{br} as shown in Fig. 4-19 is included in the previous model. In this case instead of lumping together the action of the various flexors such as the biceps brachii, the brachialis, and the brachioradialis (which would be normally active for the case being studied) into a single equivalent muscle, let us assume that there are two lines of muscular action—one for the biceps F_{bb} and one for the brachialis F_{br} as shown in

118

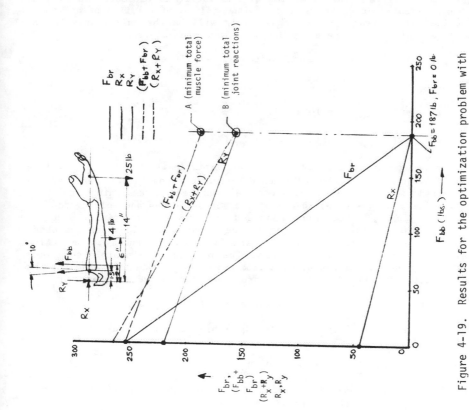

Figure 4-19. Results for the optimization problem with various criteria.

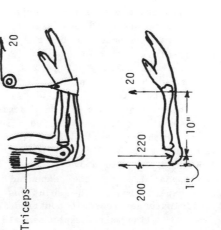

Triceps

Figure 4-18(b). Calculating the triceps force for the loading condition illustrated. After Williams and Lissner (1).

the figure. The point of action of the latter is estimated at 1.5 in. from the center of rotation at the elbow joint and its line of action is inclined (10°) from the vertical. This will be the only change in the model and all previous assumptions will remain unaltered.

Now the equations of equilibrium can be written as

$$\Sigma \ F_x = 0 = R_x - F_{br} \ (\sin 10°)$$

$$\Sigma \ F_y = 0 = F_{br} \ (\cos 10°) + F_{bb} - 4 - 25 - R_y$$

$$\Sigma \ M_z = 0 = F_{br} \ (\cos 10°)(1.5) + F_{bb}(Z) - 4(6) - 25(14)$$

The above equations now have 4 unknown variables: F_{br}, F_{bb}, R_x and R_y and consequently no determinate solution is possible. Instead an unlimited number of solutions can be obtained by assigning arbitrary values (zero or positive) to one of the muscle forces and solving for the remaining three, namely, one muscle force and two reactions. Any of these solutions would be feasible if the calculated values for the other muscle force is zero or positive since muscles are known to be active only in tension. Figure 4-19 shows the plot of all feasible solutions for this problem. As can be seen from this figure when the value for the muscle force F_{bb} is assumed to be zero, one obtains the solution as:

$$F_{bb} = 0, \ F_{br} = 253, \ R_x = 44, \ R_y = 220 \ lb$$

If on the other hand, F_{br} is assumed to be zero, the resulting solution is:

$$F_{br} = 0, \ F_{bb} = 187, \ R_x = 0, \ R_y = 158 \ lb$$

It is evident that one could arbitrarily select any value for F_{bb} between $F_{bb} = 0$ and $F_{bb} = 187$ lb to yield corresponding values for F_{br} to lie between F_{br} 253 and $F_{br} = 0$. This linear relationship between F_{bb} and F_{br} is shown in the figure. Thus there are an unlimited number of feasible solutions that can be obtained by arbitrarily assigning a value to either F_{bb} or F_{br} and solving for the remaining three.

One must therefore establish a merit criterion to select among all these solutions the one with the highest merit. If for example, the merit criterion is assumed to be the expenditure of minimum muscle effort, that is minimization of the sum $(F_{bb}+F_{br})$, the solution with the highest merit is that corresponding to a value of $F_{br} = 0$ and $F_{bb} = 187$ lb. This is shown by the point A (see Fig. 4-19) on the straight line described by the function $(F_{bb}+F_{br})$. Also if the merit criterion is assumed to be the minimization of the reaction forces on the elbow joint, that is, the sum (R_x+R_y), then the solution with the highest merit is again that corresponding to a value of $F_{br} = 0$ and $F_{bb} = 187$ lb. This is indicated by the point B (Fig. 4-19) on the straight line described by the function (R_x+R_y). The above criteria or a weighted combination of them would appear to be the expected controlling objective from physiological expectations. The solutions they produce, however, would invariably eliminate any possible contribution from the brachialis muscle which is not in line with experimental evidence such as those by

120

Basmajian and Latif [14], Stevens et al.[15], etc., that the brachialis is the "power horse" of the elbow flexors. It is expected in this type of action to function along with the biceps group and carry the bigger share of the load.

This would seem to indicate that simplified models such as the one discussed above which are developed primarily for the convenience of obtaining determinate analytical solutions may yield unrealistic results for muscle activity and load sharing.

One cannot, therefore, escape the development of a comprehensive model which takes into consideration all possible actions of the individual muscles and takes into account the true functional lines of action for the different muscles. Such representation would necessarily require the development of a comprehensive three dimensional model as described in the following section which incorporates all the potential muscle expected to be active in the investigated posture.

THE COMPREHENSIVE MODEL

A comprehensive model can be developed for the case investigated by Williams and Lissner as shown in Fig. 4-20. The body fixed axes for the segments and the muscle coordinates are the same as those described in the previous section. A total of 24 muscles (Table 1) represented by 34 action lines are considered in the model.

The problem is now one of equilibrium of a rigid body (the forearm and the hand) connected at the elbow joint to another body (the upper arm) which is in turn connected at the shoulder joint to the rest of the body which is assumed fixed. Each of the two segments under consideration can be modeled as a three-dimensional rigid body whose equilibrium is defined by six independent equations. The model for each segment is developed as follows: (See Figs. 4-21 and 4-22).

a. The forearm and the hand are assumed to be one rigid body supported at the elbow and acted upon by 18 unknown muscle forces (numbered 16 and, 18 through 34 in Table I), and its own weight [W_{FA} = 1.2 kg and W_{hand} = 0.6 kg acting as shown in Fig. 21] as well as a 25 lb load acting at a distance of 34.5 cm from the center of the elbow joint. The supporting reactions at the elbow can be modeled by three possible joint force components R_{ELBX}, R_{ELBZ} along the three respective axes as well as three joint moment components M_{ELBX}, M_{ELBY}, M_{ELBZ}. The latter represent any constraint to rotation about the three axes respectively due to the bony structure of the joint and its ligamentation. The forearm can then be described by six equations:

$$\Sigma F_x = 0 = (\Sigma_i F_{x_i}) + R_{ELBX}$$

121

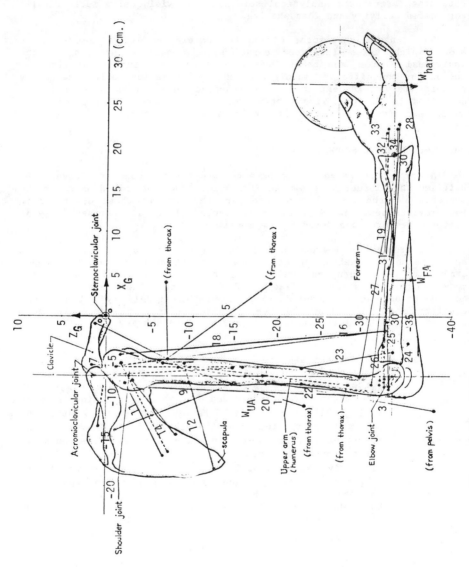

Figure 4-20. The upper extremity model.

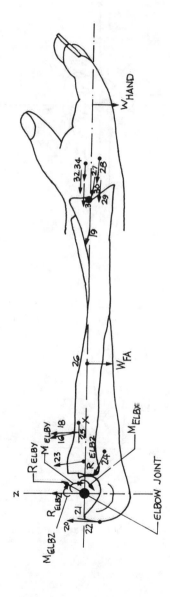

Figure 4-21. Free-body diagram of the forearm and hand.

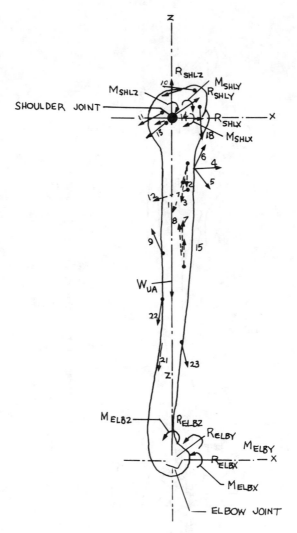

Figure 4-22.   Free-body diagram of the upper arm.

$$\Sigma\, F_y = 0 = (\underset{i}{\Sigma}\, F_{y_i}) + R_{ELBY}$$

$$\Sigma\, F_z = 0 = (\underset{i}{\Sigma}\, F_{z_i}) + R_{ELBZ} - W_{FA} - W_{hand} - L$$

$$\Sigma\, M_x = 0 = (\underset{i}{\Sigma}\, M_{x_i}) + M_{ELBX} + (R_{ELBZ} \cdot Y_{ELB} \cdot R_{ELBY} \cdot Z_{ELB})$$

$$- W_{FA} \cdot Y_{FA} - W_{hand} \cdot Y_{hand} - L \cdot Y_{hand}$$

$$\Sigma\, M_y = 0 = (\underset{i}{\Sigma}\, M_{y_i}) + M_{ELBY} + (R_{ELBX} \cdot Z_{ELB} - R_{ELBZ} \cdot X_{ELB})$$

$$+ W_{FA} \cdot X_{FA} + W_{hand} \cdot X_{hand} + L \cdot X_{hand}$$

$$\Sigma\, M_z = 0 = (\underset{i}{\Sigma}\, M_{z_i}) + M_{ELBZ} + (R_{ELBY} \cdot X_{ELB} - R_{ELBY} \cdot Y_{ELB}$$

where

$F_{x_i}, F_{y_i}, F_{z_i}$ = component of force in muscle i along the axis x,y,z respectively

$M_{x_i}, M_{y_i}, M_{z_i}$ = component of moment of force in muscle i about the axes, x,y,z respectively.

$R_{ELBX}, R_{ELBY}, R_{ELBZ}$ = reaction force components at the elbow joint.

$M_{ELBX}, M_{ELBY}, M_{ELBZ}$ = reaction moment components at the elbow joint.

$X_{ELB}, Y_{ELB}, Z_{ELB}$ = coordinates of the elbow joint with respect to the ground axes.

$W_{FA}, (X_{FA}, Y_{FA})$ = weight and the x and y coordinates of center of gravity of the forearm.

$W_{hand}, (X_{hand}, X_{hand})$ = weight and the x and y coordinates of center gravity of the hand.

L = load carried in the hand, assumed acting through the center of gravity of the hand.

These equations contain a total of 24 variables: 18 muscle forces and six joint reactions and moments.

b.   The upper arm is modeled as a rigid body supported at both the elbow and shoulder joints. It is acted upon by 15 unknown muscle forces from the main body (numbered 1 through 15 in Table I), another 15 muscle forces from the forearm (numbered 19, and 21 through 34 in Table I) equal in magnitude but opposite in direction to those shown in the freebody diagram of the forearm, and its own weight (W_{UA} = 2.2 kg).

The supporting reactions and moments at the elbow joint acting on the upper arm are equal in magnitude and opposite in direction respectively to the reactions R_{ELBX}, R_{ELBY} and R_{EBLZ} and the moments M_{ELBX}, M_{ELBY} and M_{ELBZ} which are acting on the forearm at the elbow joint. The supporting reactions at the shoulder joint can be similarly described by three reaction forces R_{SHLX}, R_{SHLY} and R_{SHLZ} along the three respective coordinate axes, and three moment components M_{SHLX}, M_{SHLY} and M_{SHLZ}, representing any constraint to rotation about the three axes respectively due to the bony structure of the joint and its ligamentation. The upper arm can now be described by its six equations of equilibrium:

$$\Sigma F_x = 0 = (\Sigma_i F_{x_i}) - (\Sigma_j F_{x_j}) + R_{SHLX} - R_{ELBX}$$

$$\Sigma F_y = 0 = (\Sigma_i F_{y_i}) - (\Sigma_j F_{y_j}) + R_{SHLY} - R_{ELBY}$$

$$\Sigma F_z = 0 = (\Sigma_i F_{z_i}) - (\Sigma_j F_{z_j}) + R_{SHLZ} - R_{ELBZ} - W_{UA}$$

$$\Sigma M_x = 0 = (\Sigma_i M_{x_i}) - (\Sigma_j M_{x_j}) + M_{SHLX} - R_{SHLZ} \cdot Y_{SHL}$$
$$- R_{SHLY} \cdot Z_{SHL}) - W_{UA} \cdot Y_{UA}$$
$$- (R_{ELBZ} \cdot Y_{ELB} - R_{ELBY} \cdot Z_{ELB}) - M_{ELBX}$$

$$\Sigma M_y = 0 = (\Sigma_i M_{y_i}) - (\Sigma_j M_{y_j}) + M_{SHLY} + (R_{SHLX} \cdot Z_{SHL}$$
$$- R_{SHLZ} \cdot X_{SHL}) + W_{UA} \cdot X_{UA}$$
$$- (R_{ELBX} \cdot Z_{ELB} - R_{ELBZ} \cdot X_{ELB}) - M_{ELBY}$$

$$\Sigma\, M_z = 0 = (\Sigma_i\, M_{z_i}) - (\Sigma_j\, M_{z_j}) + M_{SHLZ} + (R_{SHLY} \cdot X_{SHL}$$

$$-R_{SHLX} \cdot Y_{SHL}) - (R_{ELBY} \cdot X_{ELB}$$

$$-R_{ELBX} \cdot Y_{ELB}) - M_{ELBZ}$$

where

$(F_{x_i}, F_{y_i}, F_{z_i})$,

$M_{x_i}, M_{y_i}, M_{z_i})$ =   x,y,z components respectively of the force and moment of muscle i originating on the main body (clavicle, scapula and the upper torso) and inserting into the upper arm.

$(F_{x_j}, F_{y_j}, F_{z_j})$,   =   x,y,z components respectively of

$(M_{x_j}, M_{y_j}, M_{z_j})$ =   the force and moment of muscle j from the forearm acting on the humerus, equal in magnitude but opposite in direction.

$W_{UA}, (X_{UA}, Y_{UA})$ =   weight and x and y coordinates of the center of gravity of the upper arm.

$R_{SHLX}, R_{SHLY}, R_{SHLZ}$ =   reaction components at the shoulder joint along the three respective axes.

$M_{SHLX}, M_{SHLY}, M_{SHLZ}$ =   moment components at the shoulder joint about the three respective axes.

It may be mentioned here that an additional equation needs to be included. This arises from the fact that the long head of the biceps brachii is assumed to wrap around the humerus. This muscle, as described earlier, has been modelled by two lines of action——an upper part connecting the scapula to the humerus (numbered 17 in Table I) and a lower part connecting the humerus to the radius (numbered 18 in Table I). Since both these parts are actually the same muscle, the force in both must be equal. In other words, $F_{17} = F_{18}$ (= force in long hand of biceps) where F_{17} and F_{18} are the forces in the two parts. Note that in the freebody diagram of the humerus (Fig. 4-22), both the forces F_{17} and F_{18} are acting on it in the directions shown.

The 12 equilibrium equations of the forearm and the upper arm along with the equality constraint for the two parts of the long head of biceps, form a system of 13 equations and 46 unknown variables (a total of 34 independent muscle forces and a total of 12 joint reactions and moments). Evidently there is no determinate solution and unlimited solutions can be obtained by assigning arbitrary zero or positive values to 33 (the difference between the total number of variables and the number of equations) of the unknowns and solving for the remaining 13 from the 13 equations. A solution is considered feasible if all the resulting muscle forces are either zero or positive. Hence a criterion will have to be established to select a solution among the many possible alternatives that such a model will produce. Possible merit criteria can be formulated as one or a weighted combination of the following objectives:

a) minimization of forces in all the muscles.

b) minimization of the work done by the muscles to attain the given posture, that is, minimization of the product of the muscular tension and its elongation or contraction from the rest position.

c) minimization of the reaction forces at the joints.

d) minimization of the reaction moments at the different joints.

In conditions where the center of articulation of the joint (origin of body-fixed axis) is accurately defined, such moments represent additional needed support which would be carried by the ligaments.

Also in conditions where the support is possible over a defined area, the moment reactions may represent a shift in the preselected center of the area of the articulating joint surface and consequently may imply maldistribution of the resultant pressure on the surface of the joint. It can be readily seen that such shift has to be within the confines of the joint anatomical geometry. Moments in excess of what could be accommodated by the shift of the reactions to the permissible boundary of the support area would have to be carried by the pull on the joint ligaments.

e) minimization of a weighted combination of the above with differing emphasis on the various components namely, the muscle forces, the joint reactions forces and the joint reaction moments.

A possible criterion for persons with particular muscular impairments or joint malfunctions would be minimization of the forces in the muscles responsible for the abnormality or the joint disorder. Since the solution is highly dependent on the selected merit criterion and its precise mathematical form, considerable attention must be given to formulating a merit function which renders results compatible with empirical evidence and physiological expectations.

This criterion should ideally be applicable to all postures and actions and establishes the control law for the muscle load sharing in the musculoskeletal system. Because of the many biomechanical and physiological investigations available in the literature on the lower extremities, a comprehensive model of it will be utilized for the establishment of the merit criterion and to demonstrate the solution technique. This is discussed in detail in the following chapter where numerical examples illustrating the muscle load sharing for both the upper and lower extremities are considered. A comprehensive model for the lower extremities which will be used to demonstrate the solution procedure is developed in the following section.

THE MUSCULOSKELETAL MODEL FOR THE LOWER EXTREMITIES

The skeletal structure of the lower limbs consists of four elements—the pelvis, the femur (the thigh bone), the lower leg and the foot (see Fig. 4-23). The pelvis along with the rest of the upper body is modelled as one rigid segment which articulates with the thighs at the left and the right hip joints. Since the left side is identical to the right we shall limit our attention to the latter only. The model for the left side thus would be a mirror image of the right. The thigh articulates at its distal end with the leg, which is composed of two bones——the medially-placed tibia and the laterally-placed fibula, connected together by an interosseous membrane. Unlike the two bony elements of the forearm——the ulna and the radius which can move freely with respect to each other such as in the acts of supination and pronation, (see section 4.2), the tibia and the fibula have negligible motion and therefore can be treated as one single rigid body. The fourth member of the skeletal model is the foot, which is connected to the leg at the ankle joint. For the purpose of this model the multi-segmented bony structure, intrinsic musculature and complex ligamentation of the foot will not be considered. Instead, at this stage the foot will be treated as one rigid segment. A comprehensive model of the foot is described later in a special chapter.

The four rigid elements are connected together through three joints. The hip joint is formed by the cup-like acetabulum of the pelvis and the rounded head of the femur, and functions mechanically as a ball and socket type of joint. Evidently, all three rotary motions can occur at the joint——flexion and extension about a transverse axis, adduction and abduction about an anterior-posterior axis and medial and lateral rotation about a vertical axis or the longitudinal axis of the femur. At the distal end of the femur, there are two large eminences, the condyles, which articulate with their counterparts on the proximal end of the tibia, forming the knee joint. Unlike the hip joint, the knee joint is a hinge type of articulation allowing motion in one plane only, namely that of flexion and extension. It should be mentioned here that because of the curvature of the inferior surfaces of the femoral condyles and the corresponding condyles on the tibia being relatively planar surfaces, there is both translation and rotation during movements at the knee. The joint center of rotation therefore undergoes a complex motion, not easily identifiable. It will be assumed however, in this study that the flexion extension can be defined by a fixed point of rotation on the femoral condyles. The ankle joint is formed by the medial and lateral malleoli at the distal ends of the tibia and the fibula respectively and the talus bone of the foot. Normal movements permitted at this joint are those of plantar flexion and dorsi-flexion about a transverse axis, and inversion and eversion (rotation of foot inwards and outwards) about an oblique axis.

MUSCULATURE OF THE LOWER EXTREMITIES

The four segments of the lower extremities model described above are held together by 26 major muscles as listed below:

	Muscle	Figure	no. of lines required to model
1.	Gracilis	2	1
2.	Adductor longus	2	1
3.	Adductor magnus	13	2
4.	Adductor brevis	2	1

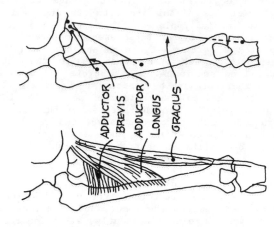

Figure 4-24. Adductor longus and brevis, and gracilis models.

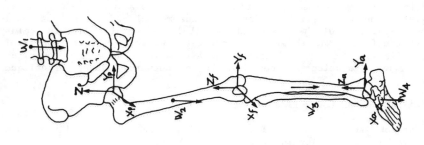

Figure 4-23. Skeletal structure of the lower extremities.

5.	Semitendinosus	3	1
6.	Semimembranosus	4	1
7.	Biceps femoris	14	2
8.	Quadriceps femoris	15abcd	4
9.			
10.	Sartorius	5	1
11.	Tensor fasciae latae	6	1
12.	Gluteus maximus	7	1
13.	Iliopsoas	16a&b	1
14.	Gluteus medius	7	1
15.	Gluteus minimus	7	1
16.	Gastrocnemius	17	2
17.	Soleus	8	1
18.	Tibialis anterior	9	1
19.	Tibialis posterior	10	1
20.	Extensor digitorum longus	11	1
21.	Extensor hallucis longus	9	1
22.	Flexor digitorum longus	12	1
23.	Flexor hallucis longus	12	1
24.	Peroneus longus	18	1
25.	Peroneus brevis	18	1
26.	Peroneus tertius	18	1

As described in the previous chapter, muscles are modelled by their functional lines of action in space, obtained by connecting their points of origin to their corresponding points of insertion. Muscles which wrap around a skeletal segment are modelled by two lines and a reaction on the interposing structure. In certain cases a muscle may have to be modelled by more than one line if it has distinct components with widely differing functions.

Several of the muscles connecting the pelvis to the femur or the leg can be conveniently modelled by straight lines. For example, the gracilis muscle which takes its origin from the lower margin of body and ramus of pubic bone and inserts into the upper part of the medial aspect of tibia is modelled by a straight line connecting the two segments as shown in Fig. 4-24. Likewise, the adductor longus and brevis, semitendinosus, semimembranosus, sartorius, tensor fasciae latae, gluteus maximus, gluteus medius and gluteus minimus, muscles which arise from various parts of the pelvis and insert either into the femur or into the tibia are simulated by spatial lines connecting their points of origin to their points of inertion (see Figs. 4-24 through 4-29). In similar fashion, among the muscles joining the leg to the foot the soleus which arises from the posterior surface of the head of the fibula, the soleal line of the tibia and the intervening interosseous septum and inserts into the calcaneous bone of the foot is represented by a straight line connecting the leg to the foot (Fig. 4-30).

Since the foot is being considered as a rigid body the extrincis muscles of the foot which originate on the tibia-fibula and insert into the foot as far as the phalanges (such as the digital flexors and extensors) need special treatment in their modeling. For these muscles, their operations along the leg are modelled separate from their portions along the foot following their course towards their final insertions. In other words, these muscles are modelled by two lines each, a line connecting the site of origin on the leg to alternate point on the foot

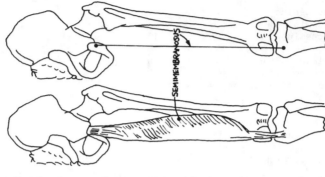

Figure 4-26.   Muscle model for semimembranosus.

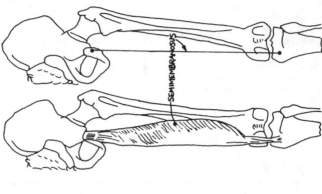

Figure 4-25.   Muscle model for semitendinosus.

132

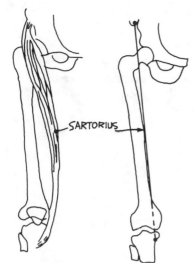

Figure 4-27. Sartorius muscle
and its model.

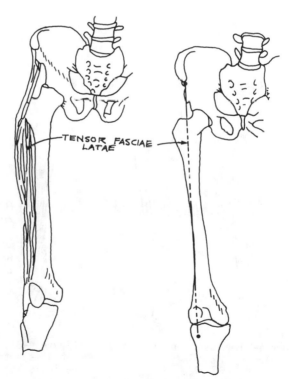

Figure 4-28. Muscle model for the tensor
fasciae latae.

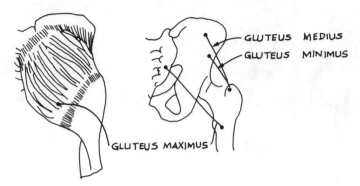

Figure 4-29. Muscle model for gluteus maximus, medius, and minimus.

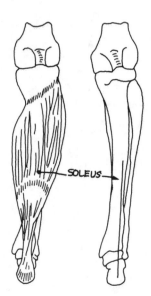

Figure 4-30. Muscle model for soleus.

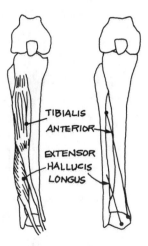

Figure 4-31. Muscle model for tibialis anterior and extensor hallucis longus.

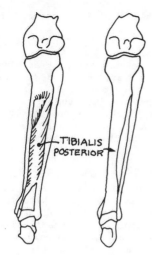

Figure 4-32. Muscle model
for tibialis posterior.

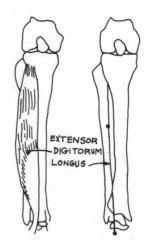

Figure 4-33. Muscle model for
extensor digitorum longus.

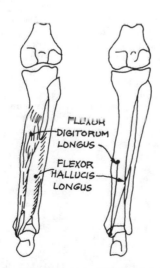

Figure 4-34. Muscle model for
flexor digitorum longus, flexor
hallucis longus.

in the vicinity of the ankle joint, and a second line representing the muscle portion on the foot. Understandably the latter portions will be excluded from actions where the toes are not brought into play.

The following muscles because of their complex anatomy require special consideration:

a) The adductor magnus:
The adductor magnus has two heads—the adductor part arising from the ischiopubic ramus and the extensor part arising from the ischial tuberosity. The former inserts into the linea aspera and the gluteal tuberosity of the femur while the latter inserts into the linea aspera, medial supracondylar ridge and adductor tubercle. Consequently it is modelled by two lines, one for the adductor component and one for the extensor component connecting the pelvis to the femur as shown in Fig. 4-35.

b) Biceps femoris:
The bicep femoris has two heads of origin—a short head from the inferior part of the linea aspera of the femur and a long head from the medial aspect of the ischial tuberosity on the pelvis, both heads inserting into the lateral side of the proximal end of the fibula. The muscle is represented by two lines; the long head from the pelvis and the short head from the femur, connected to the fibula (Fig. 4-36).

c) Quadriceps femoris:
The quadriceps group of muscles is composed of four components: the rectus femoris, the vastus medialis, the vastus intermedius and the vastus lateralis. The rectus femoris has its origin on the anterior inferior spine of the ilium and from around the margin of the acetabulum by means of a reflected tendon while the remaining three arise from the medial, anterior and the lateral sides of the femur respectively. All four head unit in a tendon that encloses the patella and inserts into the tibial tuberosity after crossing the knee joint. Since all the four components will wrap around the femur such as during acts of knee flexion they cannot be simulated by direct straight lines spanning the points of origin and insertion. Consequently the muscle is modelled by four lines, joining their respective origins to the patella and another joining the patilla to the tibia representing the patellar tendon as shown in Fig. 4-37a,b and c. Note that because of the 'wrapping around', a reaction will have to be included in the force analysis, simulating the pressure exerted from the patella to the femur.

d) Iliopsoas:
The iliopsoas is made up of the iliacus and the psoas major. The iliacus is a fan-shaped muscle which arises from the iliac fossa while the psoas major arises from the sides of the bodies and the transverse processes of the lumbar vertebrae, passes under the inguinal ligament and inserts into the lesser trochanter along with the iliacus. Since the torso is considered to be integral with the pelvis, one need select only one line to represent this extensive and complex origin. However since the muscle wraps around the pelvis, in the vicinity of the hip joint, on its way to insertion on the femur, it is modelled by two lines, with an appropriate reaction on the pelvis at the assumed point of contact or interference (see Figs. 4-38a & b).

e) Gastrocnemius:
The gastrocnemius or the calf muscle has two heads of origin from

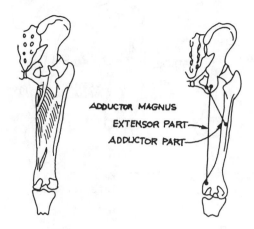

Figure 4-35. Muscle model for adductor magnus.

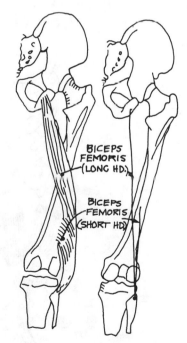

Figure 4-36. Muscle model for biceps femoris.

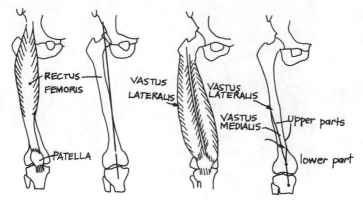

Figure 4-37a. Rectus
femoris model.

Figure 4-37b. Vastus medialis
and lateralis models.

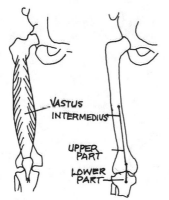

Figure 4-37c. Vastus intermedius and its model.

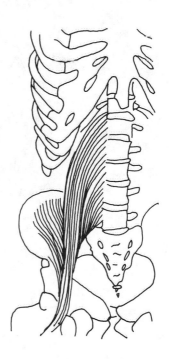

Figure 4-38a. The iliopsoas muscle.

the posterior surfaces of the medial and lateral femoral condyles and inserts into the calcaneous bone of the foot (along with the soleus muscle as described earlier) by means of a common tendon, the tendo calcaneuos (see Fig. 4-39). Consequently the muscle is modelled by two lines of action, one from the medial condyle and the second from the lateral condyle joining the foot as shown in the figure.

f) The peronei:

The peroneus longus, peroneus brevis and peroneus tertius all arise on the leg and insert into the foot. The peroneus longus passes behind the lateral malleolus, turns below the cuboid bone, and runs obliquely across the sole of the foot to insert into the proximal end of the first metatarsal and the medial cuneiform. The peroneus brevis passes behind the lateral malleolus and inserts into the dorsal surface of the fifth metatarsal. The peroneus tertius inserts on the base of the fifth metatarsal. All these muscles are modelled by single lines joining the leg to the foot (see Fig. 4-40).

SELECTION OF BODY FIXED ASEX AND COORDINATES OF MUSCLE ORIGINS AND INSERTIONS

The major muscles and the skeletal segments are represented in the manner described in the foregoing pages. The 26 muscles listed earlier are thus modeled by 31 lines of action. The next step consists of identifying the coordinates of the points of muscle origins and insertions with respect to body-fixed axes on the individual segments. These are in general selected to lie at the centers of joints formed with adjoining elements (see Fig. 4-23) such that the x axis lies in the anterior-posterior direction, the y axis is directed medio laterally and the z axis is directed upwards. The coordinates are estimated from the skeleton figure in the book by Braus (1954) (11). For example, the point of interest on the pelvis are identified with respect to a body fixed system of axes with its origin at the right hip joint. For the femur, the points of muscle origins and insertions are located with respect to a coordinate-axes system fixed at the knee joint between the two condyles. Viewed from the side, the origin of these axes is assumed to coincide with the approximate center of the condyles and treated for the sake of simplicity as the center of rotation of the femur at the knee joint. Note however, that this is not the point where the knee joint reaction forces is assumed to occur; the site for these forces is selected to lie on the inferior surfaces of the condyles, between the two tibial condyles. The body-fixed axes for the leg are chosen at the ankle joint. Since this functions as a modified ball and socket joint, the origin of the fixed axes with respect to which the coordinates of all points of interest will be evaluated using coordinate transformation matrices.

The coordinates of the points of origin and insertions of the muscles with respect to their body fixed axes are tabulated in Table III.

WEIGHTS OF SEGMENTS OF THE LOWER EXTREMITIES

The model of the lower extremities is assumed to comprise of four segments articulated by three joints. The pelvis and the remainder of the body (the torso and the upper extremities) is one rigid segment, its weight equalling the weight of the entire body minus the weight of the

139

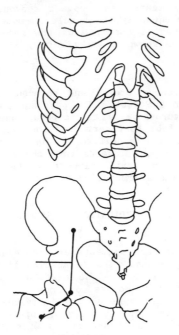

Figure 4-38b. Muscle model for iliopsoas.

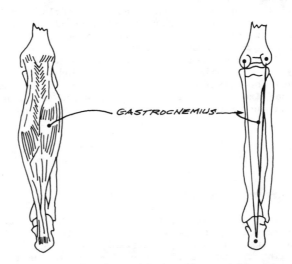

GASTROCNEMIUS

Figure 4-39. Muscle model for the gastrocnemius.

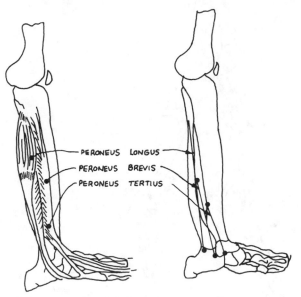

Figure 4-40. Muscle model for peroneus longus, brevis and tertius.

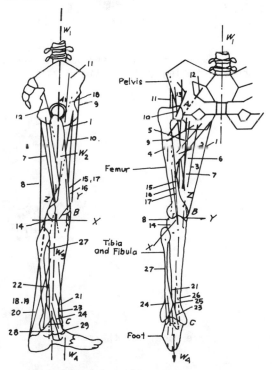

Figure 4-41-a. Musculo-skeletal model for the lower extremeties.

141

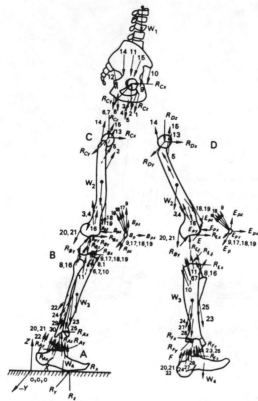

Figure 4-41b. Free-body diagrams of the various segments for a typical posture during quasi-static walking.

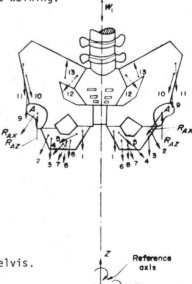

Figure 4-42. Force model for the pelvis.

142

two thighs, two legs and two feet. Referring to the information in the chapter dealing with the inertial properties of body segments, for a 72.3 kg-mass (or 151.9 lb, the 50th percentile male (see Chapter III) the weight of the pelvic segment turns out to be 48.3 kg. The weights of the thigh, the leg and the foot and the coordinates of the centers of gravity of the segments are shown in Table 2. The entire model of the lower extremities along with the weights of the segments and their corresponding points of application is schematically illustrated in Fig. 4-41a and 4-41b.

EQUILIBRIUM EQUATIONS FOR INDIVIDUAL SEGMENTS

For each of the four segments, namely, the pelvis, the femur, the leg, and the foot, which are rigid bodies in space connected by 31 muscle forces as shown in Table III six equations are needed to describe the equilibrium of each. At each of the three joints (hip, knee and ankle) connecting the four segments, the reaction forces and moments are modelled by their 3 respective x, y, and z components. As discussed in the section on the upper extremities model (Section IV.I), the joint moments, if obtained to be nonzero from the solution, represent the shift in the location of the point of application of the resultant joint reaction forces from that selected in the model. This may happen in situations where the true joint position cannot be described with reasonable accuracy. For example, in case of the knee joint where the tibio-femoral contact occurs over a wide area, the point of application of the resulting contact force cannot be pre-established. The knee joint center ha been selected to lie between the two condyles on the tibial proximal joint surface. However, if the solution indicates that the selected point of application of the reaction forces may be allowed to shift within the boundaries of the joint. If there are any residual moments even after such translation has been performed to the extreme boundary limits, this would be taken up by the joint ligaments.

   a) **The pelvis:**
For the postures which will be investigated using the model, (such as leaning and squatting) the pelvis may be considered to be symmetric about the sagittal plane. The force in a muscle on the right side is therefore equal to its counter part on the left. By selecting the origin of the ground axes in the sagittal plane one can thus eliminate three of the six equations. The remaining three are as follows:

$$\Sigma F_x = 0 = 2 (\Sigma F_{x_i}) + 2R_{AX}$$

$$\Sigma F_z = 0 = 2 (\Sigma F_{z_i}) + 2R_{AZ}$$

$$\Sigma M_y = 0 = 2 (\Sigma M_{y_i}) + 2 (R_{AX} Z_A - R_{AZ} X_A) + W_1 X_{W_1} + 2M_{AY}$$

where F_{x_i}, F_{z_i} =   force component along x,z respectively for muscle i

   M_{y_i} =   moment component about y axis of the force in muscle i

   R_{AX}, R_{AZ} =   joint reaction force along x,z respectively at the hip joint, A

143

TABLE II.   Weights and centers of gravity of segments of arm.

Body	Weight (kg)	Coordinates of C.G. (cm)			C.G. w.r.t. axes located on
		x	y	z	
Upper arm	2.2	0	0	-13.5	Humerus
Forearm	1.2	0	-1.0	-11.5	Ulna
Hand	0.6	0	0	-8.0	Hand

Length of upper arm (distance between shoulder joint and the elbow joint) = 30.5 cm.

Length of forearm (distance between elbow joint and the wrist joint) = 26.5 cm

The freebody diagram of the femur is shown in Fig. (4-43). The above three equations contain a total of 18 unknown variables: 15 muscle forces (numbered 1 through 13, 30 and 31 in Table III), 2 reaction forces and 1 reaction moment.

b) The femur:
The freebody diagram of the femur is shown in Fig. (4-43). The reactions at the hip joint on the femur R_{AX}, R_{AY}, R_{AZ} are equal and opposite to those acting on the pelvis. Similarly the force in muscle j (numbered 2 through 5, 12, 13, 30 and 31) connecting the pelvis to the femur is equal and opposite to the force shown in the free body diagram of the pelvis. In addition, the muscle forces F_i (numbered 14 through 19 Table III) connecting the femur to the leg, and the foot act in directions shown in the figure. The femur is a rigid body in space and therefore requires six equations to describe its equilibrium:

$$\Sigma F_x = 0 = (\Sigma F_{x_i}) - (\Sigma F_{x_j}) - R_{AX} + R_{BX}$$

$$\Sigma F_y = 0 = (\Sigma F_{y_i}) - (\Sigma F_{y_j}) - R_{AY} + R_{BY}$$

$$\Sigma F_y = 0 = (\Sigma F_{z_i}) - (\Sigma F_{z_j}) - W_2 + R_{AX} + R_{BX} + R_{BZ}$$

$$\Sigma M_x = 0 = (\Sigma M_{x_i}) + (R_{BZ} Y_B - R_{BY} Z_B) - (\Sigma M_{x_j}) + M_{BX} - M_{AX}$$
$$- (R_{AZ} Y_A - R_{AY} Z_A) - W_2 \quad Y_{W_2}$$

$$\Sigma M_y = 0 = (\Sigma M_{y_i}) + (R_{BX} Z_B - R_{BZ} X_B) - (\Sigma M_{y_j}) + M_{BY} - M_{AY}$$
$$- (R_{AX} Z_A - R_{AZ} X_A) - W_2$$

$$\Sigma M_z = 0 = (\Sigma M_{z_i}) + (R_{BX} Z_B - R_{BZ} Y_B) - (\Sigma M_{z_j}) + M_{BY} - M_{AY}$$
$$- (R_{AX} Z_A - R_{AZ} X_A)$$

where

R_{BX}, R_{BY}, R_{BZ}	=	components of reaction force at the knee joint B, along x,y,z axes respectively.
X_B, Y_B, Z_B	=	x,y,z coordinates respectively of the knee joint.
M_{BX}, M_{BY}, M_{BZ}	=	components of moment at the knee joint about the three axes respectively.
W_2	=	weight of the femur.
X_{W_2}, Y_{W_2}	=	X,Y coordinates of the point of application of the weight W_2.

145

TABLE III. Coordinates of muscle origins and insertions for the lower extremities.

No.	Muscle Name	Origin	x^1 (cm)	x^1 (cm)	x^1 (cm)	Insertion	x^2 (cm)	x^2 (cm)	x^2 (cm)
1	Gracilis	Pelvis	3.5	4.5	-3.5	Leg	1.0	2.5	35.5
2	Adductor longus	Pelvis	5.0	4.7	-1.5	Femur	1.5	-1.0	18.0
3	Adductor Magnus (extensor part)	Pelvis	-2.0	1.0	-5.5	Femur	0	4.0	2.0
4	Adductor Magnus (adductor part)	Pelvis	0.0	2.0	-5.5	Femur	1.0	-1.0	22.0
5	Adductor brevis	Pelvis	4.0	4.0	-2.5	Femur	0.5	-2.0	28.0
6	Semitendinosus	Pelvis	-4.0	1.0	-2.5	Leg	0	3.0	35.5
7	Semimembranosus	Pelvis	-3.8	0.0	-3.0	Leg	-1.0	3.0	39.5
8	Biceps femoris (LH)	Pelvis	-3.8	0.0	-2.5	Leg	-2.5	-3.5	38.5
9	Rectus femoris[a]	Pelvis	2.5	-2.0	2.5	Leg	4.0	-1.0	37.5
10	Sartorius	Pelvis	4.5	-3.0	8.5	Leg	1.0	2.5	37.5
11	Tensor fasciae latae	Pelvis	3.0	-4.0	9.5	Leg	3.0	-4.0	40.5
12	Glutens maximus	Pelvis	-7.0	2.0	6.5	Femur	-1.5	-4.0	33.0
13	Iliopsoas[b]	Pelvis	-1.0	0.0	10.0	Femur	-1.0	-9.5	34.5
14	Biceps femoris (SH)	Femur	0.5	-1.0	13.0	Leg	-2.5	-3.5	38.5
15	Vastus medialis[a]	Femur	2.5	-1.0	18.0	Leg	4.0	-1.0	37.5
16	Vastus intermedius[a]	Femur	3.0	-2.0	20.0	Leg	4.0	-1.0	37.5
17	Vastus Lateralis[a]	Femur	2.5	-3.0	18.0	Leg	4.0	-1.0	37.5
18	Gastrocnemius (MH)	Femur	-2.0	1.5	0	Foot	-6.5	-1.0	-4.5
19	Gastrocnemius (LH)	Femur	-2.0	-2.5	-0.5	Foot	-6.5	-1.0	-4.5
20	Soleus	Leg	-1.5	-2.5	32.5	Foot	-6.5	-1.0	-4.5
21	Tibialis anterior	Leg	2.0	0.0	22.5	Foot	3.0	1.0	2.5
22	Tibialis posterior	Leg	-0.5	-2.5	17.5	Foot	-2.5	2.8	2.5
23	Extensor digitorium longus	Leg	0	-2.5	17.5	Foot	1.0	-1.8	2.5
24	Extensor hallucis longus	Leg	0	-2.0	18.5	Foot	3.0	1.0	2.5
25	Flexor digitorum longus	Leg	-0.5	0	20.5	Foot	-2.5	2.8	2.5
26	Flexor hallucis longus	Leg	-2.0	-3.0	14.5	Foot	-1.5	-0.7	2.5
27	Peroneus longus	Leg	-4.0	-3.0	22.5	Foot	-2.0	-2.5	-0.5
28	Peroneus brevis	Leg	-3.5	-3.0	14.5	Foot	-2.0	-2.5	-0.5
29	Peroneus tertius	Leg	-1.0	-2.0	7.5	Foot	1.0	-2.0	-0.5
30	Gluteus medius	Pelvis	-3.0	-3.0	11.0	Femur	-1.5	-6.0	40.0
31	Gluteus minimus	Pelvis	2.0	-2.0	8.5	Femur	0	-6.5	40.0

a. Rectus femoris and the three vastii wrap around the femur at a point whose coordinates are x = 5.0, y = -2.0, z = 0 cm.

b. The Iliopsoas wraps around the pelvis at a point those coordinates are x = 2.5, y = 0.5, z = 0 cm.

COORDINATES OF JOINTS
Hip joint - (with respect to pelvis-fixed axes) x = 0, y = -0.5, z = 0
Knee joint - (with respect to femur-fixed axes) x = 0, y = 0, z = -2
Ankle joint - (with respect to leg-fixed axes) x = 0, y = 0, z = 0
Length of thigh - (between hip and knee joints) = 43.5 cm,
Length of leg - (between knee and ankle joints) - 42.5 cm

TABLE IV. Weight distribution for the lower extremities (Total body wt. = 72.3 kg).

| No. | Segment | Weight (kg) | Center of Gravity | | | Location of CG with respect to body-fixed axes on |
			x (cm)	y (cm)	z (cm)	
1	Pelvis & remainder of body	48.0	0	8.0	26.5	Pelvis
2	Thigh	7.45	0.0	-2.0	22.6	Femur
3	Leg	3.6	0.0	0	25.2	Leg
4	Foot	1.1	4.5	0	-3.0	Foot

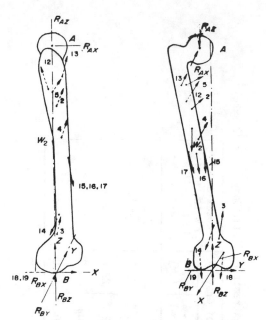

Figure 4-43. Force model for the femur.

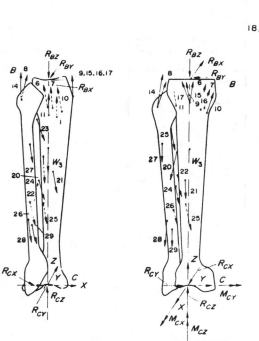

Figure 4-44. Force model for the tibia and the fibula.

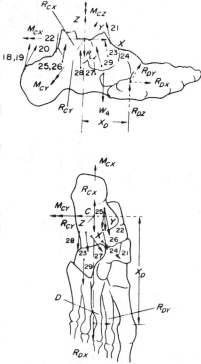

Figure 4-45. Force model for the foot.

148

Other variables have been defined for the case of the pelvis earlier. Note that R_{AY} (reaction force along Y at the hip joint) and M_{AX}, M_{AZ} (moments along X and Z respectively at the hip joint) were absent from the equations for the pelvis, which was treated as if in planar equilibrium. They must however be now included for the case of the femur which is a rigid body in space. The ground axes for moment balance for the femur are assumed to lie at the ankle joint.

The above six equations contain a total of 26 unknowns: 14 muscle forces, 6 joint reaction forces, and 6 joint moment reactions.

c) The leg:

The leg is in equilibrium under the action of the muscle forces F_i connecting it to the foot, muscles forces F_j connecting it to the femur, and the pelvis, its own weight, and force and moment reactions at the knee and ankle joints (see Fig. 4-44). The forces F_j are equal in magnitude but opposite in directions to those assigned in the analysis of the foregoing two segments. Similarly the reactions at the knee joint on the leg are equal and opposite to those on the femur. The six equations describing the equilibrium of the leg are:

$$\Sigma F_x = 0 = (\Sigma F_{xi}) - (\Sigma F_{xj}) + R_{CX} - R_{BX}$$

$$\Sigma F_y = 0 = (\Sigma F_{yi}) - (\Sigma F_{yj} \mu \mp \Theta_{\Psi T} / \Theta_{\infty T}$$

$$\Sigma F_z = 0 = (\Sigma F_{zi}) - (\Sigma F_{zj}) + R_{CX} - R_{BZ} - W_3$$

$$\Sigma M_x = 0 = (\Sigma M_{xi}) - (\Sigma M_{xj}) + (R_{CZ} Y_C - R_{CY} Z_C)$$
$$- W_3 \quad Y_{W_3} \quad - (R_{BZ} Y_B - R_{BY} Z_B)$$
$$+ M_{CY} - MBX$$

$$\Sigma M_y = 0 = (\Sigma M_{yi}) - (\Sigma M_{yi}) + (R_{CX} Z_C - R_{CZ} X_C)$$
$$+ W_3 Y_{W_3}) - (R_{BX} Z_B - R_{BZ} X_B)$$
$$= \mp M_{CY} - M_{BX}$$

$$\Sigma M_z = 0 = (\Sigma M_{zi}) - (\Sigma M_{zj}) + (R_{CY} X_C - R_{CX} Y_C)$$
$$- (R_{BX} Z_B - R_{BZ} X_B) + M_{CZ} - M_{BZ}$$

where

R_{CX}, R_{CY}, R_{CZ} = joint reaction force components along the three respective axes at the ankle joint, C.

R_{CX}, R_{CY}, R_{CZ} = joint moment components about the three respective axes at the ankle joint, C.

X_C, Y_C, Z_C = x,y,z coordinate respectively of the ankle joint.

$$X_{W_4}, Y_{W_4} \qquad = x, y \text{ coordinates of the point of application of } W_4.$$

Reactive due to wrapping of muscles:
    In addition to the above 21 equations of equilibrium, further constraints need to be included.

    These arise from conditions where a muscle wraps around a segment and therefore will cause a reaction force on the member interposed. For example, the iliopsoas muscle has been modelled to wrap around the pelvis. The resulting reaction force in terms of its 3 components must be therefore included in the equations of equilibrium for the pelvis. Also, two more equations ($\Sigma F_x = 0$, $\Sigma F_z = 0$ because of symmetry of the pelvis) will be needed to relate the forces in the two parts of the iliopsoas to the reaction components. In similar fashion, the wrapping of the quadriceps tendon through the patella at the knee joint will give rise to a reaction (modelled by its three components) and three more equations relating the forces in the four components to the reaction. It can be therefore seen that the system of lower extremities can be described by a total of 26 equations and 57 variables (31 muscle forces, 9 joint reaction moments, 9 joint reaction forces, 3 ground-to-foot forces, and 5 reactions from muscle wrapping).

    It should be noted that also in this case as in the upper extremities model, the number of unknown variables far exceeds the available number of equilibrium equations. Therefore, there is no determinate solution and an unlimited number of solutions can be obtained by arbitrarily assigning values to some of the unknown variables to make the system of equations statically determinate, and solving for the remainder. Once again, a criterion will have to be established to choose among these many possible alternatives, the one solution that best meets the desired objective.

SIMPLIFIED MODELS OF THE LOWER EXTREMITIES

    The following briefly describes some of the many simplified models which are available in the literature for evaluation of forces in the muscles of the lower extremities. In all these models, in order to overcome the problem of statical indeterminancy as discussed earlier simplifying assumptions and approximations have been used to render the problem determinate.

TWO-DIMENSIONAL MODELS

    A simple, two dimensional model of the lower leg by Williams and Lissner [1] is illustrated in Fig. (4-46). This is similar to the simplified arm model described in the preceding sections where the force in the elbow flexors group was evaluated in the load bearing posture. In this model, the knee is flexed at right angle to the thigh, and the tension developed by the hamstrings group is represented by a single line of action perpendicular to the leg. An external force of known magnitude is applied at the ankle in a direction parallel to the line of action of the muscle group. The tension in the hamstrings is then readily computed by taking moments about the knee joint. If the applied ankle force is 15

W_3 = weight of the leg.

X_{W_3}, Y_{W_3} = x,y coordinates of the point of application of W_3.

Other variables have been defined for the case of the femur in the preceding section. The muscle forces F_i are those numbered 20 through 29, and the muscle forces F_j are numbered 1, 6-11, 14-17, in Table III.

For the leg therefore, we have six equations, comprised of a total of 33 unknown variables: 21 muscle forces, 6 joint reaction forces and 6 joint reaction moments.

The foot

The foot is connected to the leg at the ankle joint, and therefore the reaction forces and moments at this joint on the foot will be equal and opposite to those on the leg. The muscle forces F_i (numbered 18 through 29) are equal and opposite to those acting on the femur and the leg. In addition, the ground-to-foot supportive reaction forces must be included in its freebody analysis. (See Fig. 4-45). The vertical supportive force R_{DZ} can be evaluated by knowing the overall equilibrium of the entire body, and will thus equal half the total body weight, since the body weight is borne by two feet, and the posture is symmetrical about the sagittal plane. The ground-to-foot forces in the medio-lateral and fore and aft directions however, are treated as unknowns and included in the analysis. The six equations of equilibrium containing 20 unknowns for the foot are as follows:

$$\Sigma F_x = - (\Sigma F_{x_j}) - R_{CX} - R_{DX}$$

$$\Sigma F_y = - (\Sigma F_{y_j}) + R_{CY} - R_{DY}$$

$$\Sigma F_z = - (\Sigma F_{z_j}) - R_{CZ} - R_{DZ} - W_4$$

$$\Sigma M_x = - (\Sigma M_{x_j}) - (R_{CZ} \, Y_C - R_{CY} \, Z_C) - M_{CX}$$
$$+ (R_{DZ} \, Y_D - R_{DY} \, Z_D) - W_4 \quad Y_{W_4}$$

$$\Sigma M_y = - (\Sigma M_{y_j}) - (R_{CX} \, Z_C - R_{CZ} \, X_C) - M_{CY}$$
$$+ (R_{DX} \, Z_D - R_{DZ} \, X_D) + W_4 \quad X_{W_4}$$

$$\Sigma M_y = - (\Sigma M_{z_j}) - (R_{CY} \, X_C - R_{CX} \, Y_C) + M_{CZ}$$
$$+ (R_{DY} \, X_D - R_{DX} \, Y_D)$$

where

R_{DX}, R_{DY}, R_{DZ} = ground-to-foot supportive forces along x,y,z axes respective active at the point D.

W_4 = weight of the foot.

kg acting at a distance of 45 cm from the fulcrum, and the lever arm of the hamstring muscles is 5 cm, then the force in the muscle group is 135 kg. Note that there are a total of three unknowns: the hamstring force F_H, and the two joint reactions R_x, R_y along two mutually perpendicular directions. These can be uniquely determined from the three equilibrium equations as follows:

$$\Sigma F_x = 0 = 15 - F_H + R_x$$
$$\Sigma F_y = 0 = R_y$$
$$\Sigma F_x = 0 = 15 - F_H + R_X$$
$$\Sigma M_o = 0 = 15 \times 45 - F_H \times 5$$

From which $R_x = 120$ kg

$$R_y = 0$$

and $R_x = 135$ kg

This analysis is the same as that done in the case of the simplified arm model (Fig. 4-20).

A similar model for the hip joint has been described by Inman[4] who performed an analysis to obtain the magnitude and direction of the forces exerted by the abductor muscles of the hip in the one-legged standing posture. Other two-dimensional models based on the same approach have been outlined by Blount[3] and Strange[5]. It should be emphasized that all of the above models are two-dimensional, that is, they deal with the equilibrium of a segment as a planar lever. In such cases three equilibrium equations can be formulated, from which only three unknowns can be determined. The inclusion of any additional muscle force or unknown variable would render the problem statically indeterminate.

THREE-DIMENSIONAL MODELS

For a three-dimensional model the problem is one of equilibrium of a rigid body in space and therefore six unknowns can be evaluated from the corresponding six equilibrium equations. Williams and Svensson[2] have discussed a three-dimensional model for the hip for the static case of standing erect on one leg. They combined seven muscles and ligament forces acting across the hip joint into three variables. Their magnitudes in addition to that of the three mutually perpendicular hip joint reactions could be determined from the six equations of equilibrium for the pelvis.

A model for studying the variation of the magnitudes of the hip joint loads during normal walking has been described by Paul[7]. In this model twenty-two hip muscles were combined into three paired groups of agonist-antagonist muscles: namely, the long and the short flexors, the long and the short extensors and the abductors and the adductors. Using myoelectric recordings of the phasic activity of muscles could be ignored. In this manner only three muscles (one muscle each from the three paired groups) would be active at any given instant. The forces in the three muscles along with the hip joint reactions along three mutually perpendicular directions can then be solved from the six equilibrium equations.

Similar analysis for the variation of the magnitudes of the reaction forces at the knee joint during normal walking has been performed by Morrison[9].

Thus it can be easily seen that in all cases discussed above various assumptions and approximations have to be made so as to render the statically indeterminate system of equations for a given segment uniquely determinate. Such simplified models may yield accurate results from a particular act under analysis but lack potential for general applications to a wide range of segmental activities.

SUMMARY

Comprehensive models for the upper and lower extremities are developed in this chapter. In all cases it has been shown that a general evaluation of the muscle load sharing during the multitudes of activities executed by skeletal system·cannot be evaluated based on equilibrium considerations alone. It is for this reason that models that have been developed in the past rely on simplifying assumptions, approximations or experimentation to select a candidate number of unknowns equal to the number of equations for any particular situations. In order to develop a complete mathematical solution for the muscle participation problem, the solution procedure should be able to predict without prior selection on the part of the modeler, the participating muscles and the extent of the forces they exert. The solution procedure for such comprehensive models is developed in the following chapter using the lower extremities during the forward and backward leaning act. It is then applied to squatting, walking as well as several representative activities of the upper extremities.

REFERENCES

1. Williams, M. and Lissner, H.R.: Biomechanics of Human Motions. W.B. Saunders, Philadelphia, 1962.

2. Williams, J.F. and Svensson, N.L.: A force analysis of the hip joint. Bio-Med. Engr.

3. Blount, W.P.: Don't throw away the cane. J. Bone Jt. Surg., 38A:695, 1956.

4. Inman, V.T.: Functional aspects of the abductor muscles of the hip. J. Bone Jt. Surg., 39:607, 1947.

5. Strange, F.C.ST.C.: The Hip. Heinemann, London, 1963.

6. Paul, J.P.: Bioengineering studies of the forces transmitted by joints. In Biomechanics and related bioengineering topics. (Ed. R.M. Kenedi). Pergamon Press, London, 1965

7. Paul, J.P.: Forces transmitted by joints in the human body. Proc. Instn. Mech. Engrs. 181:3J, 1967.

8. Paul, J.P.: Load actions on the human femur in walking and some resultant stresses. Exptl. Mech., March, 1971.

9. Morrison, J.B.: Bioengineering analysis of the forces transmitted at the knee joint. Bio. Med. Engr., 3:164, 1968.

10. Morrison, J.B.: The mechanics of the knee joint in relation to normal walking. J. Biomech., 3:51, 1970.

11. Braus, H.: Anatomic Der Menschen. Springer-Verlag, Berlin, 1954.

12. Woodburne, R.T.: Essentials of Human Anatomy. Oxford Univ. Press, N.Y., 1965.

13. Davies, D.V. and Davies, F.: Gray's Anatomy. 33rd Ed., Longmans, Green and Co., London 1962.

14. Basmajian, J.V. and Latif, A.: Integrated actions and functions of the chief flexors of the elbow. J. Bone Jt. Surg., 39A: 1106, 1957.

15. Stevens, et al: A polyelectromyographical study of the arm muscles at gradual isometric loading. EMG & Clin. Neurophysiol., 13: 465, 1973.

CHAPTER V

The Solution Procedure
for the Comprehensive
Musculoskeletal Model

"The constitution of an animal must be regarded as
resembling that of a well-governed city-state. For
when order is once established in a city, there is no
need of a special ruler with arbitrary powers to be
present at every activity, but each individual
performs his own task as he is ordered, and one act
succeeds another because of custom."

Aristotle
384-322 BC

INTRODUCTION

The comprehensive modeling of the human musculoskeletal structure
and ligaments has been illustrated in the previous chapter by the
examples of the upper and the lower extremities. It has been shown from
the mathematical formulation that the number of equations describing the
equilibrium of a segment or a system of segments is far less than the
number of unknown variables, namely, the muscle forces, joint reactions
and forces to be carried by the ligaments. The latter can be
conveniently represented by unbalanced tensile reactions and moments at
the joints.

Consequently, many feasible solutions can be obtained indicating
that different loadsharing alternatives are possible for the muscles to
maintain the equilibrium of the investigated posture. It has also been
shown that a merit criterion is necessary to select among these
alternatives the solution that is most compatible with the physiological
and anatomical expectations. In this chapter, the model of the lower
extremities is selected for the development of such a criterion and for
investigating its applicability to different actions and postures. The
upper extremity model is also used to check the generality of the merit
criterion and the solution technique.

LEANING STUDIES

The lower extremities model is first used for obtaining solutions
for the case when a subject is standing and leaning forward or
backward. These postures are assumed to be attained by sagittal rotation
of the pelvis on the femur as shown in Fig. 5-1. This implies that all
muscles (numbered 1 through 13, Table 4-2) which are attached to the
pelvis either contract or elongate according to the change in the x and z
coordinates of their origins on the pelvis. Because of the symmetry of
this act the y coordinate of the point of muscle attachment is assumed to
remain unaffected by the sagittal rotation of the pelvis. Using the
coordinate transformation procedure described in Appendix (A), the new
coordinates for the pelvic muscle origins with respect to the

155

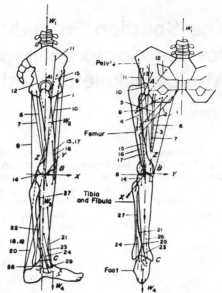

Figure 5-1(a). Musculoskeletal model for the lower extremities.

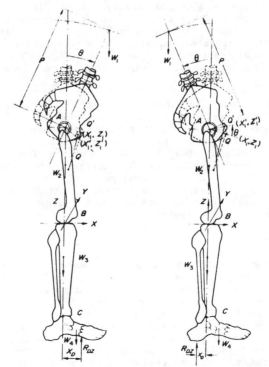

Figure 5-1(b). Theoretical model for leaning postures.

referenceground axes are calculated as a function of the angle of tilt. The new directions for the lines of action of the different muscles are then determined and utilized in formulating the equilibrium equations for the pelvis, the thigh, the leg and the foot as described in Chapter IV.

THE MERIT CRITERIA

Since the number of unknown variables vastly exceeds the number of equations available for solution, innumerable solutions are possible. One must therefore establish a merit criterion--reflecting the muscle loadsharing control law--to select from these solutions, the one with the highest merit. Several possible merit criteria can be formulated as weighted combinations of muscle forces, joint reaction forces and residual moments as follows:

1. Minimization of all the forces carried by the muscles:

$$\text{Merit criterion: } U = \sum_{i=1}^{29} F_i,$$

where F_i is the force in the muscle i. The muscles are assigned tensile forces only.

2. Minimization of the work done by the muscle to attain the given posture, that is, minimize the product of the muscular tension and its elongation or contraction:

$$\text{Merit criterion: } U = \sum_{i=1}^{13} F_i |\Delta L|_i,$$

where $|\Delta L|_i$ represents the magnitude of change in length of muscle i. For the case of forward or backward leaning (that is, forward or backward pelvic tilt) the change in muscle length for the thirteen muscles originating on the pelvis are measured with reference to their corresponding lengths during the standing erect posture.

3. Minimization of the magnitudes of the vertical reaction forces at the hip, knee and ankle joints:

$$\text{Merit criterion: } U = |R_z|_{hip} + |R_z|_{knee} + |R_z|_{ankle}$$

4. Minimization of the magnitudes of the residual moments at the joints which would have to be borne by the appropriate joint ligaments:

$$\text{Merit criterion: } y = \Sigma M = |M_{xyz}| = |M_x| + |M_y| + |M_z|$$

where M_{xyz} denotes the three respective x, y and z components of the joint moment (M_x, M_y, M_z).

5. Minimization of the magnitudes of all the reaction forces at the hip, knee and ankle joints and the residual moments at the joints merit criterion.

$$\Sigma R = \Sigma R_{xyz} = \Sigma |R_x| + |R_y| + |M_z|$$

where R_{xyz} denotes the three respective x,y,z components of the joint reaction (R_x, R_y, R_z).

6. Minimization of a combination of the muscle forces, joint reactions and ligament loads with different weighting factors:

$$\text{Merit criterion:} \quad U = K_F(\Sigma F_i) + K_R(\Sigma R_{xyz}) + K_M(\Sigma M_{xyz}) ,$$

where K_F, K_R and K_M (all >0) represent the weighting factors for the muscle forces, joint reaction forces, and residual joint moments respectively. R_{xyz} denotes the three respective x, y and z components of the joint reaction force ($\Sigma R_{xyz} = |R_x| + |R_y| + |R_z|$).

7. Minimization of a weighted sum of the muscle stresses, joint reaction forces and residual moments on the joints (ligament loads):

$$\text{Merit criterion:} \quad U = K_F \left(\Sigma \frac{F_i}{A_i} \right) + K_R(\Sigma R_{xyz}) + K_M(\Sigma M_{xyz}) ,$$

where A_i is the physiological cross sectional area of the muscle i. Since the allowable maximum stress for a skeletal muscle is normally assumed as 4 Kg/cm^2 (approx.) the weighting factor K_F was selected to be $K_F = 1/4$, so that $K_F F_i/A_i = F_i/4A_i = F_i/F_{ai}$, represents the ratio of the actual force in the muscle (F_i) to the maximum permissible force based on its area (F_{ai}).

8. Minimization of a weighted sum of the muscle stresses, joint reaction stresses and residual moments on the joints (ligament loads):

$$\text{Merit criterion:} \quad U = K_F \left(\Sigma \frac{F_i}{A_i} \right) + K_R \left(\Sigma \frac{R_{xyz}}{A_j} \right) + K_M(\Sigma M_{xyz}) ,$$

where A_j is the anatomical weight-bearing area of the joint.

9. Minimization of a weighted sum of the maximum muscle stress and residual moments on the joints (ligament loads):

$$\text{Merit criterion:} \quad U = \sigma_{max} + K_M(\Sigma M_{xyz})$$

where σ_{max} is the largest of all the F_i/A_i, $\sigma_{max} > F_i/4A_i$. The maximum stress is further constrained not to exceed 4 Kg/cm^2.

10. Minimization of the weighted sum of the maximum stresses in each of the groups of muscles crossing the hip, knee and ankle joints respectively, joint reaction stresses and ligament loads:

$$\text{Merit criterion:} \quad U = \Sigma \left. \sigma_{max} \right|_{hip} + \left. \sigma_{max} \right|_{knee} + \left. \sigma_{max} \right|_{ankle}$$

$$+ K_R \Sigma \frac{R_{xyz}}{A_j} + K_M \Sigma M$$

The maximum stresses in each of the three groups are further constrained not to exceed 4 Kg/cm^2 by the constraints:

$$s_{max}\big|_{hip} > \frac{F_i}{4A_i} \text{ for the muscles crossing the hip joint (i=1-13)}$$

$$\sigma_{max}\big|_{knee} > \frac{F_i}{4A_i} \text{ for the muscles crossing the knee joint (i=14-19)}$$

$$\sigma_{max}\big|_{ankle} > \frac{F_i}{4A_i} \text{ for the muscles crossing the ankle joint}$$

Note that a muscle crossing across more than one joint is included in only one joint group, that which is closer to the muscle origin.

11. Minimization of a weighted sum of the maximum muscle force in different groups of muscles (preselected according to their force producing ability) and the joint reaction forces and residual moments (ligament loads):

a. All the muscles are included in one group, so that

Merit criterion: $U = f + K_R(\Sigma R_{xyz}) + K_M(\Sigma M_{xyz})$,

where f is the largest of all active muscle forces, and given by $F_i < f$ for i = 1,2,...,29

b. All the muscles are grouped into three groups according to their ability to produce forces so that

Merit criterion: $U = \sum_{k=1}^{3} f_k + K_R(\Sigma R_{xyz}) + K_M(\Sigma M_{xyz})$,

the muscles being grouped as follows:

Group 1	Group 2	Group 3
$F_i < f_1$	$F_i < f_2$	$F_i < f_3$
i=3,4,6-9,12-22	i=2,5,11	i=1,10,23-29

c. The muscles are grouped into four groups according to their ability to produce forces, so that

Merit criterion $U = \sum_{k=1}^{4} f_k + K_R(\Sigma R_{xyz}) + K_M(\Sigma M_{xyz})$,

where f_k are the maximum muscle forces in the following four groups:

Group 1	Group 2	Group 3	Group 4
$F_i < f_1$	$F_i < f_2$	$F_i < f_3$	$F_i < f_3$
i= 1(4.11 cm^2)	i= 2(10.00 cm^2)	i= 8(14.25 cm^2)	i= 7(26.39 cm^2)
5(6.60 cm^2)	3(11.65 cm^2)	14(15.00 cm^2)	9(28.29 cm^2)
10(3.75 cm^2)	4(11.65 cm^2)	18(11.50 cm^2)	12(22.20 cm^2)
11(8.40 cm^2)	6(7.65 cm^2)	19(11.50 cm^2)	13(35.00 cm^2)
23(2.50 cm^2)	21(7.70 cm^2)		15(50.00 cm^2)
24(1.35 cm^2)	22(5.80 cm^2)		16(50.00 cm^2)
25(2.80 cm^2)			17(50.00 cm^2)
26(4.50 cm^2)			20(20.00 cm^2)
27(4.30 cm^2)			
28(3.80 cm^2)			
29(1.70 cm^2)			

In this case the muscles are grouped more or less according to their cross sectional areas (cm^2) as shown in parentheses, next to the respective muscle numbers.

d. The muscles are grouped into five different groups:

$$\text{Merit criterion: } U = \sum_{k=1}^{5} f_k + K_R(\Sigma R_{xyz}) + K_M(\Sigma M_{xys}) \quad ,$$

where f_k is the maximum muscle force in each of the five groups. The grouping is exactly the same as in c above, except for an additional fifth group containing two muscle forces: $F_G < f_5$ and $F_7 < f_5$. The effect of this constraint is to allow two separate muscles which are very near in origin and insertion, and thus functionally identical, to share the total group force between them.

e. Each muscle is considered as independent, thus yielding 29 different groups.

$$\text{Merit criterion: } U = \sum_{1}^{29} f_k + K_R(\Sigma R_{xyz}) + K_M(\Sigma M_{xyz})$$

Note that this criterion is in effect the same as criterion (6), which is minimization of a weighted sum of all muscle forces, joint reactions and ligament loads, and hence the new criterion which is merely another way of stating the previous criterion (6), does not entail any actual computations.

12. Minimization of a weighted sum of the maximum muscle stress in different groups of muscles (preselected according to their force producing ability), the joint reaction stresses and residual moments carried by ligaments.

a. Muscles grouped into four different groups:

Merit criterion: $U = \sum_{k=1}^{4} \sigma_k + K_R \left(\dfrac{\Sigma R_{xyz}}{A_j} \right) + K_M (\Sigma M_{xyz})$

The stress in each of the four groups is limited not to exceed 4 kg/cm^2 by means of the constraints:

$$\frac{F_i}{4A_i} \leqslant \sigma_k \qquad \text{for each group}$$

The grouping is the same as that for the case (11c) discussed above.

b. Muscles grouped into five different groups (same as in (11d)) so that

$$U = \sum_{k=1}^{5} \sigma_k + K_R \left(\frac{\Sigma R_{xyz}}{A_j} \right) + K_M (\Sigma M_{xyz})$$

METHOD OF SOLUTION OF THE SYSTEM EQUATIONS

The equations representing the equilibrium conditions for the system at a particular posture are linear algebraic equations where all the unknown variables appear in the first order and in separate terms of the equations. It can also be seen that all the merit criteria considered in the previous section are linear functions of the unknown variables. The problem can be readily formulated as a linear program and a unique solution is conveniently obtained by the simplex algorithm using a digital computer. The algorithm searches efficiently for solutions with continuously improving merit until the one with the highest possible merit is reached. Detailed discussion of the procedure is given in Appendix C.

The simplex technique requires that all the variables are greater than or equal to zero. This is always the case with the muscle forces which act in tension only. However, there are some variables, such as the reactions at the joints, which can be either positive or negative depending on the direction in which they should act. In the course of analysis of the free-body diagram of a segment, arbitrary directions can be assumed for these reactions. The solution will automatically indicate whether the assumed directions are correct or should be reversed. To allow the reactions (or any other variables, if desired), to take up either a positive or a negative sign in the program, we divide them into two parts, both of which are constrained to be positive, but the net reaction, being the difference of the two, can be either positive or negative depending on whichever part is greater. Division of a variable in this manner further increases the total number of variables.

From anatomical considerations, two muscles with points of origin and insertion close together are represented by two lines in the musculoskeletal model. Both muscles are, therefore capable of serving identical functions. However, the simplex computation may yield only one of these to be active in the final solution since their functional directions are not appreciably different. For example the semitendinosus (F_6) and the semimembranosus (F_7), the flexors of the leg, are

161

anatomically close together. The solution may define their combined activity by assigning their total load to one or the other. In order for both of them to be active simultaneously, the contribution of each muscle to the combined force can be reasonably assumed to be proportional to their cross-sectional areas. All closely connected muscles may be similarly treated in order to insure their participation in the final solution. In other words, closely connected muscles can be more reliably treated as a group. Modeling them independently can, however, illustrate preferences according to the merit criterion.

RESULTS

The equilibrium analysis of the lower extremities during leaning backward and forward is discussed in detail in Chapter IV. There are a total of 21 equilibrium equations: three for the pelvis ($\Sigma F_x = 0$, $\Sigma F_y = 0$, $\Sigma F_z = 0$) which is symmetrical about the sagittal plane for the leaning activity, and six each for the thigh, the leg and the foot ($\Sigma F_x = 0$, $\Sigma F_y = 0$, $\Sigma F_z = 0$, $\Sigma M_x = 0$, $\Sigma M_y = 0$, $\Sigma M_z = 0$). There are twenty-nine unknown muscle forces, which with the three z, y and z components of reactions at the hip, knee and ankle joints, and three x, y and z components of moments (ligament loads) at the ankle joint make up 41 unknown variables. For the foot, the x coordinate (anterior-posterior direction) of the resultant ground-to-foot reactions act on the foot at the floor. As explained earlier, when the reactions and the moments which can be either positive or negative, are divided into two parts each, the total number of unknown variables turns out to be 57.

The body weight for the subject is 72.3 Kg, and is assumed to be distributed as follows: upper torso = 48.0 Kg, thigh = 7.45 Kg, leg = 3.6 Kg and foot = 1.1 Kg. The numerical results for muscle load sharing, joint reactions and ligament loads for the leaning acts are obtained for pelvic tilt increments of five degrees, starting from the erect posture, both backwards and forwards. An illustration sample of these results, which are computed based on the various merit criteria outlined in the preceding section, are shown in Tables 5-1 through 5-8 for 10 degrees forward and backward leaning.

The results for the muscle load sharing based on the criteria U = ΣF, $U = \Sigma F_i |\Delta L|_i$, $\Xi \mp \Sigma R_z|_{\text{all joints}}$, $U = \Sigma R_{xyz} + K_m \Sigma M_{xyz}|_{\text{all joints}}$ and $U = \Sigma M_{xyz}|_{\text{ankle}}$ are shown in Table 5-1. The first three yield high unbalance or residual moments at the ankle joint. Similarly, the results when the objective is to minimize a linear combination of muscle forces and joint moments, that is, $U = \Sigma F + K_m \Sigma M$ are tabulated in Table 5-2 for the values of weighting factor K_m = 1,2,3,4 and 40. It is seen for forward leaning that when $K_m < 4$, the muscle forces change in an erratic manner with increase in forward pelvic tilt. In order for the criterion $\Sigma F + K_m(\Sigma M)$ to be applicable therefore, K_m should be selected $\geqslant 4$. When reactions at the joints are included in this criterion (U ΣF + 2 ΣR_{xyz} + 4 ∂F_{xyz}) the results for muscle activity for both forward and backward leaning are seen to be anatomically consistent.

When the general form of the criterion is modified to

$$U = K_F \sum_i \frac{F_i}{A_i} + K_R \Sigma R_{xyz} + K_M \Sigma M_{xyz}$$

the corresponding results are shown in Table 5-3. The combinations for

162

TABLE 5-1.

TABLE 5-1. 10° FORWARD LEANING

Criterion Muscle Forces	29 $U = \sum_{i=1}^{29} F_i$	13 $U = \sum_{i=1}^{13} F_i \|\Delta L\|_i$	$U = \Sigma M$	$U = \Sigma R_z$	$U = \Sigma R_{xyz} + 4\,\Sigma M$
4	0	15.39	0	0	0
7	16.70	16.65	25.53	17.82	23.88
8	7.58	[0]	14.41	5.87	11.71
10	0	0	11.44	0	8.64
12	1.38	3.39	0	2.05	0.15
13	0	0	1.48	0	-
14	0	0	12.70	14.37	17.66
18	20.02	26.48	10.82	21.64	14.06
19	12.52	15.37	5.15	0	0
22	0	0	4.08	0	3.2
24	0	[37.57]	0	0	0
26	0	-	0	0	10.94
29	-	10.25	0	0	0
Joint reactions:					
hip ⎰ x	-1.84	-9.11	-0.76	-2.02	-1.47
⎱ y	-0.48	2.27	-1.58	-0.45	-1.81
z	49.43	58.57	76.35	49.66	68.13
knee ⎰ x	1.48	2.61	2.02	2.68	2.58
⎱ y	-0.32	-1.18	0.02	1.80	1.13
z	88.00	89.64	110.88	90.63	106.72
ankle ⎰ x	2.99	-5.46	1.97	1.98	1.23
⎱ y	0.23	-6.87	-1.23	0.71	-2.09
z	67.42	122.93	54.75	56.57	62.77
ground- ⎰ x	0	0	0	0	0
to ⎱ y	-0.4	-0.93	-0.27	-0.39	-0.29
foot z	36.15	36.15	36.15	36.15	36.15
ankle ⎰ x	[-32.48]	0	0	[-20.18]	0
moments ⎱ y	91.04	0	0	23.51	0
z	-2.39	[-23.38]	0	-6.41	0
objective: U	58.21	17.18	0	196.66	248.54

10° BACKWARD LEANING

	$U = \sum F_i$	$U = \sum F_i\|\Delta L\|_i$	$U = \Sigma M$	$U = \Sigma R_z$	$U = \Sigma R_{xyz} + 4\Sigma M$
1	27.67	30.59	28.52	30.59	30.09
4	0	0	0	0	0
6	5.23	0	5.05	0	4.69
7	0	4.8	0	4.80	0
9	7.63	0	4.95	0	0
10	0	1.03	0.76	1.03	2.18
12	0.69	1.12	0.83	1.12	1.06
15	11.97	19.12	14.12	19.12	18.12
21	[0]	[0]	28.49	[0]	28.72
23	0	$F_{24} = 24.91$	15.45	0	13.51
29	0	16.62	0	0	1.41
Joint reactions:					
hip ⎰ x	0.81	1.29	0.96	1.29	1.23
⎱ y	-0.88	-0.72	-0.83	-0.72	-0.74
z	65.01	61.25	63.87	61.25	61.74
knee ⎰ x	0.37	0.59	0.44	0.59	0.57
⎱ y	-1.62	-1.89	-1.7	-1.89	-1.85
z	83.81	86.84	84.71	86.84	86.37
ankle ⎰ x	0	-8.55	-2.45	0	-2.67
⎱ y	-1.78	-6.49	-3.99	-1.98	-4.05
z	35.05	75.25	78.86	35.05	78.53
ground ⎰ x	0	0	0	0	0
to ⎱ y	-1.78	-1.98	-1.85	-1.98	-1.99
foot z	36.15	36.15	36.15	36.15	36.15
ankle ⎰ x	[-8.91]	0	0	[-9.88]	0
moments ⎱ y	110.44	0	0	110.44	0
z	5.2	[-11.33]	0	5.77	0
objective: U	53.20	25.70	0	183.14	237.74

TABLE 5-2. 10° FORWARD LEANING

Criterion / Muscle forces	$U = \Sigma F + M$	$U = F + 2M$	$U = F + 3M$	$U = F + 4M$	$U = F + 40M$	$U = F + \Sigma R + 16M$
7	22.27	22.27	23.75	25.53	25.53	23.75
8	10.15	10.15	12.09	14.41	14.41	12.09
10	6.34	6.34	8.74	11.44	11.44	8.74
12	0.66	0.66	0	0	0	-
13	0	-	0	1.48	1.48	-
14	16.78	16.78	14.98	12.70	12.70	14.98
18	16.08	16.08	13.67	10.82	10.82	13.67
19	0	0	2.36	5.15	5.15	2.36
22	3.84	3.84	3.95	4.08	4.08	3.95
Joint reactions:						
hip x	-1.62	-1.62	-1.43	-0.76	-0.76	-1.43
hip y	-1.44	-1.44	-1.83	-1.58	-1.58	1.83
hip z	63.16	63.16	68.34	76.35	76.35	68.34
x	2.61	2.61	2.36	2.02	2.02	2.36
y	1.31	1.31	0.73	0.02	0.02	0.73
z	102.44	102.44	106.41	110.88	110.88	106.41
ankle x	1.95	1.95	1.96	1.97	1.97	1.96
ankle y	-0.78	-0.78	-0.98	-1.23	-1.23	-0.98
ankle z	54.64	54.64	54.69	54.75	54.75	54.69
ground to foot x	0	0	0	0	0	0
ground to foot y	-0.32	-0.32	-0.29	-0.27	-0.27	-0.29
ground to foot z	36.15	36.15	36.15	36.15	36.15	36.15
ankle moments x	0	0	0	0	0	0
ankle moments y	0	0	0	0	0	0
ankle moments z	-3.03	-3.03	-1.63	0	0	-1.63
objective: U	79.15	82.17	84.44	85.60	85.60	583.12

10° BACKWARD LEANING

Muscle forces						
1	27.67	27.67	27.67	27.67	28.52	28.52
6	5.23	5.23	5.23	5.23	5.05	5.05
9	7.63	7.63	7.63	7.63	4.95	4.95
10	0	0	0	0	0.76	0.76
12	0.69	0.69	0.69	0.69	0.83	0.83
15	11.97	11.97	11.97	11.97	14.12	19.12
21	28.42	28.42	28.42	28.42	28.49	28.49
23	15.62	15.62	15.62	15.62	15.45	15.45
Joint reactions:						
hip x	0.81	0.81	0.81	0.81	0.96	0.96
hip y	-0.88	-0.88	-0.88	-0.88	-0.83	-0.83
hip z	65.01	65.01	65.01	65.01	63.87	63.87
knee x	0.37	0.37	0.37	0.37	0.44	0.44
knee y	-1.62	-1.62	-1.62	-1.62	-1.7	-1.7
knee z	83.81	83.81	83.81~	83.81	84.71	84.71
ankle x	-2.46	-2.46	-2.46	-2.46	-2.45	-2.45
ankle y	-3.93	-3.93	-3.93	-3.93	-3.99	-3.99
ankle z	78.97	78.97	78.97	78.97	78.86	78.86
ground to foot x	0	0	0	0	0	0
ground to foot y	-1.78	-1.78	-1.78	-1.78	-1.85	-1.85
ground to foot z	36.15	36.15	36.15	36.15	36.15	36.15
ankle moments x	0	0	0	0	0	0
ankle moments y	0	0	0	0	0	0
ankle moments z	-0.23	-0.23	-0.23	-0.23	0	0
objective U	97.47	97.69	97.93	98.16	98.17	573.79

TABLE 5-3*   TABLE 5-4

10° FORWARD LEANING

Criterion Muscle force	$k_F=k, k_R=\frac{1}{2}, k_m=1$	$k_F=k, k_R=\frac{1}{2}, k_m=2$	$k_F=k, k_R=\frac{1}{2}, k_m=4$	$k_F=k, k_R=0, k_m=1$ and $k_m=\frac{1}{2}, k$	$k_F=k, k_m=4, k_R=1$ $U=\frac{F_t}{4A}+\frac{R}{A_{jt}}+4M$
6	0	0	0	0	-
7	17.92	17.92	24.01	25.66	25.53
8	5.95	5.95	11.77	14.49	14.41
10	0	0	8.30	11.17	11.44
12	2.55	2.55	2.16	1.76	0
13	0.99	0.99	3.84	4.86	1.48
14	14.29	14.29	17.25	12.30	12.70
18	21.56	21.56	14.03	10.76	10.82
19	0	-	0	5.2	5.15
20	0	-	0	0	0
21	[9.25]	[9.25]	0	0	0
22	2.29	2.29	3.21	4.09	4.08
23	0	0	0	0	0
24	0	0	0	0	0
25	0	0	0	0	0
26	0	0	11.06	0	0
29	0	0	0	0	0
Joint Reactions					
hip x	-1.71	-1.71	-0.3	0.27	-0.76
hip y	0	0	0	0	1.58
hip z	50.91	50.91	73.08	80.78	76.35
knee x	2.65	2.65	2.48	1.92	2.02
knee y	1.79	1.79	1.09	-0.03	0.02
knee z	90.65	90.65	106.16	110.42	110.88
ankle x	1.79	1.79	1.23	1.97	1.97
ankle y	-0.52	-0.52	-2.74	-1.25	-1.23
ankle z	67.86	67.86	62.86	54.76	54.75
ankle x	0	0	0	0	0
moments y	0	0	0	0	0
z	-6.22	-6.22	0	0	0
ground x	0	0	0	0	0
to y	-0.39	-0.39	-0.31	-0.28	-0.27
foot z	36.15	36.15	36.15	36.15	36.15
Objective: U	116.58	122.79	127.35	2.03	26.79

10° BACKWARD LEANING

Muscle force					
1	29.35	29.35	29.35	26.35	28.96
2	0	0	0	3.69	0
6	0	0	0	0	1.63
7	5.12	5.12	5.12	8.12	3.48
9	3.36	3.36	3.36	8.19	4.72
12	1.57	1.57	1.57	2.04	0.88
13	2.09	2.09	2.09	0	0
15	16.58	16.58	16.58	8.49	15.04
16	0	0		0.00	0
21	28.53	28.53	28.53	28.49	28.49
22	15.09	15.09	15.09	15.45	15.45
29	0.26	0.26	0.26	0	0
Joint Reactions					
hip x	1.64	1.64	1.64	0.71	1.02
hip y	0	0	0	0	0.81
hip z	64.89	64.89	64.89	71.79	63.41
knee x	0.49	0.49	0.49	0.26	0.47
knee y	-1.79	-1.79	-1.79	-2.35	-1.74
knee z	85.69	85.69	85.69	87.50	85.13
ankle x	-2.49	-2.49	-2.49	-2.45	-2.45
ankle y	-4.00	-4.00	-4.00	-3.99	-3.99
ankle z	78.80	78.80	78.80	78.86	78.86
ankle x	0	0	0	0	0
moments y	0	0	0	0	0
z	0	0	0	0	0
ground x	0	0	0	0	0
to y	-1.88	-1.88	-1.88	-1.85	-1.85
foot z	36.15	36.15	36.15	36.15	36.15
Objective: U	124.35	124.35	124.35	4.40	27.88

$*U = \Sigma k_F \frac{F_t}{A_t} + k_R \Sigma R + k_M \Sigma M$

165

TABLE 5-5. **10°** FORWARD LEANING

$$U = \sigma_{max} + k_M M$$

Criterion Muscle force	$k_m = 0$	$k_m = 40$
1	4.38	0
2	2.94	0
6	0	0
7	28.10	25.56
8	13.54	12.61
10	3.99	4.89
12	20.15	24.04
13	37.27	45.68
14	0	12.97
18	12.25	13.52
19	11.86	0
20	0	7.41
21	8.2	10.05
22	2.18	3.72
23	0	3.26
24	1.44	0
25	2.98	0
26	4.79	5.87
29	1.81	0
Joint Reactions:		
hip { x	9.96	12.43
y	-16.37	-19.62
z	126.61	127.55
knee { x	0.17	1.33
y	-1.69	0.67
z	105.22	100.45
ankle { x	1.51	1.73
y	-3.37	-3.04
z	80.03	78.35
ankle { x	0	0
moments { y	26.32	0
z	-2.01	0
ground { x	0	0
to { y	-0.89	-0.45
foot { z	36.15	36.15
Objective: U	0.27	0.33

°**10°** BACKWARD LEANING

	$k_m = 0$	$k_m = 40$
1	3.86	13.1
2	9.39	25.25
5	6.20	0
6		23.09
7	13.09	8.66
9	27.15	0
10	3.52	11.95
12	14.02	43.32
13	32.89	111.56
15	3.86	5.12
17	0	21.98
19	0.77	0
20	6.46	0
21	0	24.54
23	0	7.97
24	1.27	4.30
27	4.04	0
28	3.57	0
29	0	5.31
Joint Reactions:		
hip { x	8.51	27.28
y	-15.79	-46.82
z	126.81	240.32
knee { x	-1.51	-2.52
y	-3.26	-7.69
z	83.57	114.94
ankle { x	0	-3.82
y	-1.66	-5.08
z	51.01	76.78
ankle { x	29.32	0
moments { y	164.01	0
z	3.72	0
ground { x	0	0
to { y	-0.95	-2.71
foot { z	36.15	36.15
σ_{max}	0.235	0.797
Objective: U	0.235	0.797

166

TABLE 5-6.   10° FORWARD LEANING

Muscle force / Criterion	$U = (\sigma_{max})_{hip} + (\sigma_{max})_{knee}$ $+ (\sigma_{max})_{ankle} + k_R \frac{R}{A_{jt}} + k_M M$ $k_R = k, \ k_m = 4$
6	0 0
7	25.19
8	13.75
10	10.42
12	1.67
13	4.20
14	13.69
18	11.65
19	3.79
20	0
21	0
22	3.85
23	0
24	0
25	0
26	2.99
29	0

Joint Reactions:		
hip	x	0
	y	0.18
	z	78.19
knee	x	-2.08
	y	0.28
	z	109.32
ankle	x	1.77
	y	-1.65
	z	59.65
ankle moments	x	0
	y	0
	z	0
ground to foot	x	0
	y	0
	z	-0.29
σ_{hip}		0.69
σ_{knee}		0.25
σ_{ankle}		0.17
Objective: U		7.33

10° BACKWARD LEANING

Muscle force	
1	15.37
2	17.58
6	13.75
7	0
9	0
10	14.03
13	6.26
15	10.55
16	10.55
21	32.57
23	10.58
27	4.09
29	0.76

Joint Reactions:		
hip	x	0.65
	y	-3.15
	z	89.09
knee	x	-0.58
	y	-4.55
	z	95.46
ankle	x	-2.87
	y	-4.29
	z	82.89
ankle moments	x	0
	y	0
	z	0
ground to foot	x	0
	y	-2.08
	z	36.15
σ_{hip}		0.94
σ_{knee}		0.053
σ_{ankle}		1.06
Objective: U		8.81

K_F, K_R and K_M used were as follows:

K_F	K_R	K_m
1/4	1/2	1
1/4	1/2	1
1/4	1/2	1
1/4	1/2	1
1/4	1/2	1
1/4	1/2	1

The results show that during forward leaning, the criterion is not applicable for the values of $K_M = 1$ and $K_M = 2$. The tibialis anterior muscle (F_{21}) seen to be active, is not likely to function during forward leaning from an anatomical point of view.

When the criterion is changed to minimize the muscle and joint stresses, $U = \Sigma F_i/A_i + K_R R_{xys}/A_j + K_M M_{xyz}$, the resulting distribution for muscle forces and joint reactions and moments is illustrated in Table 5-4, for $K_F = 1/4$, $K_R = 1$, $K_M = 4$. This criterion appears to be satisfactory for both forward and backward leaning.

The results for the case of minimization of maximum muscle stress among all the muscles, that is, $U = \sigma_{max} + K_R R_{xyz} + K_M M_{xyz}$ are tabulated in Table 5-5. Two different cases were investigated with $K_R = 0$, $K_M = 0$ and with $K_R = 0$ $K_M = 40$, both shown in the table. For both backward and forward leaning, these criteria require the muscles to carry high forces, function antagonistically and subject the ankle joint to high residual moments.

The criterion $U = (\sigma_{man})_{hip} + (\sigma_{max})_{knee} + (\sigma_{max})_{ankle} + K_R \Sigma R_{xyz}/A_{jt} + K_M \Sigma M_{xyz}$ with $K_R = 1/4$ and $K_M = 4$ (Table 5-5) may be ruled out because it shows that during backward leaning, the semitendinosus (F_6 is active which is automatically inconsistent.

The results for muscle load sharing when the muscles are divided into three groups with the objective of minimizing the maximum muscle force in each group are shown in Table 5.7. The criteria

$$U = \sum_{k=1}^{3} F_k + K_R(\Sigma R_{xyz}) + K_M(\Sigma M_{xyz})$$

is applied with $K_M = 4$ and $K_R = 0.05$, 0.1, 0.15, 0.25, 0.5 and 1.0 respectively. The criteria is inapplicable for forwarding leaning for all values of K_R with the exception of $K_R = 0.25$, as antagonistic activity in the tibialis anterior (F_{21}) is shown. It is also inapplicable for similar reasons when the criterion is modified to

$$U = \sum_{k=1}^{3} F_k + K_R \frac{R_{xyz}}{A_{jt}} K_M M \text{ with } K_R = \frac{1}{4}$$

and $K_M = 4$ (Table 5-8). When the muscle grouping is changed to four groups so that the merit criterion is to minimize

$$U = \sum_{k=1}^{4} F_k + K_R R + K_m M$$

the criteria fails again for forward leaning when $K_R = 1$ and $K_M = 8$, but appears to be feasible for backward leaning (Table 5-9). However, it appears to be satisfactory for $K_R = 2$ and $K_M = 16$ for both backward and forward leaning. When the forces in the semitendinosus and the semimembranosus are related by their areas ($F_6 : F_7 = A_6 : A_7$), the criterion

$$U = \sum_{1}^{4} F_k + 2 \Sigma R + 16 \Sigma M$$

seems feasible for leaning in either direction. For the case when the muscles are divided into five groups the fifth group consisting of the semitendinosus and the semimembranosus, the criterion

$$U = \sum_{k=1}^{5} F_k + 2 \Sigma R + 16 \Sigma M$$

seems applicable for forward and backward leaning (Table 5-9).

The criteria based on minimization of maximum muscle stresses in four groups

$$U = \sum_{1}^{4} \sigma_k + k_R \frac{R}{A_j} + k_M M \; , \quad u = \sum_{1}^{4} \sigma_k + k_R \frac{R}{A_j} + k_M M$$

with $F_6 F_7 = A_6 : A_7$ and in five groups

$$U = \sum_{1}^{4} \sigma_k + k_R \frac{R}{A_j} + k_m M)$$

seem to yield results that are compatible with anatomical findings (Table 5-10). The weighting factors were selected to be $K_R = 1/4$ and $K_m = 4$; the criteria appear valid for leaning in either direction.

When all the muscles are lumped into one group, and the objective becomes minimization of the maximum muscle force

$$U = \sum_{k=1}^{1} F_K + K_R R + K_M M$$

the selection of the magnitudes of the weighting factors seem to influence the adaptability of the criterion to the leaning act. Thus, for $K_R = 0$ and $K_M = 4$, the results show antagonistic activity in the tibialis anterior (F_{21}) as well as in the extensor muscles (F_{23}, F_{24})

TABLE 5-8

TABLE 5-7 $U = \sum\limits_{k=1}^{3} F_k + k_R R + k_M M$

10° FORWARD LEANING

Criterion / Muscle forces	$k_R = 0$ $k_m = 1$	$k_R = .05$ $k_m = 1$	$k_R = 0.1$ $k_m = 1$	$k_R = 0.15$ $k_m = 1$	$k_R = 0.25$ $k_m = 1$	$k_R = 0.5$ $k_m = 1$	$k_R = 1$ $k_m = 1$
6	13.23	13.35	13.35	13.02	6.83	0	0
7	13.71	13.56	13.56	13.69	19.33	19.52	17.82
8	13.02	13.04	13.04	12.87	12.25	7.51	5.87
10	10.28	11.38	11.38	11.57	10.86	2.42	0
12	6.56	1.84	1.84	0	0	1.52	2.05
13	13.71	4.79	4.79	1.19	0.39	0	0
14	9.81	11.20	11.20	12.08	19.33	15.29	14.37
18	13.71	13.56	13.56	13.69	13.77	19.52	21.64
19	8.49	8.22	8.22	7.69	0.27	0	0
21	[13.71]	[13.56]	[13.56]	[12.91]	0	[6.56]	[9.38]
22	5.72	1.41	1.41	1.46	3.28	2.60	2.27
23	[10.28]	0	0	0	0	0	0
26	10.28	11.38	11.38	11.57	10.86	2.42	0
Joint Reactions:							
hip x	2.69	-0.007	-0.007	-1.11	-1.37	-1.87	-2.02
hip y	-4.24	0	0	1.67	1.94	0.83	-0.45
hip z	91.60	80.79	80.79	75.88	73.32	54.68	49.46
knee x	1.78	2.05	2.05	2.19	2.74	2.65	2.68
knee y	-0.24	-0.12	-0.12	0	1.13	1.62	1.80
knee z	113.13	115.14	115.14	115.42	113.34	95.13	90.63
ankle x	0.96	1.03	1.03	1.03	1.25	1.68	1.79
ankle y	-5.02	-3.28	-3.28	-3.26	-2.78	-1.02	-0.51
ankle z	96.51	82.73	82.73	81.91	62.74	65.80	68.05
ankle moments x	0	0	0	0	0	0	0
ankle moments y	0	0	0	0	0	0	0
ankle moments z	0	0	0	0	0	-4.62	-6.26
ground to foot x	0	0	0	0	0	0	0
ground to foot y	-0.46	-0.43	-0.43	-0.42	-0.35	-0.36	-0.39
ground to foot z	36.15	36.15	36.15	36.15	36.15	36.15	36.15
F_1	13.71	13.56	13.56	13.69	19.33	19.52	21.64
F_2	0	0	0	0	0	0	0
F_3	10.28	11.38	11.38	11.57	10.86	2.42	0
Objective: U	31.63	37.19	53.46	67.62	95.33	153.04	264.07

10° BACKWARD LEANING

	$k_R = 0$ $k_m = 1$	$k_R = .05$ $k_m = 1$	$k_R = 0.1$ $k_m = 1$	$k_R = 0.15$ $k_m = 1$	$k_R = 0.25$ $k_m = 1$	$k_R = 0.5$ $k_m = 1$	$k_R = 1$ $k_m = 1$
1	24.23	24.47	24.47	23.69	23.44	27.67	28.52
2	3.14	0	0	0	0	0	0
3	9.72	0	0	0	0	0	0
4	9.72	13.66	13.66	14.86	9.88	0	0
5	3.14	0	0	0	0	0	0
6	9.72	6.91	6.91	6.77	6.61	5.23	5.05
7	0	0	0	0	0	0	0
9	[0.06]	3.65	3.65	3.52	9.88	7.63	4.95
10	13.41	9.01	9.01	9.89	5.15	0	0.76
12	7.81	2.41	2.41	2.48	1.65	0.69	0.83
13	9.72	0	0	0	0	0	0
15	9.72	13.66	13.66	14.86	9.88	11.97	14.12
16	7.56	1.03	1.03	0	0	0	0
21	9.72	13.66	13.66	14.86	9.88	11.97	28.49
23	24.23	24.47	24.47	23.69	17.98	17.81	15.45
24	22.22	18.75	18.75	17.19	15.51	13.73	0
26	[24.23]	[24.47]	[24.47]	[22.35]	0	0	0
Joint Reactions:							
hip x	2.84	0.21	0.21	0.18	0.12	0.81	0.96
hip y	5.71	1.40	1.40	1.59	0.67	-0.88	-0.83
hip z	111.32	83.27	83.27	89.34	79.99	65.01	63.87
knee x	0.42	0.36	0.36	0.32	0.22	0.37	0.44
knee y	-3.98	-2.95	-2.95	-3.01	-2.39	-1.62	-1.70
knee z	95.93	89.98	89.98	89.99	86.25	83.81	84.71
ankle x	-7.11	-6.71	-6.71	-6.35	-4.49	-4.27	-2.45
ankle y	-12.56	-11.88	-11.88	-11.22	-5.97	-5.09	-3.99
ankle z	114.15	115.21	115.21	112.04	77.81	78.01	78.86
ankle moments x	0	0	0	0	0	0	0
ankle moments y	0	0	0	0	0	0	0
ankle moments z	0	0	0	0	-4.26	-3.93	0
ground to foot x	0	0	0	0	0	0	0
ground to foot y	-2.37	-2.06	-2.06	-2.05	-1.83	-1.78	-1.85
ground to foot z	36.15	36.15	36.15	36.15	36.15	36.15	36.15
F_1	9.72	13.66	13.66	14.86	9.88	11.97	28.49
F_2	3.14	0	0	0	0	0	0
F_3	24.23	24.47	24.47	23.09	23.44	27.67	28.52
Objective: U	45.29	53.73	69.33	84.90	114.83	175.59	294.82

TABLE 5-9. 10° FORWARD LEANING

Criterion / Muscle forces	$U=\sum F_k+k_m M+k_R R$ (4 groups) $k_R=1, k_m=8$	$k_R=2, k_m=16$	$U=\sum F_k+k_m M+k_R R,\ \frac{F_8}{F_7}=\frac{A_8}{A_7}$ $k_m=16,\ k_R=2$	$U=\sum F_k+2R+16N$
6	0	0	5.81	0
7	23.86	24.15	20.05	24.15
8	12.27	12.15	12.23	12.15
10	8.92	9.12	10.59	9.12
12	0	0	0	0
13	0.11	0.01	0.33	0.01
14	14.69	16.86	18.96	16.86
18	13.46	13.53	13.73	13.53
19	2.67	0.85	0.36	0.85
21	2.94	0	0	0
22	2.94	3.35	3.29	3.35
26	8.92	9.12	10.59	9.12
Joint Reactions:				
hip x	-1.38	-1.42	-1.38	-1.42
hip y	-1.81	-1.88	1.93	1.88
hip z	68.91	69.19	72.7	69.19
knee x	2.32	2.49	2.7	2.49
knee y	0.66	0.95	1.10	0.95
knee z	106.69	107.44	112.46	107.44
ankle x	1.33	1.36	1.27	1.36
ankle y	-2.49	-2.45	-2.73	-2.45
ankle z	65.53	61.43	62.54	61.43
ankle moments x	0	0	0	0
ankle moments y	0	0	0	0
ankle moments z	0	0	0	0
ground to foot x	0	0	0	0
ground to foot y	-0.29	-0.29	-0.34	-0.29
ground to foot z	36.15	36.15	36.15	36.15
F_1	8.92	9.12	10.59	9.12
F_2	2.94	3.35	5.81	3.35
F_3	14.69	16.86	18.96	16.86
F_4	23.86	24.15	20.05	24.15
F_5	-	-	-	24.15
Objective: U	301.55	550.71	573.04	574.86

10° BACKWARD LEANING

1	27.96	28.52	29.13	28.84
2	0.79	0	0	0
6	6.43	5.05	1.14	2.55
7	0	0	3.95	2.55
9	8.11	4.95	4.33	4.78
10	0	0.76	0	0.21
12	1.08	0.83	0.9	0.86
15	8.11	14.12	15.46	14.79
16	3.32	0	0	0
21	28.49	28.49	28.36	28.49
23	15.45	15.45	15.44	15.45
24	0	0	0.11	0
Joint Reactions:				
hip x	0.69	0.96	.05	1.00
hip y	-0.68	-0.83	-0.80	-0.82
hip z	68.06	63.87	63.19	63.53
knee x	0.39	0.44	0.48	0.46
knee y	-1.84	-1.70	-1.75	-1.73
knee z	85.23	84.71	85.30	85.01
ankle x	-2.45	-2.45	-2.46	-2.45
ankle y	-3.99	-3.99	-4.02	-3.99
ankle z	78.86	78.86	78.84	78.86
ankle moments x	0	0	0	0
ankle moments y	0	0	0	0
ankle moments z	0	0	0	0
ground to foot x	0	0	0	0
ground to foot y	-1.85	-1.85	-1.86	-1.85
ground to foot z	36.15	36.15	36.15	36.15
F_1	27.96	28.52	29.13	28.84
F_2	28.49	28.49	28.36	28.49
F_3	0	0	0	0
F_4	8.11	14.12	15.46	14.79
F_5	-	-	-	2.55
Objective: U	306.75	546.76	548.75	550.39

TABLE 5-10. 10° FORWARD LEANING

Criterion Muscle forces	$U=\sum_1^4 \sigma_k+k_R R+k_m M^2$ $k_m=4,\ k_R=\frac{1}{4(\text{joint area})}$	$U=\sum_1^4 \sigma_k+k_R R+k_m M,\ \frac{F_6}{F_7}=\frac{A_6}{A_7}$ $k_m=4,\ k_R=\frac{1}{4A_{\text{joint}}}$	$U=\sum_{k=1}^5 k+\frac{R}{4A_{\text{joint}}}+4M$
6	0	5.79	0
7	24.14	19.98	24.14
8	11.98	12.02	11.98
10	8.52	9.91	8.52
12	2.13	2.19	2.13
13	3.92	4.33	3.92
14	16.88	18.82	16.88
18	13.79	14.05	13.79
19	0.39	0	0.39
21	0	0.32	0
22	3.28	3.16	3.28
26	10.22	11.89	10.22

Joint Reactions:

hip	x	-0.26	-0.19	-0.26
	y	0	0	0
	z	73.67	77.12	73.67
knee	x	2.43	2.63	2.43
	y	1.01	1.15	1.01
	z	106.48	111.30	106.48
ankle	x	1.29	1.18	1.29
	y	-2.63	-2.94	-2.63
	z	62.25	63.96	62.25
ankle moments	x	0	0	0
	y	0	0	0
	z	0	0	0
ground to foot	x	0	0	0
	y	-0.31	-0.36	-0.31
	z	36.15	36.15	36.15
	σ_1	0.57	0.66	0.57
	σ_2	0.14	0.19	0.14
	σ_3	0.29	0.31	0.29
	σ_4	0.23	0.19	0.23
	σ_5	-	-	0.23
Objective:	U	7.48	7.86	7.71

1	25.39	25.39	25.39
2	3.34	4.0	4.0
6	6.13	1.45	1.45
7	0	4.99	4.99
9	3.56	2.94	2.94
10	3.69	3.18	3.18
12	1.76	2.02	2.02
15	16.02	17.72	17.72
21	28.49	28.49	28.49
23	15.45	15.45	15.45

Joint Reactions:

hip	x	0.97	1.06	1.06
	y	-0.09	0.08	0.08
	z	67.33	67.37	67.37
knee	x	0.2	0.19	0.19
	y	-2.18	-2.32	-2.32
	z	86.08	86.95	86.95
ankle	x	-2.45	-2.45	-2.45
	y	-3.99	-3.99	-3.99
	z	78.86	78.86	78.86
ankle moments	x	0	0	0
	y	0	0	0
	z	0	0	0
ground to foot	x	0	0	0
	y	-1.85	-1.85	-1.85
	z	36.15	36.15	36.15
	σ_1	1.55	1.55	1.55
	σ_2	0.93	0.93	0.93
	σ_3	0	0	0.09
	σ_4	0.08	0.08	0.05
		-	-	0.93
Objective:	U	8.49	-	8.58

which are also located on the anterior side of the leg and hence unlikely to be active during this posture (see Table 5-11). However, when $K_R = 2$ and $K_M = 16$, the criterion appears to be suitable for leaning in either direction.

EXPERIMENTAL ANALYSIS

An experimental investigation was undertaken in order to investigate the correlation between the computed results and the isometric EMG activity of the muscles for the selected postures.

Electromyographic potentials were recorded from six superficial thigh and leg muscles (vastus medialis, rectus femoris, biceps femoris (long head), gastrocnemius (medial head), semitendinosus and tibialis anterior) by means of surface electrodes placed on a well trained subject (height = 5'5", weight = 72.3 Kg). The subject stood on a force platform and performed different leaning postures by tilting the pelvis on the femur. EMG results from these muscles are plotted as recorded chart lines versus the pelvic tilt in Figs. 5-2 through 5-6, thus indicating the trend in the observed signal with increasing sagittal rotation in either direction. The electrodes are kept intact and the recording amplification is kept unaltered throughout the experiments for all the measured EMG signals.

The EMG signals for each of the isometric posture qualitatively represent the activity of a muscle. Its pattern or trend in a series of activities is therefore indicative of change in muscle tension with change of posture. The relationship between integrated EMG and muscle tension though commonly agreed to be linear for static isometric actions is less certain for dynamic activities when the EMG is influenced by such factors as muscle speed, length, tension, state of contraction (elongation of shortening), etc.

It can be seen from the EMG data that the semitendinosus, the biceps femoris and the gastrocnemius are active only during forward leaning while the other three, namely, the rectus femoris, the vastus medialis and the tibialis anterior are active only during backward leaning. Moreover, the EMG patterns for these active muscles show a consistent increase with increasing angle of rotation (Figs. 5-2 through 5-6).

THE ESTABLISHMENT OF AN APPROPRIATE MERIT CRITERION

It is widely agreed that there is no convenient non-invasive experimental method for direct quantitative measurement of muscle forces. Hence the correlation between the patterns of the EMG signals and the calculated muscle force patterns can serve as a qualitative indicator of the appropriateness of using a particular merit criterion to render an acceptable solution. The experimental data plotted in Figs. 5-2 through 5-6 as obtained from the six superficial muscles are used for this purpose.

When these experimental findings are compared with the predicted theoretical muscle forces, several of the various criteria which were proposed can be readily eliminated.

1. Merit criteria: $U = \sum_{i=1}^{4} F_i$

   This is readily ruled out because of the unusually large unbalance (residual) moments at the ankle joint for both backward and forward leaning.

2. Merit criteria: $U = \sum_{i=1}^{13} F_i |\Delta L|_i$

   The criteria to minimize work done by the muscles appears to be invalid for both backward and forward leaning. According to this criterion, the biceps femoris (F_8) is inactive in the forward leaning posture and the tibialis anterior (F_{21}) is inactive in the backward leaning posture. Both of these muscles however are know to be active as confirmed by the EMG measurement.

3. Merit criteria: $U = R_z\Big|_{hip} + R_z\Big|_{knee} + R_z\Big|_{ankle}$

   This criterion yields excessive residual moments at the ankle for either directions of leaning and hence can be ruled out. Moreover, for backward leaning the tibialis anterior (F_{21}) force is computed to be zero, which is inconsistent with the observed EMG signals from this muscle which show it to be active for the investigated postures.

4. Merit criteria: $U = \Sigma D + K_M M, \ K_M < 4$

   Although the criterion is valid for backward leaning when $K_m < 4$, it does not appear to be satisfactory for the opposite direction until $K_M > 4$. When $K_M < 4$, it is seen that the tibialis anterior (F_{21}) is active during forward leaning contrary to expected physiological behavior. Also, the muscle forces show erratic change with increasing pelvic inclination, until the weighting factor for the moments, K_M is made > 4. It seems therefore that $\Sigma F + 4$ would be valid for both directions of leaning.

5. Merit criteria: $U = \Sigma K_F \dfrac{F_i}{A_i} + K_R R + K_M M$

   For the two sets of values of the weighting factors, namely, $K_f = 1/4$, $K_M = 1/2$ and $K_F = 1$, and $K_R = 1/4$, $K_R = 1/2$ and $K_M = 2$, this criteria fails for the forward leaning as activity in the tibialis anterior (F_{21}) is shown. The other sets of the same weighting factors that were also investigated, however, appear to be applicable for both directions of leaning.

6. Merit criteria: $U = s_{max} + K_M \Sigma M$

For the two values of K_M that were used ($K_M = 0$ and $K_M = 40$ respectively) this criteria fails for both backward and forward leaning. For forward leaning, the tibialis anterior muscle (F_{21}) is seen to be active, while during backward leaning the iliopsoas (F_{13}) and the semitendinosus (F_6) are seem to produce excessively large forces, hence this criteria can be rejected.

7. Merit criteria: $U = s\sigma_{max}\big|_{hip} + \sigma_{max}\big|_{knee} + \sigma_{max}\big|_{ankle} + K_R \dfrac{R}{A_{jt}} + K_M M$

This criteria seems to be valid for forward leaning. However inspection of the results for the backward leaning make it a candidate for elimination___the force in the semitendinosus (F_6) is appreciably large.

8. Merit criteria: $U = \sum\limits_{k=1}^{3} F_k + K_R R + K_M M$

Over the wide range of magnitudes of the weighting factors that were investigated, the criteria fails for forward leaning when $K_R = 0.05$, 0.1, 0.15, 0.5 or 1.0 and $K_M = 4$, because of antagonistic behavior revealed by the tibialis anterior (F_{21}). It also fails for backward leaning when $K_R = 0.05$, 0.10 or 0.15 and $K_M = 4$ because of antagonistic behavior as seen in the toe flexor (F_{26}). Since this muscle is located on the posterior side of the ankle joint, it is normally expected to be inactive during this posture as a primary or even as an assistive muscle. This apparently leaves $\sum\limits_{k=1}^{3} + 0.25 R + 4 M$ as the only valid possibility in this type of criteria for both directions of leaning. Closer inspection of the results for backward leaning however reveal that the tibialis anterior muscle force (F_{21}) is considerably lower than that obtained from other valid criteria and hence it can also be eliminated.

9. Merit criteria: $U = \sum\limits_{k=1}^{4} F_k + K_M M + K_M$

When the muscles are divided into four different groups with the objective of minimization of maximum muscle force in each, the set of weighting factors $K_R = 1$, $K_M = 8$ cause the criterion to be rejected for the forward leaning act. The results indicate that the tibialis anterior (F_{21}) and the toe extensor (F_{23}) are required to be active—muscles which by reason of their locations anterior to the ankle joint are not expected to act in this position. Similarly, for leaning in the opposite direction, the toe flexor (F_{26}) is shown to exert force—behavior which because of its position posterior to the ankle joint is indicative of antagonism to the anteriorly located muscles normally active in this posture.

10. Merit criteria: $U = \sum\limits_{k=1}^{4} F_k + K_M M + K_M M$

When the muscles are divided into four different groups with the objective of minimization of maximum muscle force in each, the set of weighting factors $K_R = 1$, $K_M = 8$ cause the criterion to be rejected for the forward leaning act. The results indicate the tibialis anterior (F_{21}) is force bearing, which is anatomically inconsistent.

11. Merit criteria: $U = \sum\limits_{k=1}^{1} F_k + K_R \Sigma R + K_M \Sigma M$

This criterion was investigated with two sets of values of the weighting factors $K_R = 0$, $K_M = 4$ and $K_R = 2$ and $K_M = 16$. When the reactions are not included in the criterion (that is, the former set) it is not applicable to leaning in either direction as the results exhibit antagonistic activity. For forward leaning, the tibialis anterior (F_{21}) and the toe extensors (F_{23}, F_{24}), and for the reverse direction, the significant forces in the adductors (F_2, F_4, F_5) the semitendinous (F_6), the semitendinosus (F_7) and the toe flexor (F_{26}) suggest elimination of this criterion.

All the remaining merit criteria appear to be compatible with the measured muscle activities in leaning forward and backward. The theoretical results obtained by using these criteria are plotted in Figs. 5-2a and 5-2b through 5-6a and 5-6b for comparison. It can be seen that all the following criteria are equally applicable and give practically the same results:

$$\Sigma F + 4 \ \Sigma M$$

$$\Sigma F + 2 \ \Sigma R + 16 \ \Sigma M$$

$$\Sigma R_{xyz} + 4 \ \Sigma M$$

$$\sum \frac{F_i}{4A_i} + \frac{\Sigma R}{2} + 4 \ \Sigma M$$

$$\sum \frac{F}{4A_i} + \dot{K}_M \ \Sigma M, \ K_M = 1; \ \frac{1}{2} \text{ or } \frac{1}{4}$$

$$\sum \frac{F}{4A_i} + \frac{\Sigma R}{A_j} + 4 \ \Sigma M$$

$$\sum\limits_{k=1}^{4} f_k + 2 \ \Sigma R + 16 \ \Sigma M$$

$$\sum\limits_{1}^{4} f_k + 2 \ \Sigma R + 16 \ \Sigma M, \ F_6 : F_7 = A_6 : A_7$$

176

$$\sum_1^5 f_k + 2\ \Sigma R + 16\ \Sigma M$$

$$\sum_1^4 \sigma_k + \frac{1}{4}\frac{\Sigma R}{A_j} + 4\ \Sigma M, \quad F_6:F_7 = A_6:A_7$$

$$\sum_{11}^5 \sigma_k + \frac{1}{4}\frac{\Sigma R}{A_j} + 4\ \Sigma M$$

$$\sum_{k=1}^1 f_k + 2\ \Sigma R + 16\ \Sigma M$$

These criterion therefore appear to be plausible and consistent for all considered postures. It should be noted here that the unique optimum solutions obtain show that the considered postures can be attained without support from the ligaments at the hip, knee and ankle. The first criterion in the foregoing list is one of the most convenient and economical to use and is therefore recommended for further applications. The theoretical results for muscle load sharing in the leaning acts (Fig. 5-2b through 5-6b and 5-6c) based on this criterion were then used to calibrate the EMG data. This provided a means of converting the EMG signals for the studied muscles (Fig. 5-2a through 5-6a and 5-6b) to equivalent force values.

GENERATION OF SOLUTIONS WITH REDUNDANT MUSCLE FORCES

As repeatedly stated, the general solution for equations of equilibrium of the musculoskeletal system or any of its segment is statically indeterminate. This is because the number of equilibrium equations is always smaller than the number of unknowns. The linear programming formulation with a particular merit criterion would produce a unique solution for the equilibrium equations with the highest possible merit. Theoretically even if all the unknown variables appear in the merit criterion the optimum solution is expected to contain a maximum of nontrivial (non zero) unknowns as there are equations.

In other words, because of the nature of the linear programming formulation for such a case, the number of unknown variables (with non zero magnitude) in the final solution can be either equal or less than the number of equations. This can raise a physiological question that this approach is too restrictive. It would not allow more muscles than the minimum number needed for static equilibrium to participate in sharing the load even if they are anatomically capable to do so because of their location.

In order to insure that such a situation does not occur, the linear programming problem can be stated to give all the muscles or all the muscles in a similar functional group an opportunity to share the load if a higher merit can be attained. This type of formulation is considered in the formulation using the merit criteria 11 (through e) and 12 (a and b). As can be seen, the final results did not produce any significantly different patterns of muscle load sharing for the postures considered.

THE SQUATTING POSTURE

177

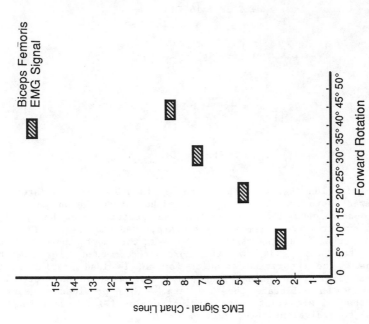

FIGURE 5-3(a). EMG response from the Biceps Femoris (Long Head).

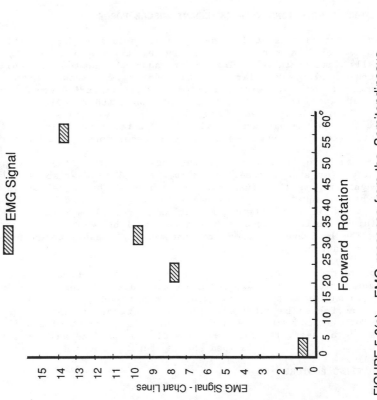

FIGURE 5-2(a). EMG response from the Semitendinosus.

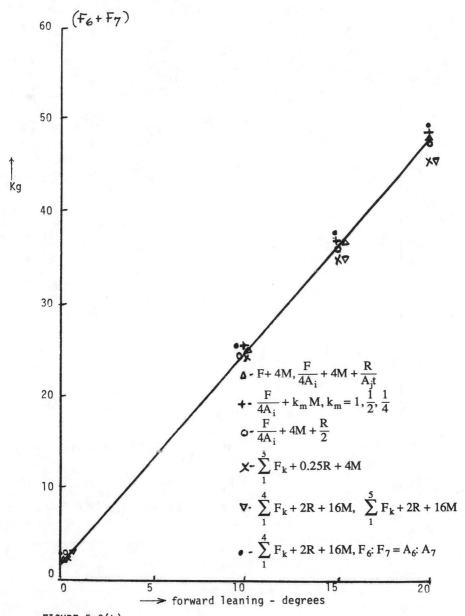

FIGURE 5-2(b).

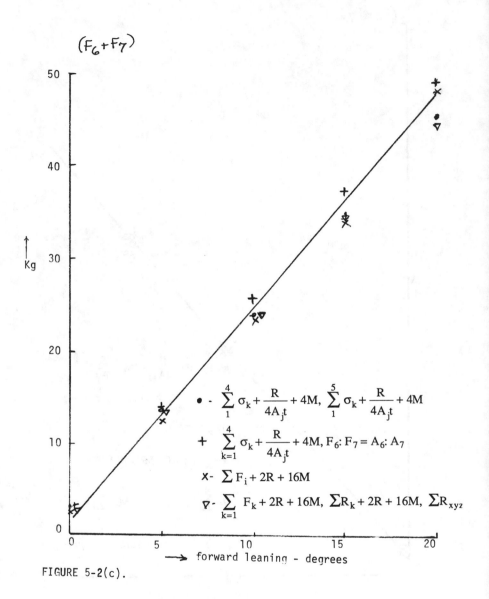

$(F_6 + F_7)$

FIGURE 5-2(c).

The figure shows a plot with Kg on the vertical axis (marked 0, 10, 20, 30, 40, 50) and "forward leaning - degrees" on the horizontal axis (marked 0, 5, 10, 15, 20).

Legend:

$\bullet$ - $\displaystyle\sum_1^4 \sigma_k + \frac{R}{4A_j t} + 4M, \ \sum_1^5 \sigma_k + \frac{R}{4A_j t} + 4M$

$+$ $\displaystyle\sum_{k=1}^4 \sigma_k + \frac{R}{4A_j t} + 4M, \ F_6 : F_7 = A_6 : A_7$

x - $\displaystyle\sum F_i + 2R + 16M$

$\triangledown$ - $\displaystyle\sum_{k=1} F_k + 2R + 16M, \ \sum R_k + 2R + 16M, \ \sum R_{xyz}$

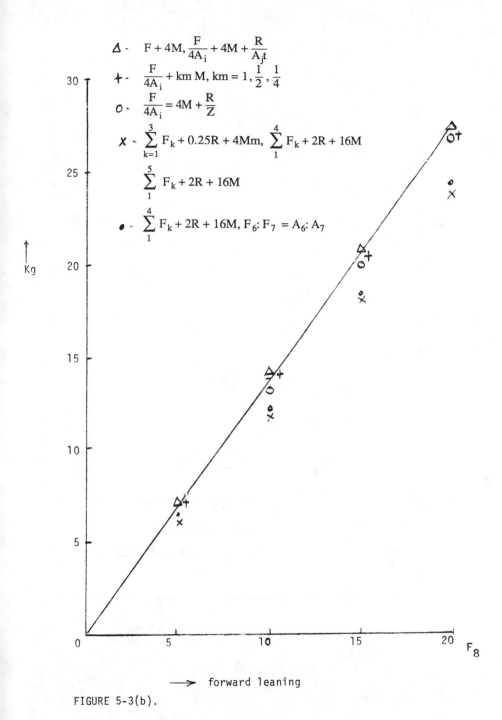

FIGURE 5-3(b).

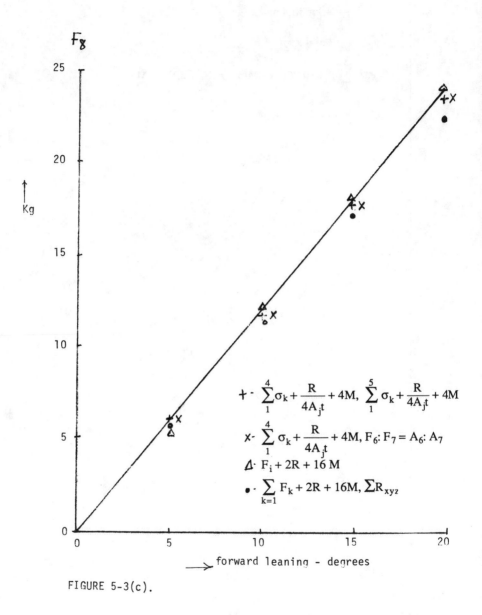

FIGURE 5-3(c).

182

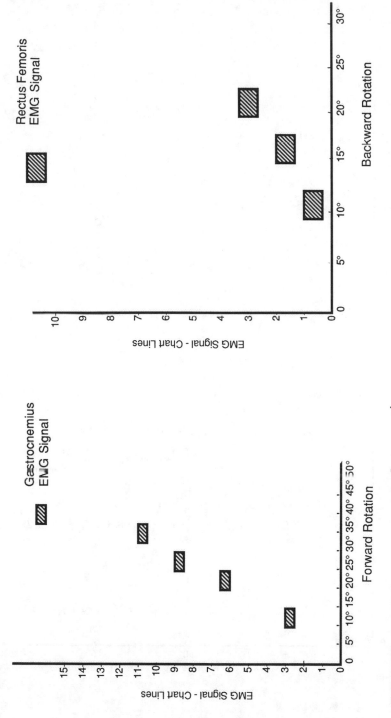

FIGURE 5-5(a). EMG response from the Rectus Femoris.

FIGURE 5-4(a). EMG response from the Gastrocnemius.

183

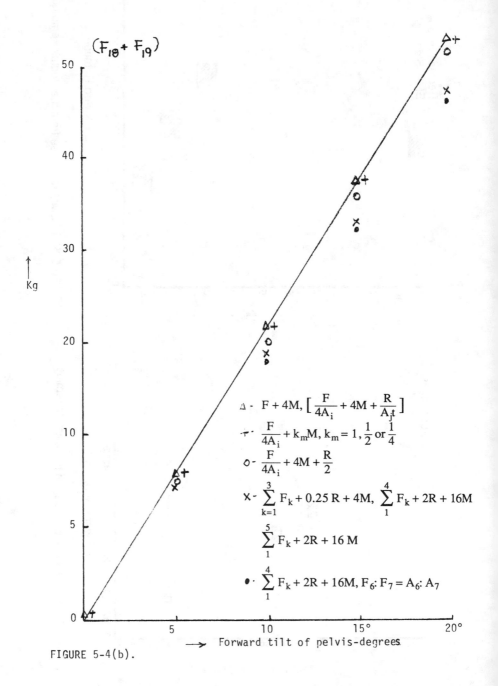

FIGURE 5-4(b).

$(F_{18} + F_{19})$

Kg

Forward tilt of pelvis-degrees

$\triangle$ - $F + 4M, \left[\dfrac{F}{4A_i} + 4M + \dfrac{R}{A_j t} \right]$

$+$ - $\dfrac{F}{4A_i} + k_m M, k_m = 1, \dfrac{1}{2}$ or $\dfrac{1}{4}$

$\circ$ - $\dfrac{F}{4A_i} + 4M + \dfrac{R}{2}$

$\times$ - $\displaystyle\sum_{k=1}^{3} F_k + 0.25\,R + 4M, \ \sum_{1}^{4} F_k + 2R + 16M$

$\displaystyle\sum_{1}^{5} F_k + 2R + 16\,M$

$\bullet$ - $\displaystyle\sum_{1}^{4} F_k + 2R + 16M, F_6 : F_7 = A_6 : A_7$

184

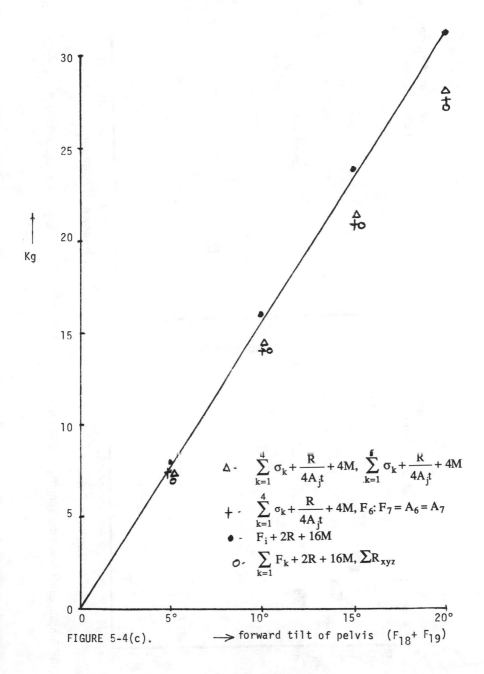

FIGURE 5-4(c).　$\longrightarrow$ forward tilt of pelvis　$(F_{18} + F_{19})$

Legend:

$\triangle$ -　$\displaystyle\sum_{k=1}^{4} \sigma_k + \frac{R}{4A_j t} + 4M, \sum_{k=1}^{6} \sigma_k + \frac{R}{4A_j t} + 4M$

$+$ -　$\displaystyle\sum_{k=1}^{4} \sigma_k + \frac{R}{4A_j t} + 4M, \; F_6 : F_7 = A_6 = A_7$

$\bullet$ -　$F_i + 2R + 16M$

$\circ$ -　$\displaystyle\sum_{k=1} F_k + 2R + 16M, \; \Sigma R_{xyz}$

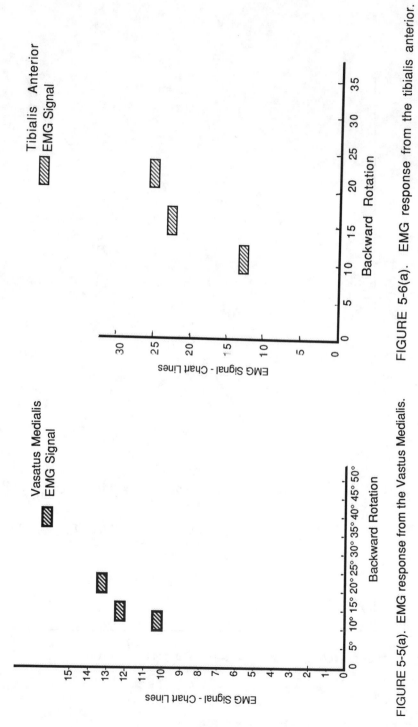

FIGURE 5-6(a).   EMG response from the tibialis anterior.

FIGURE 5-5(a). EMG response from the Vastus Medialis.

186

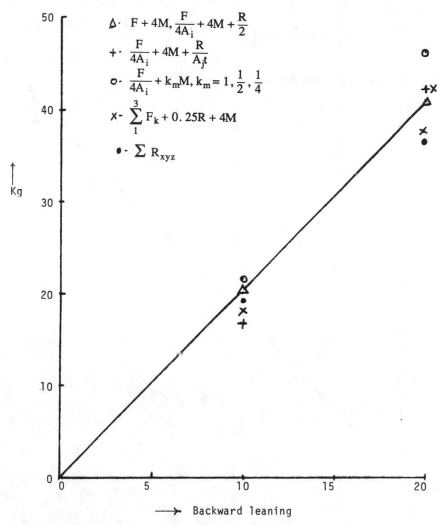

$$F_9 + F_{15-17}$$

Kg

Δ· $F + 4M, \dfrac{F}{4A_i} + 4M + \dfrac{R}{2}$

+· $\dfrac{F}{4A_i} + 4M + \dfrac{R}{A_j t}$

o· $\dfrac{F}{4A_i} + k_m M, \ k_m = 1, \dfrac{1}{2}, \dfrac{1}{4}$

x· $\displaystyle\sum_1^3 F_k + 0.25R + 4M$

•· $\displaystyle\sum R_{xyz}$

→ Backward leaning

FIGURE 5-5(b).

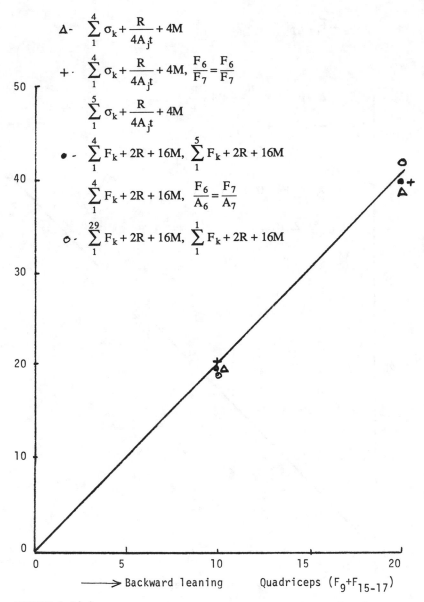

FIGURE 5-5(c).

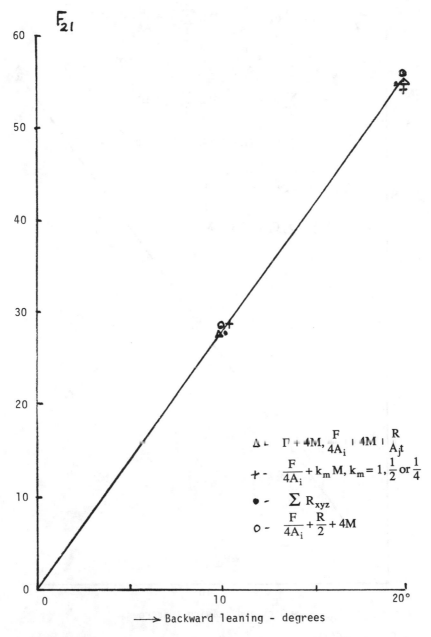

FIGURE 5-6(b).

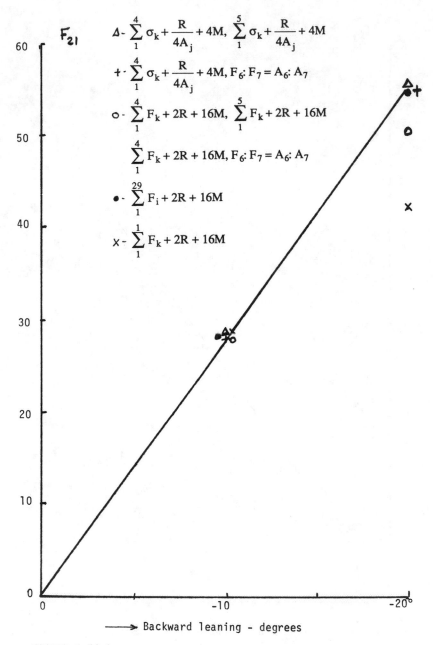

FIGURE 5-6(c).

TABLE 5-11.   10° FORWARD LEANING

Criterion Muscle forces	$U = \sum_{k=1} F_k + k_R R + k_m M$	
	$k_R=0, k_m=4$	$k_R=2, k_m=16$
3	0.19	0
6	13.28	0
7	13.28	23.88
8	13.28	11.71
10	11.77	8.64
12	0	0.15
13	1.47	0
14	8.88	17.66
18	13.28	14.06
19	10.65	0
20	0	0
21	13.28	0
22	5.96	3.19
23	13.28	-
24	3.54	-
25	0	-
26	13.28	10.94
29	0	-

Joint Reactions:

		$k_R=0, k_m=4$	$k_R=2, k_m=16$
hip	x	-0.99	-1.47
	y	1.58	1.81
	z	76.78	68.13
knee	x	1.94	2.58
	y	-0.50	1.13
	z	115.28	106.72
ankle	x	0.21	1.24
	y	-6.46	-2.69
	z	107.36	62.77
ankle moments	x	0	0
	y	0	0
	z	0	0
ground to foot	x	0	0
	y	-0.42	-0.29
	z	36.15	36.15
F_1		13.28	23.88
Objective:	U	13.28	520.96

10° BACKWARD LEANING

Muscle forces	$k_R=0, k_m=4$	$k_R=2, k_m=16$
1	15.22	28.52
2	15.22	-
4	2.99	-
5	15.22	-
6	11.33	5.05
7	15.22	0
9	5.91	4.95
10	15.22	0.76
12	15.22	0.83
13	7.32	0
15	0	14.12
16	4.16	0
17	15.22	0
21	15.22	28.49
23	15.22	15.45
24	15.22	0
26	12.77	0
29	3.42	0

Joint Reactions:

		$k_R=0, k_m=4$	$k_R=2, k_m=16$
hip	x	2.91	0.96
	y	13.15	-0.83
	z	136.49	63.87
knee	x	-1.42	0.44
	y	-7.43	-1.7
	z	113.36	84.71
ankle	x	-5.88	-2.45
	y	-9.37	-3.99
	z	95.97	78.86
ankle moments	x	0	0
	y	0	0
	z	0	0
ground to foot	x	0	0
	y	-2.74	-1.85
	z	36.15	36.15
F_1		15.22	28.52
Objective:	U	15.22	504.14

The squatting configuration is stated in three angels as shown in Fig. 5-7, namely:

θ = angle between the femoral axis and the vertical

ϕ = angle between the femoral and pelvic axes

ζ = angle between the tibial axis and the vertical

The coordinates of muscles origin and insertion points are computed as functions of these angles.

The gluteus maximus (F_{12}) requires a different representation in this case. It can no longer be represented by a straight line joining the points of origin and insertion as it wraps around the pelvis. As indicated in Fig. 5-7, this muscle requires to be divided into two parts along with a reaction acting on one or both of the bodies which it connects as determined by the way it wraps around. The reaction is the resultant of the two parts; and for the gluteus maximum static equilibrium yields:

$$2F_{12} \cos \alpha = R_1$$

It is assumed that the reaction at the hip R_1 acts on the pelvis. Numerical results are obtained for the posture with the kinematic configuration defined by $\theta = 60°$, $\phi = 90°$ and $\zeta = 50°$. Electromyographic results are also recorded by placing surface electrodes on the semitendinosus, biceps femoris, rectus femoris and vastus medialis muscles. The theoretical results are given in Table 5-12 for the merit criteria U = ΣF + 4ΣM which gives results for the leaning acts in agreement with EMG signals. The experimental EMG signals which were obtained at the same recording magnification as the leaning experiments are given as polygraph chart lines in Table 5-13.

The calibration of EMG signal/muscle force as obtained from the leaning studies for the muscles rectus femoris, vastus medialis, semitendinosus and biceps femoris is shown in Table 5-13. When this conversion is applied to the EMG signals recorded during the squatting posture and the forces calculated, it showed good correlation with the results predicted by the mathematical model for this act as shown in that table.

The criterion used for evaluation of muscle forces and joint reactions for the leaning studies therefore seem to be applicable to the squatting posture. The predicted theoretical results for muscle forces show good agreement with the experimental electromyographic evidence.

It will be further demonstrated in the following section that this merit criterion used for leaning and squatting can be used for dynamic activities too. The quasi-static walking will be analysed using the musculoskeletal model for the lower extremities described in the previous sections.

QUASI-STATIC WALKING

The applicability of the merit criterion, minimize U ΣF + 4 ΣM established for the leaning studies and the squatting posture will be further evaluated for determining the muscle load sharing during normal walking. Several excellent experimental studies such as those at the

192

TABLE 12(a). Calculated muscle forces and joint reactions for forward rotation.

Force No.	Rotation		Force magnitude (kg)		
	0°	5°	10°	15°	20°
1	0	0	0	0	0
3	0	0	0	0	0
5	0	0	0	0	0
6	1·57	8·44	15·56	22·90	30·46
7	1·57	8·44	15·56	22·90	30·46
8	0·22	7·64	14·69	21·39	27·75
9	0	0			
10	3·12	9·50	15·83	21·85	27·47
11	0	0	0	0	0
12	0·64	1·32	2·08	2·94	3·88
13	1·79	4.01	6·25	8·41	10·45
14	0·37	11·19	22·60	34·29	46·06
15	0	0	0	0	0
16	0	0	0	0	0
17	0	0	0	0	0
18	0·01	5·79	11·16	16·09	20·49
19	0	0	0	0	0
20	0	2·17	4·85	7·84	11·15
21	0	0	0	0	0
22	0·25	2·27	4·26	6·25	8·23
23	0	0	0	0	0
24	0·22	0	0	0	0
29	0	0	0	0	0
R_{AX}	0·48	0·47	0·56	0·70	0·91
R_{AY}	0	0	0	0	0
R_{AZ}	32·57	62·48	92·52	122·29	151·71
R_{BX}	0·01	1·39	2·82	4·24	5·64
R_{BY}	-0·67	0·15	0·97	1·77	2·54
R_{BZ}	38·27	82·03	125·99	169·56	212·37
R_{CX}	0	1·10	2·19	3·29	4·39
R_{CY}	-0·39	-0·88	-1·44	-2·04	-2·69
R_{CZ}	35·51	45·09	54·94	64·66	74·17
R_{DX}	0	0	0	0	0
R_{DY}	-0·27	-0·34	-0·04	-0·48	-0·56
R_{DZ}	36·15	36·15	36·15	36·15	36·15
U	0	0	0	0	0
M_{CX}	0	0	0	0	0
M_{CY}	0	0	0	0	0
M_{CZ}	0	0	0	0	0

TABLE 12(b). Calculated muscle forces and joint reactions for backward
rotation.

| | Rotation | | Force magnitude (kg) | | |
Force No.	0°	5°	10°	15°	20°
1	0	33·39	43·58	49·94	55·58
5	1·41	2·57	2·97	3·31	3·69
6	1·93	8·79	9·56	8·69	6·55
7	1·93	8·79	9·56	8·69	6·55
8	0·23	7·42	11·56	15·38	19·56
9	0	3·67	7·81	11·84	15·58
10	3·36	1·26	4·95	10·89	19·51
11	0	0	0	0	0
12	0·59	1·42	1·94	2·35	2·69
13	0	0	0	0	0
14	0	0	0	0	0
15	0	7·35	15·62	23·68	31·17
16	0	0	0	0	0
17	0	0	0	0	0
18	0·16	0	0	0	0
21	0·58	15·98	30·33	44·25	57·77
22	0·29	0	0	0	0
23	0	0	0	0	0
24	0	0	0	0	0
29	0	3·65	11·23	19·27	27·31
R_{AX}	0·11	0·72	1·31	1·87	2·38
R_{AY}	0	0	0	0	0
R_{AZ}	33·11	90·54	114·87	133·85	152·23
R_{BX}	-0·07	0·08	0·31	0·53	0·74
R_{BY}	-0·85	-1·95	-2·63	-3·24	-3·82
R_{BZ}	39·02	101·92	133·74	160·16	185·45
R_{CX}	0·02	-1·68	-4·24	-6·88	-9·51
R_{CY}	-0·46	-3·21	-4·42	-5·32	-6·11
R_{CZ}	36·07	54·53	76·2	97·88	119·17
R_{DX}	0	0	0	0	0
R_{DY}	-0·34	-2·41	-2·9	-3·12	-3·23
R_{DZ}	36·15	-36·15	36·15	36·15	36·15
M_{CX}	0	0	0	0	0
M_{CY}	0	0	0	0	0
M_{CZ}	0	0	0	0	0
U	0	0	0	0	0

Muscle	Experimental force values, kg using extrapolated results of learning studies with U = ΣM for calibration (mm)	Theoretical forces based on U = ΣM (kg)
Rectus femoris	15 ≈ 90	89
Vastus medialis	40 ≈ 100	108
Semitendinosus	1 1/2-5	0
Biceps femoris	1 1/2-6	0

TABLE 13. Comparison between experimental and theoretical results for stooping posture.

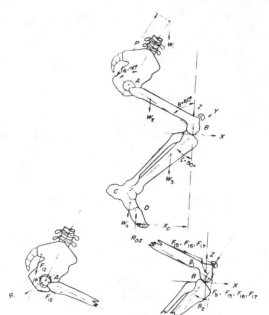

Figure 5-7(a). Theoretical model for stooping postures.

Figure 5-7(b). Experimental study of stooping.

University of California (1947) on the activity of the muscles of the lower extremities during walking are readily available for corroborating the theoretical results for forces exerted by these muscle as obtained from the musculoskeletal model.

For the purpose of investigation of this act, the lower extremities are now treated as a system of seven rigid bodies instead of the four as modelled for the leaning studies. In addition to the previous four (the pelvis, the upper leg, the lower leg and the foot) one must now include the three remaining segments of the right side—the right thigh, the right leg and the right foot. The lower extremities are now a system of seven rigid bodies connected together by a total of 62 major muscles (there are 31 muscles on each side, as listed in Table 5-14 and articulated at six joints, the left and the right hip, knee and ankle joints respectively). The data for the origins and the insertions of the various muscles and the inertial parameters of the segments remain unaltered.

For quasi-static walking (so termed because of the assumption that the act is done slowly enough to generate negligible inertia forces) the cycle is treated as a progression of static postures at successive intervals. The data for the angles between segments in the plane of progression (sagittal plane) and in the plane perpendicular to the direction of walking (coronal plane) are obtained from the University of California report (1947). In order to calculate the coordinates of the points of origin and insertion of the muscles (on various segments) with respect to fixed ground axes (Fig. 5-8), reference axes are chosen on each segment, similar to the ones established for the leaning studies. The coordinates with respect to the ground axes are then obtained using a set of coordinate transformation matrices, as described in Appendix A.

EQUILIBRIUM EQUATIONS OF THE SEGMENTS

The muscular force system keeps the lower extremities in equilibrium at all times. Each of the seven segments of the lower extremities is assumed as a rigid body in space and is therefore described by six equations of equilibrium: three for force balance along the three coordinate axes and three for moment balance about the three coordinate axes. The general form of these equations and the technique of their formulations is identical to the analysis carried out for the previous studies. The force equations include the muscle force components acting on each segment, inertia forces, gravity forces, joint reactions and ground reactions where applicable.

The moment equations include the moments of all forces about the three reference axes as well as the inertia moments, and any unbalanced moments at the joints which would be carried by the ligaments, or employed to shift the selected joint reaction force locations to their true sites within the boundary limits of the joints.

The inertia forces and moments are neglected in the case considered in this chapter as the walking is assumed to be performed quasistatically. For faster rates of walking, terms representing the inertia forces and moments can be readily included in the equations when all linear and angular accelerations of all the body segments are know. The free body diagrams of the various segments at a particular instant of the cycle are shown in Fig. 5-9.

197

Muscle

Number (i)

1	Gracilis
2	Adductor longus
3	Adductor magnus (extensor part)
4	Adductor magnus (adductor part)
5	Adductor brevis
6	Semitendinosus
7	Semimembranosus
8	Biceps femoris (long head)
9	Rectus femoris
10	Sartorius
11	Tensor fasciae latae
12	Gluteus maximus
13	Iliacus
14	Biceps femoris (short head)
15	Vastus medialis
16	Vastus intermediuss
17	Vastus lateralis
18	Gastrocnemius (med. head)
19	Gastrocnemius (lateral head)
20	Soleus
21	Tibialis anterior
22	Tibialis posterior
23	Extensor digitorum longus
24	Extensor hallucis longus
25	Flexor digitorum longus
26	Flexor hallucis longus
27	Peroneus longus
28	Peroneus Brevus
29	Peroneus Tertius
30	Gluteus medius
31	Gluteus minimus

TABLE 14. Important muscles for the lower extremities.

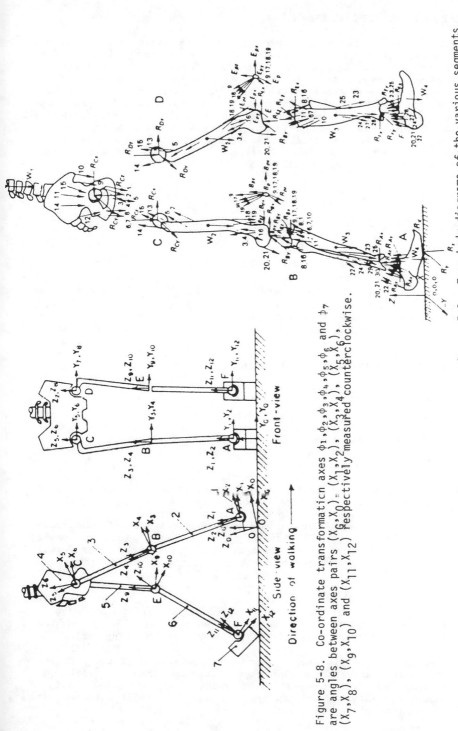

Figure 5-8. Co-ordinate transformation axes $\phi_1, \phi_2, \phi_3, \phi_4, \phi_5, \phi_6$ and ϕ_7 are angles between axes pairs $(X_G, X_0) = (X_1, X_2)$, (X_3, X_4), (X_5, X_6), (X_7, X_8), (X_9, X_{10}) and (X_{11}, X_{12}) respectively measured counterclockwise.

Figure 5-9. Free-body diagrams of the various segments for a typical posture during quasi-static walking.

199

EVALUATION OF MUSCLE LOAD SHARING

The 42 equilibrium equations of all the seven rigid bodies contain a large number of unknown variables, which far exceeds this number of equations. The unknown variables consist of the 62 muscle forces, 3 joint reaction components along the three reference axes at each of the six joints, 3 moment components at each joint and any other variables such as the reactions occurring from the wrapping of the quadriceps tendon around the femur or from the wrapping of iliopsoas around the pelvis. To solve for these variables therefore the same merit criterion as that used for leaning studies and squatting will be utilized. This implies minimization of the sum of all the muscle forces plus four times the sum of the moments at all the joints, i.e.,

$$U = \sum_{i=1}^{i=62} F_i + 4(M_{jX}+M_{jY}+M_{jZ})_R$$

$$+ 4(M_{jX}+M_{jY}+M_{jZ})_L,$$

where u = objective function; F_i = force in muscle i; $(M_{jX}+M_{jY}+M_{jZ})_R$ = right-side moments at the joint j (hip, knee or ankle) about the X, Y or Z axis, respectively; $(M_{jX}+M_{jY}+M_{jZ})_L$ = left-side moments at the joint j (hip, knee or ankle) about the X, Y, or Z axis, respectively. As before, the objective function and the constraints being linear the simplex technique can be conveniently employed to obtain the solution.

MUSCLE FORCES AND JOINT REACTIONS DURING QUASISTATIC WALKING

a) Muscle forces

Since the model is used to evaluate muscular forces during quasistatic walking (i.e., the inertia forces and moments are assumed to be relatively insignificant compared to the gravity forces), the net ground-to-foot reactions in X and Y directions are zero. This also implies that net ankle moment about the vertical axis (Z-axis) is also zero. The variation in the ground-to-foot vertical supportive force is assumed to be distributed over the walking cycle as given in Fig. 5-10. The force on each foot is taken to vary linearly from a zero value at heel strike to a value equal to weight to the body at the beginning of one-legged stance phase. At this time the other leg is off the ground and swinging free while the entire weight of the body is borne by the leg on the ground. As soon as the free-swinging extremity reaches forward for its heel strike the weight starts shifting from the other leg. Consequently, the supportive force starts decreasing linearly from a value equal to weight of the body at the instant of heel strike of the other leg to a zero value at the instant of take-off, and stays zero till the next heel strike occurs. The corresponding path of center of support is shown in Fig. 5-11 as calculated from conditions of total body equilibrium at different intervals of the quasistatic walking cycle.

Figures 5-12 - 5-16 show a representative sample of the forces in various muscles during the walking cycle obtained from the model. The figures also show the corresponding EMG results as reported by Institute of Engineering Research, University of California (1953) obtained from tests on 5 subjects.

200

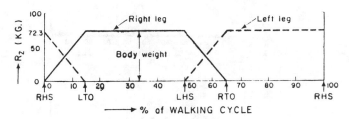

Muscular load sharing during walking

Figure 5-10. Simplified ground reaction R_z during quasi-static walking. (RHS: right heel strike; LTO: left toe off; LHS: left heel strike; RTO: right toe off.)

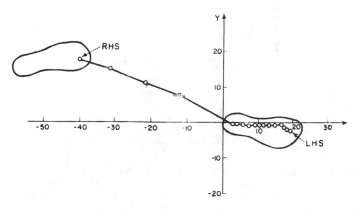

Figure 5-11. Path of center of support during quasi-static walking. (Each interval represents three percent of the walking cycle.

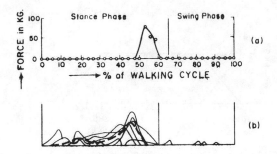

Figure 5-12. (a) Theoretical results for the tibialis posterior.
(b) EMG response from the tibialis posterior during normal level walking.

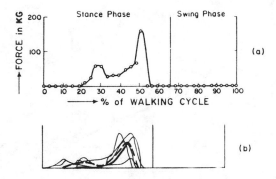

Figure 5-13. (a) Theoretical result for the flexor digitorum longus.
(b) EMG response from the flexor digitorum longus during normal level walking.

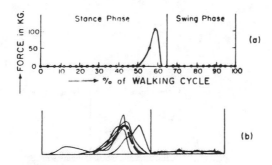

Figure 5-14. (a) Theoretical result for the flexor hallucis longus. (b) EMG response from the flexor hallucis longus during normal level walking.

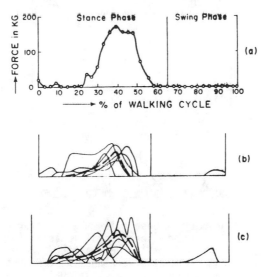

Figure 5-15. (a) Theoretical result for the calf group (gastrocnemius and soleus). (b) EMG response from the gastrocnemius during normal level walking, (c) EMG response from the soleus during normal level walking.

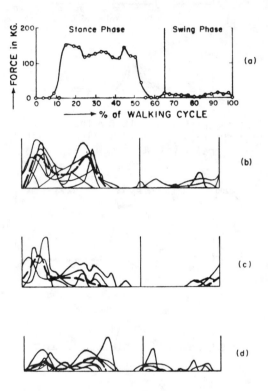

Figure 5-16. (a) Theoretical result for the abductor group (gluteus minimus, gluteus medius and tensor fascial latae). (b) EMG response from the g. minimus during normal level walking. (c) EMG response from the g. medius during normal level walking. (d) EMG response from the the tensor fasciae latae during normal level walking.

The calculated force patterns of the tibialis posterior, the flexor digitorum longus, the flexor hallucis longus, and the calf group (soleus and gastrocnemius) are shown in Figs. 5-12a through 5-15a, respectively. The experimental EMG responses from these muscles are shown in Figs. 5-12b through 5-15b and c, respectively. All these muscles are situated in the back of the leg and with the exception of gastrocnemius which is a two-joint muscle, originate on tibia and insert on the foot. All seem to act exclusively in the stance phase.

The theoretical results for the abductor group are shown in Fig. 5-16a. The result plotted is the sum of the individual muscle forces—the gluteus medius, the gluteus minimus and the tensor fasciae latae. The experimental EMG patterns (Figs. 5-16b-d) appear to closely match the results obtained from the model.

In general the theoretical results for every muscle of the lower extremities show good correlation with the reported averages of the EMG pattern obtained experimentally from different subjects. Any deviations between the theoretical results and the reported averages is of a smaller magnitude than the deviations between the data from the different subjects. It should also be emphasized that magnitude variations and phase shifts in the EMG patterns can be expected due to dynamic effects which have been neglected in this quasi-static investigation.

b) Joint reaction forces

A plot of the resultant joint reaction and its respective components during a complete walking cycle, along X (the direction of walking), Y (medio-lateral direction) and Z (the vertical direction) for the hip, knee and ankle joints is shown in Figs. 5-17 through 5-19, respectively. The maximum values for the joint reaction are 5.4, 7.1 and 5.2 times the body weight respectively for the hip, knee and ankle joint.

The variation in magnitude and direction of the resultant hip joint reaction is demonstrated by a three-dimensional view of the femoral head (Fig. 5-20).

c) Results obtained when the objective function includes terms requiring the minimization of a joint reaction

To illustrate the versatility of the optimization technique employed in the development of the model the objective function is modified by adding a term requiring the minimization of the reaction on one of the joints. Since the vertical component at each of the three joints is predominantly higher than the lateral components it is decided to include this component only with a suitable weighting factor to describe the desire for alleviating the load on the joint.

The objective function, for the hip joint, takes the form:

$$\text{minimize } U = \left| \sum_{i=1}^{i=62} {}_i + 4(M_{jX}+M_{jY}+M_{jZ})_R + 4(M_{jX}+M_{jY}+M_{jZ})_L + k\, R_{Z_{hip}} \right|$$

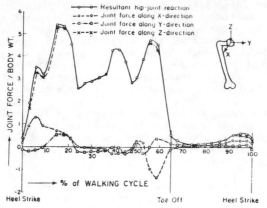

Figure 5-17. Hip joint reaction and components.

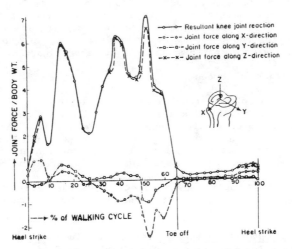

Figure 5-18. Knee joint reaction and components.

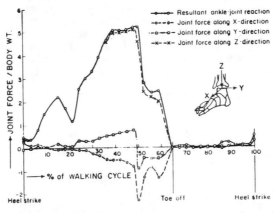

Figure 5-19. Ankle joint reaction and components.

206

where $^Rz_{hip}$ = hip joint force along Z axis and k a weighting factor. The solutions corresponding to values of k = 2,3,4,5 and 10 for the hip joint at 15 per cent of the cycle (where the maximum force occurs) are shown in Table 5-15. Similarly, solutions are obtained with the same values of k for the knee joint (both at 39 and 53 per cent of the cycle where the two large peaks occur) and the ankle joint (at 48 per cent of the cycle). The results are shown in Tables 5-16a and b and Table 5-17, respectively. When the objective function is modified to include minimization of the vertical component of the hip joint force, the same is reduced from its peak value of 384.5 kg with k = 0 to 299.7 kg with k = 2 (see Table 7). With subsequent increase in k, the force is further reduced to 241.6 kg with k = 3. However, any further increase in the value of k does not alter the solution suggesting that no further reduction in the hip force is possible without changing the walking pattern. The corresponding changes in muscle forces can also be seen in the table.

For the knee joint, inclusion of the vertical force on the knee in the objective function does not produce any changes in the muscle forces or the joint reaction at 53 per cent of the cycle for values of k between 2 and 10. The joint force and the muscle force patterns are altered however, at 39 per cent of the cycle with the modified criterion as shown in Table 8a where a reduction in the vertical force on the knee from 445.3 to 326.8 kg is obtained for a weighting factor k of 3, 4, 5, and 10. Similar results are obtained for the ankle joint as shown in Table 5-17.

It should be noted here that in Tables 5-15 through 5-17 some of the muscles have been grouped in the tabulation of the result in order to quantify their combined effort. These are: abductor groups (gluteus medius, gluteus minimus and tensor fasciae latae); adductor group (adductor brevis, longus and magnus); calf group (gastrocnemius and soleus); hamstrings (semimembranosus, semitendinosus and biceps femoris); quadriceps (rectus femoris, vastus medialis, intermedius and lateralis).

We have so far used the criteria U = ΣF + 4 ΣM for the acts of leaning, squatting and walking. It should be reiterated however that this merit criterion does not necessarily imply universal applicability. To determine that, it has to be applied to a wide range of body activities and postures. Within the scope of the investigations undertaken however, for the activities reported in this chapter, the criterion that includes minimization of all muscle forces and ligament actions at all the joints (or its modified version in certain situations) nevertheless seems to warrant a wide range of applications. This is further demonstrated in the following by its application to the upper extremities.

APPLICATIONS OF THE UPPER-EXTREMITY MODEL

The comprehensive musculoskeletal model of the arm which has been formulated in the preceding chapter will now be used to investigate the universality of the developed criterion and solution technique. The arm is considered as two rigid bodies—the upper arm or the humerus, and the forearm. The hand is assumed to be a intergral part of the forearm. The remaining segments of the upper extremity gridle (such as the clavicle and the scapula) and the upper torso are treated as a fixed reference.

The upper arm is connected at its proximal end to the fixed torso at

207

TABLE 5-15. Muscle forces (kg) and joint reaction (kg) at 15 percent of the cycle with modified objective function: minimize $U = \Sigma F_i + 4(\text{joint moments}) + kR_{z_{hip}}$.

Muscle	$k = 0$ Right leg	$k = 0$ Left leg	$k = 2$ Right leg	$k = 2$ Left leg	$k = 3,4,5$ and 10 Right leg	$k = 3,4,5$ and 10 Left hip
Abductor group*	154.9	13.1	83.7	13.1	97.0	13.1
Adductor group	0	0	0	0	0	0
Calf group	0	0.4	70.3	0.4	92.4	0.4
Hamstrings	156.2	36.1	58.1	36.1	41.7	36.1
Quadriceps	78.2	9.9	148.6	9.9	146.9	9.9
Gracilis	0	0	0	0	0	0
Sartorius	0	0	0	0	47.8	0
Iliopsoas	0	0	0	0	0	0
Gluteus maximus	0	1.0	0	1.0	0	1.0
Tibialis anterior	49.3	0	172.6	0	211.2	0
Tibialis posterior	0	0	0	0	0	0
Extensor digitorum longus	40.8	0	86.6	0	100.9	0
Extensor hallucis longus	0	0.3	0	0.3	0	0.3
Flexor digitorum longus	0	0	0	0	0	0
Flexor hallucis longus	0	0	0	0	0	0
Peroneus longus	0	0	0	0	0	0
Peroneus brevis	0	0	0	0	0	0
Peroneus tertius	0	0.2	0	0.2	0	0.2
Hip joint reaction: X	50.8	-7.2	29.9	-7.5	5.7	-7.5
Y	35.7	9.4	1.5	9.4	-4.2	9.43
Z	384.5	35.5	299.7	35.5	241.6	35.5
Knee joint reaction: X	31.9	-12.9	-27.3	-12.9	-32.7	-12.9
Y	52.6	2.7	47.5	2.7	55.9	2.7
Z	431.2	37.2	335.4	37.2	386.4	37.2
Ankle joint reaction: X	-3.4	-0.72	-22.3	0.72	-28.2	-0.72
Y	13.6	0.02	46.8	0.02	57.2	0.02
Z	160.2	0.6	396.1	0.6	469.9	0.6
Hip moment: X, Y, Z	0	0	0	0	0	0
Knee moment: X, Y, Z	0	0	0	0	0	0
Ankle moment: (kg-cm) X and Y	0	0	0	0	0	0
Z	-28.2	0	-30.9	0	-31.8	0

TABLE 5-16a. Muscle forces (kg) and joint reactions (kg) at 39 percent of the cycle with modified objective function: minimize $U = \Sigma F_i + 4(\text{joint moments}) + kR_z$

Muscle	$k = 0$ Right leg	$k = 0$ Left leg	$k = 0$ Right leg	$k = 0$ Left leg	$k = 3,4,5$ and 10 Right leg	$k = 3,4,5$ and 10 Left leg
Abductor group*	118.7	11.1	169.7	11.1	179.7	11.1
Adductor group	0	4.0	33.4	4.0	57.4	4.0
Calf group	172.5	3.7	172.5	3.7	129.8	3.7
Hamstrings	116.9	0	85.1	0	110.4	0
Quadriceps	0	2.6	1.8	2.6	23.8	2.6
Gracilis	0	0	0	0	0	0
Sartorius	49.1	16.4	0	16.4	0	16.4
Iliopsoas	0	0	0	0	0	0
Gluteus maximus	0	0	0	0	0	0
Tibialis anterior	0	5.9	0	5.9	0	5.9
Tibialis posterior	0	0	0	0	0	0
Extensor digitorum longus	0	2.9	0	2.9	0	2.9
Extensor hallucis longus	90.5	1.5	90.5	1.5	38.9	1.5
Flexor digitorum longus	35.4	0	35.4	0	85.1	0
Flexor hallucis longus	0	0	0	0	0	0
Peroneus longus	0	0	0	0	0	0
Peroneus brevis	0	0	0	0	0	0
Peroneus tertius	0	0	0	0	63.4	0
Hip joint reaction: X	-22.2	12.4	-26.5	12.4	-37.9	12.4
Y	-2.2	3.8	-30.1	3.8	-50.4	3.8
Z	304.6	18.4	341.7	18.4	419.1	18.4
Knee joint reaction: X	-64.3	9.0	-67.9	9.0	-43.6	9.0
Y	20.7	0.8	12.8	0.8	12.5	0.8
Z	445.3	23.8	366.4	23.8	326.8	23.8
Ankle joint reaction: X	-38.0	0.7	-38.0	0.7	-35.2	0.7
Y	41.8	0.6	41.8	0.6	43.7	0.6
Z	359.7	12.8	359.7	12.8	377.9	12.8
Hip moment X,Y,Z	0	0	0	0	0	0
Knee moment X,Y,Z	0	0	0	0	0	0
Ankle moment (kg-cm) X,Y,Z	0	0	0	0	0	0

TABLE 5-16b. Muscle forces (kg) and joint reaction (kg) at 53 percent of the cycle with modified objective function: minimize $U = \Sigma F_i + 4(\text{joint moments}) + kR_{z_{knee}}$.

Muscle	$k = 0$		$k = 2,3,4,5$ and 10	
	Right leg	Left leg	Right leg	Left leg
Abductor group*	0.8	44.1	0.8	43.5
Adductor group	4.2	65.3	4.2	80.0
Calf group	2.7	77.9	2.7	74.2
Hamstrings	106.8	4.9	106.8	0
Quadriceps	0	272.1	0	271.4
Gracilis	5.8	41.7	5.8	44.4
Sartorius	19.6	84.0	19.6	88.5
Iliopsoas	0	0	0	0
Gluteus maximus	0	0	0	0
Tibialis anterior	0	0	0	0
Tibialis posterior	0	78.9	0	77.2
Extensor digitorum longus	2.7	0	2.7	0
Extensor hallucis longus	12.0	0	12.0	0
Flexor digitorum longus	0	0	0	0
Flexor hallucis longus	0	0	0	9.2
Peroneus longus	0	0	0	0
Peroneus brevis	0	0	0	0
Peroneus tertius	0	0	0	0
Hip joint reaction:				
X	65.1	12.9	65.1	16.2
Y	-13.5	-22.4	-13.5	-26.1
Z	121.6	276.2	121.6	291.9
Knee joint reaction:				
X	64.7	-176.9	64.7	-175.3
Y	-11.9	-71.9	-11.9	-71.8
Z	127.2	476.2	127.2	475.2
Ankle joint reaction:				
X	9.4	-93.2	9.4	-94.9
Y	0.2	-32.6	0.2	-34.4
Z	28.4	174.1	28.4	177.3
Hip moment X,Y,Z	0	0	0	0
Knee moment X,Y,Z	0	0	0	0
Ankle moment (kg-cm)				
X and Y	0	0	0	0
Z	0	16.9	0	12.1

TABLE 5-17. Muscle forces (kg) and joint reactions (kg) at 48 percent of the cycle with modified objective function: minimize $U = \Sigma F_i + 4(\text{joint moments}) + kR_{z_{knee}}$.

| Muscle | $k = 0$ | | $k = 2,3,4,5$ and 10 | |
	Right leg	Left leg	Right leg	Left leg
Abductor group*	122.3	15.6	154.2	15.6
Adductor group	0	4.5	32.5	4.5
Calf group	155.7	0	101.7	0
Hamstrings	60.3	13.7	70.4	13.7
Quadriceps	0	0	0	0
Gracilis	0	0	0	0
Sartoruis	0	21.3	7.2	21.3
Iliopsoas	0	6.6	0	6.6
Gluteus maximus	0	0	0	0
Tibialis anterior	0	0.4	0	0.4
Tibialis posterior	0	0	0	0
Extensor digitorum longus	0	1.3	0	1.3
Extensor hallucis longus	86.5	1.0	0	1.0
Flexor digitorum longus	68.3	0	88.1	0
Flexor hallucis longus	0	0	0	0
Peroneus longus	0	0	0	0
Peroneus brevis	0	0	0	0
Peroneus tertius	0	0	0	0
Hip joint reaction:				
X	-5.0	21.0	-21.4	21
Y	-13.5	6.4	-33.2	6.4
Z	201.9	38.1	316.1	38.1
Knee joint reaction:				
X	-56.4	21.4	-35.8	21.4
Y	13.7	3.4	11.7	3.4
Z	318.8	40.8	260.9	40.8
Ankle joint reaction:				
X	-57.4	1.6	-50.0	1.6
Y	53.6	0	29.9	0
Z	366.8	1.14	250.6	1.14
Hip moment: X,Y,Z	0	0	0	0
Knee moment X,Y,Z	0	0	0	0
Ankle moment (kg-cm)				
X and Y	0	0	0	0
Z	0	0	48.6	0

the shoulder joint and the forearm is connected at its proximal extremity to the upper arm at the elbow joint. A total of 34 muscular lines of action representing 24 major muscles produce the motions of the two segments and maintain their equilibrium during any particular activity. Taking into consideration the resulting reaction forces and moments at the two joints, a system of 13 equations with 46 unknown variables is obtained. In this section we will discuss the application of the arm model to several different situations, specifically when the forearm is held flexed with respect to the upper arm and external loads are applied at the hand (representing either weights held in the hand or traction force on the wrist). The merit criterion developed earlier for the lower extremities will be used to investigate the muscle participation.

ARM CONFIGURATION

In normal arm motion, movements of all the elements of the shoulder gridle are coupled with each other. In order to accurately establish a particular configuration of the arm it is necessary therefore, to provide information regarding the kinematics of all the elements, namely clavicle, scapula, arm, ulna, radius and hand. If the thorax is held fixed, one may select the sternoclavicular joint for the establishment of the reference ground (global) axes, with respect to which the coordinates of all points of interest on any of the segments of the shoulder girdle can be related. We can assume however that for the action investigated, namely forearm flexion with respect to the vertical arm does not cause significant motion in the remainder of the shoulder girdle. The weight carried in the hand is assumed to act through the center of gravity of the hand, and any traction forces applied on the forearm is assumed to act at the wrist joint.

CARRYING LOAD IN THE HAND

The pattern of muscle load sharing for the position when the forearm is flexed with respect to the vertical arm at a right angle and supporting different weights (0,5,10,15,20 lb respectively) in the hand (Table [5-18]). The criterion used is a weighted linear combination of muscle forces, ligament moments and tensile reactions at the joints:

$$U = \Sigma F + K_R \Sigma R_t + K_M \Sigma M$$

where K_R = 2 and K_M = 4. Since the position of the arm is not altered throughout the test, this is an instance of isometric contraction for the muscles involved. The theoretical results indicate that in this posture, the elbow flexors are strongly active and this is confirmed by the experimental results shown in Fig. 5-21. Electromyographic activity from the biceps (long head) and the triceps (medial head) are recorded using surface electrodes. The subject was an adult male weighing 61.5 Kg and was seated in a chair in an upright position. The weights were held in the hand with the forearm supine, and the resulting muscle electrical activity was amplified, integrated and rectified before recording. In another part of the experiment, the arm was raised in the sagittal plane to a horizontal level and the forearm held flexed supine at ninety degrees. The external loading in this situation consisted of a horizontal pull on the forearm at the wrist applied by means of suitably placed pulleys and cables (Fig. 5-22). The muscle activity was recorded once again for the same five magnitudes of applied pull, that is, 0, 5, 10, 15 and 20 lb. The calculated theoretical results for muscle load

212

sharing in this posture and external loading are in Table [5-19]. All these results were based on the criterion that minimizes all muscle forces, joint reaction forces and ligament actions (U = $\Sigma F + 2R + 4M$).

For the first case of carrying weights in the hand, the calculated flexor muscle forces are shown in Fig. 5-23. It is seen that the force in the biceps (both heads) and in the brachialis increase linearly with increasing loads lifted. This agrees excellently with the experimental EMG results (Fig. 5-21) wherein a similar linear relationship is observed. Calibration of the EMG signals in terms of the expected muscle tension as obtained from the model shows that on the average, 1 unit of EMG chart lines equals 1.208 Kg of muscle force. This can be further checked against the calibration factor obtained from the second part of the experiment when horizontal pulls are applied on the wrist. In this situation, the flexor group is again active with hardly any support from the antagonistic triceps group (Fig. (5-22) - a finding matched by the theoretical results from the model. As shown in Figs. 5-24a and b with increasing magnitudes of applied traction forces, the participation by the flexor groups, namely, the biceps and the brachialis, increases in a linear fashion. This is also the case of the corresponding experimental EMG activity where the recorded signal intensity increasing linearly with heavier pulls. The calibration of the EMG signals from the biceps to the calculated biceps force (sum of forces in the two heads) yields a factor of 1.2 KG/EMG line which is identical to the value obtained from the weight carrying exercise described above. It is further seen that at the shoulder joint, the muscles predicted to be active using the model agree with anatomical expectations. When the arm is flexed and loads are held in the hand, the clavicular part of the pectoralis major, the supraspinatus and the infraspinatus are required to be active (Table 5-18). In the position considered these muscles should be active to stabilize the glenohumeral joint. When the arm is horizontal and traction forces are applied on the forearm, the stability of the shoulder joint is maintained mostly by the participation of the deltoid, both anterior and posterior portions (Table 5-19).

Another finding that deserves particular attention is that among the components of the elbow flexors, the brachialis is always seen to be more strongly active than the other members of this group. From an analytical point of view, the brachialis is closer to the elbow joint than the others, and hence the consequent shortness of the lever arm would unquestionably place it in a position of lower mechanical advantage. This entails that to resist a given amount of moment at the elbow joint, the brachialis will be more severely stressed than the biceps or the brachioradialis if each acted independently. It has been recognized though that despite this inherent mechanical disadvantage the brachialis is the workhorse among the elbow flexors. (Basmajian,ref.13,Ch.7) This clearly demonstrated from the predicted model results, the brachialis always carries a greater force than the biceps in the two postures investigated.

In a third application of the arm model, the arm is held vertical, the forearm horizontal at right angles to the arm and vertical pull is applied at the wrist. This posture is identical to that during carrying loads in the hand except that the loading now is vertically upward instead of downward. The predicted muscle forces, joint reactions and ligament moments for the different magnitudes of the vertical pull (0, 5, 10, 15 and 20 lb respectively) are given in Table 5-20. In this situation the triceps would be expected to be active and this is

213

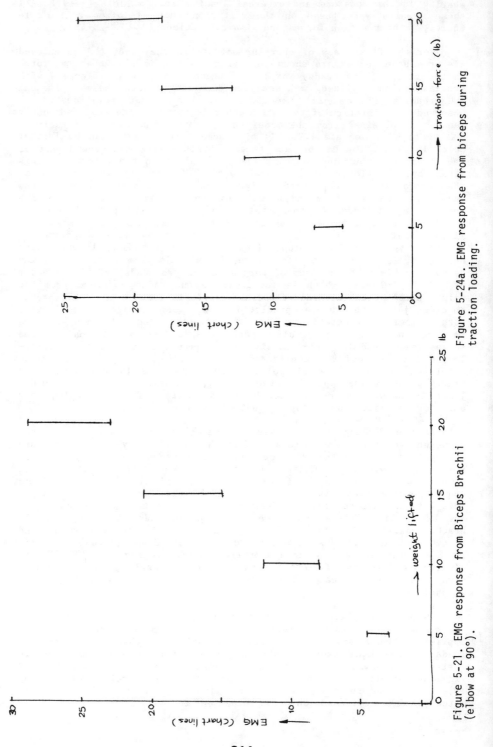

Figure 5-24a. EMG response from biceps during traction loading.

Figure 5-21. EMG response from Biceps Brachii (elbow at 90°).

214

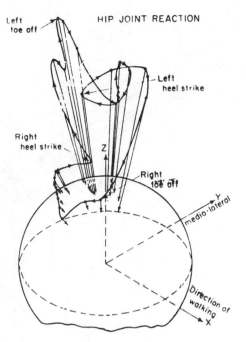

Figure 5-20. Variation of total hip joint force vector in a walking cycle.

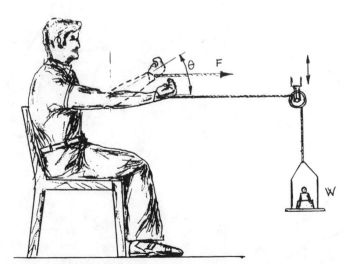

Figure 5-22. Horizontal traction on the arm.

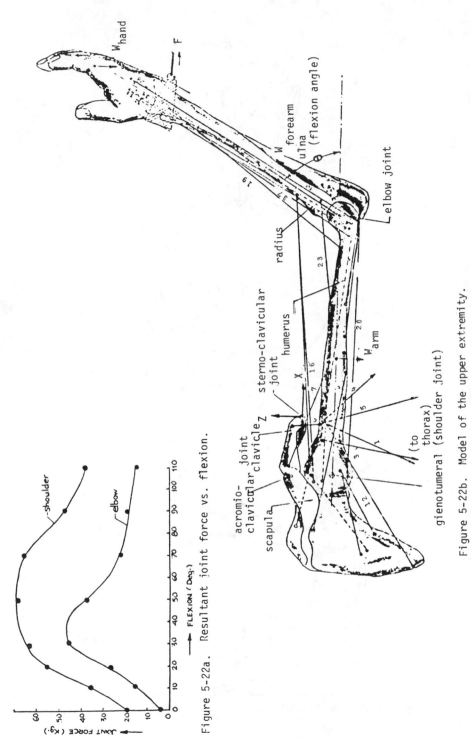

Figure 5-22a. Resultant joint force vs. flexion.

Figure 5-22b. Model of the upper extremity.

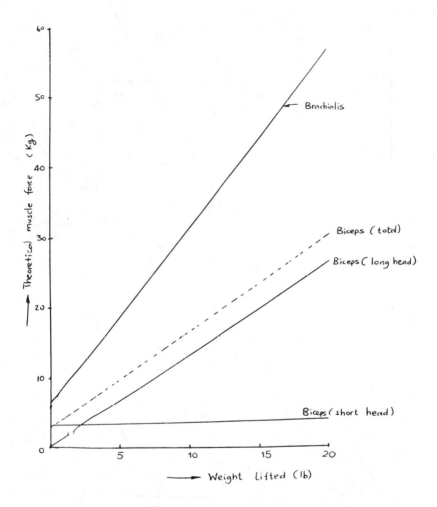

Figure 5-23. Calculated force in arm muscles during weight lifting (elbow at 90°).

217

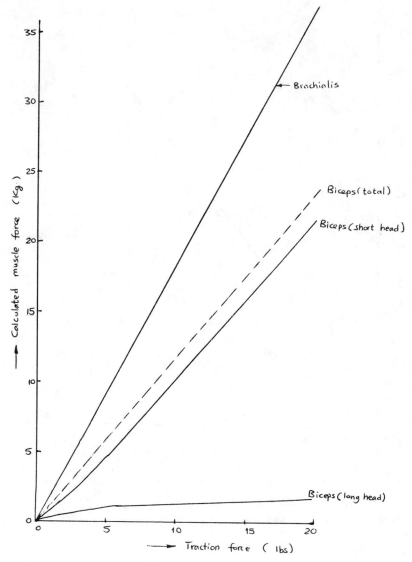

Figure 5-24b. Calculated muscle force during traction loading.

TABLE 5-18. Results for carrying weight in hand.

W(lb) Muscle No.	0	5	10	15	20
6	3.51	16.46	29.41	42.36	55.30
10	11.25	28.79	46.33	63.87	81.42
11	4.95	19.66	34.37	49.08	63.79
16	2.93	3.12	3.31	3.51	3.69
17,18	0.18	6.76	13.35	19.93	26.51
23	5.73	18.46	31.18	43.90	56.63
26	0.69	2.85	7.00	10.16	13.31
Biceps group	3.11	9.88	16.66	23.44	30.20
Shoulder Joint Reactions (x	10.06	31.12	52.18	73.25	94.31
y	-15.08	-53.18	-91.28	-129.38	-167.48
z	0	0	0	0	0
Elbow Joint Reactions (x	-2.10	-7.93	-13.75	-19.58	-25.40
y	-0.51	-1.64	-2.78	-3.92	-5.06
z	6.79	23.66	40.53	57.39	74.27
Objective U	98.49	338.92	579.35	819.77	1060.19

TABLE 5-19. Results for horizontal pull on the wrist.

W(lb) Muscle No.	0	5	10	15	20
7	12.38	18.54	23.62	28.71	33.79
8	14.98	28.76	43.21	57.65	72.09
13	4.13	9.02	14.11	19.19	24.29
16	0	4.57	10.33	16.08	21.84
17,18	0	1.00	1.22	1.43	1.64
23	0	9.15	18.39	27.64	36.89
26	0	2.43	3.06	3.69	4.32
20	0	0	0	0	0
21	0	0	0	0	0
22	0.51	0	0	0	0
34	1.20	0	0	0	0
Biceps group	0	5.57	11.55	17.51	23.48
Shoulder Joint Reactions $\begin{cases} x \\ y \\ z \end{cases}$	28.93 -4.42 0.86	53.15 -8.45 0	78.91 -11.89 0	104.67 -15.35 0	130.43 -18.79 0
Elbow Joint Reactions $\begin{cases} x \\ y \\ z \end{cases}$	-0.51 0.66 -2.93	-12.13 -0.17 -6.38	-24.59 -0.97 -9.44	-37.07 -1.78 -12.49	-49.53 -2.58 -15.55
$M_z = -0.28$					
Objective U	109.96	235.02	366.78	498.53	630.28

TABLE 5-20. $U = \Sigma F + 2R + 4M$

W(lb) Muscle No.	0	5	10	15	20
6	3.51	0	0	0	0
2	0	7.08	26.71	46.34	65.97
8	0.35	1.84	3.33	3.33	4.82
9	0	0	0	0	0
10	11.25	0	-	-	-
11	4.95	0.93	3.42	5.91	8.40
12	0	0	0	-	-
14	0	0	0	-	-
20	0	8.67	25.11	41.55	57.99
21	0	1.94	10.63	19.33	28.02
Triceps	0	10.61	35.74	60.88	86.01
22	0	0	0	-	-
32	0	0	0	-	-
34	0	0	0	-	-
16	2.93	0	0	=	
17,18	0.18	0	0	-	-
23	5.73	0	0	-	-
26	0.69	0	0	-	-
25	0	0.08	0.16	0.25	0.33
Shoulder x	10.06	1.19	4.76	8.31	11.87
Shoulder y	-15.08	-1.36	-5.16	-8.97	-12.77
Shoulder z	0	0	0	0	0
Elbow x	-2.10	<0.46 M=-3.97	<1.99 M=-10.97	<3.53 M=-17.97	<5.07 M=-24.97
Elbow y	-0.51	-0.11	-0.25	-0.38	-0.51
Elbow z	6.79	11.06	38.38	65.69	73.02
Objective U	98.49	63.32	212.84	362.36	511.89

221

TABLE 5-21.   Arm Flexion $U = \sum_{i=1}^{4} f_i + 2R + 16M$

Flexion angle / Muscle No.	0	10	20	30	50	70	90	110
7	8.74	11.15	13.46	22.39	26.83	20.62	17.08	12.35
8	12.89	19.08	30.87	31.94	39.35	40.99	28.69	22.19
10	0	0	0	0	0	0	0	0
11	1.32	0	0	0.12	0	0	0	0
14	0.35	13=1.58	13=4.66	0	4.37	13=7.91	5.5	6.21
16	2.67	4.77	6.80	7.61	6.96	6.13	4.47	3.81
17,18	2.67	4.77	5.21	5.46	3.67	2.62	1.62	1.33
19	0.14	0	0	0	0	0	0	0
23	0	1.58	8.44	14.07	14.82	12.77	10.08	9.13
24	0.68	0	0	26=0.95	26=1.12	0	26=0.19	26=0.48
32	0	0	0	0.95	1.12	0	0.46	0.48
33	0	0	0	0.95	1.12	0	0.46	0.48
34	0	2.34	1.70	14.07	14.82	4.29	10.08	9.13
Shoulder Joint x	19.92	28.71	46.16	52.81	67.82	66.08	47.84	37.39
y	0	-0.06	0	0	0	0	0	-0.66
z	-1.25	-2.99	-2.96	-7.49	-5.45	-2.79	-1.64	0
Elbow Joint x	-3.25	-10.41	-18.73	-37.87	-33.62	-19.34	-13.02	-7.96
y	-0.53	-0.53	-2.15	-1.08	-0.49	-2.26	0	0.36
z	-0.76	-1.99	-2.78	-12.07	-19.11	-9.27	14.93	-12.54
f_1	12.89	19.08	30.87	31.94	39.35	40.98	28.69	22.19
f_2	0	1.58	4.66	0	0	7.91	0	0
f_3	2.67	4.77	8.44	14.07	14.82	12.77	10.08	9.13
f_4	0.68	0	0	0.95	1.12	0	0.46	0.48
Objective U	67.64	114.79	189.53	269.60	308.26	261.14	194.09	149.62

indicated both by the EMG findings (Fig. 5-25) and by the theoretical results plotted in Fig. 5-26.

In all the above examples, the muscles are in a state of isometric contraction. This is for all the magnitudes of the loads lifted or traction forces applied are gradually increased from lower to high values, the arm posture is maintained the same, so that the lengths of the muscles between the points of origin and insertion do not change. The arm model is further applied to study a situation where the participating muscles are in a state of simulated (quasistatic) isotonic contraction. The forearm is flexed through a range of flexion angles, starting from the horizontal position for the arm and forearm. The arm is always kept horizontal while the forearm is flexed in increments of 10 degrees to a final position of about 110 degrees of flexion from the extended arm position.

A constant force of 7 lb (2.95 Kg), horizontally directed is applied at the wrist by means of a cable and pulley system. The subject in this test is seated upright in a chair with posterior back support and EMG recordings are obtained from the long head and the short head of the biceps respectively using surface electrodes throughout this activity.

The theoretical results for the muscle load sharing for the various angles of flexion are shown in Table 5-21. These are obtained using the criterion

$$U = \sum_{k=1}^{4} f_k + 2 \Sigma R_t + 16 \Sigma M$$

where f_k is the maximum muscle forces in the four pre-selected muscle groups. Note that this criterion gives the same results as the criterion $U = \Sigma F + 4M$. The groups are formulated as follows based on the capability of muscles to perform similar functions or to exert similar tensile forces according to their physiological cross sectional areas (shown in parentheses, cm^2).

f_1	f_2	f_3	f_4
1(20.0)	2(3.0)	16(3.22)	(24(3.18)
7(25.3)	3(3.0)	17(3.33)	(25(2.20)
8(25.3)	4(10.0)	18(3.33)	(26(3.24)
9(25.3)	5(10.0)	19(1.86)	(27 3.14)
11(14.1)	6(4.8)	20(4.75)	(28(2.22)
14(25.2)	10(7.7)	21(5.66)	(29(4.30)
	12(9.8)	22(6.78)	(30(5.30)
	13(2.4)	23(6.40)	(31(5.00)
	15(5.8)	34(10.70)	(32(2.16)
			(33(2.90)

Thus the respective muscles in each group are constrained to exert a force less than or equal to the maximum force in that group ($F_i \leqslant f_k$).

The experimental electromyographic findings from the long and the

223

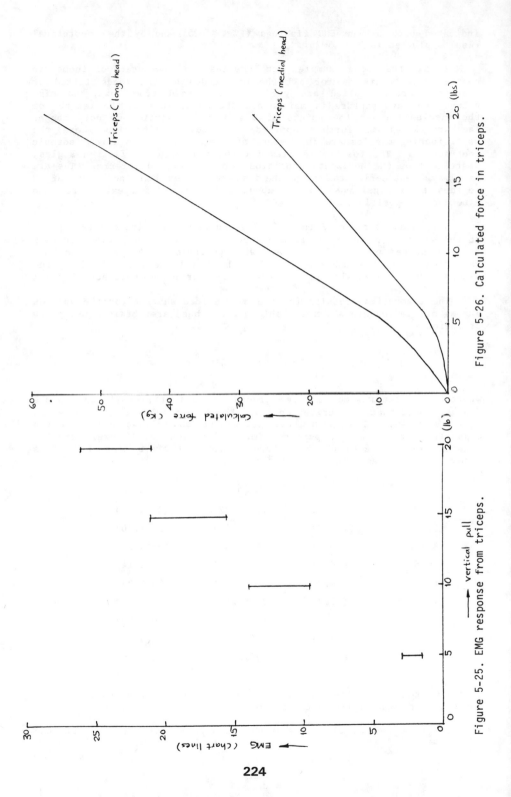

Figure 5-26. Calculated force in triceps.

Figure 5-25. EMG response from triceps.

224

short head respectively are plotted against the flexion angles in Figs. 27 and 5-28. The corresponding theoretical forces in the two muscles are plotted in Figs. 5-29 and 5-30 respectively. As can be seen, the trends or patterns of muscle forces with flexion angles for theoretical results and the experimental EMG plots show good agreement.

CONCLUSIONS

In this chapter an attempt has been made to establish a merit criterion or several criteria that can be used to obtain the muscle load sharing from the musculoskeletal models for the various segments of the body. The three basic parameters or variables which are found to constitute the linear criterion are muscle forces, joint reaction forces and residual joint reaction moments to be carried by the ligaments. The general form of this type of criterion can be mathematically expressed as $U = K_F \Sigma F_i + K_R \Sigma_R + K_M \Sigma M$ and by manipulating the arbitrary weighting factors for the three variables a plethora of criteria can be easily obtained. Many of these, however can be eliminated when the predicted muscle forces or ligament loads are apparently contrary to expected anatomical findings such as activity in a particular muscle which is for certain known not to function in the given posture or act; or inactivity in a particular muscle which is known to be definitely bear tension for the given act. Or in some cases the ligament loads as assigned by the joint reaction moments in the final solution may be excessively high and thus eliminating it as a viable criterion.

While the criteria based on weighted linear combinations of F, R and M can be readily formulated other criteria which involve minimization of stresses in muscles or in joints may not be as convenient to devise. For such criteria, a knowledge of physiological cross sectional areas of muscles or of areas of weight bearing surfaces in the joints becomes necessary. Because of the complex patterns of origin and insertion of most of the muscles it may not always be possible to readily ascertain the areas of the muscles or their various functional components. Moreover, it is likely that the area of a muscle may change depending on the posture or act or depending on the anatomical function which a particular part or component of the muscle may actually be performing in a given posture or act.

Similarly the division of muscles into different groups, wherein muscles with similar functions or similar capacity to exert tension are placed in individual groups requires good judgement. This division cannot be done according to the areas of muscles alone—since the cross sectional area of a muscle may not necessarily reflect its force bearing property. The strength capacity of a muscle is more dependent on its fibre structure than on its area. Thus, the tension that a muscle can produce can be taken to be equal to the product of its area and a stress factor. The latter indicates how many units of force the muscle can exert per unit cross sectional area, and is generally found to vary between 3-10 Kg/cm^2. Since the stress factor is not the same for all muscles, obviously, the area of the muscle may not therefore reflect its maximal strength. The grouping of the muscles therefore cannot be based solely on their cross sectional area. It should also be mentioned that the criteria that involve grouping of muscles is a convenient way to force redundancy in the optimal solution of a higher merit value can be achieved.

225

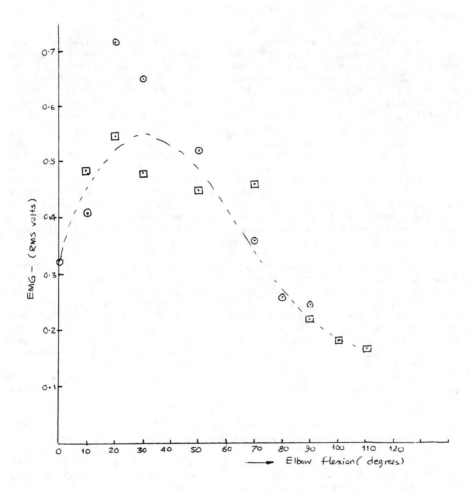

Figure 5-27. EMG response from long head of biceps.

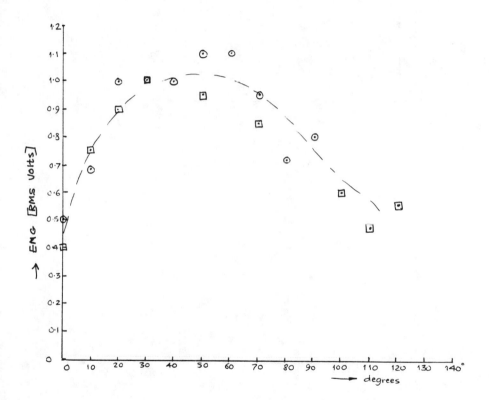

Figure 5-28. EMG response of biceps (short head) with elbow flexion.

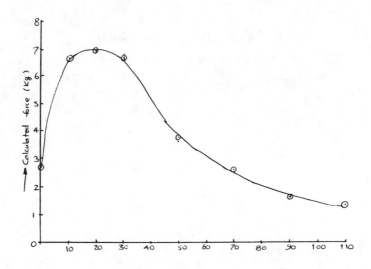

Figure 5-29. Calculated force in biceps   (long head).

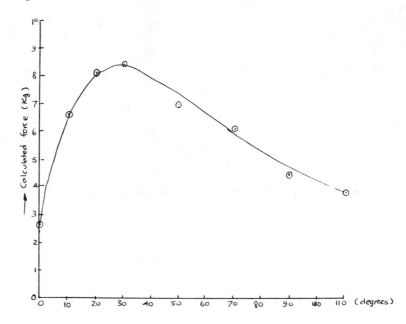

Figure 5-30. Calculated force in biceps (short head).

228

From the various merit criteria that were formulated and tried for the leaning acts, it was concluded that the criterion that minimizes all the muscle forces and four times the residual joint moments, that is, $U = \Sigma F + 4\ \Sigma M$ gives results that seem to agree with the experimental observations. This criteria is one of the most convenient to formulate and economical to use. It has also been shown that many other criteria which are more difficult to formulate given essentially the same solution.

In situations where tensile forces are likely to occur at the joints, such as for the upper extremities they should also be included in the criteria to reflect minimization of tensile components of joint reaction forces (which have to be carried by the ligaments).

It should be noted that the developed merit criterion is proposed for normal subject. Criteria governing muscle load sharing for different types of disabilities can be similarly developed based on correlation with test data.

CHAPTER VI

The Spinal Column

"This convergence of muscles in the spine
keeps it upright just as the ropes of ship
supports its mast, and the same ropes bound
to the mast also support in part the edges
of the ship to which they are joined."

Leonardo da Vinci
1452-1519

INTRODUCTION

The spinal column is functionally one of the most important parts of the human musculoskeletal structure. Formed of alternating rigid vertebrae and elastic fibrocartilaginous discs held together by ligament and muscles, the spine can undergo complex patterns of motion during the different human activities. Its structure thus allows it to effectively transmit the load that it is consequently subjected to, providing maximum comfort to the individual, protection for the life sustaining physiological organs and systems and vital support to other elements of the skeletal system to which it is connected.

Unlike the lower and upper extremities which consist of easily identifiable skeletal bodies, articulating joints and muscles connecting them, the spine is comprised of many rigid bodies, each of which articulates with its adjoining members at several joints. It is, however, the complex anatomy of the muscles connecting the vertebrae that would seem to make the task of modelling difficult. For the sake of illustration consider the case of the erectores spinae or the back muscle. This muscle begins below in a broad, thick tendon attached to the posterior aspect of the sacrum, posterior part of the iliac crest, and the spinous processes of the lumbar vertebrae. At the lumbar levels, the muscular fibers split into three columns—the lateral iliacostalis, the intermediate longissimus, and the medial spinalis. The erectores spinae, as a whole, ascend throughout the length of the back but the columns are composed of fascicles of shorter length. Each column is composed of a rope-like series of fascicles, various bundles arising as others are inserting, each fascicle spanning six to ten vertebral segments between bony attachments. Thus the iliocostalis fascicles originate on lower ribs of the thorax and insert in upper ribs and, at cervical levels, insert into the posterior tubercles of cervical transverse processes as high as the fourth cervical vertebra. Similarly, the fascicles of the longissimus arise on the transverse processes of the vertebrae at lower levels and insert into the transverse processes of the vertebrae at higher levels. Likewise, the fascicles of the spinalis arise from the spinous processes of the vertebrae and insert into the spinous processes of the vertebrae at higher levels. Each of the three columns can be further divided into their individual regional portions. The

231

iliocostalis has three different portions—lumborum, thoracis and cervicis while the longissimus and the spinalis columns have thoracis, cervicis and capitis portions. (See Fig. 1a & 1b). It is therefore evident that modelling of these muscles will not be as simple as, for example, modelling the muscles of the lower extremities where a muscle could be identified by its distinct origin and insertion. If one were to model each fascicle individually to take into account the anatomic minutiae, innumerable lines of force will have to be included. However, by selecting the most significant functional line or lines or action, a complex muscle like the erector spinae can be represented by a more reasonable number of lines.

SKELETAL SEGMENTS OF THE SPINE

The vertebral column is composed of twenty-four mobile vertebrae— seven cervical, twelve thoracic, and five lumbar. Typical vertebrae from the three regions are shown in Fig. 2. To simplify the model it will be assumed that the bony thorax, its organs, and the twelve thoracic vertebrae all constitute one single rigid body. Note that this will considerably facilitate modelling of several muscles which originate and insert on the ribs of the thorax (such as the thoraxic portion of the iliocostalis of the erectores spinae) or on the thoracic vertebrae (such as the thoracis portion of the longissimus or the spinalis columns of the erectores spinae)—since such muscles now originate and insert on the same segment, they can understandably be excluded from the model. The vertebral column is attached to the head through the atlanto occipital joint (between the atlas or the first cervical vertebra and the occipital condyles of the skull) and to the pelvis through the lumboscaral junction (between the fifth lumbar vertebra and the sacrum of the pelvis). Since there are muscular attachments between various parts of the vertebral column and the upper extremities, such as to the scapula, to the clavicle and to the humerus, it seems necessary to include these segments into the model of the spine. Similarly many muscles connect the vertebral column to the skeletal bones of the lower extremities, such as the pelvis and the femur and therefore for meaningful modelling it is imperative that the lower extremities be also included in the model of the spine. The model of the spine therefore invariably means modelling the entire musculoskeletal structure consisting of 23 rigid bodies as listed in Table I, and shown in Figure 3. Elements of the upper and lower

TABLE I. Segments of the human musculoskeletal structure.

1.	Head	13.	Fourth lumbar vertebra
2.	First cervical vertebra	14.	Fifth lumbar vertebra
3.	Second cervical vertebra	15.	Clavicle
4.	Third cervical vertebra	16.	Scapula
5.	Fourth cervical vertebra	17.	Humerus
6.	Fifth cervical vertebra	18.	Ulna-radius
7.	Sixth cervical vertebra	19.	Hand
8.	Seventh cervical vertebra	20.	Pelvis
9.	Thorax	21.	Femur
10.	First lumbar vertebra	22.	Tibia-fibula
11.	Second lumbar vertebra	23.	Foot
12.	First cervical vertebra		

extremities are shown in the table for one side of the body only.

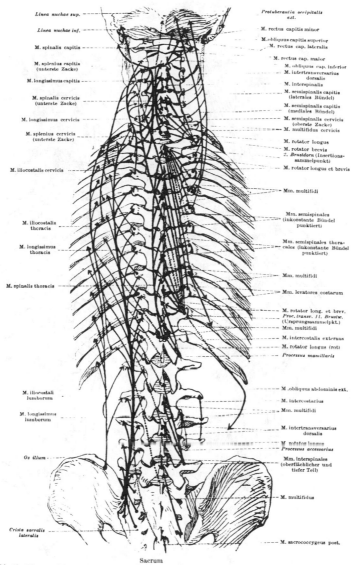

Abb. 48. Lineares Schema für die tiefen Rückenmuskeln. Rechts: kurze Muskeln rot, Multifidus und Semispinalis thoracis et cervicis blau, Semispinalis capitis und Muskeln ventraler Abkunft schwarz. Links: Spinalis und Iliocostalis blau, Longissimus schwarz und Splenius rot. Die Insertion der Muskeln als Pfeilspitze.

FIGURE 6-1a.　Anatomical plan of the muscles of the back.

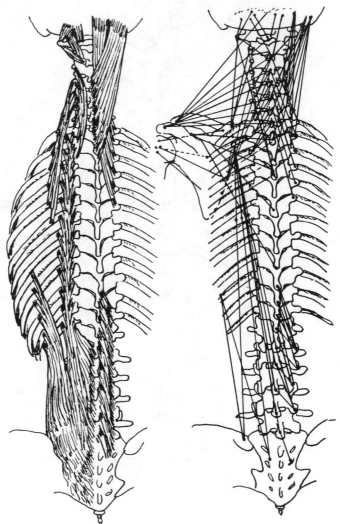

FIGURE 6-1b. Muscles of the vertebral column.

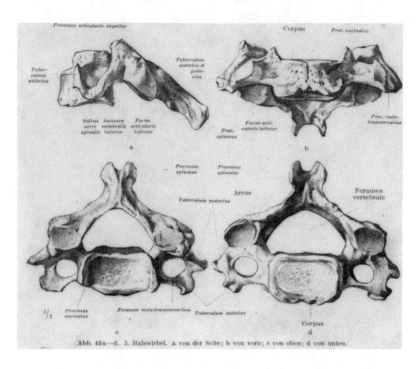

Abb. 43a—d. 5. Halswirbel. a von der Seite; b von vorn; c von oben; d von unten.

FIGURE 6-2a.   A typical cervical vertebra.

235

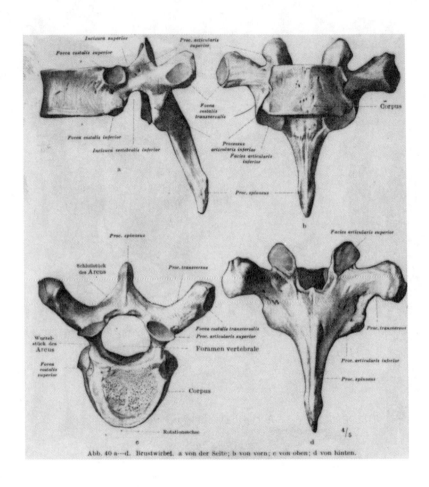

Abb. 40 a—d. Brustwirbel. a von der Seite; b von vorn; c von oben; d von hinten.

FIGURE 6-2b.   A typical thoracic vertebra.

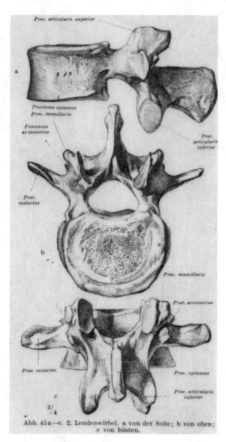

Proc. articularis superior

Processus spinosus
Proc. mamillaris

Processus accessorius

Proc. articularis inferior

Proc. costarius

Proc. mamillaris

Proc. accessorius

Proc. costarius

Proc. spinosus

Proc. articularis inferior

3/4

Abb. 45a—c. 2. Lendenwirbel. a von der Seite; b von oben; c von hinten.

FIGURE 6-2c. A typical lumbar vertebra.

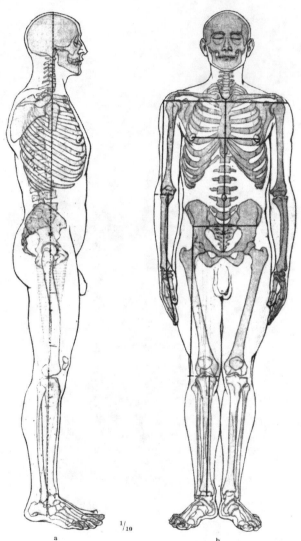

Abb. 1 a u. b. Ansicht eines kräftigen Mannes mit eingezeichnetem Skelet. a von der Seite, b von vorn. Das Individuum ist das gleiche wie das zu den Abbildungen des Bewegungsapparates benutzte (S. 6). In a ist der Gesamtschwerpunkt des Körpers (×) und das Lot für die „Normalstellung" (S. 17) eingetragen. (Nach BRAUNE und FISCHER.) Das schraffierte Dreieck entspricht der Gesamtheit der Adductoren des Oberschenkels. In b sind die Knochenmeßpunkte für den Kundigen abzulesen (man vgl. für die Meßtechnik MARTIN, Lehrbuch der Anthropologie). Das Becken und der Brustkorb zeigen normale Asymmetrie. × Gesamtschwerpunkt.

FIGURE 6-3.   The human skeleton.

Inclusion of left and right side bones will therefore increase the number to 32. The ulna and the radius which make up the forearm, and the tibia and the fibula which make up the lower leg are both treated as single rigid bodies for the purpose of force analysis in the model. The motion between the ulna and the radius such as during the acts of pronation and supination, however, is duly accounted whenever required for the purpose of establishment of the forearm configuration.

The twenty-three rigid bodies listed in Table I are articulated through various joints permitting diverse types and ranges of motion. The atlanto occipital joints are formed between the convex occipital condyles and the concave superior articular facets of the atlas (first cervical vertebra). The articular surfaces of the joints may be viewed as forming parts of a single ellipsoidal surface with the longer dimension situated transversely. The movements permitted at the atlanto occipital joints are those of flexion and extension about a transverse axis (such as in nodding of the head forward and backward) and lateral bending about an anterior-posterior axis. The construction of the joints naturally prevents any rotary movement about the vertical axis.

The articulations between the atlas and the axis (the second cervical vertebra) occur through two joints. The lateral atlanto axial joints are formed by the reciprocal surfaces of the articular processes of the atlas and the axis and are bilateral gliding joints. The median atlanto axial joint is formed by the dens of the axis passing through the space between the anterior arch of the atlas and its transverse ligament and is a rotary joint. At the atlanto axial joints therefore the only motion that can occur is rotation of the skull and atlas together in moving the face from side to side. In this action, the dens of the axis acts as a vertical pivot enclosed in the ring formed by the anterior arch of the atlas and its transverse ligament. The rotary motion is accompanied by forward and backward gliding between the articular surfaces of the atlas and the superior articular facets of the axis.

The articulations of the vertebral column occur at the cartilaginous joints between the adjacent vertebrae and at the synovial joints between the vertebral arches. The bodies of the vertebrae are united by fibrocartilaginous intervertebral disks interposed between the bodies of adjacent vertebrae from axis to sacrum. The size and shape of the disks varies depending on which part of the column they belong to. In the cervical and lumbar region they are thicker anteriorly than posteriorly, thus contributing to the cervical and lumbar curvatures of the spine. The disks are of uniform thickness at thoracic levels while they are thickest posteriorly in the lumbar region of the column. The vertebral arches on the other hand, are united by plane synovial joints between the anticular processes—two superior processes and two inferior processes for each vertebrae. The joint is formed by the inferior articular processes of the vertebra above fitting into the superior articular processes of the vertebra below, the extent and type of intersegmental motion at the joint thus being dependent on the shape and orientation of the contacting processes. The movement between vertebrae takes place in the intervertebral disks and at the joints of the articular processes. The displacement between adjacent vertebrae though small, accumulates over the entire column to provide a considerable range of motion. In general, the permissible movements of the column are flexion-extension in the sagittal plane, lateral bending about an anterior-posterior axis and rotation about a vertical axis. The degree of motion, however, varies in different parts of the column. In the neck all movements can freely occur. In the thoracic region, the obliquely set articular facets permit slight flexion and extension, bending and rotation, but the presence of the ribs and the overlapping of the long spinous processes tend to limit the vertebral motion. In the lumbar region, characterized by thick disks and vertical orientation of the articular surfaces, flexion and extension can occur freely and lateral bending is allowed but rotation in the lumbar column is restricted because of the sagittal orientation of the articular surface, flexion and extension can occur freely and lateral bending is allowed but rotation in the lumbar column is restricted because of the sagittal orientation of the articular surfaces.

The only connection between the trunk (thorax) and the upper limbs occurs at the sternoclavicular joint. This joint is formed by the sternal end of the clavicle entering into the fossa created by the manubrium sterni and the cartilage of the first rib (see Fig. 3). The movements at the sternoclavicular joint are accompanied by movements of the scapula which glides on the surface of the thorax. The scapula is connected to the clavicle at the acromioclavicular joint formed between the acromial end of the clavicle and the medial margin of the acromion of the scapula. The motions that can occur at the sternoclavicular joint are those of elevation and depression of the shoulder, forward and backward movement of the shoulder and circumduction of the shoulder. At the acromioclavicular joint the allowable movements are those of gliding of the acromial end of the clavicle on the acromion and rotation of the scapula upon the clavicle.

The joints between the other segments of the shoulder girdle (shoulder joint, elbow joint and wrist joint) have already been dealt with in Chapter 4. The joints between the segments of the pelvic girdle (hip joint, knee joint and ankle joint) have also been described in the same chapter. Since the muscoskeletal model of the lower extremities and part of the upper extremities have been developed in these chapters IV and V) a repetition of these systems would be unnecessary. We shall therefore devote our attention to the modeling of the muscles of the head and the neck, and thorax and abdomen and those connecting the spine to the shoulder and the pelvic girdle.

MAJOR MUSCLES OF THE SPINE

The next step after identifying the rigid bodies and their connecting joints is modelling of the major muscles by their lines of action. The major muscles to be modelled are shown in Table II. Among the muscles of the head and neck, and the back, certain muscles are described by their different regional parts even though there is truly only one muscle. For example, the semi-spinalis or the splenius etc. are described by their cervicis and capitis portions separately. Note that as explained earlier, the thoracis portions of the back muscles which originate from and insert into the thoracic vertebrae or the ribs are omitted from this list because the thorax has been modelled as a single rigid body. Similarly muscles of the thorax which primarily function as respiratory muscles are also excluded from the modelling process. Most of the muscles of the upper and lower extremities have been described in detail in proceeding chapters and hence will not be discussed herein. Only those muscles belonging to these segments will be dealt with which need modification in their models due to the inclusion of the spine.

240

Muscles of The Head and Neck

Most muscles in this region can be conveniently represented by straight lines connecting their points of origin to their points of insertion.   For example, the suboccipital muscles rectus capitis posterior major and minor, and obliquus capitis superior which arise on the first or the second cervical vertebra and insert into the skull, and the obliquus capitis inferior which arises on the second cervical vertebra and inserts into the first are all modelled by straight lines joining the corresponding points of origin and insertion (Fig. 4). Likewise, the anterior vertebral muscles rectus capitis anterior and rectus capitis lateralis which originate on the atlas and insert into the skull can be represented by straight lines (Fig. 4).   The sternocleidomastoid which passes obliquely across the side of the neck arises by two heads—the medial or sternal head form the ventral surface of the manubrium sterni and the lateral or clavicular head from the superior border and anterior surface of the medial third of the clavicle and inserts into the mastoid process of the skull.   This muscle is modelled by two lines—one from the thorax and one from the clavicle both inserting into the skull (Fig. 5).

The infrahyoid muscles, sternothyoidus and sternothyroideus spanning the hyoid bone and the thorax may assist the anterior vertebra neck muscles by keeping the hyoid bone fixed.   It was decided therefore, to include these muscles in the model even though they are primarily responsible for the actions of the larynx and hyoid.   The sternohyoideus arises from the posterior aspect of the sternal end of the clavicle and from the back of the manubrium sterni and inserts into the lower border of the body of the hyoid bone.   It is modelled by two lines—one from the thorax and one from the clavicle both inserting into the hyoid bone which is assumed to be intergral with the skull.   Similarly the sterno- thyroideus which arises from the dorsal surface of the manubrium sterni and the cartilage of the first rib and inserts into the lamina of the thyroid cartilage is modelled by a line connecting the thorax to the thyroid cartilage also assumed integral with the skull.

Though most muscles of the head and neck can be simulated by straight lines as described above, several present a problem in identifying their points of origin and insertion.   They have a wide area of origin and insertion spanning several vertebrae and therefore require considerable judgement in modelling.

a)   Longus colli:

The longus colli lies on the anterior surface of the vertebral column between the first cervical vertebra and the third thoracic vertebra and consists of three portions.   The superior oblique portion arises from the anterior tubercles of the transverse processes of vertebrae C3 to C5 and inserts on the anterior tubercle of the atlas (C1) and the body of the axis (C2).   Consequently this part of the muscle is modelled by six lines—a set of three from C3, C4 and C5 respectively connected to C1 and another set of three from the same three vertebrae connected to C2.

The inferior oblique portion arises from the bodies of vertebrae T1 to T3 and inserts into the anterior tubercles of the transverse processes of the fifth and sixth cervical vertebrae.   This part is modelled by two lines, a line from the thorax between T1 and T3 going to C5 and another line from the same location going to C6.

241

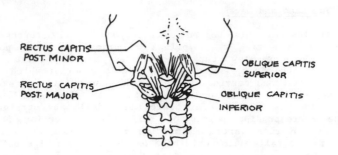

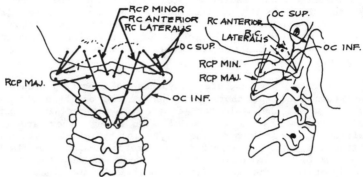

FIGURE 6-4. Muscle models for rectus capitis posterior major,
rectus capitis posterior minor, oblique capitis superior and
inferior, rectus capitis anterior and lateralis.

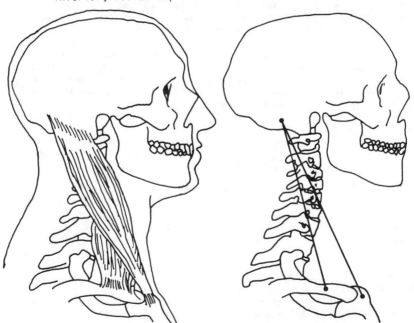

FIGURE 6-5. Muscle model for the sternocleidomastoid.

242

The vertical portion of the longus colli arises from the anterior surface of the bodies of the upper three thoracic and the last three cervical vertebrae and inserts into the anterior surface of the bodies of the first four cervical vertebrae. Consequently the muscle is represented by sixteen lines: four sets of four each from C5, C6, C7 and the thorax (between T1 and T3) respectively connected to C1, C2, C3 and C4 respectively. The muscle model for the longus colli is illustrated in Fig. (6).

b) Longus capitis

The longus capitis (Fig (6)) arises from the anterior tubercles of the transverse processes of the vertebrae C3 to C6 inclusive and inserts into the inferior surface of the basilar part of the occipital bone in the skull. The muscle can be modelled by four lines from C3, C4, C5 and C6 vertebrae respectively connected to the skull.

c) Scalenus:

The scaleni are lateral vertebral muscles and consists of three: anterior, medius and posterior (Fig. 7). The scalenus anterior originates from the anterior tubercles of the transverse processes of the vertebrae C3 to C6 and inserts into the scalene tubercle on the inner border of the first rib. It is represented by four lines from the four cervical vertebrae (C3-C6) joining the thorax. The scalenus medius which arises from the posterior tubercles of the transverse processes of all the cervical vertebrae and inserts into the first rib is modelled by seven lines, one each from the seven cervical vertebrae joining the thorax. Finally, the scalenus posterior which takes its origin from the posterior tubercles of the transverse processes of the last three cervical vertebrae and inserts into the outer surface of the second rib is conveniently modelled by three lines connecting the three cervical vertebrae (C5-C7) to the thorax.

d) Semispinalis capitis, splenius capitis, longissimus capitis

These muscles in general originate on the cervical and thoracic vertebrae and insert into the skull. For example the semispinalis capitis takes origin from the transverse processes of the upper six thoracic vertebrae and the articular processes of the lower four cervical vertebrae and inserts into the occipital bone of the skull. The part of the muscle with wide thoracic origin is represented by one line connecting the thorax to the skull, best representative of the expansive origin from the upper six thoracic vertebrae. In this manner, inclusion of excessive lines of action (from each of the thoracic vertebrae) is not necessary and the number of lines in the muscle model may be kept to a minimum without mitigating the function of the entire group. This technique will be adopted for modelling of all muscles with extensive origin from the thoracic vertebrae. The remainder of the semispinalis capitis arising from the lower four cervical vertebrae is modelled by four lines connecting these bodies to the skull (Fig. 8).

The splenius capitis and the longissimus capitis which arise on the cervical and thoracic vertebrae and insert into different regions of the skull are also modelled in the same fashion (see Figs. 9 and 10).

e) Iliocostalis cervicis, splenius cervicis, semispinalis cervicis and longissimus cervicis

With the exception of the iliocostalis cervicis which arises from the angles of the ribs, all of these muscles arise from different parts of the cervical and thoracic vertebral regions of the spinal column

243

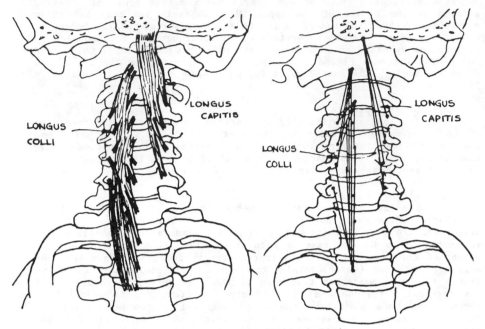

FIGURE 6-6. Muscle models for longus colli and capitis.

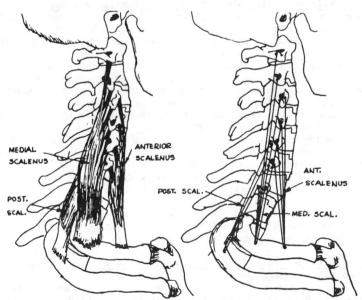

FIGURE 6-7. Muscle models for scalenus anterior, medial and posterior.

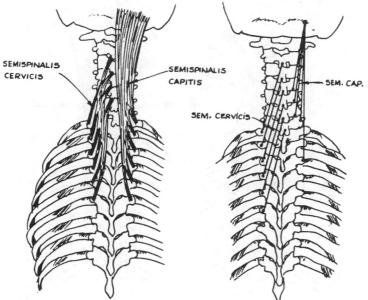

FIGURE 6-8. Muscle models for semispinalis capitis and
cervicis.

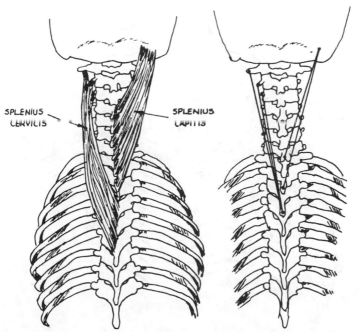

FIGURE 6-9. Muscle models for splenius cervicis and
capitis.

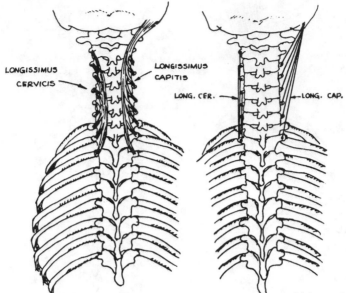

FIGURE 6-10. Muscle models for longissimus cervicis and capitis.

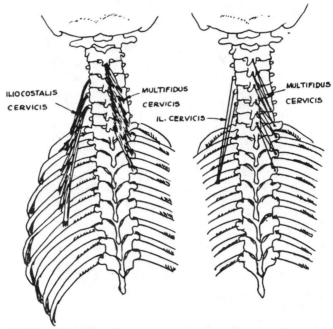

FIGURE 6-11. Muscle models for the iliocostalis cervicis and the multifidus cervicis.

and insert either into the spinous processes or into the posterior tubercles of the transverse processes of the cervical vertebrae. They are all modelled by several lines connecting their origins on the thorax (or on the cervical vertebrae) to their insertions on the cervical vertebrae (see Figs. 8 through 11).

f) Multifidus and rotatores
The multifidus cervicis and lumborum, and rotatores cervicis and lumborum are actually prolongations of the multifidus and rotartores muscles. They belong to the transversospinal muscle group with origins in transverse processes and insertions in spinous processes of the vertebrae. The multifidus extends throughout the length of the vertebral column, its transverse attachments being to transverse processes in the thoracic region to articular processes at cervical levels, and to mammillary processes at lumbar levels. Its lowest fibers originate from the posterior surface of the sacrum and its highest origin is on the fourth cervical vertebra. The fibers of multifidus ascend obliquely to insert two to four segments above their origin into the spinous processes of all vertebrae from the last lumbar to the axis. Accordingly, the cervicis portion of this muscle is modelled by twelve lines, two each inserting into the spines of each of the six cervical vertebrae C2 through C7 inclusive. Their corresponding origins lie two and three segments below these insertions (Fig. 11). The lumbar portion of the multifidus is modelled by eleven lines. There is a pair of line from each of the five lumbar vertebrae L1 through L5 inclusive (from their mammillary processes) inserting into the spines of the second and third segment above their origins. In addition the sacral part of the multifidus is modelled by a line connecting the sacrum to the spine of L5 (Fig. 12).

The rotatores muscles are the shortest representative of the transversospinal group. Their general plan of attachment is similar to the multifidus described above their origin (rotatores breves) or the second vertebral spine above their origin (rotatores longis). Consequently, the cervicis portion of the rotatores muscles is modelled by eleven lines. There is a pair of lines from each of the four cervical vertebrae C4 to C7 inclusive, and from the first thoracic vertebera T1, inserting into the spines of the first and second segment above their origins. The inferior portion of the muscle is represented by a line connecting the second thoracic vertebra T2 to the last cervical vertebra C7. The rotores lumborum, the lumbar part of the rotatores muscles, is simulated by five lines, a line from each of the five lumbar vertebrae connecting the spine of the second segment above the origin. (See Fig. 12 and 13).

Muscles of The Back and Abdomen:
The abdominal and the back muscles with extensive origins and insertions and complex anatomy pose considerable challenge to their modelling. The muscles of the abdominal walls have extensive aponeurotic origins and insertions, compounded with multi-directional muscular fascicles arranged in layers of flat sheets. While functioning in movements of the trunk, such as in flexing the trunk and the pelvis, side to side bending and rotation of the trunk they also assist in compressing the abdominal viscera, in elevating the diaphragm in respiration, and in other physiological processes. The erectores spinae, the major back muscles extend over the entire vertebral column and are primary extensors of the back.

247

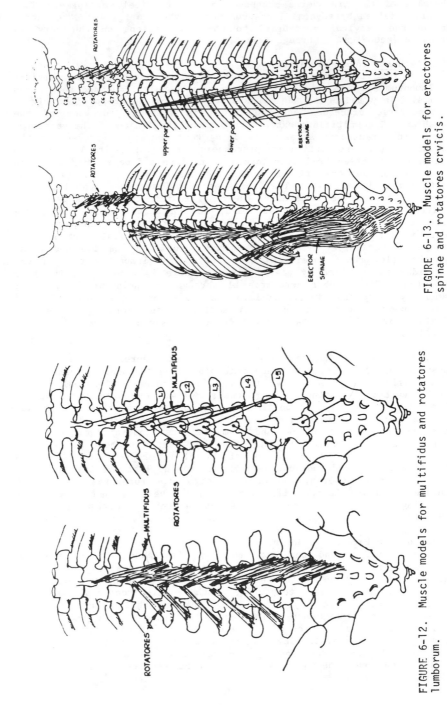

FIGURE 6-13. Muscle models for erectores spinae and rotatores crvicis.

FIGURE 6-12. Muscle models for multifidus and rotatores lumborum.

a) The erectores spinae:

The erector spinae takes its origin below in a broad, thick tendon attached to the posterior aspect of the sacrum, posterior part of the iliac crest, and the lumbar spinous processes. At the lumbar levels, the muscular fibers split into three columns—the medial spinalis, the intermedius longissimus and the more lateral iliocostalis. The iliocostalis is inserted into the angles of the lower five or six ribs. The longissimus is the most massive and powerful division of the erector spinae and has a double set of inertions or the transverse processes of the thoracic vertebrae and the backs of the ribs. The spinalis is inserted into the spines of the thoracic vertebrae. The erector spinae as a whole ascends throughout the length of the back, the columns, however, are composed of short, rope-like series of fascicles which span six to ten vertebral segments between bony attachments. Each column therefore has its own regional portion, such as lumborum (in the lumbar region), thoracis (in the thoracic region), cervicis (in the neck region) and capitis (inserting on the skull). Relevant portions of each column have already been discussed and modelled as muscles of the head and neck in the foregoing section. The thoracis portion of each column was omitted from the model since the thorax is being treated as a single rigid body.

The iliocostalis lumborum which inserts into the angles of the lower six ribs is modelled as two lines—one from the posterior iliac crest and the other from the posterior surface of the sacrum both joining the thorax (see Fig. 13). The longissimus lumborum is modelled as having seven components—one from each of the five lumbar vertebrae L1 through L5, and the remaining two from the same locations on the iliac crest and the sacrum respectively as those selected for the iliocostalis lumbroum model.

None of the seven components can directly connect the points of origin below to the point of insertion, (assumed to be common to all seven and situated at the level of the first thoracic vertebrae) due to the concave curvature of the rib cage. All seven lines are therefore assumed to wrap around the rib cage and are consequently modelled by two parts each, a lower part representing the direction of the muscle fibers in the proximity of their origin and an upper part representing the direction of the muscle fibers in the proximity of their insertion. Due to this wrapping around, reactions on the ribcage simulating pressure exerted by the seven components will have to be included in the force analysis of the thorax.

b) External obliquus abdomini:

The internal abdominal oblique arises from the anterior two-thirds of the iliac creas and passes upward and medially to insert into the costal cartilages of the lower three ribs and the linea alba and crest of the pubis. This muscle is represented by a line joining the iliac crest to the xiphoid process of the sternum (Fig. 14a and b). The muscular fasciculi from the remaining six ribs directed obliquely downward and anteriorly to end in the aponeurosis are represented by a line joining the thorax to the pubic crest of the pelvis.

c) Internal obliquus abdominis:

The internal abdominal oblique arises from the anterior two-thirds of the iliac creas and passes upward and medially to insert into the costal cartilages of the lower three ribs and the linea alba and crest of the pubis. This muscle is represented by a line joining the

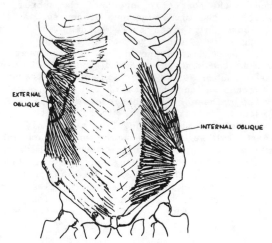

FIGURE 6-14a. The external and internal oblique abdominis.

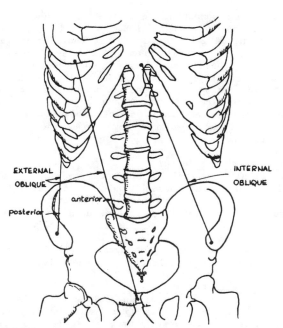

FIGURE 6-14b. Muscle model for the internal and external abdominal muscles.

iliac crest to the xiphoid process of the sternum (Fig. 14a and b).

d) Rectus abdominis:

The rectus abdominis takes its origin from the crest of the pubis and inserts in the xiphoid process of the sternum and the costal cartilages of the fifth, sixth and seventh ribs. It can be conveniently represented by a line connecting the origin on the pubic crest to its approximate point of insertion on the ribcage (see Fig. 15).

e) Quadratus lumborum:

The quadratus lumborum forms part of the posterior wall of the abdomen and arises from the posterior aspect of the crest of the ilium and the transverse processes of the lower four lumbar vertebrae and is inserted into the transverse processes of the upper four lumber vertebrae and the last rib. It is modelled by five lines, each from the posterior iliac crest, and connected to the upper four lumbar vetebrae L1 through L4 and the thorax at the last rib (Fig. 16a and b).

f) Psoas major:

The psoas major has been described as part of the iliopsaoas during modelling of the muscles of the lower extremities. The psoas arises from the sides of the bodies and transverse processes of the five lumbar vertebrae, passes obliquely lateralward beneath the inguinal ligament to insert into the lesser trochanter of the femur along with the iliacus. The muscle is modelled by five lines—one each from the five lumbar vertebrae L1 through L5, connected to the femur. Since the muscle distinctly wraps around the pelvis on its way to insertion on the femur, each component is assumed to be integral with the forearm and therefore muscles connecting the forearm to the hand were excluded. Here, however, we will describe these muscles and also those connecting the spine to the other segments of the upper extremities.

a) Trapezius

The trapezius is a flat, triangular muscle placed superficially on the upper back and the back of the neck. It takes origin on the skull from the external occipital protuberance and the medial part of the superior nuchal line from the ligamentum nuchae (ligament joining the spines of the cervical vertebrae), from the spine of the seventh cervical vertebra and the spinous processes of all thoracic vertebrae. The fibers converge lateral-ward towards the point of the shoulder for insertion. The occipital and upper cervical fibers are inserted in the posterior margin of the outer third of the clavicle; the lower cervical and the crest of the spine of the scapula, and the lower thoracic fibers converge to insert into the tubercle of the crest of the spine of the scapula. The model for the trapezius muscle is shown in Fig. 17. The clavicular part of the muscle is shown by lines from the skull and the upper four cevical vertebrae going to the clavicle. The scapular part of the muscle is modelled as:

i)   two lines from the vertebrae C5 and C6 respectively connected to the acromion of the scapula.

ii)   a line from C7 a line representing fibers originating from T1 through T3, both joining the middle of the spine of the scapula.

iii)   a line representing fibers arising from T4 through T7 connected to the tubercle of the crest of the spine of the scapula. This

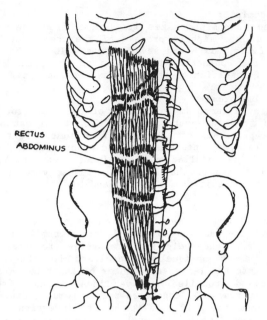

FIGURE 6-15. The rectus abdominis muscle and its model.

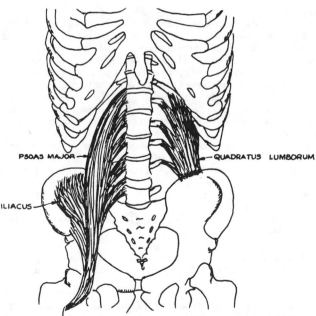

FIGURE 16a. The psoas major, iliacus and quadratus lumborum muscles.

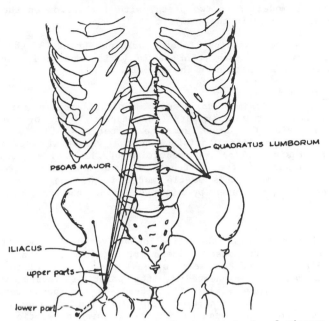

FIGURE 6-16b. Muscle models for quadratus lumborum, iliacus and psoas major.

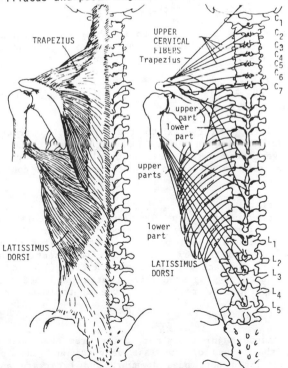

FIGURE 6-17. Latissimus dorsi and its model; Trapezius and its model.

line is modelled as two parts, with a reaction on the scapula to be included in the analysis.

iv)   a line representing fibers arising from T8 through T12 connecting the same point as in (iii), modelled also as two parts and a reaction on the scapula.

b)   Latissimus dorsi:
        The latissimus dorsi was modelled earlier for the two segmented arm model but only in part (see Chapter IV). For the posture investigated therein, the trunk and the vertebral column was assumed fixed so that muscle fibers arising on the vertebral segments were represented by a single line representative of the entire group. With the lumbar segments of the vertebral column now being treated as individual rigid bodies, this muscle model is no longer valid. The muscle is therefore simulated by eight lines in the following manner (Fig. 17).

i)   a line each from the five lumbar vertebrae L1 through L5 representing the muscle fibers arising from the thora-columbar fascia attached to the spious processes of the lumbar vertebrae, and connected to the point of insertion on the humerus. However, since the ribcage is interposed between the origins and the insertion, this portion of the muscle is assumed to wrap around the thorax, causing a reaction force on this segment. Consequently, the five lines are modelled by two parts each, with appropriate reaction force on the thorax at the assumed points of contact.

ii)   a line connecting the pelvis to the humerus, representing the portion of the muscle originating on the posterior aspect of the iliac crest.

iii)   a line joining the ribcage to the humerus, simulating the portion of the muscle that takes origin from the outer surface of the lower four ribs.

iv)   a line joining the thorax to the humerus, representing the muscle part arising on the spinous processes of the lower six thoracic vertebrae. This line is modelled as two parts, along with a reaction force at an assumed wrap-around contact point on the thorax.

c)   Serratus anterior:
        The serratus anterior is functionally antagonist of the trapezius, arising by fleshy digitations from the upper eight or nine ribs and inserting into the medial border of the costal surface of the scapula. The muscle is modelled by three lines, representing the upper, middle and lower fibers respectively as shown in Fig. (18a and b). The middle fibers are assumed to wrap around the thorax and hence modelled by two lines with an appropriate reaction at the approximate wrap-around point on the thorax.

d)   Rhomboideus::
        The rhomboideus major and minor, arise from the lower part of ligamentum nuchae, the spines of the last cervical and the upper five thoracic vertebrae. The fibers pass downward and laterward and insert

254

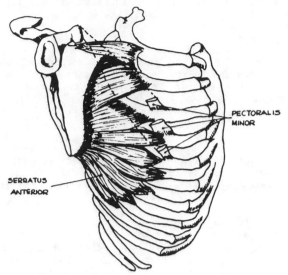

FIGURE 18a. The serratus anterior and pectoralis
minor muscles.

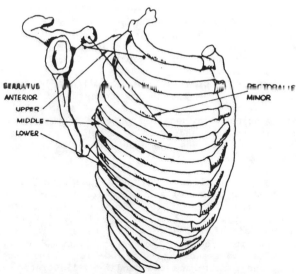

FIGURE 18b. Muscle models for the serratus anterior
and pectoralis minor.

into the medial border of the scapula below the root of the spine. The muscles are modelled by five lines—two from the spines of the cervical vertebrae C6 and C76 and three from the thoracic spines, representing the origin from the thoracic vertebrae (see Fig. 19).

e) Levator scapulae:
The levator scapulae arises from the transverse processes of the atlas and axis and from the posterior tubercles of the transverse processes of the third and fourth cervical vertebrae and inserts into the vertebral border of the scapula from the superior angle to the spine. The muscle is modelled by four lines—a line each from the four cervical vertebrae C1 through C4, joining the scapula (Fig. 19).

f) Subclavius:
The subclavius is a small muscle placed between the clavicle and the first rib. It arises on the first rib and its cartilage at their junction and inserts into the inferior surface of the clavicle. It is modelled conveniently by a straight line connecting the thorax to the clavicle.

g) Pectoralis major and minor:
The pectoralis major (see Chapter IV) has an extensive origin from the sternal half of the clavicle, from the ventral surface of the sternum and from the costal cartilages of the upper seven ribs. It is modelled by three lines—one from the clavicle and two from the thorax representing the origin from the sternum and the costal cartilages, all going to the humerus. (See Fig. in Chapter IV).

The pectoralis minor arises from the outer surfaces of the third, fourth and fifth ribs near their cartilages and inserts into the coracoid process of the scapula. It is represented by a straight line connecting the thorax to the scapula. (See Figs. 18a and b).

Details of the models of the other muscles of the upper extremities can be found in Chapter IV. Since the forearm and the hand together were considered to be a single rigid body muscles connecting the forearm to the hand were excluded from the model for the action investigated. Since the hand is now considered as a segment separated from the forearm at the wrist joint, these muscles will be included. Thus, portions of extensor carpi ulnaris, flexor carpi ulnaris, flexor pollicis longus and flexor digitorum superficialis which arise on the forearm and insert into the hand will now be considered for modelling. Note that only the humeral portions of these muscles were modelled before. In addition, muscles connecting the forearm to the hand, such as extensor pollicis longus and brevis, flexor digitorum profundus and abductor pollicis longus formerly altogether omitted will also be included.

Since the hand is treated as a single segment, that is, action of the fingers is omitted, the flexors and extensors of the fingers require special treatment in their modelling. There muscles are modelled as straight lines connecting their points of origin on the arm or on the forearm to "alternate" sites of insertion in the vicinity of the wrist joint, and their portions beyond the wrist joint which are mainly responsible for digital movements are not modelled. The models of the muscles of the arm, the forearm and the hand are illustrated in Figs. 20 through 23.

Muscles of the Lower Extremities

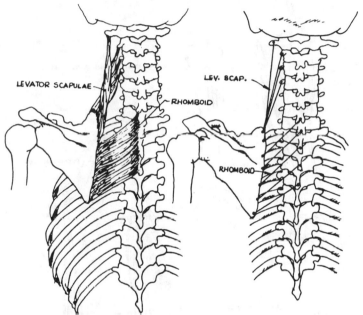

FIGURE 6-19. Muscle models for levator scapulae and
the rhomboids.

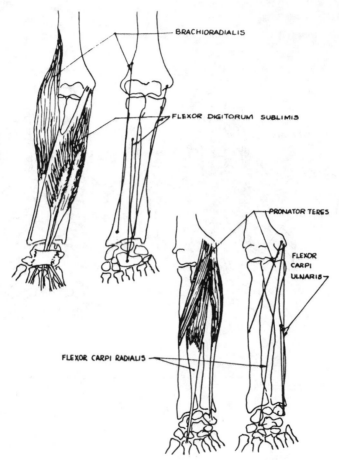

FIGURE 6-20. Muscle model for brachioradialis, flexor digitorum sublimis, pronator teres, flexor carpi radialis, flexor carpi ulnaris.

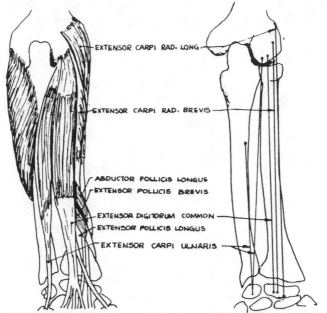

FIGURE 6-21. Muscle models for extensor carpi radialis longus and brevis, extensor digitorum communis, and extensor carpi ulnaris.

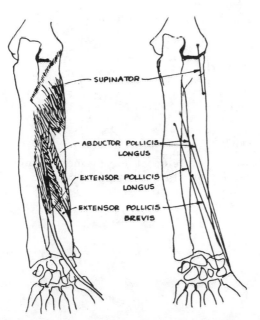

FIGURE 6-22. Muscle models for supinator, abductor pollicis longus, extensor pollicis longus and brevis.

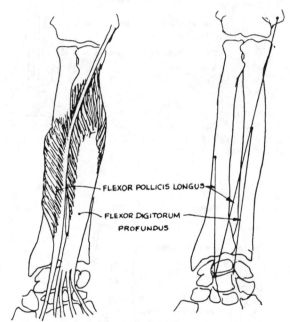

FIGURE 6-23. Muscle model for flexor pollicis longus and flexor digitorum profundus.

With the exception of the psoas which connects the spine to the upper leg, and which is now included in the model of the spine, there is no modifications required in the models of the muscles of the lower extremities. The musculoskeletal model of the lower extremities is therefore exactly identical to the one developed in Chapter IV.

In all, the 88 muscles listed in Table II are represented by 254 functional lines of application.

COORDINATES OF POINTS OF MUSCLE ORIGINS AND INSERTIONS

The third step in musculoskeletal modeling after identifying the segments and muscles connecting them deals with determining the coordinates of points of muscle origins and insertions with respect to suitable body-fixed coordinate axes. These axes are fixed with any particular segment and move along with them. The origin of the axes is generally selected to lie on or outside the segment, coincident with the joint the segment makes with adjoining member. The location of the body-fixed axes for the upper and lower extremities have been described in the foregoing chapters. For the five cervical vertebrae (C3 to C7) and the five lumbar vertebrae (L1 to L5) the coordinate axes are selected with their origin at the center of the intervertebral disc immediately below the specific segment. It may be mentioned here that these vertebral segments are assumed to comprise of the body of the vertebra, and half of the disc above and half of the disc below, so that the center of the articulation between two segments would be at the center of the disc in between (see Fig. 24a). The axes are chosen so that xz plane coincides with the sagittal plane and yz plane coincides with the coronal plane. The axes are directed in the following manner: x axis is posterior to anterior, parallel to the inferior surface of the vertebral body; z axis is vertically upwards and y axis lies mediolaterally.

The vertebrae common to a particular region are assumed for convenience to be of the same size. Thus all the five lumbar vertebrae are taken to be of the same dimensions, and all the lower five vertebrae are assumed of same size. This will greatly simplify the task of measurement of the coordinates on these segments. For example, the coordinates of a point of interest located on the transverse process of a lumbar vertebrae would be identical for the five segments belonging to the lumbar region.

Since the first and second cervical vertebrae differ in shape from the remaining five, their body-fixed coordinate axes are selected in alternate manner. For the first cervical vertebra, the origin of the axes is selected to be in the sagittal plane, at the level of the inferior articulating surfaces (Fig. 24a). The origin of the axes for the second cervical vertebra is chosen to be in the sagittal plane at the center of the C2/C3 disk. The x axis is directed posterior to anterior parallel to the inferior surface of the vertebral body and the z axis is vertically upwards perpendicular to the former (Fig. 24a).

For the skull, the points of interest are measured with respect to an axis system with its origin at the level of the atlanto occipital joint, midway between the two occipital condyles. Finally, the coordinates axes on the thorax are selected with their origin at the center of the T12/L1 disk with the x-axis directed anteriorly and parallel to the inferior surface of the body of T12 (Fig. 24a.).

261

TABLE II. Major muscles of the spine.

A. Muscles of the Head and Neck

1. Semispinalis capitis
2. Splenius capitis
3. Longissimus capitis
4. Rectus capitis posterior major
5. Rectus capitis posterior minor
6. Obliquus capitis superior
7. Obliquus capitis inferior
8. Rectus capitis anterior
9. Rectus capitis lateralis
10. Longus capitis
11. Sternocleidomastoid
12. Longissimus cervicis
13. Iliocostalis cervicis
14. Splenius cervicis
15. Semispinalis cervicis
16. Multifidus cervicis
17. Rotatores cervicis
18. Longus colli
19. Sclenus
20. Sternothyroid
21. Sternohyoid

B. Muscles of the Back and Abdomen
1. Rectus abdominis
2. External obliquus abdominis
3. Internal obliquus abdominis
4. Quadratus lumborum
5. Psoas major
6. Erector spinae

C. Muscles of the Upper Extremeties

1. Trapezius
2. Rhomboids
3. Levator scapulae
4. Latissimus dorsi
5. Subclavius
6. Serratus anterior
7. Pectoralis major
8. Pectoralis minor
9. Deltoideus
10. Supraspinatus
11. Infraspinatus
12. Teres major
13. Teres minor
14. Subscapularis
15. Coracobrachialis
16. Biceps brachii
17. Brachioradialis
18. Triceps brachii
19. Brachlalls
20. Anconeus
21. Supinator
22. Pronator teres
23. Extensor pollicis longus
24. Extensor pollicis brevis
25. Extensor carpi radialis longus
26. Extensor carpi radialis brevis
27. Extensor digitorum communis
28. Extensor carpi ulnaris
29. Flexor carpi ulnaris
30. Flexor carpi radialis
31. Flexor pollicis longus
32. Flexor digitorum superficialis
33. Flexor digitorum superficialis
34. Flexor digitorum profundus
35. Abductor pollicis longus

D. Muscles of the Lower Extremities

1. Gracilis
2. Adductor longus
3. Adductor magnus
4. Adductor brevis
5. Semitendinosus
6. Semimembranosus
7. Biceps femoris
8. Rectus femoris
9. Sartorius
10. Tensor fasciae latae
11. Gluteus maximus
12. Iliopsoas
13. Gluteus medius
14. Gluteus minimus
15. Gastrocnemius
16. Soleus
17. Tibialis anterior
18. Tibialis posterior
19. Extensor digitorum longus
20. Extensor hallucis longus
21. Flexor digitorum longus
22. Flexor hallucis longus
23. Peroneus longus
24. Peroneus brevis
25. Peroneus tertius

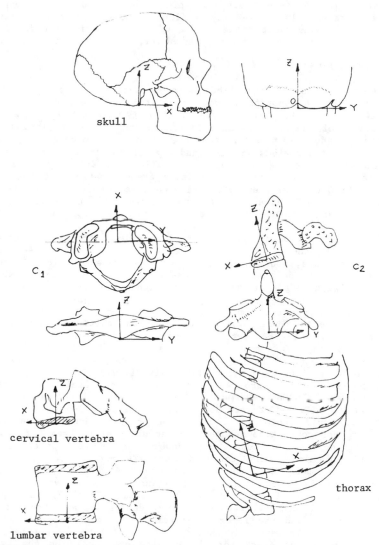

skull

C_1

C_2

cervical vertebra

lumbar vertebra

thorax

FIGURE 6-24a. Body-fixed coordinate axes for the segments.

The coordinates of the various muscle origins and insertions were obtained from the skeletal figures in the book by Braus (1) (listed in Table IIIa). The coordinates of the joints are listed in Table IIIb.

The selected weights and locations of the centers of gravity the segments are listed in Table IV for an individual of 72.3 Kg body weight. Since the information regarding the nature of distribution of the weight of the trunk over the spinal column and the corresponding centers of gravity is not readily available, the vertebrae are assumed to be massless. Consequently the weight of the trunk is considered to be distributed among three segments: the skull, the thorax and the pelvis.

MODEL APPLICATIONS

The model was first applied to investigate the muscle load-sharing during the act of stooping forward to lift weights off the ground (see Fig. 24b). The kinematic configuration is defined by knowing the orientations of the segments in relation to each other (Table Va). This resulting change in coordinates of points of muscle origins and insertions on the various segments and any other points of interest (such as locations of centers of masses, locations of intersegmental joints etc.) is then computed using coordinate transformation matrices. This transformation will relate the coordinates of a point on any segment with respect to ground-fixed-axes which are selected with their origin at the level of the ankle joints and situated in the sagittal plane.

In the posture being studied, the left side of the body is identical with respect to the right side. Since the ground axes are selected in the sagittal plane, about which the two sides are symmetrical it follows that one may write only three equations for each of the segment which is symmetrical with respect to the sagittal plane. The computer program was initially developed in two parts: a part to investigate the muscle load sharing for the spinal column and the upper extremities only and a second part to analyze the lower extremities, with the pertinent input (namely, the forces in muscles connecting the spine to the lower extremities and the joint reaction forces at the lumbosacral (L5/Pelvis joint) obtained from the first part. The programs are thus designed for application to situations of sagittal symmetry only. This division is primarily done to reduce the size of the program so that it could efficiently adapted to the available SIMPLEX subroutine, and to meet the computer space requirements. If the program is of vast size, round-off errors during computation could prolong the iteration process requiring substantial expenditure and could also encumber the accuracy of the results. Later a program was developed to study more general actions and postures, those that would require truly three dimensional modelling. This would understandably make the program size enormous and only then can be handled by special subroutine packages (such as FMPS, available on the UNIVAC 1110, University of Wisconsin Computing Center). This program incorporates all the segments of the musculoskeletal structure, the spine, the upper and the lower extermities, and will be referred to as the "Total Model" in later discussions. The 'Total Model' can be used to provide a convenient check for the results obtained from the two-part model.

THE TWO-PART MODEL

There are a total of 20 segments, of which 15 are symmetrically placed about the sagittal plane for the stooping posture. The five

266

TABLE IIIa.   Coordinates of muscle attachment points.

MUSCLE NO.	MUSCLE NAME	Body1	X1 (CM.)	Y1 (CM.)	Z1 (CM.)	Body2	X2 (CM.)	Y2 (CM.)	Z2 (CM.)
1	Semipinalis capitis	Skull	-5.30	-1.80	2.50	C4	-1.20	-2.20	2.00
2	"	"	-5.30	-1.80	2.50	C5	-1.20	-2.20	2.00
3	"	"	-5.30	-1.80	2.50	C6	-1.20	-2.20	2.00
4	"	"	-5.30	-1.80	2.50	C7	0.00	-2.80	1.00
5	"	"	-5.30	-1.80	2.50	THX	4.00	-3.00	24.00
6	Splenius capitis	"	-3.00	-4.40	1.60	C7	-4.00	-.40	-1.00
7	"	"	-3.00	-4.40	1.60	THX	3.00	0.00	25.00
8	Longissimus capitis	"	-1.80	-4.30	.50	C3	0.00	-2.50	1.00
9	"	"	-1.80	-4.30	.50	C4	0.00	-2.50	1.00
10	"	"	-1.80	-4.30	.50	C5	0.00	-2.50	1.00
11	"	"	-1.80	-4.30	.50	C6	0.00	-2.50	1.00
12	"	"	-1.80	-4.30	.50	C7	0.00	-2.50	1.00
13	"	"	-1.80	-4.30	.50	THX	6.50	-3.50	26.00
14	Rectus cap. post. major	"	-4.00	-2.30	1.20	C2	-2.70	-.40	-.80
15	" minor	"	-3.70	-1.00	.80	C1	-2.70	0.00	.60
15	Oblique cap. superior	"	-3.70	-3.20	1.70	C1	0.00	-3.60	.70
17	Rectus cap. anterior	"	1.00	-.60	.80	C1	0.00	-2.70	.80
18	" lateral	"	-.90	-3.00	.90	C1	0.00	-3.60	.70
19	Longus capitis	"	1.70	-.40	.80	C3	.50	-2.30	1.00
20	"	"	1.70	-.40	.80	C4	.50	-2.30	1.00
21	"	"	1.70	-.40	.80	C5	.50	-2.30	1.00
22	"	"	1.70	-.40	.80	C6	.50	-2.30	1.00
23	Sternocleidomastoid	"	-3.50	-5.00	2.30	THX	13.00	-1.00	23.00

TABLE IIIA (continued)

24	"	Skull	-3.50	-5.00	2.30	Clav.	.50	3.50	1.20
25	Longissimus Cervicis	C2	-.40	-2.80	.50	THX	4.50	-3.50	25.00
26	"	C3	-.40	-2.80	.50	"	4.50	-3.50	25.00
27	"	C4	-.40	-2.80	.50	"	4.50	-3.50	25.00
28	"	C5	-.40	-2.80	.50	"	4.50	-3.50	25.00
29	"	C6	-.40	-2.80	.50	"	4.50	-3.50	25.00
30	"	C2	-.40	-2.80	.50	C7	-.50	-2.80	.70
31	"	C3	-.40	-2.80	.50	"	-.50	-2.80	.70
32	"	C4	-.40	-2.80	.50	"	-.50	-2.80	.70
33	"	C5	-.40	-2.80	.50	"	-.50	-2.80	.70
34	"	C6	-.40	-2.80	.50	"	-.50	-2.80	.70
35	Ilio-costalis cervicis	C4	-.40	-2.80	.50	THX	1.50	-4.00	21.50
36	"	C5	-.40	-2.80	.50	"	1.50	-4.00	21.60
37	"	C6	-.40	-2.80	.50	"	1.50	-4.00	21.50
38	Splenius cervicis	C1	0.00	-3.60	.70	"	1.50	0.00	22.50
39	"	C2	-.40	-2.80	.50	"	1.50	0.00	22.50
40	"	C3	-.40	-2.80	.50	"	1.50	0.00	22.50
41	Semispinalis cervicis	C2	-2.70	-.40	-.80	"	1.50	-3.00	22.00
42	"	C3	-3.50	-.40	-.70	"	1.50	-3.00	22.00
43	"	C4	-3.50	-.40	-.70	"	1.50	-3.00	22.00
44	"	C5	-3.50	-.40	-.70	"	1.50	-3.00	22.00
45	"	C6	-3.50	-.40	-.70	"	-1.00	-3.00	16.00
46	"	C7	-4.00	-.40	-1.00	"	-1.00	-3.00	16.00
47	Multifidus cervicis	C2	-2.70	-.40	-.80	C5	-1.20	-2.20	2.00
48	"	C2	-2.70	-.40	-.80	C6	-1.20	-2.20	2.00
49	"	C3	-3.50	-.40	-.70	C6	-1.20	-2.20	2.00
50	"	C3	-3.50	-.40	-.70	C7	-1.20	-2.20	2.00

TABLE IIIA (continued)

51	"	C4	-3.50	-.40	-.70	C7	-1.20	-2.20	2.00
52	"	C4	-3.50	-.40	-.70	THX	7.00	-3.50	28.00
53	"	C5	-3.50	-.40	-.70	"	7.00	-3.50	28.00
54	"	C5	-3.50	-.40	-.70	"	6.50	-3.50	26.00
55	"	C6	-3.50	-.40	-.70	"	6.50	-3.50	26.00
56	"	C6	-3.50	-.40	-.70	"	5.50	-3.50	25.00
57	"	C7	-4.00	-.40	-1.00	"	5.50	-3.50	25.00
58	"	C7	-4.00	-.40	-1.00	"	4.00	-3.50	24.00
59	Rotatores cervicis	C2	-2.70	-.40	-.80	C4	0.00	-2.50	1.00
60	"	C3	-3.50	-.40	-.70	C4	0.00	-2.50	1.00
61	"	C3	-3.50	-.40	-.70	C5	0.00	-2.50	1.00
62	"	C4	-3.50	-.40	-.70	C5	0.00	-2.50	1.00
63	"	C4	-3.50	-.40	-.70	C6	0.00	-2.50	1.00
64	"	C5	-3.50	-.40	-.70	C6	0.00	-2.50	1.00
65	"	C5	-3.50	-.40	-.70	C7	0.00	-2.50	1.00
66	"	C6	-3.50	-.40	-.70	C7	0.00	-2.50	1.00
67	"	C6	-3.50	-.40	-.70	THX	7.00	-3.50	28.00
68	"	C7	-4.00	-.40	-1.00	"	7.00	-3.50	28.00
69	"	C7	-4.00	-.40	-1.00	"	6.50	-3.50	27.00
70	Longus colii (vertical)	C1	1.60	0.00	1.00	C5	1.00	0.00	.80
71	"	C2	1.00	0.00	.80	"	1.00	0.00	.80
72	"	C3	1.00	0.00	.80	"	1.00	0.00	.80
73	"	C4	1.00	0.00	.80	"	1.00	0.00	.80
74	"	C1	1.60	0.00	1.00	C6	1.00	0.00	.80
75	"	C2	1.00	0.00	.80	"	1.00	0.00	.80
76	"	C3	1.00	0.00	.80	"	1.00	0.00	.80
77	"	C4	1.00	0.00	.80	"	1.00	0.00	.80

78	"	C1	1.60	0.00	1.00	C7	1.00	0.00	.80
79	"	C2	1.00	0.00	.80	"	1.00	0.00	.80
80	"	C3	1.00	0.00	.80	"	1.00	0.00	.80
81	"	C4	1.00	0.00	.80	"	1.00	0.00	.80
82	"	C1	1.60	0.00	1.00	THX	10.30	0.00	27.00
83	"	C2	1.00	0.00	.80	"	10.30	0.00	27.00
84	"	C3	1.00	0.00	.80	"	10.30	0.00	27.00
85	"	C4	1.00	0.00	.80	"	10.30	0.00	27.00
86	Longus colli (Inf. obliq.)	C5	.50	-2.30	1.00	"	10.30	0.00	27.00
87	"	C6	.50	-2.30	1.00	"	10.30	0.00	27.00
88	Longus colli (sup. obliq.)	C1	1.60	0.00	1.00	C3	.50	-2.30	1.00
89	"	C3	1.00	0.00	.80	C4	.50	-2.30	1.00
90	"	C1	1.60	0.00	1.00	C5	.50	-2.30	1.00
91	"	C2	1.00	0.00	.80	C3	.50	-2.30	1.00
92	"	C1	1.60	0.00	1.00	C4	.50	-2.30	1.00
93	"	C2	1.00	0.00	.80	C5	.50	-2.30	1.00
94	Scalenus (anterior)	C3	.50	-2.30	1.00	THX	10.50	-5.30	24.00
95	"	C4	.50	-2.30	1.00	"	10.50	-5.30	24.00
96	"	C5	.50	-2.30	1.00	"	10.50	-5.30	24.00
97	"	C6	.50	-2.30	1.00	"	10.50	-5.30	24.00
98	Scalenus (medial)	C1	0.00	-3.60	.70	"	7.50	-5.30	27.00
99	"	C2	-.40	-2.80	.50	"	7.50	-5.30	27.00
100	"	C3	-.40	-2.80	.50	"	7.50	-5.30	27.00
101	"	C4	-.40	-2.80	.50	"	7.50	-5.30	27.00
102	"	C5	-.40	-2.80	.50	"	7.50	-5.30	27.00
103	"	C6	-.40	-2.80	.50	"	7.50	-5.30	27.00
104	"	C7	-.40	-2.80	.50	"	7.50	-5.30	27.00

105	Scalenus (posterior)	C5	-.40	-2.80	.50	THX	4.50	-5.30	25.50
106	"	C6	-.40	-2.80	.50	"	4.50	-5.30	25.50
107	"	C7	-.40	-2.80	.50	"	4.50	-5.30	25.50
108	Oblique cap. inferior	C1	0.00	-3.60	.70	C2	-2.70	-.40	-.80
109	Trapezius	Skull	-6.60	-1.00	2.70	Clav	-1.00	13.00	.50
110	"	C1	-2.70	0.00	.60	"	-1.00	13.00	.50
111	" "	C2	-2.70	-.40	-.80	"	-1.00	13.00	.50
112	"	C3	-3.50	-.40	-.70	"	-1.00	13.00	.50
113	"	C5	-3.50	-.40	-.70	"	-1.00	13.00	.50
114	"	C5	-3.50	-.40	-.70	"	-1.00	13.00	.50
115	"	C6	-3.50	-.40	-.70	"	-1.00	13.00	.50
116	"	Skull	-6.60	-1.00	2.70	Scap	-2.00	-1.00	-1.50
117	"	C1	-2.70	0.00	.60	"	-2.00	-1.00	-1.50
118	"	C2	-2.70	-.40	-.80	"	-2.00	-1.00	-1.50
119	"	C3	-3.50	-.40	-.70	"	-2.00	-1.00	-1.50
120	"	C4	-3.50	-.40	-.70	"	-2.00	-1.00	-1.50
121	"	C5	-3.50	-.40	-.70	"	-2.00	-1.00	-1.50
122	"	C6	-3.50	-.40	-.70	"	-2.00	-1.00	-1.50
123	"	C7	-4.00	-.40	-1.00	"	-6.50	5.00	-2.00
124	"	THX	3.00	0.00	25.50	"	-6.50	5.00	-7.00
125	Rhomboid	C6	-3.50	-.40	-.70	Scap	-11.00	9.20	-4.00
126	"	C7	-4.00	-.40	-1.00	"	-11.50	9.00	-6.50
127	"	CHX	3.00	0.00	27.50	"	-11.80	8.50	-9.00
128	"	THX	3.00	0.00	26.00	"	-12.00	8.00	-11.00
129	"	THX	3.00	0.00	25.00	"	-12.00	7.50	-13.50
130	Levator Scapulae	C1	0.00	-3.60	.70	"	-5.70	9.00	-.40
131	"	C2	-.40	-2.80	.50	"	-6.80	9.00	-1.20
132	"	C3	-.40	-2.80	.50	"	-8.00	8.70	-2.80

TABLE IIIA (continued)

134	Rectus abdominis	THX	15.00	-4.00	5.00	Plvs	5.50	-.80	-1.00
135	Ext. abd. oblique	THX	-4.00	-12.00	-4.50	"	4.00	-11.50	11.50
136	"	THX	7.00	-13.00	0.00	"	6.00	0.00	-1.00
137	Int. abd. oblique	THX	14.50	-.50	2.00	"	2.50	-12.50	12.50
138	Quadratus lumb.	THX	-5.00	-5.50	1.00	"	-4.00	-7.50	16.00
139	"	L1	-3.50	-3.50	2.00	"	-4.00	-7.50	16.00
140	"	L1	-3.50	-3.70	2.00	"	-4.00	-7.50	16.00
141	"	L3	-3.50	-4.20	2.00	"	-4.00	-7.50	16.00
142	"	L4	-3.50	-4.50	2.00	"	-4.00	-7.50	16.00
143	Multifidus lumb.	THX	-5.00	0.00	9.50	L1	-4.50	-2.00	3.30
144	"	"	-5.50	0.00	6.50	L1	-4.50	-2.00	3.30
145		"	-5.50	0.00	6.50	L2	-4.50	-2.00	3.30
146	"	"	-5.50	0.00	3.00	L2	-4.50	-2.00	3.30
147	"	"	-5.50	0.00	3.00	L3	-4.50	-2.00	3.30
148	"	"	-5.50	0.00	0.00	L3	-4.50	-2.00	3.30
149	"	"	-5.50	0.00	0.00	L4	-4.50	-2.00	3.30
150	"	L1	-6.50	0.00	1.00	L4	-4.50	-2.00	3.30
151	"	L1	-6.50	0.00	1.00	L5	-4.50	-2.00	3.30
152	"	L2	-6.50	0.00	1.00	L5	-4.50	-2.00	3.30
153	"	L5	-6.50	0.00	1.00	Plvs	-8.00	-3.00	6.50
154	Rotatores lumb.	THX	-5.50	0.00	3.00	L2	-4.50	-2.00	3.30
155	"	"	-5.50	0.00	0.00	L2	-4.50	-2.00	3.30
156	"	L1	-6.50	0.00	1.00	L3	-4.50	-2.00	3.30
157	"	L2	-6.50	0.00	1.00	L4	-4.50	-2.00	3.30
158	"	L3	-6.50	0.00	1.00	L5	-4.50	-2.00	3.30
159A	Iliopsoas	L1	-.50	-2.00	2.00	Plvs	2.50	-7.50	0.00
B	"	Femur	-1.00	-9.50	34.50	"	2.50	-7.50	0.00
160A	"	L2	-.50	-2.30	2.00	"	2.50	-7.50	0.00

TABLE IIIA   (continued)

161A	"	L3	-.50	-2.50	2.00	Plvs .	2.50	-7.50	0.00
B	"	Femr	-1.00	-9.50	34.50	"	2.50	-7.50	0.00
162A	"	L4	-.50	-2.70	2.00	"	2.50	-7.50	0.00
B	"	Femr	-1.00	-9.50	34.50	"	2.50	-7.50	0.00
163A	"	L5	-.50	-2.80	2.00	"	2.50	-7.50	0.00
B	"	Femr	-1.00	-9.50	34.50	"	2.50	-7.50	0.00
164	Erectores spinae	THX	-6.00	-7.00	10.00	"	-6.50	-6.00	14.00
165	"	THX	-6.00	-7.00	10.00	"	-6.00	-3.00	8.00
166A	"	L1	-6.50	0.00	1.00	THX	-4.00	-2.00	12.50
B	"	THX	2.00	-3.50	22.00	"	-4.00	-3.50	12.50
167A	"	L2	-6.50	0.00	1.00	"	-4.00	-2.30	12.50
B	"	THX	2.00	-3.50	22.00	"	-4.00	-2.30	12.50
168A	"	L3	-6.50	0.00	1.00	"	-4.00	-2.50	12.50
B	"	THX	2.00	-3.50	22.00	"	-4.00	-2.50	12.50
169A	"	L4	-6.50	0.00	1.00	"	-4.00	-2.70	12.50
B	"	THX	2.00	-3.50	22.00	"	-4.00	-2.70	12.50
170A	"	L5	-6.50	0.00	1.00	"	-4.00	-2.90	12.50
B	"	THX	2.00	-3.50	22.00	"	-4.00	-2.90	12.50
171A	"	Plvs	-6.00	-3.00	8.00	"	-4.00	-3.50	12.50
B	"	THX	2.00	-3.50	22.00	"	-4.00	-3.50	12.50
172A	"	Plvs	-6.50	-6.00	14.00	"	-4.00	-4.20	12.50
B	"	THX	2.00	-3.50	22.00	"	-4.00	-4.20	12.50
173A	Latissimus dorsi	L1	-6.50	0.00	1.00	"	-4.00	-11.50	5.00
B	"	arm	1.00	0.00	5.80	"	-4.00	-11.50	5.00
174A	"	L2	-6.50	0.00	1.00	"	-4.00	-11.50	4.00
B	"	arm	1.00	0.00	5.80	"	-4.00	-11.50	4.00
175A	"	L3	-6.50	0.00	1.00	"	-4.00	-11.50	3.00
B	"	arm	1.00	0.00	5.80	"	-4.00	-11.50	3.00

TABLE IIIA (continued)

176A	"	L4	-6.50	0.00	1.00	"	-3.00	-11.50	2.00
B	"	arm	1.00	0.00	5.80	"	-3.00	-11.50	2.00
177A	"	L5	-6.50	0.00	1.00	"	-3.00	-11.50	1.00
B	"	arm	1.00	0.00	5.80	"	-3.00	-11.50	1.00
178	"	THX	-2.50	-14.00	-3.50	arm	1.50	.60	4.00
179	"	Plvs	-6.00	-8.00	14.00	"	1.50	.60	4.00
180	Subclavius	THX	11.50	-5.00	21.30	clav.	0.00	9.50	.30
181	serratis anterior	THX	8.50	-8.50	22.50	scap	-5.50	9.00	.50
182A	"	THX	8.50	-10.00	19.00	THX	3.00	-11.00	19.00
B	"	scap	-12.50	9.50	-8.50	THX	3.00	-11.00	19.00
183	"	THX	8.00	-14.00	7.00	scap	-13.00	6.80	-14.80
184	Pectoralis major	Clav	1.80	5.00	.70	arm	2.00	-.90	4.50
185	"	THX	13.50	-1.50	15.00	"	2.00	-.90	4.50
186	"	THX	12.00	-4.50	4.00	"	2.00	-.90	4.50
187	" minor	THX	13.00	-9.00	11.00	scap	1.70	4.50	-2.00
188A	Trapezius	THX	-.70	0.00	15.00	scap.	-11.50	9.50	-5.00
B	"	scap	-9.00	6.50	-2.70	"	-11.50	9.50	-5.00
189	Deltoid	clav	.80	13.00	.50	arm	1.00	-.80	12.00
190	"	scap	0.00	-2.50	-1.50	"	1.00	-1.00	12.00
191	"	"	-6.50	6.00	-2.50	"	-1.00	0.00	12.00
192	Supraspinatus	"	-6.00	7.50	-.60	"	1.90	-2.00	-2.50
193	Infraspinatus	"	-9.50	6.00	-8.00	"	-.80	-2.30	-.40
194	Teres major	"	-11.00	6.00	-14.00	"	1.40	.80	6.40
195	" minor	"	-7.00	2.50	-9.50	"	-.60	-2.30	.50
196	Subscapularis	"	-9.00	6.50	-8.50	"	1.50	1.20	-.80
197	Coracobrachialis	"	1.00	3.60	-2.50	"	1.00	.80	13.20
198	Biceps (SH)	"	1.00	3.60	-2.50	Rad	.80	.50	4.00
199A	" (LH)	"	.80	.50	4.00	arm	2.50	0.00	-1.00
B	"	"	-.50	2.00	-2.00	scap	.50	.50	-2.50

No.	Name							
200	Triceps	scap -1.50	.50	-6.00	ulna -2.30	0.00	.30
201	"	arm -1.00	1.20	20.00	" -2.30	0.00	.30
202	"	" -.80	-1.00	16.00	" -2.30	0.00	.30
203	Brachialis	" 1.00	-.60	20.00	" 1.00	.30	3.00
204	Anconeus	" -1.60	-2.50	30.00	" -1.80	-.60	4.00
205	Supinator	" -1.30	-2.80	30.00	rad. .60	-.60	6.50
207	Prona teres	" -.50	3.50	29.00	rad. 0.00	-1.00	13.00
210	Ext. poll. longus	ulna -1.00	-.90	13.50	hand -.70	-2.50	3.00
211	Ext. poll. brevis	rad. -1.00	.80	16.00	" 0.00	-3.00	3.00
212	Ext. carp. rad. long.	arm -1.00	-2.80	28.00	" -.60	-1.30	3.40
213	Ext. carp. rad. brev.	arm -1.00	-3.00	29.00	" -.60	-.90	3.60
214	Ext. digit. common	arm -1.20	-3.00	29.50	" -.70	0.00	1.50
215	Ext. Carpi ulnaris	arm -1.40	-2.70	30.00	" -.30	2.50	3.00
216	"	ulna -1.00	-.60	12.00	" -.30	2.50	3.00
217	Flex. carpi ulnaris	arm -1.20	-3.00	29.50	" .70	1.80	1.20
218	"	ulna -.30	-1.00	9.00	" .70	1.80	1.20
219	Flex. carpi rad.	arm -.30	3.60	29.50	" .80	-1.80	2.50
220	Flex. poll. long.	arm -.50	3.50	29.50	" .70	-3.00	3.00
221	"	rad. .50	-.80	14.00	" .70	-3.00	3.00
222	Flex. digit. superf.	arm -.50	3.50	29.50	" .70	0.00	2.50
223	"	ulna .30	1.50	3.50	" .70	0.00	2.50
224	"	rad .80	0.00	7.50	" .70	0.00	2.50
225	Flex digit prof.	ulna .80	-.50	12.00	" .70	0.00	2.50
226	Abd. poll. long.	ulna -.90	-.90	12.00	" 1.00	-3.00	3.00
227	"	rad -.90	0.00	13.50	" 1.00	-3.00	3.00
228A	Trapezius	THX -4.50	0.00	8.00	scap -12.50	9.00	-8.00
B	"	scap -9.00	6.50	-2.70	" -12.50	9.00	-8.00
237	Brachio radialis	arm 0.00	-1.50	20.00	rad 0.00	-.60	21.00

TABLE IIIb. Joint coordinates.

no.	w.r.t.	$V1_x$	$V2_y$	$V3_z$	Description
		0	0	0	R. ankle jt.
1	R. leg	0	0	44.5	R. knee jt.
2	R. femur	0	0	41.5	R. hip jt.
3	Pelvis	0	16	0	L. hip jt.
4	L. femur	0	0	-41.5	L. knee jt.
5	L. leg	0	0	-44.5	L. ankle jt.
6	Pelvis	-1.0	8.0	10.4	Sacrum/L5 jt.
7	L5	0	0	3.5	L5/L4 jt.
8	L4	0	0	3.5	L4/L3 jt.
9	L3	0	0	3.5	L3/L2 jt.
10	L2	0	0	3.5	L2/L1 jt.
11	L1	0	0	3.5	L1/Thx jt.
12	Thorax	8.7	0	27.5	Thx/C7 jt.
13	C7	0	0	2.	C7/C6 jt.
14	C6	0	0	1.8	C6/C5 jt.
15	C5	0	0	1.8	C5/C4 jt.
16	C4	0	0	1.8	C4/C3 jt.
17	C3	0	0	1.8	C3/C2 jt.
18	C2	-1.0	0	1.8	C2/C1 jt.
19	C1	0	0	1.2	C1/Skull jt.
20	Thorax	12	-2.5	22.5	R. St/Cl jt.
21	Clavicle	1.5	-16.5	0	R. Acr/Cl jt.
22	Scapula	0	0	-4.0	R. Glenohum jt.
23	Humerus	0	1.0	-30.5	R. Elbow jt.
24	Humerus	0	-1.5	-31.0	R. Radio-hum jt.
25	Radius	0	0	-26.5	R. wrist jt.
26	Thorax	12.	2.5	22.5	L. St/Cl jt.
27	Clavicle	1.5	16.5	0	L. Acr/Cl jt.
28	Scapula	0	0	-4.0	L. glenohum. jt.
29	Humerus	0	-1.0	-30.5	L. elbow jt.
30	Humerus	0	1.5	-31.0	L. radio-hum. jt.
31	Radius	0	0	-26.5	L. wrist jt.

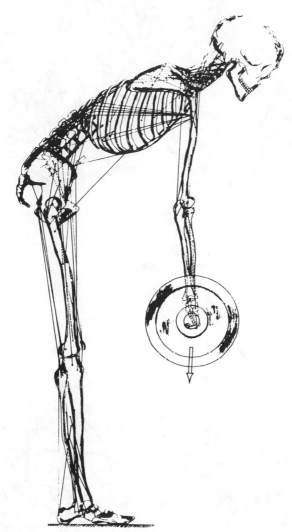

FIGURE 6-24b. The stooping posture.

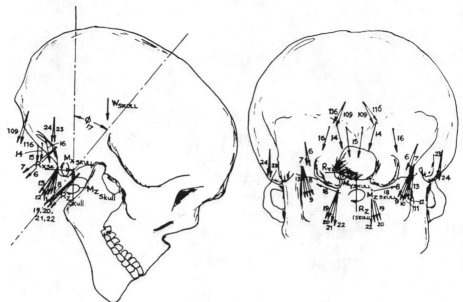

FIGURE 6-25.   Free body diagram of the skull.

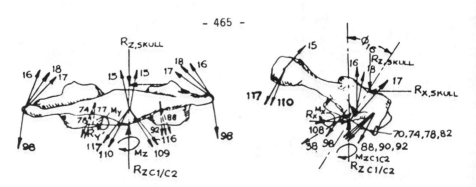

FIGURE 6-26.   Free body diagram of the first cervical vertebra.

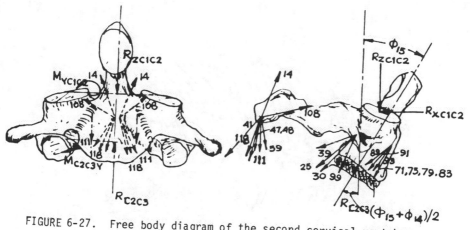

FIGURE 6-27. Free body diagram of the second cervical vertebra.

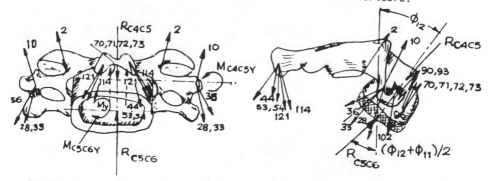

FIGURE 6-28. Free body diagram of the fifth cervical vertebra.

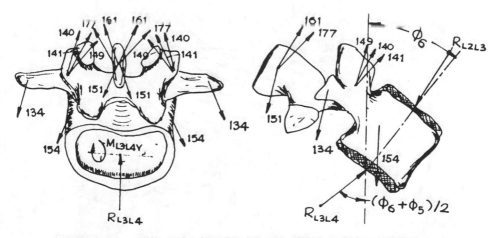

FIGURE 6-29. Free body diagram of the third lumbar vertebra.

279

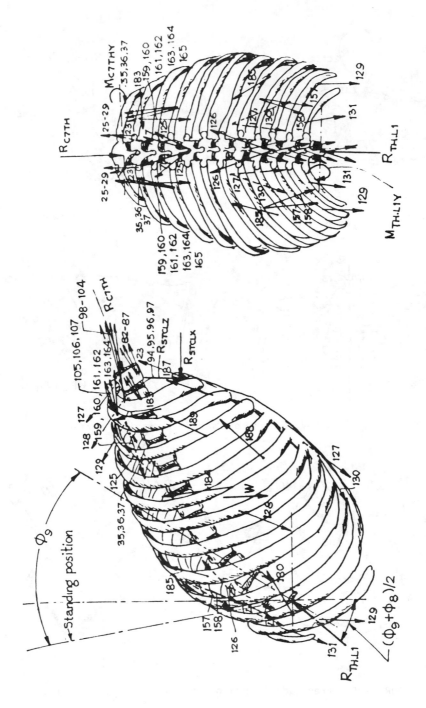

FIGURE 6-30. Free body diagram of the thorax.

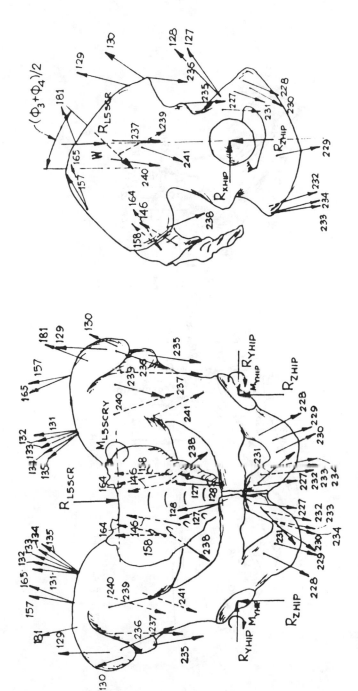

FIGURE 6-31. Free body diagram of the pelvis.

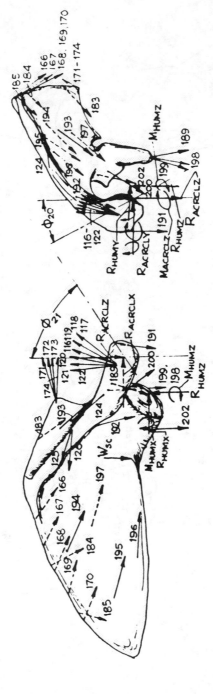

FIGURE 6-32. Free body diagram of the scapula.

TABLE IV. Weights and C.G.s of segments.

no.	G_x	G_y	G_z	Wt.	Description
1	4.5	0	-3.0	1.0	R. foot
2	-1.0	0	25.2	3.8	R. leg
3	2.0	-2.0	23.6	7.2	R. thigh
4	0	8.0	12.0	18.0	Pelvis
5	2.	2.	-17.9	7.2	L. thigh
6	-1.	0.	-19.3	3.8	L. leg
7	4.5	0	-3.0	1.0	L. foot
8	2.0	0.	9.5	8.9	Thorax
9	1.0	0.	3.	5.4	Skull
10	-2.	2.	-6.5	4.0	R. Scapula
11	0.	0.	-13.5	2.2	R. humerus
12	0.	-1.0	-11.5	1.2	R. Ulna + radius (w.r.t. ulna)
13	0.	0.	-8.	0.6	R. Hand (extended fingers)
14	-2	-2	-6.5	4.0	L. Scapula
15	0.	0.	-13.5	2.2	L. humerus
16	0.	1.0	-11.5	1.2	L. (U+R)
17	0.	0.	-8.	0.6	L. hand

Note: From Dempster*, W.T.,:
Hand: wrist axis to knuckle III, 50.6% to wrist axis
Forearm: Elbow axis to wrist axis, 43.0% of elbow axis
Upper arm: Glenohumeral axis to elbow, 43.6% to Glenohumeral
           axis

* Dempster, W.T. (1955) Space requirements of the seated operator.
Wright-Patterson base, Ohio.

TABLE Va. Segmental angles: stooping to lift weight.

No.	Body	ϕ_x	ϕ_y	ϕ_z
1	Right foot	0	0	0
2	Right leg	0	-6	0
3	Right thigh	0	-6	0
4	Pelvis	0	0	0
5	Left thigh	0	-6	0
6	Left leg	0	-6	0
7	Left foot	0	0	0
8	L5	0	40	0
9	L4	0	40	0
10	L3	0	45	0
11	L2	0	48	0
12	L1	0	50	0
13	Thorax	0	52	0
14	C7	0	65	0
15	C6	0	55	0
16	C5	0	55	0
17	C4	0	47	0
18	C3	0	47	0
19	C2	0	45	0
20	C1	0	45	0
21	Skull	0	45	0
22	R. clavicle	-25	0	10
23	R. scapula	10	52	-10
24	R. humerus	0	0	0
25	R. ulna	-12	0	0
26	R. radius	0	0	105
27	R. hand	-20	0	105
28	L. clavicle	25	0	-10
29	L. clavicle	-10	52	+10
30	L. humerus	0	0	0
31	L. ulna	12	0	0
32	L. radius	0	0	-105
33	L. hand	20	0	-105

TABLE Vb. Results for lifting weights in the stooping posture.

		Weight lifted		
		0 KG.	22.7 KG	45.4 KG
1	LONGISSIMUS CAPITIS	0.00	0.00	0.00
2	RECTUS CAPITIS POSTERIOR MAJOR	2.58	2.58	2.53
3	LONGUS CAPITIS	0.00	0.00	0.00
4	STERNOCLEIDOMASTOID	0.00	0.00	0.00
5	LONGISSIMUS CERVICIS	0.00	0.00	0.00
6	ILICCOSTALIS CERVICIS	4.61	12.05	19.53
7	SPLENIUS CERVICIS	4.57	4.50	4.44
8	SEMISPINALIS CERVICIS	0.00	0.00	0.00
9	MULTIFIDUS CERVICIS	0.00	0.00	0.00
10	ROTATORES CERVICIS	0.00	0.00	0.00
11	LONGUS COLLI	7.18	28.14	51.20
12	SCALENUS	11.67	30.50	49.41
13	TRAPEZIUS	10.78	28.75	46.79
14	RHOMBOIDEUS	0.00	0.00	0.00
15	DELTICIDEUS	9.68	19.53	29.42
16	LATISSIMUS DORSI	2.89	5.83	8.73
17	SERRATUS ANTERIOR	0.00	0.00	0.00
18	PECTORALIS MINOR	0.00	0.00	0.00
19	INFRASPINATUS	0.00	0.00	0.00
20	TERES MINOR	1.19	2.39	3.61
21	CORACOBRACHIALIS	0.00	0.00	0.00
22	BICEPS BRACHII	0.00	0.00	0.00
23	TRICEPS BRACHII	0.00	0.00	0.00
24	BRACHILIALIS	.27	2.63	4.91
25	ANCONEUS	.08	0.00	0.00
26	BRACHIORADIALIS	.007	4.88	0.11
27	SUPINATOR	.01	.33	.62
28	FLEXOR CARPI ULNARIS	.10	2.06	4.13
29	FLEXOR CARPI RADIALIS	.23	3.60	6.71
30	FLEXOR ROLLICIS LONGUS	0.00	.34	.92
31	FLEXOR DISITORUM SUPERFICIALIS	.27	6.08	11.77
32	FLEXOR DIGITORUM PROFUNDUS	0.00	0.00	0.00
33	ADDUCTOR POLLICIS LONGUS	0.00	0.00	0.00
34	EXTERNAL OBLIQUE	0.00	0.00	0.00
35	INTERNAL OBLIQUE	11.21	20.13	29.95
36	QUADRATUS LUMBORUM	0.00	0.00	0.00

TABLE Vb. (continued)

37	MULTIFIDUS LUMBORUM	37.59	60.10	78.06
38	ROTATORES LUMBORUM	12.4	20.01	26.11
39	ILIOPSOAS	10.34	16.71	21.87
40	ERECTOR SPINAE	89.22	151.11	205.93
41	GRACILIS	0.00	0.00	0.00
42	ADDUCTOR LONGUS	0.00	0.00	0.00
43	ADDUCTOR MAGNUS	0.00	0.00	0.00
44	SEMITENDINOSUS	0.00	0.00	0.00
45	SEMIMEMBRANOSUS	80.17	134.49	189.23
46	BICEPS FEMORIS (LONG HEAD)	72.68	132.61	192.73
47	BICEPS FEMORIS (SHORT HEAD)	33.82	67.05	100.77
48	QUADRICEPS FEMORIS	0.00	0.00	0.00
49	SARTORIUS	24.85	38.04	51.78
50	TENSOR FASCIAE LATAE	0.00	0.00	0.00
51	GLUTEUS MAXIMUS	0.00	0.00	0.00
52	ILIACUS	0.00	0.00	0.00
53	GLUTEUS MEDIUS	9.22	26.73	43.04
54	GLUTEUS MINIMUS	0.00	0.00	0.00
55	GASTROCNEMIUS	37.59	87.45	137.53
56	SOLEUS	0.00	0.00	0.00
57	TIBIALIS ANTERIOR	0.00	0.00	0.00
58	TIBIALIS POSTERIOR	0.00	0.00	0.00
59	EXTENSOR DIGITORUM LONGUS	0.00	0.00	0.00
60	EXTENSOR HALLUCIS LONGUS	0.00	0.00	0.00
61	FLEXOR DIGITORUM LONGUS	0.00	0.00	0.00
62	FLEXOR HALLUCIS LONGUS	0.00	0.00	0.00
63	PERONEUS LONGUS	0.00	0.00	0.00
64	PERONEUS BREVIS	0.00	0.00	0.00
65	PERONEUS TERTIUS	0.00	0.00	0.00

JOINT REACTIONS AND MOMENTS

SKULL/CI JOINT	RX	1.69	1.69	1.69
	RZ	9.97	9.97	9.97
	MY	0.00	0.00	0.00
C1/C2 JOINT	RX	3.33	3.33	3.33
	RZ	12.14	12.14	12.14
	MY	0.00	0.00	0.00

TABLE Vb. (continued)

C2/C3 JOINT	R	19.00	20.10	21.20
	MY	0.00	0.00	0.00
C3/C4 JOINT	R	19.60	20.70	21.80
	MY	0.00	0.00	0.00
C4/C5 JOINT	R	37.80	96.70	155.90
	MY	0.00	0.00	0.00
C5/C6 JOINT	R	40.40	105.70	171.20
	MY	0.00	0.00	0.00
C6/C7 JOINT	R	49.90	130.40	211.30
	MY	2.40	6.30	10.30
C7/THORAX JOINT	R	56.90	148.80	241.00
	MY	0.00	0.00	0.00
THORAX/L1 JOINT	R	205.80	331.40	432.40
	MY	0.00	0.00	0.00
L1/L2	R	205.00	330.10	430.60
	MY	0.00	0.00	0.00
L2/L3 JOINT	R	204.40	329.20	429.60
	MY	0.00	0.00	0.00
L3/L4 JOINT	R	203.90	328.30	428.40
	MY	0.00	0.00	0.00
L4/L5 JOINT	R	210.10	338.30	441.40
	MY	0.00	0.00	0.00
L5/SACRUM JOINT	R	195.20	314.60	410.70
	MY	0.00	0.00	0.00
STERNOCLAVICULAR JOINT	RX	.80	2.50	4.20
	RY	-9.50	-25.90	-42.30
	RZ	4.30	11.20	18.20
	MX	0.00	0.00	0.00
	MY	-1.30	-2.40	-3.40
	MZ	0.00	0.00	0.00
ACROMIOCLAVICULAR JOINT	RX	-3.20	-7.20	-11.30
	RY	2.50	6.60	10.70
	RZ	-3.70	-11.00	-13.30
	MX	0.00	0.00	0.00
	MY	-15.80	-40.00	-64.30
	MZ	0.00	0.00	0.00

287

TABLE Vb. (continued)

GLENOHUMERAL JOINT	RX	-3.50	-7.00	-10.60
	RY	1.10	2.20	3.40
	RZ	6.20	3.10	4.10
	MX	0.00	0.00	0.00
	MY	0.00	0.00	0.00
ELBOW JOINT	RX	0.00	-.20	-.40
	RY	0.00	0.00	0.00
	RZ	0.00	0.00	0.00
	MX	0.00	0.00	0.00
	MY	0.00	0.00	0.00
	MZ	0.00	0.00	0.00
WRIST JOINT	RX	0.00	-.40	-.70
	RY	.10	1.20	2.30
	RZ	0.00	0.00	0.00
	MX	-.70	-15.20	-29.00
	MY	0.00	0.00	0.00
	MZ	0.00	0.00	0.00
HIP JOINT	RX	-22.16	-40.19	-58.69
	RY	1.83	7.17	11.94
	RZ	217.62	377.27	536.47
KNEE JOINT	RX	-22.99	-39.48	-56.07
	RY	3.43	9.83	16.43
	RZ	277.64	497.36	718.99
ANKLE JOINT	RX	-.212	-.25	-.39
	RY	1.34	3.93	6.51
	RZ	72.69	133.83	195.23
	MX	0.00	0.00	0.00
	MY	0.00	0.00	0.00
	MZ	0.00	0.00	0.00
GROUND TO FOOT REACTIONS	RX	0.00	0.00	0.00
	RY	0.00	0.00	0.00
	RZ	72.50	94.98	117.78
X-COORDINATE OF BODY C.G.		6.89	12.07	15.26

288

segments of the upper extremities which are not so positioned would require six equations of equilibrium. The total number of equations is therefore 15 x 3 + 5 x 6 = 75. Inclusion of equations relating forces in muscles that wrap around segments and the reactions forces caused thereby would increase the total number of equations to 120. Free-body diagrams of the skull, C1, C2 and C5, thorax, L3, pelvis and the scapula are shown in Figs. 25 through 32 for illustration. There are 14 joints in the vertebral column formed by the skull, thorax and vertebrae. At each joint, the joint reaction force can be expressed in terms of its x and z components (due to symmetry there is no lateral component along y). However, the intervertebral joints constructed with fibrocartilaginous disks interposed between two adjacent vertebrae require special treatment. It will be assumed that the joint reaction force at such joints acts along the axis of the disk since these compressible elements act like springs transmitting forces along their axes. Since a vertebra has been treated as a rigid body with half the disk above and half the disk below, we will therefore assume that the intervertebral joint force is inclined at an angle equal to the average of the sagittal tilts of the two segments forming the joint (see Fig. 33). Since there are no disks at the joints between the skull and C1 and between C1 and C2, at these joints the reaction force will be assigned in terms of the two components, along x and along z respectively. At the remaining twelve intervertebral joints, however, the reaction force will be constrained to act as explained above.

At the 5 joints of the upper extremities (sternoclavicular, acromioclavicular, glenohumeral, elbow and wrist) the reaction forces will be expressed in terms of their three components along the three respective axes. Note that on the thorax, there will be two joints formed by the attachments of the left and right side shoulder girdles. Since only the right side girdle is analyzed (as the left side is identical to the right), the reactions at the right sternoclavicular joint will be simply doubled and directions reversed when their effect on the thorax is considered in its free body diagram.

At each joint we will also include the reaction moments in terms of the respective orthogonal components. Since the spine is like a planar problem for the stooping posture, the only moment that the joints can transmit will be about the y axis.

At the five joints of the upper extremities, the joint moments will be expressed in terms of all the three components. As in the case of the joint reaction, the sternoclavicular joint moment components will be doubled and their signs reversed when the free body diagram of the thorax is analyzed to account for the left and right side shoulder girdles. The total number of unknown variables including the muscle forces, joint reaction forces and moments, and reaction forces on the segments due to wrapping around of muscles thus turns out to be 324, while the number of equations available for solution is only 120. This implies there are innumerable feasible solutions, and as before we will attempt to obtain an optimal solution to this problem based on a preselected merit criterion. A complete description of the number of equations and the variables is given for the development of the "Total Model" in the following pages.

Merit criterion

For the lower extremities (Chapter V) it was shown that a criterion

Constraint on the direction of intervertebral joint reaction

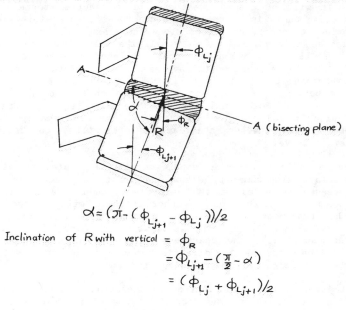

$$\alpha = (\pi - (\phi_{L_{j+1}} - \phi_{L_j}))/2$$

Inclination of R with vertical $= \phi_R$

$$= \phi_{L_{j+1}} - (\tfrac{\pi}{2} - \alpha)$$

$$= (\phi_{L_j} + \phi_{L_{j+1}})/2$$

FIGURE 6-33. Figure showing how the intervertebral reaction (joint load) R is constrained to act between the vertebra L_j and the vertebra L_{j+1} immediately below it.

290

based on minimizing the total muscle force and ligament effort, with suitable weighting factors, seems to yield results for muscle activity that closely matches the experimental electro-myographic data. The minimization of a merit function $U = \Sigma F + 4M$ when applied to the model gives good correlation between the theoretical muscle force results and the corresponding experimental EMG activity for such acts as forward and backward leaning, squatting and quasi-static walking.

We will therefore use this criterion for the application of the musculoskeletal model to the stooping posture, with slight modification. It is worth noting that in case of the lower extremities the joints are subjected to mostly compressive forces. In the upper extremities, however this is not necessarily the case since the extremities can be subjected to tensile forces. It was therefore, decided to include the reaction forces at the five joints of the upper extremities (with an arbitrary weighting factor of 2) in the merit criterion.

Results for the stooping posture

The muscle load sharing was studied for different magnitudes of weights lifted—0, 22.7 (50 lb), and 45.4 (100 lb) kg respectively. A strenuous exercise of this type tends to increase the intra-abdominal and the intra-thoracic pressures. Several investigators (see references 2-4) have hypothesized that the increase in the pressures in the thoracic and the lumbar cavities, which become like rigid-walled cylinders of air, and of liquid and semisolid materials respectively under the action of the trunk muscles is responsible for reduction of loads on the disks of the vertebral column. We will use the values for the abdominal pressures as reported by Morris et al (4) for the different loads lifted in this posture and analyze how much reduction in the disk loads occurs. The pressure is assumed to act on the diaphragm and its force value is obtained by multiplying the pressure with the area of the diaphragm (=80 sq. in.). This force acts equal and opposite on the thorax and on the pelvis.

The results for the forces required in the different muscles for the posture studied, as calculated from the model based on the preselected criterion are shown in Table Vb. It is worth noting that the variation in the forces developed in the back muscles and in the abdominal muscles with increasing weights lifted, shows excellent agreement with the EMG results in these muscles for the same act (Ref. 4). This is illustrated in Figs. 34 and 35 for the back muscles where it is evident that increase in the EMG output and in the theoretical force as obtained from the model, are both linear, and consequently the EMG-Force graph as shown in Fig. 36 is almost linear. Such relationship is also observed for the abdominal muscles as plotted in Figs. 34 through 36.

The intervertebral disk loads:

A typical variation of the intervertebral disk loads and the resulting stress (disc load/disc area) over the entire spinal column is illustrated in Fig. 37. The loads on the cervical disks are low compared to those on the lumbar disks, the maximum load of 439.6 kg occurring on the L4/L5 disk when the lifted weight is 45.4 kg. The high pressures on the lumbar disks correspond to the activity of the abdominal and the back muscles which are always more active than the head and neck muscles, the latter responsible for the loads on the cervical disks.

291

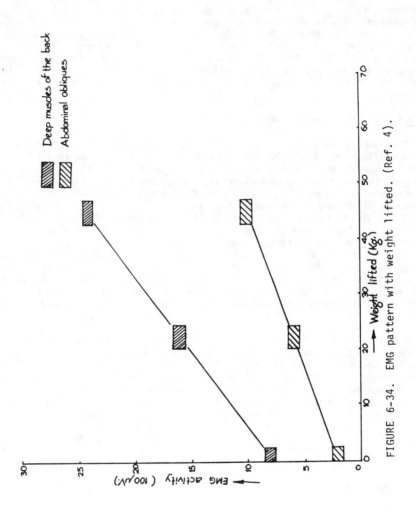

FIGURE 6-34. EMG pattern with weight lifted. (Ref. 4).

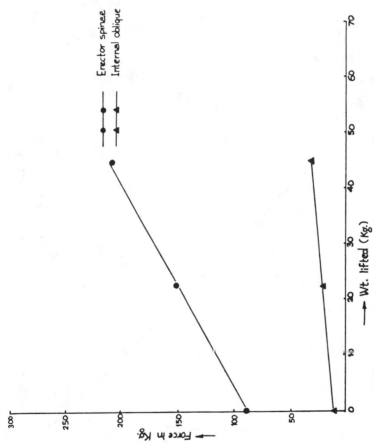

FIGURE 6-35. Results for internal oblique and erectores spinae from the model.

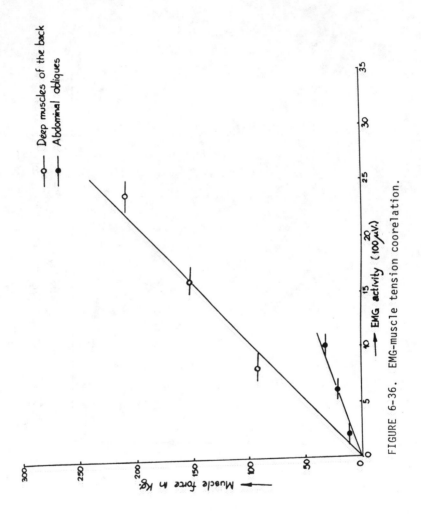

FIGURE 6-36. EMG-muscle tension coorelation.

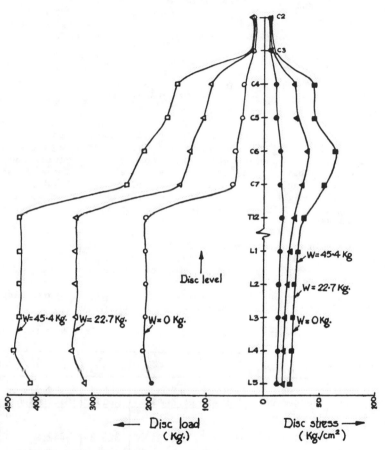

FIGURE 6-37. Variation of disc loads and pressure along the spinal column for the three weights lifted.

The values for these disk loads obtained from the model seem to agree well with the reported experimental measurements. Using discometry techniques Nachemson has [5-7] made in-vivo measurements on the intradiscal pressure (within the nucleus pulposis) in the lower lumbar disks (L3 and L4) in several subjects during a variety of activities, such as sitting standing, leaning forward while standing and sitting, reclining, carrying loads in the arms. For the standing posture this is expressed as:

$$P_{standing} \ (kg) = 15 + 2.1 \ W + 3.6 \ W \quad \sin \phi$$

where $P_{standing}$ = total load on the lumbar disk (L3 or L4)

ϕ = angle of forward leaning

W = body weight above the level measured,

including load lifted in the arms, if any.

In our case the angle of flexion in the stooping posture (angle between the vertical and the thorax) is ϕ = 52 degrees. Notwithstanding the fact that this is considerably large than the 20 degrees range as investigated by Nachemson, one can use the above expression and compute the loads on the L3 and L4 disks for the stooping act. The distribution of the total body weight above these vertebral levels can be obtained by using the results of Ruff [8], who suggests that 59% of the body weight acts on the L4 disk, and 57% on the L3 disk.

Accordingly, for an individual of 72.3 kg body weight, the weight W acting on the L4 disk will be 42.66 kg, and the weight acting on the L3 disk will be 41.21 kg. The results for the loads on these two disks corresponding to the three weights lifted are shown in Table VI. The calculated loads using Nachemson's formula and predicted loads using the musculoskeletal model show very close agreement (Fig. 40).

TABLE VI. Lower lumbar disc loads from the model and from Nachemson during stooping.

Weight lifted:	0		22.7 kg		45.4 kg	
Disc	From Nachemson	From Model	From Nachemson	From Model	From Nachemson	From Model
L3	214.3	203.9	330.0	328.3	442.5	428.4
L4	225.5	210.1	337.0	338.3	449.2	441.4

DEVELOPMENT OF "TOTAL THREE DIMENSIONAL MODEL"

After having modelled the various parts of the human musculoskeletal system individually and independently, it seems natural to develop one single model of the entire body. This will incorporate the muscles of the upper and lower extremities, the head and neck, the thorax and the abdomen. The model is essentially the same as that developed as "Two-part Model" described before. However, in order to make the model applicable to simulation of any general activity that the human body can perform, muscles to the left and right sides of the sagittal plane are modelled individually with the six general equations of equilibrium formulated for each segment. Certain alternations, however, were made in the model of the vertebral column to allow for six degrees of freedom for each vertebrae. Due to the presence of viscoelastic discs between the vertebral, which can undergo complex translational and rotational movements, the motion of the entire column (modelled as seven cervical segments, a single, rigid thoracic cage with its twelve thoracic vertebrae and five lumbar segments) is cumbersome to analyze. In addition, some muscles of the lower extremities were remodelled for better representation of their anatomical functions.

Modifications in the model of the spine

1. For each single vertebral unit the body-fixed set of axes is selected with its origin at the center of its inferior surface. Previously, however, a vertebral unit was assumed to be made up of half the superior disc, the rigid body in between and half the lower disc. The body-fixed coordinate axes were selected to have an origin at the center of the lower disc (see Fig. 38). All previously measured coordinates therefore are modified to conform to the new set of coordinate axes—simply by reducing the old z coordinate by an amount equal to half the disc thickness. Figure 39 shows how this adjustment is considered for each vertebral segment.

2. To obtain the locations of the intervertebral joints, with respect to ground axes a vertebra is first assumed to undergo three rotations about the x, y, z axes respectively (Fig. 39). This will yield the position of the intervertebral joint immediately above, provided, the superior disc was incompressible and an integral part of the vertebra below. However, since the disc is compressible, the true position of the intervertebral joint is obtained by superimposing incremental amounts Δx, Δy, Δz, which represent the components of the translational vector for the disc.

3. The reaction forces at the intervertebral joints are assumed to have three components along x, y, z respectively and not constrained to act along the axis of the disc as done for the case of stooping analyzed using the two part-model.

Modifications in the model of the lower extremities

Certain muscles of the lower extremities were remodelled for better representation of their anatomical functions. The sartorius, the tensor faxial latae, the gluteus maximus, and the semitendinosus which were previously modelled by individual lines connecting the points of origin to the points of insertions are now modelled as wrapping around the femur. Each is assumed to consist of two parts along with a reaction on the interposing structure at the approximate point of contact. The gluteus maximus is now modelled as two lines—a line connecting the pelvis to the femur (this is identical to the one in the previous model) and another joining the pelvis to the tibia at its lateral side representing the superior fibers of the muscle which insert into the iliotibial tract. The latter line is modelled to wrap around the femur in the proximity of the greater trochanter. The tensor fascial latae is also modelled to connect the pelvis to the tibia with a reaction on the

297

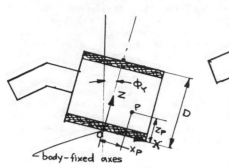

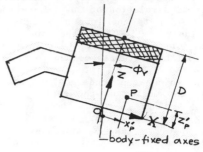

OLD MODEL
(TWO-PART MODEL)

X_P, Z_P = coordinates of point P

FIGURE 38

NEW MODEL
(Three-Dimensional Model)

X_P', Z_P' = coordinates of point P

where $X_P' = X_P$

$Z_P' = Z_P - h$

$2h$ = disc thickness

$h = 0.25$ for C2 through C7

$h = 0.40$ for L1 thru L5, and thorax

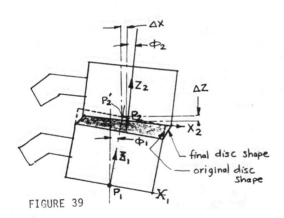

final disc shape

original disc shape

FIGURE 39

X_{P_1}, Z_{P_1} = coordinates of P_1 w.r.t. ground

X_{P_2}', Z_{P_2}' = coordinates of P_2' w.r.t. ground

X_{P_2}, Z_{P_2} = coordinates of P_2 w.r.t. ground

$$X_{P_2} = X_{P_2}' + \Delta X$$

$$Z_{P_2} = Z_{P_2}' - \Delta Z$$

where $\Delta X, \Delta Z$ = components of disc-compression vector

298

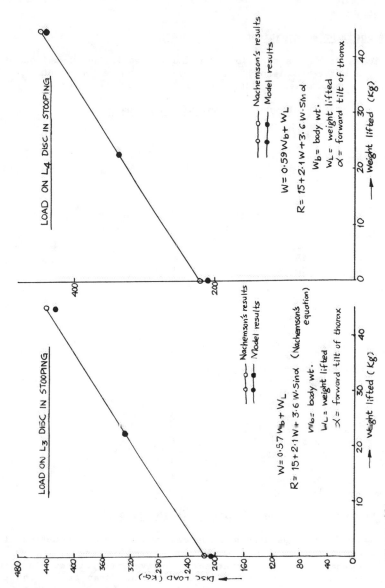

Figure 40.

LOAD ON L_4 DISC IN STOOPING

$W = 0.59 W_b + W_L$

$R = 15 + 2.1 W + 3.6 W \cdot \sin \alpha$

W_b = body wt.
W_L = weight lifted
α = forward tilt of thorax

○—○ Nachemson's results
●—● Model results

— Weight lifted (Kg)

LOAD ON L_3 DISC IN STOOPING

$W = 0.57 W_b + W_L$

$R = 15 + 2.1 W + 3.6 W \cdot \sin \alpha$ (Nachemson's equation)

W_b = body wt.
W_L = weight lifted
α = forward tilt of thorax

○—○ Nachemson's results
●—● Model results

— Weight lifted (Kg)

DISC LOAD (Kg.)

femur in the proximity of the greater trochanger of the femur. Its lower part connecting the femur to the tibia is assumed to be the same as that for the glutens maximus fibers inserting into the cliotibial tract. The sartorius and semitendinosus are also assumed to wrap around the distal and of the femur in their course to their insertions on the proximal part of the tibia.

Number of equilibrium equations:

a. There are now a total of 29 rigid bodies in the musculoskeletal model of the entire skeleton, each of which requires six equations of equilibrium to be completely described. The bodies are as follows:

Body	No. of Equations
Skull	6
C1	6
C2	6
C3	6
C4	6
C5	6
C6	6
C7	6
Thorax	6
Skull	6
L1	6
L2	6
L3	6
L4	6
L5	6
Pelvis	6
Right clavicle	6
Right scapula	6
Right humerus	6
Right ulna + radius	6
Right hand	6
Left clavicle	6
Left scapula	6
Left humerus	6
Left ulna + radius	6
Left hand	6
Right thigh	6
Right leg	6
Left thigh	6
Left leg	6

b. **Reactions from wrapping of muscles (right side);**

Each muscle that wraps around a segment requires three equations ($\Sigma F_x = 0$, $\Sigma F_y = 0$, $\Sigma F_z = 0$) to related the force in the muscle to the reactions at the point of interference. However, certain muscle groups may need only three if all the muscle components wrap around the same point on a particular segment. For example, the iliopsoas muscle has six components——5 from the psoas and one from the iliacus all of which are assumed to wrap about the same point on the pelvis. The reaction on the pelvis from each component can therefore be lumped together, so that for the entire group only three equations are needed to describe the wrapping. If a muscle group has n components, then in general one would need n x 3 equations if the components contact a different points.

Erectores spinae (7 components)	21
Latissimus dorsi (5 components)	15
Serratus anterior (middle part)	3
Trapezius (2 components)	6
Biceps brachii (long head)	3
Semitendinosus	3
Quadriceps femoris (4 components)	3
Sartorius	3
Tensor fasciae latae	3
Gluteus maximus	3
Psoas major (5 components) and iliacus	3
Total	66

c. **Reactions from wrapping of muscles (left side):**
These are identical to those of the right side, giving us an additional 66 equations.

The total number of equations thus is (29 x 6 + 66 + 66) = 306.
Number of variables

The total number of variables is 1132, the breakdown of which is as follows:

a. **Muscles of right side:**
There are 254 muscle forces on right side:

Muscle No.	Muscle or Muscle Group
1-5	Semispinalis capitis
6-7	Splenius capitis
8-13	Longissimuss capitis
14	Rectus capitis posterior major
15	Rectus capitis posterior minor
16	Obliquus capitis superior
17	Rectus capitis anterior
18	Rectus capitis lateralis
19-22	Longus capitis
23-24	Stenocleidomastoid
25-34	Longissimus cervicis
35-37	Iliocostalis cervicis
38-40	Splenius cervicis
41-46	Semispinalis cervicis
47-58	Multifidus cervicis
59-69	Rotatores cervicis
70-93	Longus colli
94-107	Scalenus
108	Obliquus capitis inferior
109-117	Trapezius
118-122	Rhomboideus
123-126	Levator scapulae
127	Rectus abdominis
128-129	External obliquus abdominis
130	Internal obliquus abdominis
131-135	Quadratus lumborum
136-146	Multifidus lumborum

147–151	Rotatores lumborum
152–156	Psoas major
157–165	Erectores spinae
166–172	Latissimus dorsi
173	Subclavius
174–176	Serratus anterior
177–179	Pectoralis major
180	Pectoralis minor
181	Trapezius
182–184	Deltoideus
185	Supraspinatus
186	Infraspinatus
187	Teres major
188	Teres minor
189	Subscapularis
190	Coracobrachialis
191,192	Biceps brachii
193–195	Triceps brachii
196	Brachialis
197	Anconeus
198	Supinator
199	Pronator teres
200	Extensor pollicis longus
201	Extensor pollicis brevis
202	Extensor carpi radiali
203	Extensor carpi radialis brevis
204	Extensor digitorum communis
205–6	Extensor carpi ulnaris
207–8	Flexor carpi ulnaris
209	Flexor carpi radialis
210–211	Flexor pollici longus
212–214	Flexor digitorum superficialis
215	Flexor digitorum profundus
216–217	Abductor pollicis longus
218	Trapezius
219	Brachioradialis
220	Sternothyroid
221–222	Sternohyoid
223	Gracilis
224	Adductor longus
225–226	Adductor magnus
227	Adductor brevis
228	Semitendinosus
229	Semimembranosus
230	Biceps femoris (long head)
231	Rectus femoris
232	Sartorius
233	Tensor fasciae latae
234	Gluteus maximus
235	Iliacus
236	Gluteus medius
237	Gluteus minimus
238	Biceps femoris (short head)
239	Vastus medialis
240	Vastus intermedius
241	Vastus lateralis
242	Gastrocnemius (medial)
243	Gastrocnemius (lateral)

244	Soleus
245	Tibialis anterior
246	Tibialis posterior
247	Extensor digitorum longus
248	Extensor hallucis longus
249	Flexor digitorum longus
250	Flexor hallucis longus
251	Peroneus longus
252	Peroneus brevis
253	Peroneus tertius
254	Tensor fasciae latae

b. Muscles of left side:

There are 254 muscles on the left side, identical to those on the right side as listed above.

c. Reactions from muscles wrapping (right side)

The reaction from a muscle wrapping around a body segment is assumed to have three components—R_x, R_y, R_z along x, y, z axes respectively. The muscle force F in the muscle is related to the reactions due to wrapping in the following manner:

$$F \ell_i + F \ell_j + R_x = 0$$

$$F m_i + F m_j + R_y = 0$$

$$F n_i + F n_j + R_z = 0$$

where   F= muscle force

ℓ_i, m_i, n_i= direction cosines of part i of muscle

ℓ_j, m_j, n_j= direction cosines of part j of muscle

R_x, R_y, R_z= components of reaction R from wrapping

Since R_x, R_y, R_z can take either +ve or −ve sign depending on their directions, one will need 6 variables to simulate the reaction. A muscle group with more than one component will require correspondingly more variables unless all the components wrap about the same point in which case only six variables will suffice. The variables from the muscles that wrap are as follows:

Muscle	No. of Variables
Psoas, iliacus	6
Erectores spinae (7 components)	42
Latissimus dorsi (5 components)	30
Serratus Anterior	6
Trapezius (2 components)	12
Biceps brachii	6
Semitendinosus	6
Quadriceps femoris (4 components)	6
Sartorius	6
Tensor fasciae latae	6
Gluteus maximus	6
Total	132

303

d. Reactions from muscles wrapping (left side

Since the muscles that wrap around the left side are identical to those on the right side, 132 additional variables are required as listed above.

e. Joint reactions forces (extremities)

At each joint the reaction force is assumed to have three components (R_x, R_y, R_z). Since these can take either a +ve sign or a =ve sign, six variables are needed at each joint to simulate the joint reaction force. A total of 96 variables is required thus for the following 16 joints.

Joint	No. of Variables
Right ankle	6
Right knee	6
Right hip	6
Left ankle	6
Left hip	6
Right sternoclavicular	6
Right acromioclavicular	6
Right glenohumeral	6
Right elbow	6
Right wrist	6
Left sternoclavicular	6
Left acromioclavicular	6
Left glenohumeral	6
Left elbow	6
Left wrist	6
Total:	96

f. Joint reaction moments (extremities)

Similar to the six joint reaction force variables for each joint listed above, a total of 96 variables representing joint reaction moments will be required.

g. Joint reaction force (intervertebral)

The joint reaction force at each of the intervertebral joints is represented by the three orthogonal components similar to the reactions at the joints of the extremities. Corresponding to the fourteen joints between the vertebral segments a total of 84 variables need to be included in the analysis:

Joint	No. of Variables
Skull/C1	6
C1/C2	6
C2/C3	6
C3/C4	6
C4/C5	6
C5/C6	6
C6/C7	6
C7/Thorax	6
Thorax/L1	6
L1/L2	6
L2/L3	6
L3/L4	6
L4/L5z	6
L5/sacrum	6
Total	84

304

h. Joint reaction moments (vertebral)

For each of the fourteen joints of the vertebral column listed above, we will need 6 variables to represent the joint moments, yielding a total of 84.

The number of unknown variables for the Total Model (508 muscle forces, 264 reaction forces from wrapping of muscles, 192 reaction forces and moments at the joints of the upper and lower extremities, and 168 reaction forces and moments at the joints of the vertebral column) thus turns out to be 1132.

The Merit Criterion

The criterion selected for the two-part model was:

$$\text{minimize } U = \Sigma F + 2(R_x + R_y + R_z)_{UE} + 4(M_x + M_y + M_z)_{UE} + 4(M_x + M_y + M_z)_{vert}$$
$$+ 4(M_x + M_y + M_z)_{LE}$$

that is, minimize sum of all muscle forces, twice the reaction components and four times the moment components at the joints of the upper extremities, four times the moments at the vertebral joints and four times the moments at the joints of the lower extremities.

THE COMPUTER PROGRAM

The model is solved using the FMPS, a subroutine package suitable for large problems. The cost of a single run with no basis specified can be enormous--the case of stooping forward during weight lifting for example, may run over $100.00. However, once the solution is obtained, and a basis can be specified, further computations, provided model application is not altered substantially (for example, the case of lifting different loads in the same posture through minor changes to the RHS or to the cost vector) may be more economical, with a cost reduction of around 80%.

MODEL APPLICATION TO STOOPING

It was decided to use the model to investigate the case when a person stoops forward in order to lift weights off the ground. Since a solution for this situation has been obtained previously using the Two Part Program, one can use these results to check the accuracy of the "Total Model". The results for the muscle load sharing during the stooping posture carrying a load of 100 lb (45.4 kg) in the hands, using the "Total Model and the old 'Two-part Model' are compared in Table VII. The segmental orientations that define the kinematic configuration of the body in this posture are described in the Table Ia.

I. It can be seen from Table VII comparing the results from both models that barring a few minor discrepancies the results are identical. Results from the total model show slight differences between the magnitudes of muscle forces on right and left sides of the body, possibly due to accumulation of round off error during computation. The longus colli on the right side is shown to be totally inactive, while on the left side it carries a force of 68.9 kg. In view of the fact that the longus colli muscle on the left and right side of the neck are placed entirely in the sagittal plane, so that in the model they are simulated

TABLE VII.	Right Side	Left Side	Old*	
longissimus capitis	0	0	0	
Rect. Cap. Post maj.	0	0	19.66	
Longus capitis	0	0	0	
SCM	13.3	13.3	0	NOTE
Longis cervicis	0	0	0	In old program
Ilio. cerv.	0.7	0	17.3	vertebral reactions constrained to act along vertebral axes.
Splenius cerv.	0	0	9.5	
Semispin cerv.	20.7	20.8	0	
Multif cerv.	0	0	0	
Rotat cerv.	0	0.2	0	
longus colii	0	68.9	4.9	
Scalenus	0	0	44.7	
Trapezium	43.3	43.3	40.3	Lifting weights (100 lb.)
Rhomboid	0	0	0	
Deltoid	29.9	29.9	31.7	
Lat Dorsi	8.1	8.1	9.5	
Serv. Ant.	0		0	
Pect. minor	6.9	6.9	0	*Old refers to the results obtained using the two-part model.
Infraspinatus	2.9	2.8	0	
Teres minor	0	0	3.9	

	Right Side	Left Side	Old
Coracobrachrilis	0	0	0
B.B.	10.5	10.6	0
T.B.	0	0	0
Brachialis	6.2	6.2	4.9
Brachorad	1.9	1.9	9.1
Supinator	2.1	2.1	0.6
Flex carp uln	6.2	6.3	4.2
Flex carp rad	4.1	4.0	6.7
Flex poll long	3.5	3.6	0.9
F.D. Superf	9.7	9.7	11.8
F.D. Prof	0	0	0
A.P. Longus	0	0	0
Int. oblique	0	0	29.9
Q. Lumb	61.9	62.5	0
Multif. lumb	0	0	77.8
Rotat lumb	7.1	7.1	26.1
Iliopsoas	0		21.91
Erector Spinae	119.9	119.04	20 6.
Sternohyoid	18.1	18.1	27.8
Adductor longus	1.8	1.6	0
Semimebranosus	151.8	154.3	168.1
B.F. (long)	159.3	159.9	163.3
B.F. (short)	43.8	43.3	27.4
O.F.	0	0	0
Sartorius	25.4	24.9	26.5
TFL	0		0
Gastr. (med.)	134.9	132.9	125.9
Gastr. (lat.)	0	0	11.6

Table VII (contd.)

		Right Side	Left Side	Old
Skull/C1	jt x	53.1		73.3
Ry = 0	z	70.3		102.2
C1/C2	x	56.5		79.0
	z	74.8		109.8
C2/C3	x	108.5		R=112.4
	z	119.9		
C3/C4	R_x	112.1		R=115.7
	z	125.9		
C4/C5	x	112.6		R=137.7
	z	126.0		
C5/C6	x	74.4		R=155
	z	94.9		
C6/C7	x	78.8		R=191.2
	z	87.7		
	M_y	0		9.3
C7/T	x	101.2		R=218.1
	z	92.7		
T/L1	x	131.6		R=432.2
	z	190.2		
L1/L2	x	183.2		R=430.5
	z	230.2		
L2/L3	x	219.2		R=429.4
	z	254.4		
L3/L4	x	224.1		R=428.2
	z	267.1		
L4/L5	x	224.2		R=441.2
	z	267.2		
L5/Sacrum	x	214.4		410.7
	z	250.5		

Table VII (contd.)

		Right Side	Left Side	Old
Ankle	x	-0.7	-0.3	-0.4
	y	1.6	-1.5	0.4
	z	192.7	190.6	195.3
M_x				
	y	0	0	0
	z			
Knee	x	-54.3	-54.3	-58.3
	y	18.1	-17.8	14.4
	z	561.9	561.9	569.4
M_x				
	y	0	0	0
	z			
Hip	x	-48.1	-48.2	-43.2
	y	-2.7	2.8	1.2
	z	379.9	382.2	418.9
M_x				
	y	0	0	0
	z			
STCL	x	-0.0	0	2.8
	y	36.4	-36.4	-28.9
	z	-6.8	-6.8	11.6
M_x				0
	y	0	0	-3.2
	z			0
ACrCL	x	-14.4	-14.4	-12.3
	y	19.9	-19.9	11.9
	z	-18.42	-18.5	-17.4
M_x		0	0	0
	y	-61.4	-61.4	-63.1
	z	U	U	U
G.H.	x	-2.4	-2.3	-11.4
	y	10.7	-10.7	3.7
	z	0	0	6.5
M_x				
	y	0	0	0
	z			
Elbow	x	0.4	0.4	-0.4
	y	0	0	0
	z	0	0	0
Wrist	x	-0.2	-0.2	-0.7
	y	-1.9	-1.9	2.3
	z	0	0	0
M_x		-30.7	30.7	-29.8
	y	0	0	0
	z	0	0	0

as two coincident spatial lines, these results appear reasonable. For practical purposes therefore one may assume that the longus colli of the left side and that of the right side are both active, each carrying a force of 68.9/2 or 34.45 kg.

II. It may be interesting to note that in the Total Model the reaction forces at the intervertebral joints have the freedom to select three arbitrary components along x, y and z respectively. In the two part program however, the reactions were constrained to have no lateral (y) component and to act along the "average inclination" of the axis of the intervertebral disc interposed between two adjacent vertebrae. This implies that, if the vertebra L_i is tilted about the medio-lateral y axis through an angle ϕ_{L_i} and the vertebra L_j above it is tilted through an angle ϕ_{L_j}, then the reaction $R_{L_i L_j}$ at the L_i/L_j joint has an inclination to the vertical of $(\phi_{L_i} + \phi_{L_j})/2$.

The accuracy of the Total Model is also substantiated when considering the results in Table VII where the intervertebral joint reactions are shown to have only a x and a z component and no y component. Also, there are no moments at any of these joints. One may compute the direction of the resultant joint reaction force and compare it to the actual sagittal tilts of the vertebrae. This is shown in Table VIII. The calculated sagittal orientations of the resultant interverbral joint force and their actual body tilts show excellent agreement leading to the conclusion that the directions of the resultant intervertebral joint reactions tend to follow the actual sagittal tilts of the bodies upon which they act, even in absence of any constraints to force them to follow such orientation.

III. The pattern of variation of the intervertebral joint reactions at different vertebral levels using the total model appears slightly different than that obtained using the 2-part program. As mentioned earlier, the body-fixed coordinate axes have been shifted to the inferior surface of the vertebra, omitting half the disc below. Since the sagittal inclination of the vertebrae are identical to those used previously, this will tend to alter the final positions of the points of muscle origins and insertions. The stooping posture simulated using the Total Model therefore is not truly identical to the two part model. This can effect the muscle force pattern as well as the joint reactions and moments.

MODEL APPLICATION TO STANDING

The geometrical configuration of erect standing is defined by the segmental angles in Table IXa. The results for muscle forces, joint reaction forces and moments are shown in Table IXb.

The round-off error during computation appears to be more prominent in this case than in the case of stooping. The y-components of the intervertebral joint reactions show small values instead of being zero. These range from a low of 0.11 kg to a high of 0.17 kg at the upper lumbar and lower cervical levels respectively. The moments, however are zero for all vertebral levels.

The values for joint reaction forces at the lower extremities for the right side are appreciably different than those for the left side. For example, at the right ankle joint the x, y, z components of reaction

310

TABLE VIII.. Calculated vertebral joint reaction orientations.

actual — calculated

No.	Body	$\phi_y°$	Joint	R_x	R_z	$R=\sqrt{R_x{}^2+R_z{}^2}$	$\phi_y'=\tan^{-1}\dfrac{R_x}{R_z}$
1	C2	45°	C2/C3	108.5	119.9	161.7	42.1
2	C3	47°	C3/4	112.1	125.9	168.6	41.7
3	C4	47°	C4/5	112.6	126.0	168.9	41.8
4	C5	55°	C5/6	74.4	94.9	120.6	38.1
5	C6	55°	C6/7	78.8	87.7	117.9	41.9
6	C7	65°	C7/Thorax	101.2	92.7	137.2	47.5
7	Thorax	52°	Thorax/L1	131.6	190.2	231.3	34.7
8	L1	50°	L1/L2	183.2	230.2	294.2	38.5
9	L2	48°	L2/3	219.2	254.4	335.8	40.8
10	L3	45°	L3/4	224.1	267.1	348.7	39.9
11	L4	40°	L4/L5	224.2	267.2	348.8	39.9
12	L5	40°	L5/Sacrum	214.4	250.5	329.7	40.6
13	C1	45°	C1/C2	56.5	74.8	93.8	38.5
14	Skull	45°	Skull/C1	53.1	70.3	88.0	34.7

TABLE IXa. Standing erect posture.

Body	no.	ϕ_x	ϕ_y	ϕ_z
Rt. foot	1	0	0	0
Rt. leg	2	0	5°	0
Rt. thigh	0	0	5°	0
Pelvis	4	0	0°	0
L. thigh	5	0	5°	0
L. leg	6	0	5°	0
L. foot	7	0	0	0
L5	8	0	18	0
L4	9	0	5	0
L3	10	0	-10	0
L2	11	0	-14	0
L1	12	0	-15	0
Thx	13	0	-15	0
C7	14	0	20°	0
C6	15	0	15°	0
C5	16	0	15°	0
C4	17	0	15°	0
C3	18	0	10°	0
C2	19	0	10°	0
C1	20	0	-5°	0
Skull	21	0	0	0
R. Cl	22	-5°	0	-30°
R. Scap	23	0	0	0
R. Hum	24	0	0	0
R. Ulna	25	-10°	0	0
R. Radius	26	5°	0	90°
R. wrist	27	0	0	90°
L. clav	28	5°	0	30°
L. scap	29	0	0	0
L. Hum	30	0	0	0
L. Ulna	31	10°	0	0
L. Radius	32	-5°	0	-90°
L. wrist	33	0	0	-90°
	34	0	0	0
	35	0	0	0

TABLE IXb. Standing erect with no weights in hand.

	Right side	Left side
Longissimus capitis	0	0
Rect cap post minor	9.44	9.44
Longus capitis	0	0
SCM	6.02	6.02
Longici cervicis	0	0
Ilio cerv.	0.1	0
Splenius cerv.	0	0
Semispin cerv.	6.46	5.39
Multif cerv	0	0.86
Rotat cerv	0	0.05
Longus colli	22.08	11.66
Scalenus	0.28	0
Trapezius	10.68	10.68
Lev. scap.	0.32	0.32
Deltoid	4.13	4.13
Lat. Dorsi	-	-
Serv Ant	0.53	0.53
Pect minor	-	-
Infraspinatus	-	-
Teres major	0.15	0.15
Coracobrachiels	0	0
B.B.	0	0
T.B.	0	0
Brachialis	0.35	0.35
Anconeus	0.10	0.10
Brachiorachiels	0.95	0.95
Ext carpi rad longus	0.25	0.25
Flex carpi uln	0.16	0.16
Flex carp rad	0	0
Flex roll. long	0	0
F.D. Superf	0.19	0.19
F.D. prof.	0	0
A.P. longus	0	0
Int. oblique	6.01	5.87

Stdg. Erect
Carrying 0 wt.

$\Sigma F + 2(Rx\ yz)UE.$
  + 4M at all jts.

U = 860.5

		Right side	Left side
Q. lumb			0.11
Multif lumb		6.02	5.91
Rotat lumb			
Iliopsoas			
Erectorspine		11.99	9.65
Sternohyoid		7.48	7.48
Adductor longus		11.37	8.33
Semimembranosus		0	0
B.F. (long)		0.1	5.69
B.F. (short)		0	0
D.F.		0	0
Santorius		2.49	4.68
TFL		2.94	1.89
Gastr. (med)		27.89	30.3
Gastr. (lat)		6.39	7.76
Rectus abdominis		3.91	4.18
Skull/Cl jt	x	-10.49	
	z	47.74	
C/C2	x	1.09	
	z	47.49	
C2/C3	x	8.69	
	z	80.37	
C3/C4	x	9.58	
	z	83.62	
C4/C5	x	9.41	
	z	84.19	$R_y = -0.27$
C5/C6	x	4.63	
	z	62.79	$R_y = -0.19$
C6/C7	x	2.6	
	z	54.4	$R_y = -0.13$
C7/T	x	3.36	
	z	60.0	$R_y = -0.13$
T/L1	x	0.5	
	z	75.6	$R_y = 0.12$

		Right side	Left side
L1/L2	x	1.56	
	z	64.98	
L2/L3	x	2.72	
	z	56.06	
L3/L4	x	5.43	
	z	47.85	
L4/L5	x	5.43	
	z	47.85	
L5/Sacrum	x	6.09	
	z	52.13	
Ankle	x	5.63	6.43
	y	.05	-0.13
	z	68.68	72.85
	M_x		
	y	0	0
	z		
Knee	x	5.39	6.18
	y	0.23	-1.46
	z	70.32	81.12
	M_x		
	y	0	0
	z		
Hip	x	1.77	2.28
	y	2.86	-1.84
	z	39.31	43.81
	M_x		
	y	0	0
	z		
STCL	x	0	0
	y	7.54	-7.54
	z	5.73	5.73
	M_x	0	0
	y	0	0
	z	-43.53	43.53
AcRCL	x	0.69	0.69
	y	0.29	-0.29
	z	-6.38	-6.38
	M_x	0	
	y	-11.89	-11.89
	z	-4.94	4.94

$R_y = 0.11$

$R_y = 0.11$

		Right side	Left side
G.H.	x	-0.37	-0.37
	y	0.18	-0.18
	z	0	0
	M_x		
	y	0	0
	z		
Elbow	x	-0.03	-0.03
	y	-0.18	0.18
	z	0	0
	M_x		
	y	0	0
	z		
Wrist	x	-0.013	-0.13
	y	-0.043	0.043
	z	0	0
	M_x		
	y	0	0
	z		

TABLE X.

	Right side	Left side
Rect Cap P. Minor	9.44	9.44
SCM	6.02	6.02
Ilio cerv.	0.1	0
Spl. cerv.	0	0
Rect cerv.	0	0.05
Semispin cerv	6.46	5.40
Multif cerv.	0	0.85
Longus colli	22.08	11.66
Scalenus	0.28	0
Trapezius	10.68	10.68
Lev. scap	0.32	0.32
Deltoid	4.13	4.13
Serv. Ant.	0.53	0.53
Teres major	0.15	0.15
Brachielis	0.35	0.35
Anconeus	0.10	0.10
Brachiorad ilis	0.95	0.95

ΣF_z constraint for right foot

$U = \Sigma F + 2 R_{xyz} \; UE$ Jts. $+ 4M$ at all jts

$U = 860.79$

	Right side	Left side
Ext. carp. red longus	0.25	0.25
Flex car Uln.	0.16	0.16
F.D. Superf.	0.19	0.19
Int. Oblique	6.01	5.87
Multif. Lumb	6.02	5.91
Erector Spinae	11.99	9.65
Sternohyoid	7.48	7.48
Rectus Abd.	3.91	4.18
O. Lumb	0	0.11
Adductor longus	9.21	10.23
Semimemb	0	0
BF (long)	4.11	1.69
(short)	0	0
Clf	0	0
Sartorius	4.45	3.06
TFL	2.04	2.58
Gastr (Med)	29.64	27.99
Gastr (Lat)	7.81	6.53
T. Post.	0.75	0

		Right side	Left side
Ankle	x	6.30	5.79
	y	-0.13	-0.07
	z	72.72	69.13
Knee	x	5.89	5.56
	y	1.22	-0.58
	z	78.66	72.56
Hip	x	2.19	1.94
	y	2.02	-2.50
	z	42.75	40.34

R.

m=0

m=0

m=0

OTHER RESULTS SAME AS IN TABLE IX.

are 5.63, 0.05 and 68.68 kg respectively as against corresponding values of 6.43, -0.13 and 72.85 kg at the left ankle joint. Analysis of the free body diagrams for the left and right foot show that the body C.G. falls at a point with x = 5.769 cm, and y = 8.002 cm. These results seem to agree with the values of C.G. calculated through static force equilibrium, x = 5.7691 cm, y = 8.000 cm. The ground-to-foot reaction in the vertical direction is 35.9307 kg for the right foot and 36.3694 kg for the left foot; both, however, sum up to the total body weight of 72.3 kg. This involves addition of one more constraint to the model. The results obtained are shown in Table X. It appears that despite the constraint, the muscle force patterns for left and right lower extremities are still not completely identical though some minor improvement is seen.

EFFECT OF CHANGE IN MERIT CRITERION

The criterion used so far has been to minimize the sum of all muscle forces, four times the moments at all joints (lower extremities, upper extremities and intervertebral joints) and twice the reaction forces at the upper extremities joints. It was decided to modify this criterion by including the reactions at the joints of the lower extremities (ankle, knee and hip on both sides) with a weighting factor of 2. The results for the standing erect posture with the new criterion are shown in Table XI. As expected, the results for the muscle forces for the upper extremities and the vertebral column do not change. What is worth noting however, is that notwithstanding the inclusion of the reaction forces at the joints of the lower extremities in the objective function, the muscle forces for the lower extremities and the joint reactions do not change significantly. Moreover, the individual muscle force pattern for the left and right extremity and the corresponding reaction forces at the joints of each side continue to exhibit the disparity that resulted from the old criterion. Modification of the objective function therefore does not seem to alleviate the bisagittal discrepancy.

STANDING ERECT WITH WEIGHTS IN HANDS

The case of standing erect with 22.75 kg in each hand (total weight of 45.5 kg or 100 lb.) was investigated using the model. The criterion was to minimize all muscle forces, four times the moments at all joints and twice the reactions at the upper extremities joints. The results are tabulated in Table XII. Since the abdominal pressure during such a maneuver is not known, it was assumed to be zero, the same as in standing erect with no weights in the hands. Significant increase in the forces in the back and abdominal muscles, in the muscles of the upper extremities, in the muscles of the neck, and in the muscles of the lower extremities can be seen from Table XII. Interestingly enough, the patterns of muscle force activity and joint reactions and moments are completely identical for the left and right sides of the body.

UNILATERAL WEIGHT LIFTING

The results for standing erect with 22.75 kg in right hand only and using the same criterion are shown in Table XIII. It was assumed that there was no change in the geometrical configuration during this maneuver——a factor, which in light of the tremendous magnitude of the weight lifted (50 lb) may need to be seriously regarded. An interesting muscle force pattern can be seen in Table XIII. When the load is in the right hand, the neck muscles and the abdominal obliques of the left side are more strongly active than the corresponding muscles on the right side. On the other hand, the muscles of the upper extremity and of the back are significantly more active on the side of the weight-carrying extremity s opposed to their counterparts on the contralateral side. The latter observation also holds in the case of the lower extremities where the muscles of the right extremity seems to be more stressed than those on the left side. Accordingly, the reaction forces on the hip, knee and ankle joints on the right side are substantially larger than those on the left side.

ACCELERATION EFFECTS ON MUSCLE LOAD SHARING AND DISC PRESSURE

The model developed to study the forces in the muscles of the head and neck, abdomen and back, and the upper extremities and used previously to study the case of forward stooping (that is, the 2-part model) while weight-lifting is applied to investigate the situation of sudden impact on a seated person in motion. Various instances of studies detailing the impact phenomena and their analyses through rigorous mathematics of lumped-mass oscillatory systems can be readily pointed out. The present musculoskeletal model however could be the only tool of its kind to provide better insight into the response of the muscles in maintaining the stability and integrity of the skeletal structure through simulation of the impact situation. In the seated posture it is assumed that the part of the body below, and inclusive of the pelvis is grounded. The arms could be either grounded, if the forearm and hand are gripping the armrests or hanging free at the sides. An analysis of both situations will be attempted.

The forces caused by sudden stopping when in motion in a horizontal direction, are the inertia forces acting on the various segments. These are obtained my multiplying the mass (w_i/g) of the segment i with the acceleration prior to impact, and they act at the centers of gravity of the segments in a direction opposite to that of the motion. Three acceleration were selected for analysis:

a. acceleration = 0
b. acceleration = +g
c. acceleration = -g

The program used is essentially identical to the one used for the study of stooping posture. However, since the pelvis is grounded, its three equations of equilibrium are ($\Sigma F_x = 0$, $\Sigma F_z = 0$, $\Sigma M_y = 0$) excluded. The criterion selected was to minimize the sum of all muscle forces, four times the moments at all the joints (intervertebral, and upper extremities) and twice the reaction forces at the joints of the upper extremities. The geometrical configuration for the spine and the upper extremities is shown in Fig. 41 and is the same as that in the standing erect posture (Table IXa).

RESULTS

The results for a seated subject with his arms hanging by the side, for no motion (F_{xi} on segment i = 0), for backward motion ($F_{xi} = w_i$) and for forward motion ($F_{xi} = -w_i$) are shown in Tables (XIV-XVI), respectively. Table XVII shows the results for the static stooping posture (weight lifted = 0) from the same model.

319

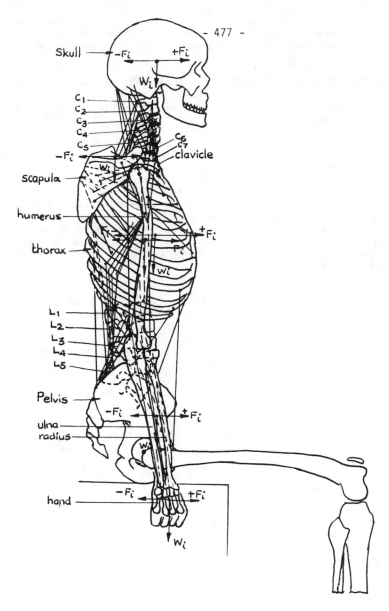

Figure 6-41. The seated posture.

TABLE XI.

	Right side	Left side
Rect cap post minor	9.43	9.44
SCM	6.03	6.02
longisimus cerv.	0.23	0
Ilio. cerv.	0.19	0
Semi. cerv.	5.45	6.00
Mult. cerv.	0	0.31
Rotat cerv.	0.51	0
longus colli	21.32	12.44
scalenus	0.27	0
Trapezium	10.67	10.67
Lev. scap	0.32	0.32
Deltoid	4.13	4.12
Lat dorsi	0	0
Serv Ant	0.53	0.53
Teres Maj	0.15	0.15
Brachielis	0.35	0.38
Anconeus	0.09	0.09
Brachioradrelis	0.95	0.92
Ext. carpi rad longus	0.25	0.25
Flex carpi uln	0.16	0.16
F.D. Superf.	0.19	0.19
Int. Oblique	6.03	5.87
Multif lumb	6.04	5.91
Q. lumborum	0	0.12
Erect spinae	12.06	9.68
Rect. abdominis	3.98	4.27
Sternohyoid	7.51	7.46

Standing erect

0 wt.

$\Sigma F + 2(R_{xyz})$ UE, LE

+ $4M$ at all jts.

$U = 1665.56$

TABLE XI

(continued)

		Right side	Left side
Adductor longus		13.72	7.06
Semimembranous		0	0
B.F. (long)		0	1.42
B.F. (short)		0	0
O.F.		0	0
Santorius		0	4.25
TFL		0	0
Gastr. Med.		27.87	22.19
Gastr. Lat.		11.24	11.04
Glut. max		0	1.61
Glut. med.		8.39	
Ankle	x	6.24	5.82
	y	0	0.28
	z	73.93	67.61
	M_z	2.50	
Knee	x	6.24	5.37
	y	0	-0.74
	z	70.13	69.34
Hip	x	1.99	2.22
	y	4.25	-0.89
	z	45.43	37.48

OTHER RESULTS SAME AS FOR $\Sigma F + 2 R_{xyz}$ $UE + 4M$

322

TABLE XII.

	Right side	Left side
R.C. Post min	35.16	35.16
SCM	25.68	25.68
Semispinal cerv	36.96	36.97
Longus colli	82.69	36.15
Trapezius	40.33	40.33
Rect scelenus	0	0
lev scap	0	0
Deltoid	23.45	23.45
Serv Ant	3.57	3.57
Teres major	0.75	0.75
Brachielis	9.24	9.24
Anconeus	0	0
Ext. carp. Uln.	5.73	5.73
Brachioradlis	0	
Ext. carp R. Brevis	1.86	1.86
Flex carp Ul .	0	0
Flex carpi Rad.	2.75	2.75
F.D. Superf.	9.12	9.12
D. Lumb	0	0
B. Brachll	4.22	4.22
Ext. Poll. long	4.09	4.09
Erector spinae	28.92	23.86
Multif lumb	15.02	14.99
Int. oblique	14.32	14.32
Rect. Abd.	9.79	9.79
Sternohyoid	31.22	31.22
Gracilis	9.45	9.45
Adductor L.	12.69	12.69
Semimemb.	0	0

Standing Erect

22.75 kg in each hand
(or 45.4 kg total wt.)

$\Sigma F + 2 R_{xyz}$ UE

+ 4M at all jts.

$U = 2931.31$

323

TABLE XII

(continued)

		Right side	Left side
B.F. (long hd)		0	0
(short hd)		0	0
Sartorius		4.13	4.13
Gastr. (M)		36.72	36.71
(lat)		16.96	16.96
TFL		2.52	2.52
Ankle	x	9.06	9.06
	y	-0.81	0.81
	z	110.75	110.75
Knee	x	10.01	10.01
	y	.06	-.06
	z	122.8	122.80
Hip	x	4.08	4.08
	y	2.06	-2.06
	z	74.06	74.06
StCL	x	0	0
	y	27.89	-27.89
	z	23.09	23.09
AcRCL	x	3.52	3.52
	y	1.36	-1.36
	z	-21.79	-21.79
G.H.	x	-2.05	-2.05
	y	2.38	-2.38
	z	0	0

StCL:
$$\overset{R}{M_z} = -151.33, \quad \overset{L}{M_z} = 151.33$$

AcRCL:
$$\overset{R}{M_y} = -43.16, \quad \overset{L}{M_y} = -43.16$$
$$M_z = -19.39, \quad M_z = 19.39$$

TABLE XII

(continued)

		Right side	Left side
Elbow	x	-0.1	-0.1
	y	-0.49	0.49
	z	0	0
Wrist	x	-0.78	-0.78
	y	-0.87	-0.87
	z	0	0
Sk/C1	x	-39.99	
	y	0	
	z	168.49	
C1/C2	x	1.98	
	y	0	
	z	173.96	
C2/C3	x	28.68	
	y	0	
	z	289.77	
C3/C4	x	32.24	
	y	0	
	z	302.83	
C4/C5	x	38.45	
	y	0	
	z	326.29	
C5/C6	x	20.09	
	y	0	
	z	245.66	
C6/C7	x	14.51	
	y	0	
	z	223.59	

M=0 for all vert jts.

TABLE XII

(continued)

		Right side	Left side
C7/T	x	17.67	
	y		
	z	244.90	
Th/L1	x	1.14	
	y	0.14	
	z	186.56	
L1/L2	x	3.706	
	y	0.13	
	z	160.50	
L2/L3	x	6.49	
	y	0.12	
	z	138.59	
L3/L4	x	13.39	
	y	0	
	z	118.03	
L4/L5	x	13.39	
	y	0	
	z	118.03	
L5/Sacrum	x	15.04	
	y	0	
	z	128.59	

TABLE XIII.

	Right side	Left side
R.C. Post. Maj.	34.58	0
R.C. Post. Min.	0	39.99
Obliq. C. Sup	0.56	0
R.C. Lat.	4.31	0
S.C.M.	3.74	13.82
Semispin. Cervicis	0	26.63
Multif. Cervicis	3.36	33.65
Rotatores Cervicis	0	1.03
Longus colli	36.05	0
Scalenus	0	40.03
Trapezius	39.78	8.8
levator scapulae	0.28	0.84
Int. oblique	7.09	22.82
Ext. oblique	0.00	32.49
Q. Lumborum	9.45	16.11
Multif lumborum	22.67	16.24
Erectores spinae	37.16	8.06
Subclavius	0	4.46
Deltoid	21.38	2.85
Teres Major	1.15	0.25
Coracobrachialis	0	1.18
D. Brachii	0.27	0.00
Brachialis	8.63	0.35
Anconeus	0	0.09
Ext. poll. longus	4.31	0
Ext. carp. Rad. Brevis	1.47	0.24
Ext. carpi ulnaris	5.86	0
Flexi carpi unlarius	0	0.14

Standing
Erect

22.75 kg in right
hand only

$\Sigma F + 2R_{xyz} UE + 4 M_{xyz}$

$U = 2152.12$

TABLE XIII

(continued)

		Right Side	Left Side
Flexicarpi radialis		2.78	0
Flex poll longus		0	0.03
Flex digit superg		9.14	0.19
Brachioradielis		0	0.93
St hyoid		51.01	0
Gracilis		0	3.67
Adductor longus		15.04	6.49
Sartorius		13.49	2.06
Gastrocnemius (m)		36.85	18.46
(L)		17.70	7.11
Tibialis Posterior		19.29	0
T.F.L.		0	1.54
Sk/C1	x	-31.89	
	y	12.44	
	z	136.2	
C1/C2	x	-4.44	
	y	13.59	
	z	138.17	
C2/C3	x	15.69	
	y	32.83	
	z	156.84	
C3/C4	x	19.06	
	y	22.82	
	z	175.56	
C4/C5	x	20.65	
	y	15.4	
	z	182.66	

TABLE XIII

(continued)

		Right side	Left side
C5/C6	x	14.63	
	y	3.96	
	z	159.43	
C6/C7	x	20.68	
	y	1.82	
	z	173.82	
C7/Thx	x	21.93	
	y	1.63	
	z	186.54	
Thx/L1	x	-4.97	
	y	4.14	
	z	175.79	
L1/L2	x	-1.48	
	y	3.87	
	z	152.68	
L2/L3	x	2.42	
	y	3.68	
	z	132.81	
L3/L4	x	13.71	
	y	0	
	z	120.73	
L4/L5	x	13.71	
	y	0	
	z	120.73	
L5/Sacrum	x	15.38	
	y	.0	
	z	131.53	
Hip	x	5.94	1.49
	y	0.79	-1.23
	z	84.07	27.18

TABLE XIII

(continued)

		Right side	Left side
Knee	x	7.67	4.33
	y	3.06	-0.06
	z	130.77	55.83
Ankle	x	13.49	4.0
	y	-5.99	0.31
	z	139.22	50.48
St Cl	x	2.98	0
	y	27.67	-10.04
	z	1.57	10.69
	M_x	79.45	0
	y	0	0
	z	-164.19	42.49
Acr Cl	x	0.21	0.45
	y	1.82	-0.23
	z	-23.83	-7.33
	M_x	0	0
	y	-14.43	-4.37
	z	0	5.34
Glenohum	x	-2.19	-0.43
	y	2.0	-0.01
	z	0	0
	M_x	0	0
	y	0	0
	z	0	0
Elbow	x	-0.001	-0.02
	y	-0.16	0.16
	z	0	0
	M_x	0	0
	y	0	0
	z	0	0
Wrist	x	-0.76	-0.01
	y	-0.91	0.04
	z	0	0
	M_x	0	0
	y	0	0
	z	0	0

$F_i = 0$ $U = 1370.87$

TABLE XIV*   Sitting
hands at side
with stdg erect
posture

F_i	value	F_i	value	Joint			Joint		
6	1.28	153	4.62	St/C1 x	-6.59		C2/C3	R	44.10
14	13.34	130	0.27	z	47.50			M	
27	5.92	110	0.67	C1/C2 x	-6.17		C3/C4	R	55.93
39	0.31	111	13.08	z	48.90			M	
40	3.99	189	0.85				C4/C5	R	68.89
84	2.13	154	38.42	St/C1 x	-3.98	0		M	
87	0.73	229	1.92	y	-9.82	0	C5/C6	R	68.89
106	5.08	204	0.18	z	-1.42	34.82		M	
107	10.05	194	0.16	Ar/C1 x	2.05		C6/C7	R	77.88
208	8.49	210	0.33	y	-0.14	-20.26		M	5.26
182	1.95	218	0.24	z	-6.68	-13.10	C7/Thx	R	90.55
166	1.68	216	0.04	GH x	-0.33	0		M	
155	19.98	137	5.96	y	0	0	Thx/L1	R	169.47
152	10.67	134	17.45	z	0	0		M	
141	16.04	156	23.66	Elbow x	-0.08	0	L1/L2	R	142.94
142	55.55	184	0.08	y	0.03	0		M	
158	9.86	190	3.20	z	0.28	-0.07	L2/L3	R	128.87
				Wrist x	-0.08	0		M	
				y	0.08	0	L3/L4	R	121.93
				z	0	-0.02		M	36.87
							L4/L5	R	110.73
								M	
							L5/Sacrum	R	96.83
								M	

TABLE XV

Sitting $F_i = w_i$

$U = 4347.84$

No.	Value	No.	Value	Joint		Value		Joint		Value
6	3.04	154	6.43	Sk/C1	x	-14.84		C2/C3	R	81.59
14	22.41	229	0		z	74.09			M	
27	10.66	204	0	C1/C2	x	-14.25		C3/C4	R	100.79
39	6.15	194	19.28		z	77.79			M	
40	6.34	210			R	M		C4/C5	R	124.15
89	0	218	0.23	St/CL	x	-14.07			M	
81	1.32	216	1.32		y	-22.34	95.90	C5/C6	R	124.15
106	9.16	137	42.74		z	-15.15			M	
107	16.21	134	0	Ar/C1	x	7.56		C6/C7	R	140.34
208	0	156	0		y	0.81	-153.65		M	9.48
182	0	184	0		z	-13.92	-2.48	C7/Thx	R	151.48
166	0	190	23.78	Hum	x	-23.75			M	
155	70.03	80	3.59		y	13.60		Thx/L1	R	305.47
152	51.62	168	26.98		z	4.34			M	
141	105.48	167	13.13	Elbow	x	2.97		L1/L2	R	300.66
142	139.87	201	13.61		y	0.04			M	
158	0	205	1.67		z	13.29	4.44	L2/L3	R	251.62
153	18.32	151	2.95	Wrist	x	0.63			M	
130	0.09	148	0.67		y	0.21		L3/L4	R	307.02
110	2.47	193	12.44		z	0.93	-0.51		M	92.84
111	23.29	178	6.07					L4/L5	R	278.79
189	0	209	12.76						M	
								L5/Sacrum	R	238.15
									M	

TABLE XVI Sitting $F_i = -W_i$

$U = 3981.49$

No.	Value	No.	Value	Joint		Value		Joint		Value
6	0.78	137	0	Sk/C1	x	-2.66		C2/C3	R	43.41
14	10.60	134	104.97		z	53.98			M	
27	2.66	156	130.33	C1/C2	x	-2.02		C3/C4	R	43.04
39	0	184	2.53		z	52.99			M	
40	0	190	0				M	C4/C5	R	47.50
89	0.94	80	0	St/C1	x	5.43	-33.76		M	
81	0	168	0		y	-1.78	0	C5/C6	R	47.50
106	3.50	167	0		z	1.57	0		M	
107	7.18	201	0	Ar/C1	x	5.89	0	C6/C7	R	53.69
208	10.36	205	0		y	0	0		M	3.63
182	8.42	151	0		z	-2.22	-65.35	C7/Thx	R	64.04
166	10.99	148	0	Hum	x	-37.96	0		M	
155	24.19	193	6.57		y	47.78	0	Thx/L1	R	631.65
152	0	178	0		z	24.71	0		M	
141	16.85	209	0	Elbow	x	-1.8	0	L1/L2	R	488.64
142	178.02	15	0.66		y	-1.24	21.64		M	
158	65.81	1	0.53		z	10.36	-7.70	L2/L3	R	488.24
153	16.87	8	1.56	Wrist	x	-1.09	0		M	
130	0	140	25.35		y	0.01	0	L3/L4	R	390.76
110	0	135	8.78		z	1.09	0		M	
111	6.19	24	2.12		y	0.01	0	L4/L5	R	354.84
189	4.02	197	22.53		z	1.09	0		M	
154	207.63	183	5.16					L5/Sacrum	R	332.50
229	0	192	47.43						M	
204	0	203	12.22							
194	0	221	0.11							
210	0	211	1.32							
218	0	227	0.34							
216	0									

TABLE XVIII. Intervertebral disk loads (kg) in a 72.3 kg subject.

Level of Disk	Seated Posture			Stooping	
	No acceleration $(F_i=0)$	Forward acceleration, 1 g $(F_i=w_i)$	Backward acceleration, 1 g $(F_i=-w_i)$	Weight lifted=0 kg	Weight lifted=45.5 kg
C2/C3	44.10	81.59	43.41	33.37	112.40
C3/C4	55.93	100.79	43.04	34.36	115.73
C4/C5	68.89	124.15	47.50	40.89	137.67
C5/C6	68.89	124.15	47.50	44.34	154.96
C6/C7	77.88	140.34	53.69	54.71	191.21
C7/Thorax	90.55	151.48	64.04	62.39	218.08
Thorax/L1	109.47	305.47	631.65	205.82	432.23
L1/L2	142.94	300.66	488.64	204.99	430.49
L2/L3	128.87	251.62	488.24	204.48	429.41
L3/L4	121.93	307.02	390.76	203.91	428.22
L4/L5	110.73	278.79	354.84	210.11	441.24
L5/Sacrum	96.83	238.15	332.50	195.22	410.71

The effect of alteration of the objective function so as to include twice the reactions at the intervertebral joints (Table XVIII) and changing it simply to ΣM at all joints (Table XIX) was also tried for the case of forward motion ($F_{xi} = -w_i$).

The variations in the reactions at the intervertebral joints along the length of the spine is plotted (Fig. 41) for different cases studied. The maximum load falls on the disc between the thorax and the first lumbar vertebra with the lower loads occurring at discs above and below this level. It may be worth noting that the reactions have been constrained to act along the axis of the disc, or inclined in the sagittal plane at an angle average of the sagittal tilts (ϕ_y's) of the vertebrae between which the disc lies. In general, the cervical spine is characterized by low reaction forces, compared to those in the lumbar spines the maximum occurring at the transition from the thoracic to the lumbar. The load on the disc between the fifth lumbar vertebra and the sacrum (pelvis) is seen to be usually lower than those on the upper lumbar levels. In the case of stooping however, the distribution of the loads on the lumbar discs is most uniform (Fig. 41), the discs below the thorax being subjected to more or less a constant load.

Sudden stopping during forward motion ($F_{xi} = -w_i$) seems to cause higher disc loads than during either sitting still ($F_{xi} = 0$) or backward motion ($F_{xi} = w_i$). For example, the load on the Thorax/L1 disc when sitting still in a seated posture is only 169.5 Kg. Sudden stopping during backward motion increases this value to almost twice (305.5 Kg) while similar impact during motion in opposite direction increases the load almost fourfold to 631.7 Kg.

An interesting result seen from Fig. 41 is that while the loads on the discs below the thorax (that is, lumbar discs) are generally highest for forward motion, the loads on the cervical discs are generally lowest for the same case. In fact the cervical discs appear to be subjected to lesser extent during stopping while in forward motion than during either sitting still or stopping in backward motion. In conclusion, while sudden impact to motion in one direction can be pernicious to the neck, impact during reverse motion can be harmful to the lower back.

The inclusion of intervertebral reaction forces in the objective function does not effectuate significant reduction. As can be seen in Fig. 42 the maximum reduction occurs in the value of the reaction force on the Thorax/L1 disc (for the case of forward motion). The load is reduced from its original value of 631.7 kg to 529.9 kg or a reduction of only 16.1%. The reduction in the value of the reaction force on the C6/C7 disc (which was originally 53.7 kg) to the new value of 33.5 kg or through 37.5% is more substantial. Any attempt to reduce the reactions however implies consequential increase in the joint moments as evidenced in Table XVIII.

The objective function $U = \Sigma M$ at all joints may given a fair indication of the highest possible joint reaction forces. Understandably, in Table XIX the intervertebral disc forces are boosted to astronomical figures: the maximum (1150.4 kg) occurring at the C7/Thorax disc rather than at the Thorax/L1 disc as obtained in earlier results. The loads on the cervical discs are unbearably high and significantly higher than those at the lumbar levels. Since the moments at the joints of the upper extremities are now totally eliminated through

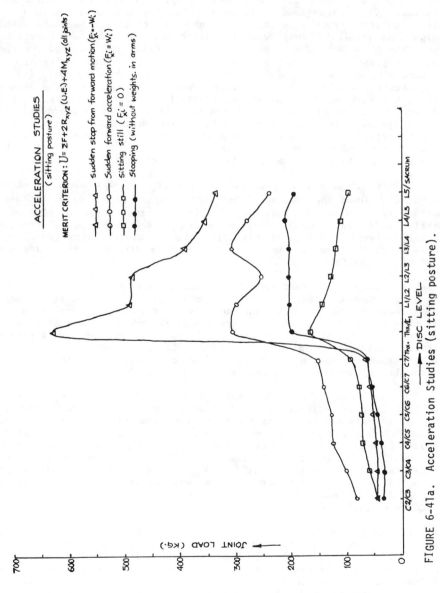

FIGURE 6-41a. Acceleration Studies (sitting posture).

336

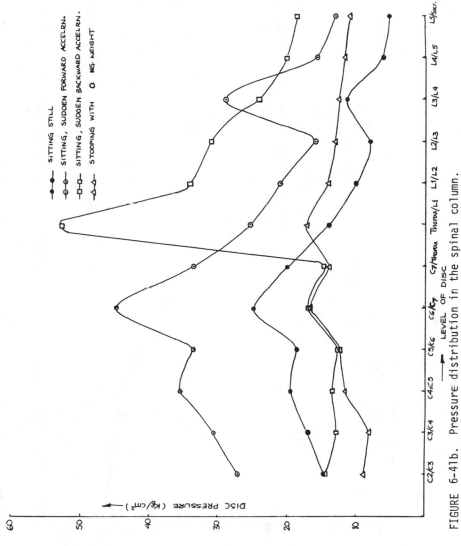

FIGURE 6-41b. Pressure distribution in the spinal column.

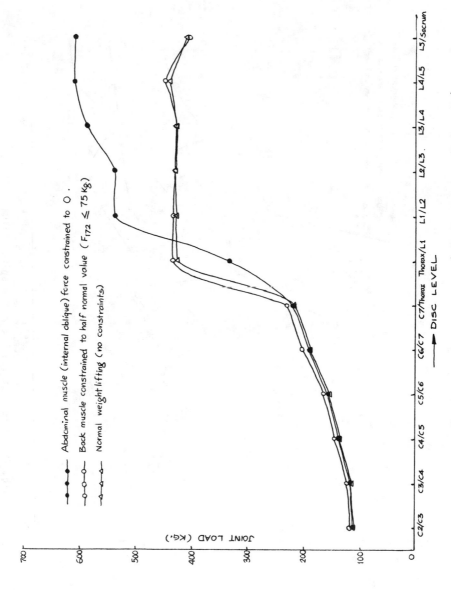

FIGURE 6-42. Effect of changing merit criterion, (sitting posture, $F_x^{\,i} = -W_i$).

338

$$F = -w_i$$

TABLE XVIII

$$U = 9347.65$$
$$U = \Sigma F_i + 2R_{vert} + 4M_{all}$$

No.	Value	No.	Value	Joint		Value		Joint		Value
6	0	137	0	Sk/C1	x	-2.27		C2/C3	R	38.09
14	11.09	134	89.92		z	48.29			M	-1.58
27	0	156	106.74	C1/C2	x	-1.66		C3/C4	R	36.16
59	1.32	184	4.65		z	47.35				
63	0.23	190	0		R		M		M	
84	0	80	0	St/C1	x	-5.73	-42.29	C4/C5	R	35.23
81	0	168			y	-1.59				
106	0	167			z	3.66			M	
107	4.76	201		Ar/C1	x	6.19		C5/C6	R	35.23
208	9.48	205			y	0.64			M	
182	8.30	151			z	-4.87	-61.27	C6/C7	R	33.47
166		148		Hum	x	-37.63				
155	30.17	193	7.79		y	48.45			M	
152		178			z	21.67		C7/Thx	R	41.32
141	13.79	209		Elbow	x	-1.80			M	2.49
142	145.79	15	0.63		y	-1.24	21.64			
158	53.89	3	0.97		z	10.36	-7.7	Thx/L1	R	529.95
153	13.82	8	1.34	Wrist	x	-1.09			M	
130		140	7.54		y	0.01				
110		135	12.66		z	1.09		L1/L2	R	438.76
111	5.97	24	0.08						M	146.27
189		197	22.64					L2/L3	R	399.87
154	164.98	183	5.01						M	
229		192	46.48					L3/L4	R	320.04
204		203	12.22						M	
194		201	0.11					L4/L5	R	290.62
210	0	211	1.32						M	-96.78
218	0	227	0.34					L5/Sacrum	R	272.32
216	0								M	

TABLE XIX $F_i = -W_i$

$$U = \Sigma M$$

$$U = 117.02$$

$U = \Sigma M$ at
UE, all vert. jts.

No.	Value	No.	Value	Joint	Value		Joint	Value
6	15.58	208	121.18	St/CL x	-86.96	0	Sk/C1 X	-76.94
14	132.49	154	205.79	y	-148.95	0	z	599.49
24	51.35	166	9.95	z	-7.89	0	C1/C2 x	-75.51
27	92.41	155	87.29	Ankle x	-3.02	0	z	609.97
39	120.1	141	32.95	y	3.43	0	C2/C3 R	707.36
40	54.99	158	56.77	z	-5.32	0	C3/C4 R	873.75
48	40.13	142	309.04	GH x	-34.53	0	C4/C5 R	1076.25
80	31.20	152	46.61	y	47.46	0	C5/C6 R	1076.25
81	11.40	218	0.18	z	0.56	0	C6/C7 R	1059.75
106	29.71	219	0.23	Elbow x	-6.13	0	C7/Thx R	1150.37
107	124.53	153	31.26	y	16.11	0	Thx/L1 R	766.59
110	7.31	156	127.96	z	132.82	0	L1/L2 R	624.67
111	20.87	191	6.61	Wrist x	-1.17	0	L2/L3 R	563.18
134	131.55	193	15.31	y	0.09	0	L3/L4 R	474.03
162	70.18	199	36.14	z	1.55	0	L4/L5 R	542.47
180	171.12	203	48.53			0	M	-117.02
197	10.89	205	1.23				L5/Sacrum R	490.79
183	3.56	207	53.34					
184	5.29	211	1.82					

the increased activity of the muscles responsible, it is obvious why the reactions at the cervical levels are so immensely high. The increase in the values of the reactions at the lumbar vertebral joints appears more reasonable. However, the L4/L5 joint is required to carry a moment of 117 cm for equilibrium—a number that in light of the high reaction force of 542.5 kg on the same disc appears negligible. This would merely imply a shift in the location of the resultant joint force through approximately 117.0/542.5 or 0.22 cm.

SITTING WITH ARMS GROUNDED

In this case the forearm and the hand are assumed grounded so that the six equations of equilibrium for each of these segments along with the three for the pelvis are excluded from the original model. Another major change performed was to break up the intervertebral joint reactions into their individual x and z components omitting the previous constraint on the direction of the resultant. This increases the number of unknown variables. Since the forearm and the hand are grounded, muscles connecting the forearm to the hand (22 in all) are eliminated. Also coordinates of certain muscles of the arm (brachioradialis, triceps brachii) were changed. The posture studied is somewhat different than that described earlier and the angles selected are defined in Table XXa. The criterion selected was to minimize the sum of all muscle forces, four times the moments at all joints, and twice the reactions at all joints (U.E., intervertebral). The results for $F_{xi} = 0$ (sitting still), $F_{xi} = w_i$ (impact during backward motion) and $F_{xi} = -w_i$ (impact during forward motion) are shown in Tables XXb,c,d.

In general, the results show that muscular activity is significant only during impact. During backward motion, the force of impact is apparently borne by the muscles of the arm, notably the triceps brachii, while the muscles of the vertibral column are not strongly active. During forward motion, the abdominal muscles become strongly active and tricpes brachii are inactive. For all three cases though, there are no reaction forces whatsoever at the intervertebral joints below the level of the thorax, suggesting that if the forearm and hand are grounded, the upper extremity and the neck are the only skeletal structures resisting impact and the lower back does not play any role. This fact is confirmed by the results in Table XXb. When the arms hand by the side in a motionless seated posture, reaction forces appear at the lumbar intervertebral joints. The values for these reactions are lower than those in Table I, primarily because (a) reactions at intervertebral joints have not been constrained to act along a particular direction: (b) intervertebral reactions are included in the objective function.

EFFECT OF CONSTRAINING MUSCLE FORCES OR ABDOMINAL PRESSURES

Stooping Forward to Lift Weight
1. It was decided to investigate the effect of constraining certain muscles to carry forces below and up to a certain level only. This situation is equivalent to simulating the case when due to impairment a muscle can develop only a limited amount of tension, causing other muscles to take up its activity and change the normal muscle-force pattern thereby. When lifting a load of 100 lb (45.5 kg, abdominal pressure = 10.5 cm of Hg) the results obtained are shown in Table XXI. It is seen that the erectores spinae (F_{172}) is carrying a force of 148 kg. It was decided to constrain this muscle so that $F_{172} \leqslant 75$ kg or

TABLE XXa. Configuration for sitting with arms grounded.

No.	Body	ϕ_x	ϕ_y	ϕ_z
1	Right foot	0	0	0
2	Right leg	0	10	0
3	Right thigh	0	-80	0
4	Pelvis	0	-30	0
5	Left thigh	0	-80	0
6	Left leg	0	10	0
7	Left foot	0	0	0
8	L5	0	-5	0
9	L4	0	-5	0
10	L3	0	-8	0
11	L2	0	-11	0
12	L1	0	-14	0
13	Thorax	0	-14	0
14	C7	0	2	0
15	C6	0	15	0
16	C5	0	15	0
17	CX	0	15	0
18	C3	0	10	0
19	C2	0	10	0
20	C1	0	-5	0
21	Skull	0	0	0
22	R. Clavicle	-30	0	10
23	R. Scapula	0	0	0
24	R. humerus	0	0	0
25	R. ulna	0	90	10
26	R. radius	0	-90	-180
27	R. hand	0	-90	-180
28	L. scapula	30	0	-10
29	L. scapula	0	0	0
30	L. humerus	0	0	0
31	L. ulna	0	-90	10
32	L. radius	0	-90	180
33	L. hand	0	-90	180

TABLE XXb $F_i = 0$

U = 211.59

No.	value	name	joint		value	joint		value
14	0.86	R.C. Post major				Sk/Cl	x	-0.68
15	.05	R.C. Post minor					z	-6.95
42	0.14	⎫				C1/C2	x	-0.31
43	0.29	⎬ Semispinalis cervicis					z	7.36
44	0.29					C2/C3	x	0.32
45	0.26	⎭					z	5.88
60	0.25	⎫				C3/C4	x	0
62	0.35	⎬ Rotatores cervicis					z	6.24
57	0.61	multif cervicis				C4/C5	x	0
64	0.45						z	6.88
66	0.54	Rotatores cervicis				C5/C6	x	0
69	0.45						z	7.47
132	0.02	Lev. scapulae				C6/C7	x	0
215	0.17	Ext. carpi ulnaris	St/Cl	x	-0.71		z	8.06
164	0.28	Erectores		y	-1.54	C7/Thx	x	0
187	1.13	Pect. major		z	0.13		z	8.96
192	2.67	Latissi mass dorsi	Acr/C	x	0.71	Thx/L1	x	0
130	0.29			y	1.54		z	
193	2.6			z	-0.13	L1/L2	x	
201	4.23		Hum	x	-3.58		z	
186	1.14			y	4.80	L2/L3	x	
178	6.47			z	3.76		z	
			Elbow	x		L3/L4	x	
				y	-0.22		z	
				z	17.84	L4/L5	x	
							z	

TABLE XXc $\qquad F = W_{\dot{1}}$

$$U = 1223.19$$

$$U = 1\Sigma F + (2R+4M)UE$$
$$+ 4Mv + 2Rxz$$

No.	Value	Name	Joint		Value	Joint		Value
5	1.32	Semispin. Capitis				Sk/C1	x	-4.93
7	1.21	Splenius capitis					z	10.33
42	1.39					C1/C2	x	
43	1.63						z	17.49
44	1.62	Semispinalis cervicis					M_y	-0.94
45	1.51					C2/C3	x	-2.93
46	1.22						z	13.29
57	1.23	Multif. cervicis				C3/C4	x	-2.15
66	0.22	Rotatores cervicis					z	+15.93
108	3.42	Oblig. cpa. Inferior				C4/C5	x	-1.25
193	16.66	Infraspinetis					z	18.99
187	1.72	Pectoralis minor				C5/C6	x	-0.35
130	1.85	Levator scapulae					z	22.04
202	64.91	T. Brachii	ST/C1	x	-0.73	C6/C7	x	0
178	4.39	Latissimus dorsi		y	-1.57		z	25.07
190	2.37	Deltoideus		z	0.14	C7/Thx	x	0
186	9.75	Pect. major	Acr/C1	x	0.73		z	29.06
201	180.08	T. Brachii		y	1.57	Thx/L1	x	0
182	1.87	Serratus Anterior		z	-0.14		z	0
172	1.82	Erectores spinae	Hum	x	-18.64	L1/L2	x	0
				y	11.62		z	0
				z	0	L2/L3	x	0
							z	0
			Elbow	x	-1.42	L3/L4	x	0
				y	0		z	0
				z	259.26	L4/L5	x	0
							z	0

TABLE XXd

$$F = W_i \qquad \boxed{F_i = -W_i}$$

$$U = 766.6 \text{ kg}$$
$$U = \Sigma F + (2R_{xyz} + 4M_{vert})_{UE} + 4M_{vert} + 2(R_{xyz})_{vert}$$

No.	Value	Name		axis	Value		axis	Value
1	1.60	Semispinelis capatis				Sk/Cl	x	
14	4.57	R.C. Post major					z	32.89
15	0.92	R.C. Post minor				C1/C2	x	0.89
62	1.02						z	31.52
64	0.51	Rotatores cervicis				C2/C3	x	4.18
69	0.80						z	23.69
68	9.34					C3/C4	x	4.18
135	12.12	Ext. oblique					z	23.69
199	3.69	B.B. long head				C4/C5	x	4.35
190	2.24	Deltoideus					z	21.49
206	7.45	Sternocleidomastoid				C5/C6	x	5.19
208	0.37	Thoracic part (Stenohyoid)					z	21.24
59	0.05	Rotatores cervicis				C6/C7	x	6.03
178	37.22	Latissimus dorsi	St/CL	x	-1.05		z	20.99
182	5.85	Serratus Anterior		y	-2.26	C7/Thx	x	
193	15.87	Infraspinelis		z	0.19		z	22.36
134	21.58	Rectus abdominis	Ankle	x	1.05	Thx/L1	x	
187	1.66	Pectoralis minor		y	2.26		z	
229	10.98	Brachioradialis		z	-0.19	L1/L2	x	
			Hum	x	-19.52		z	
				y	16.27	L2/L3	x	
				z			z	
			Elbow	x	-6.22	L3/L4	x	
				y	-2.14		z	
				z	48.29	L4/L5	x	
							z	

about half the initial force value. The results obtained in this case are shown in Table XXII and Fig. 43. It can be seen that F_{172} = 51.75 kg for equilibrium and the overall pressure pattern at the intervertebral joints is only negligibly altered. Reduction of F_{172} is accompanied by activity in F_{164} which can suitably replace F_{172} from an anatomical point of view (both arise from the iliac crest of the pelvis and insert into the thorax and are posteriorly placed). A significant change is seen in the internal oblique abdominis muscle F_{137} which is increased from 29.88 Kg initially top 40.08 kg (or an increase of 31%) with the erectores spinae F_{172} constrained. This implies that a person with weak back muscles stresses his abdominal muscles to a greater extent to perform the same activity than a healthy individual.

2. If the abdominal muscle (internal oblique) is altogether removed, i.e., F_{137} = 0, the loads on all the lumbar disc are tremendously increased (shown in Table XXIII). Note that the muscle force pattern for the upper extremities and for the head and the neck is unaltered and therefore the loads on the cervical vertebrae are completely unchanged. The load on the Thxl LL disc is lower than before, possibly because the abdominal muscle activity has been reduced due to the constraint. However, the rectus abdominis muscle which was previously inactive, is now required to carry a force of 18.50 kg. The back muscles show reduced activity ($F_{164-172}$ = 160 Kg vs 205 kg) which may be compensated by the previously inactive quadratus lumborum muscle F_{139} which is increased to 241.25 kg. Anatomically these muscles are placed behind the lumbar vertebral joints and may replace the erectores spinae.

3. **Abdominal Force as a Variable**

In this case the abdominal force is treated as a variable free to select its own value. The objective function is selected to be

$$U = \Sigma F + 2R_{xyz} + 4M_{all} + \frac{k_p}{\text{abd. force}}$$

where p = abdominal press in cm of mercury

$$k = \frac{13.6 \times 80 \times (2.54)^2}{1000} = \text{factor to convert pressure into force}$$

The results are shown in Table XXIV. The abd. pressure turns out to be 38.9 cm of Hg. However this is much higher than the experimentally measured value for this exercise (Morris et al) which is 10.5 cm of Hg.

EFFECT OF ABNORMAL LUMBAR CURVATURE ON MUSCLE LOAD SHARING AND DISC LOADS

In this study the effect of arbitrarily changing the lumbar curvature to represent abnormal or pathological cases is studied. As shown in Fig. 44, the lumbar curve was displaced both upward (convex anteriorly) and downward (concave anteriorly) from its normal position in the stooping posture. These alterations were done without changing the

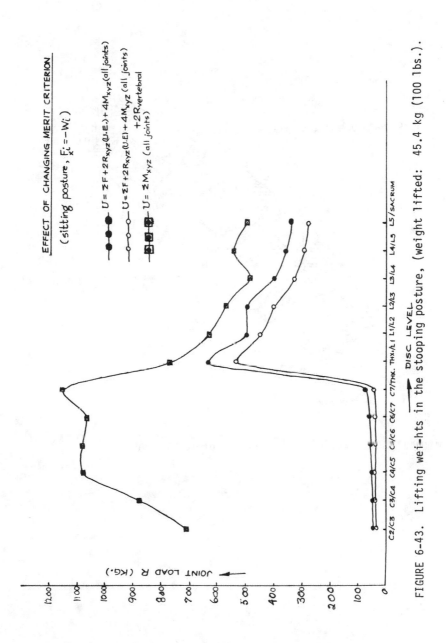

FIGURE 6-43. Lifting wei-hts in the stooping posture, (weight lifted: 45.4 kg (100 lbs.).

347

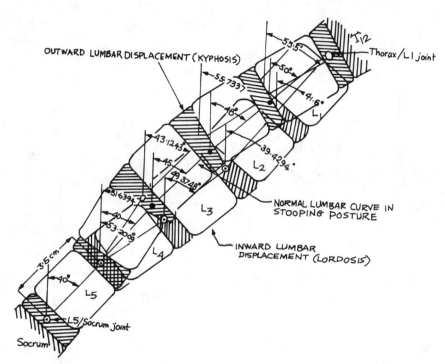

FIGURE 6-44. Abnormal displacement of lumbar curve.

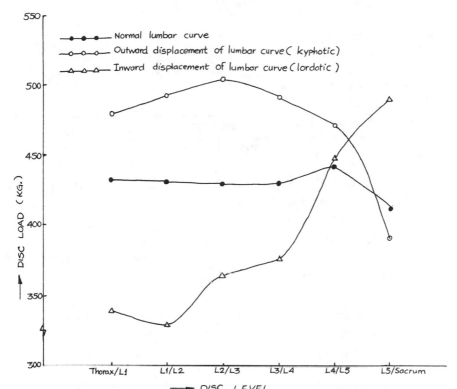

FIGURE 6-45. Lifting weights in the stooping posture. (Weight lifted: 45·4 kg (100 lbs.)

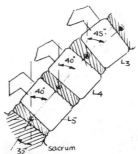

FIGURE 6-46a. Normal stooping curvature during weightlifting.

TABLE XXI

$WL = 45.5$

$P = 10.5$ cm

$U = 2461.76$

F_i	value	F_i	value	Upper	Extremities		Joint		Value
					R	M			
14	19.65	166	10.69	St/Cl x	2.76	0	Skull/Cl	x	73.33
35	7.4	167	19.63	y	-22.88	-3.22		z	102.21
36	9.87	168	23.65	z	11.64	0	Cl/C2	x	79.02
39	7.64	169	3.13	ACr/Cl x	-12.3			z	109.75
40	1.85	172	148.48	y	11.90	-63.07	C2/3		112.40
85	4.81	173	9.46 L.D.	z	-17.36		C3/4		115.73
97	7.51	189	14.19 D.	GH x	-11.38		C4/5		137.67
106	15.74	190	17.49 D.	y	3.65		C5/6		154.96
107	21.47	195	3.88 T.minor	z	6.51		C6/7		191.21 m=9.28
109	22.23	203	4.91	Elbow x	-0.43		C7/Thx		218.08
110	7.35	205	0.62	y	0		Thx/L1		432.23
112	0.4	208	27.77	z	0		L1/L2		430.69
121	5.23	217	3.37	Wrist x	-0.69	-29.8	L2/L3		429.41
137	29.88	218	0.81	y	2.29	0	L3/L4		428.22
151	15.31	219	6.71	z	0	0	L4/L5		441.24
152	19.91	221	0.92				L5/Sacrum		410.71
153	42.58	223	11.77						
158	26.09	228	5.02						
162	13.66	229	9.12						
163	8.25	Semimemb	189.23	168.07					
		BF(LH)	192.73	163.34	$\begin{cases} -43.22 \\ 1.2 \\ 418.99 \end{cases}$ hip	$\begin{cases} x \\ y \\ z \end{cases}$	-58.69 11.94 536.47		
		BF(SH)	100.77	27.41					
		Sartonius	51.78	28.49	$\begin{cases} -58.3 \\ 14.39 \\ 569.41 \end{cases}$ knee	$\begin{cases} x \\ y \\ z \end{cases}$	-56.07 16.43 718.99		
		Glut Med	43.64	15.0					
		Gastroc.	137.53 0	125.95 11.58	$\begin{cases} -0.38 \\ 0.41 \\ 195.26 \end{cases}$ ankle	$\begin{cases} x \\ y \\ z \end{cases}$	-0.39 6.51 195.25		

$$F_{172} \leqslant 75 \text{ kg}$$

TABLE XXII

$$U = 2491.02$$

F_i	Value	F_i	Value	Upper Extremities		R	M	JOINT	VALUE	
14	20.85	167	19.81	St/Cl	x	2.97		Skull/Cl	x	77.99
35	7.81	168	23.88		y	-30.73	-0.15		z	108.25
36	10.46	169	3.16		z	10.56		C1/C2	x	83.96
39	7.82	172	51.75	ACr/Cl	x	-13.07			z	116.16
40	1.95	173	9.96		y	12.77	-62.13	C2/3		118.66
85	5.18	189	14.96		z	-16.69		C3/4		122.18
97	7.94	190	18.43	GH	x	-11.99		C4/5		145.34
106	16.65	195	4.09		y	3.85		C5/6		163.89
107	22.70	203	4.91		z	8.29		C6/7		202.23 m=9.82
109	23.59	205	0.62	Elbow	x	-0.43		C7/Thx		230.66
110	7.70	208	29.57		y	0		Thx/L1		436.28
112	0.42	217	3.37		z	0		LL1/2		434.52
121	5.68	218	0.81	Wrist	x	-0.69	-29.77	L2/3		433.43
137	40.08	219	6.71		y	2.29	0	L3/4		432.23
151	15.37	221	0.92		z	0	0	L4/5		445.36
152	20.03	220	11.77					L5/Sacrum		403.88
153	53.29	228	5.44							
158	26.34	229	9.12							
162	13.79	164	81.59							
163		170	9.98							
166	10.47	Slack	24.0							

(Constrain back muscle)

By forcing $F_{172} \leqslant 75$, F_{137} (Int. oblq) is increased from 29.88 to 40.08. That is, weakness in E. spinae is compensated by abdominal muscles.

351

$$U = 2831.54$$
$$F_{137} = 0$$

TABLE XXIII.

F_i	value	F_i	value	Upper Extremities		R	M	Joint	Value	
14	19.65	166	135.95	Stcl	x	2.76		Skull/Cl	x	73.33
35	7.40	167	24.68		y	-28.88	-3.22		z	102.20
36	9.87	168	0		z	11.64		Cl/C2	x	79.02
39	7.64	169	0	ACr/Cl	x	-12.30			z	109.75
40	1.85	172	0		y	11.90	-63.07	C2/3		112.39
85	4.81	173	9.46		z	-17.36		C3/4		115.72
97	7.51	189	14.19	GH	x	-11.38		C4/5		137.67
106	15.74	190	17.49		y	3.65		C5/6		154.95
107	21.47	195	3.89		z	6.51		C6/7		191.20 m=9.28
109	22.23	203	4.91	Elbow	x	-0.43		C7/Thx		218.08
110	7.35	205	0.62		y	0		Thx/L1		335.13
112	0.39	208	27.77		z	0		L1/L2		541.05
121	5.23	217	3.37	Wrist	x	-0.69	-29.77	L2/L3		539.69
137	0	218	0.81		y	2.29	0	L3/L4		591.21
151	3.23	219	6.71		z	0	0	L4/L5		608.98
152	25.03	221	0.92					L5/Sacrum		609.33
153	23.48	223	11.77							
158	0.44	228	5.02							
162	21.54	229	9.12	(Constrain abdominal muscle)						
163	12.35	73	0.09	L. collis						
		150	6.36	Multi Lumbar						
		134	18.50	Rectum Abd.						
		139	241.25	Qudr Lumb						
		161	33.36	Iliopsos						

XXIV.

Constrain Abdp as a variable $U = \Sigma F + 2R + 4M + kp$ $k = \dfrac{13.6 \times 80 \times (2.54)^2}{1000}$

F_i	value	F_i	value	Upper Extremities				Joint	Value
						R	M		
14	19.67	162	0.25	ACr/Cl	x	-12.32	0	Skull/Cl	x	+73.42
35	7.41	163	0.23		y	11.92	-63.08		z	102.32
36	9.88	169	0.06		z	-17.36	0	Cl/C2	x	79.12
39	7.64	173	8.49	ST/CL	x	2.76	0.		z	109.88
40	1.85	172	77.12		y	-28.90	-3.28	C2/C3	112.52
73	0.09	174	0.45		z	11.66	0	C3/4	115.85
85	4.82	175	0.52	GH	x	-11.39	0	C4/5	137.82
97	7.51	189	14.24		y	3.65	0	C5/6	155.13
109	22.26	190	17.51		z	6.52	0	C6/7	191.42 m=9.29
110	7.36	205	0.62	Elbow	x	-0.43	0	C7/Thx	218.33
107	21.49	195	3.89		y	0	0	Thx/L1	0
112	0.39	203	4.91		z	0	0	L1/L2	7.82
106	15.76	208	27.80	Wrist	x	-0.69	-29.77	L2/L3	7.80
121	5.24	217	3.37		y	2.29	0	L3/4	7.78
137	43.39	218	0.81		z	0	0	L4/L5	8.02
139	10.99	221	0.92					L5/sacrum	7.83
152	0.30	219	6.71				P =	38.90 cm of Hg
154	2.40	223	11.77					(actual = 10.5 cm)
153	0.51	229	9.12				U =	2285.19
158	0.44	228	5.03					

position of the thorax/L1 joint substantially. In other words, the total bending moment from externally applied forces (such as the weights in the arms) and the body weights about the thorax/L1 joint is almost identical for the three curves. The L5 vertebra was kept in its normal position. Using the same criterion (minimize the sum of all forces, joint moments at all the joints and reactions at the upper extremities joints with suitable weighting factors) which was used for investigation of muscle load sharing in the stooping posture results were obtained and are shown in Table XXVa. Note that the muscle load sharing and joint reactions in the upper and lower extremities and the crevical vertebral region do not change from the normal stooping posture and hence have not been shown in this table. The distribution of the forces in the abdominal muscles, and in both the long and short extensors of the back are as tabulated. It is worth noting that the changes in the lumbar curve significantly affect the original muscle force distribution (left column of Table XXVa. Upward disturbance in the lumbar configuration (middle column, Table XXVa) causes increased strain on the short rotarores, namely the multifidu and the rotatores lumborum. In this case the long extensors (erectores spinae) do not seem to be affected. However with downward displacement of the lumbar curve (right column, Table XXVa) the iliopsas and the quadratus lumborum seem to be called into stronger action. Note that in this case while a slight reduction in the erectores spinae force is seen as compared to their activity in the normal posture, the multifidus are all together inactive.

Vertebral Joint Reactions

The variation in the reactions at the thorax/L1, L1/L2, L2/L3, L3/L4, L4/L5 and L5/sacrum joints is illustrated in Fig. 45 for the three cases of lumbar configurations. Upward displacement of the lumbar curve causes the forces on the upper lumbar joints to increase. The only joint to show reduced loading is the L5/sacrum joint. The maximum decrease is ≈−26% occurring at the L5/sacrum disc while the maximum increase is about ≈19% occurring at L2/L3 joint. On the other hand for the downward displacement of the lumbar curve, the reactions at the upper lumbar joints are considerably lower than those in the normal stooping exercise while the L5/sacrum joint shows an increase. The maximum increase occurs at the L5/sacrum joint (≈+20%) while the maximum decrease occurs at the L1/L2 joint (26) compared to the normal posture.

It should be emphasized that comprehensive musculoskeletal models of the spine on the lines described in this chapter serve as valuable tools in the study of effect of muscle load sharing and disc pressures due to change in vertebral configuration.

EFFECT OF DISC FUSION ON SPINAL PRESSURE DISTRIBUTION

In many pathological cases, two or more adjacent lower lumbar vertebrae are mechanically fused or fixed. The object of this study is to investigate the effect on spinal pressure distribution during weight lifting when one or more of the lower lumbar vertebrae are fused. Two cases will be investigated, when the fifth lumbar vertebra is fused to the sacrum (L5-S1 fusion), and when the fourth and the fifth lumbar vertebrae and the sacrum are all fused together (L4-L5-S fusion).

When two or more segments are fused, the combination is treated as one single rigid body. This implies that muscles connecting the segments

XXVa.

Displacement of
lumbar curve

Muscle name	Attachment points	Normal (kg)	Upward (kg)	Downward (kg)
Internal Oblique	Thorax-Pelvis	29.88	48.64	11.36
Multifidus	L5-L1	15.31	0	0
	L5-L2	19.91	52.19	0
	Pelvis-L5	42.58	82.55	0
Quadratus Lumborum	L1-Pelvis	0	25.19	0
	L2-Pelvis	0	0	62.41
	L3-Pelvis	0	0	30.3
	L4-Pelvis	0	0	19.34
Rotatores Lumborum	L1-Thorax	0	24.35	0
	L3-L1	0	0	23.78
	L5-L3	26.09	74.1	0
Iliopsoas	L4-femur	13.66	4.33	36.05
	L5-femur	8.25	1.61	47.69
Erectores Spinae	L1-Thorax	10.69	0	20.51
	L2-Thorax	19.63	35.89	37.91
	L3-Thorax	23.65	70.64	2.69
	L4-Thorax	3.13	13.03	0
	L5-Thorax	0	0	5.05
	Pelvis-Thorax	148.48	79.10	103.76

XXVb. Vertebral joint loads with abnormal lumbar displacements in the stooping posture.

Joint		Normal	Inward (Lordotic)	Outward (Kyphotic)
Skull/C1		73.33	73.33	73.93
	Z	102.20	102.20	102.98
C1/C2	X	79.02	79.03	79.66
	Z	109.75	109.75	110.57
C2/C3	R	112.4	112.4	113.20
C3/C4	R	115.73	115.72	116.55
C4/C5	R	137.67	137.67	138.66
C5/C6	R	154.96	154.95	156.10
C6/C7	R	191.21	191.20	192.62
		My=9.28	My=9.28	My=9.35
C7/Thx	R	218.08	218.08	219.69
Thx/L1	R	432.23	338.83	481.33
L1/L2	R	430.49	326.63	491.87
L2/L3	R	429.41	363.44	503.82
L3/L4	R	428.22	373.83	489.48
L4/L5	R	441.24	446.52	471.16
L5/sacrum	R	410.71	448.53	390.45

Weight lifted:

45.4 kg (100 lbs.)

should be eliminated along with the reactions at the joint or joints between the segments. For the fusion of the lower lumbar discs it was assumed that the vertebrae were fused with the same orientations as those adopted during the erect standing posture. Thus, for L5-S1 fusion, the L5 is fused to the sacrum with a sagittal anterior tilt of 18°, which occurs in the standing erect posture. The remaining upper lumbar segments are treated as mobile and allowed to orient themselves in the same configuration as that selected for forward flexion during weight lifting (Fig. 46b). Similarly, for L4-L5-S1 fusion, the L4 and L5 vertebrae are fixed with anterior tilts of 5° and 18° respectively (Fig. 47), while the upper lumbar spine maintains the same configuration as that for weight-lifting with forward flexion.

When L5 is fused to the sacrum muscles connecting the two segments as well as the reaction force and moment at the L5/sacrum joint are omitted from the model. The L5 is now treated as an integral part of the pelvis (sacrum), so that the muscles connecting L5 to other parts of the body are considered as attached between the pelvis and the same parts of the body. This includes five muscles, namely, multifidus lumborum (three parts: L1 to L5, L2 to L5 and L5 to pelvis, which is omitted), rotarores lumborum (13 to 15), iliopsoas (L5 to femur with a reaction on the pelvis, which is omitted), erectores spinae (L5 to thorax) and latissimus dorsi (L5 to humerus).

With L4-L5 -sacrum union, both L4 and L5 are treated as being intergral with the pelvis. In this case the muscles multifidus lumborum (L5 to pelvis), iliopsoas (two parts: L4 to femur, L5 to femur with reactions on the pelvis) and quadratus lumborum (L4 to pelvis) and the reactions at the L4/L5 and the L5/sacrum joints are eliminated from the model. Also, the muscles which originate on the L4, such as multifidus lumborum (two parts: thorax to L4, L1 to L4), rotatores lumborum (L2 to L4) erectores spinae (L4 to thorax) and latissimus dorsi (L4 to humerus) in addition to the muscles described above which originate on the L5 are now treated as connecting the pelvis with their respective insertions on other segments.

RESULTS

The results for muscle load sharing and disc pressures for the normal spine (52° forward leaning) during weight lifting (W = 45.4 kg or 100 lb) shown in Table XXVI. The criterion used was to minimize all muscle forces, four times the joint moments and twice the reactions at the joints of the upper extremities. The intra-abdominal pressure for this activity was included in the analysis as a force acting between the pelvis and the diaphragm (on the thorax). The disc pressures at the spinal joints with discs, have been constrained to act at an inclination average of the sagittal tilts of the two adjoining vertebrae forming the joint.

The results for the L5-sacrum fusion and L4-L5-sacrum fusion with the same 52° forward leaning posture and the same weight in the arms are shown in Table XXVI for comparison. The disc loads at the various levels of the spine are plotted for the three cases in Fig. 48. With L5 fusion, the L4/L5 joint pressure goes up to 608.29 Kg as compared with 441.24 Kg for the normal spine with no fusion. The rest of the lumbar discs and the cervical discs are only negligibly affected . However, with L4-L5-sacrum fusion, while the cervical loads remain unchanged, all the lumbar disc pressures are greatly increased as compared to the normal unfused

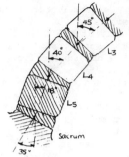

FIGURE 6-46b. L₅ fused to sacrum
during weightlifting.

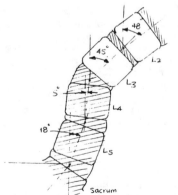

FIGURE 6-47. L₅ and L₄ fused to sacrum.

TABLE XXVI. Muscle forces and joint reactions during weight lifting (w=45.4 kg).

No.	Muscle Force (Kg)	Normal	L5 fused	L4 & L5 fused
1	Abdominal obliques	29.88	15.99	25.42
2	Erectores	206.00	213.57	267.64
3	Multifidus	77.80	34.55	17.51
4	Rotatores	26.09	26.79	21.19
5	Quadratus lumborum	0	16.84	0
6	Iliopsoas	21.91	106.82	205.1
	Joint Reactions* (Kg)			
	L5/Sacrum	410.71	–	–
	L4/L5	441.24	608.29	–
	L3/L4	428.22	434.43	840.9
	L2/L3	429.41	435.64	560.0
	L1/L2	430.49	436.73	561.0
1	Thorax/L1	432.23	438.50	563.7
	C7/Thorax	218.08	218.08	214.8
	C6/C7	191.21	191.21	188.3
	C5/C6	154.96	154.96	152.6
	C4/C5	137.67	137.67	135.6
	C3/C4	115.73	115.73	113.8
	C2/C3	112.40	112.40	110.5

*The disc reaction forces at these joints with discs are constrained to be inclined the average of the sagittal tilts of the upper and lower vertebra between which the disc lies.

Stooping forward, wt. lifted = 45.5 Kg (110 lb)

spine. The L3/L4 joint pressure reaches a value of 840.9 Kg as compared to 428.22 Kg for the normal case. It appears therefore that as the fusion is performed to include more of the lower lumbar vertebrae, the pressure between the fused segment and the next immediate free vertebra builds up.

In all the above results the disc loads are constrained in their orientations in the sagittal plane. When these joint reactions are modelled by their x and z components and therefore not constrained, the results for pressure distribution appear to significantly change.

For the normal unfused spine during weight lifting, if the reactions at the L4/L5 and L5/sacrum joints are assigned x and z components while the other joints reactions are kept constrained the resulting disc pressures at the lumbar joints are much lower compared to those obtained when all the disc reactions were constrained (Table XXVII). Now if the L5 were fused to the pelvis, the reactions are increased as shown in the same table and plotted in Fig. 48. The results for joint pressures during weight lifting when three of the lower lumbar disc forces are unconstrained (that is allowed to have x and z components), namely, the L3/L4, L4/L5, and L5/sacrum joints, shown in Table XXVII. The indicate that the disc loads for the lumbar discs are considerably lower than those for the case when the reactions at these joints were constrained, and also lower than those for the case when only two disc loads (at L4/L5 and L5/sacrum joints) are constrained.
The orientations of the resultant joint pressure

$$\left(R = \sqrt{R_x^2 + R_z^2}, \ \theta = \tan^{-1} \frac{R_z}{R_x} \right)$$

as calculated for the lower two unconstrained joints shown in Table XXIX. It appears likely that forcing the disc reactions to align with the inclinations of the vertebrae may cause their magnitudes to be higher. Since higher disc pressures result from higher muscle forces, it is evident that such constraints may also cause increase muscle forces for equilibrium (Table XXVIII).

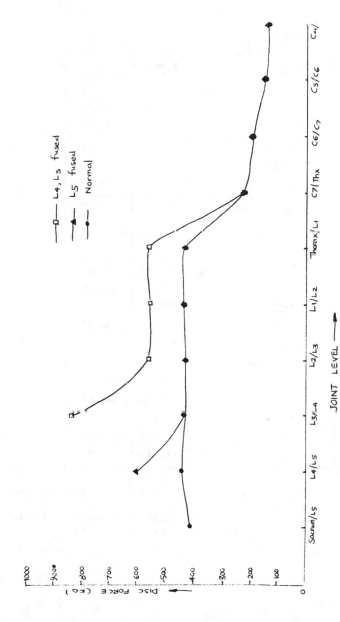

FIGURE 6-48. Disc loads during weight lifting. (W = 45.5 kg (100 lb)).

361

TABLE XXVII.

Muscle force (KG)	Normal	L5 fused
1. Abdominal Obliques	10.56	15.22
2. Erectores	179.20	210.46
3. Multifidus	12.5	33.96
4. Rotatores	14.97	26.35
5. Quadratus lumborum	5.85	0
6. Iliopsoas	22.08	0
Joint Disc Loads (Kg)		
L5/Sacrum* $\quad x$	248.33	-
$\qquad\qquad\quad z$	245.18	-
L4/L5* $\qquad x$	263.02	280.59
$\qquad\qquad z$	265.87	309.21
L3/L4	392.97	427.29
L2/L3	358.9	428.48
L1/L2	359.8	429.56
Thorax/L1	361.23	431.29
C7/Thorax	218.08	218.08
C6/C7	191.21	191.21
C5/6	154.96	154.96
C4/5	137.67	137.67
C3/4	115.73	115.73
C2/3	112.40	112.40

*The reaction forces at these two joints are unconstrained, that is, allowed x and z components. The resultant ($R = \sqrt{R_x^2 + R_z^2}$) at the L5/sacrum and L4/L5 joints for normal spine is 348.97 and 373.98 respectively. For the fused spine, the resultant disc force at the L4/L5 joint is 417.54 Kg.

TABLE XXVIII. Normal unfused spine: effect of joint reaction constraint on disc loads.

Muscle Force (kg)	All Reactions Constrained	Two Reactions Unconstrained (L5/S,L4/L5)	Three Reactions Unconstrained (L5/S,L4/L5,L3/L4)
1. Abdominal obliques	29.88	10.56	9.99
2. Erectores	206.00	179.2	177.0
3. Multifidius	77.80	12.5	17.48
4. Rotatores	26.09	14.97	9.81
5. Quadratus lumborum	0	5.85	0
6. Iliopsoas	21.91	22.08	0
Joint reactions (kg): $R = \sqrt{R_x^2 + R_z^2}$ for unconstrained joint			
L5/sacrum	410.71	$\begin{cases} x=248.33 \\ z=245.18 \end{cases} R=348.97$	$\begin{cases} x=216.43 \\ z=189.18 \end{cases} R=287.46$
L4/L5	441.24	$\begin{cases} x=263.02 \\ z=265.87 \end{cases} R=373.98$	$\begin{cases} x=235.80 \\ z=217.59 \end{cases} R=320.85$
L3/L4	428.22	392.97	$\begin{cases} x=252.38 \\ z=240.67 \end{cases} R=348.73$
L2/L3	429.41	358.9	353.59
L1/L2	430.49	359.8	354.48
Thorax/L1	432.23	361.23	355.92
C7/Thorax	218.08	218.08	218.08
C6/C7	191.21	191.21	191.21
C5/C6	154.96	154.96	154.96
C4/C5	137.67	137.67	137.67
C3/C4	115.73	115.73	115.73
C2/C3	112.40	112.40	112.40
Joint reaction orientation: (Degrees) $\theta = \tan^{-1}\frac{R_z}{R_x}$ for unconstrained joints			
L5/Sacrum	52.5	44.7	41.2
L4/L5	50.0	45.3	42.7
L3/L4	47.5	47.5	43.7
L2/L3	43.5	43.5	43.5
L1/L2	41.0	41.0	41.0

TABLE XXIX. Direction of resultant reaction (ϕ) at unconstrained joints for normal and fused spine.

I. All Reactions Constrained

$$\phi = \tan^{-1} \frac{R_z}{R_x} \text{ (degrees)}$$

Disc level	Normal	L5 fused
L5/sacrum	52.5°	
L4/L5	50.0°	61.0°
L3/L4	47.5°	47.5°
L2/L3	43.5°	43.5°
L1/L2	41.0°	41.0°

II. L4, L5 Reactions Unconstrained

| Disc Level | Normal | | L5 fused | |
	Angles input	Calculated $\phi = \tan^{-1} \frac{R_z}{R_x}$	Angles input	Calculated
L5/sacrum	52.5°	44.7°	-	-
L4/L5	50.0°	45.3°	61°	47.75
L3/L4	47.5°	47.5°	47.5°	47.5
L2/L3	43.5°	43.5°	43.5°	43.5°
L1/L2	41.0°	41.0°	41.0°	41.0°

REFERENCES

1. Braus, H. Anatomie Des Menschen Bewegungs. Apparatus. J. Springer, Berlin, Vol. I., 1954.

2. Nachemson, A. Lumbar intradiscal pressure measurements in vivo. Lancet, 1:1140-1142, 1963.

3. Nachemson, A. and Morris, J. In vivo measurements of intra discal pressure. J. Bone Jt. Surg.: 46A:1077, 1964.

4. Morris et al. The role of the trunk in stability of the spine. J. Bone Jt. Surg: 43A, 327, 1961

CHAPTER VII

The Jaw

"The tooth has less power in gripping
which is more remote from the center of
its movement."

Leonardo da Vinci
1452-1519

INTRODUCTION

The human jaw has received wide attention from many investigators
working in diverse areas such as anatomy, orthodontics, and anthropology
who have viewed the jaw function from various vantage points. The
anatomist is primarily interested in the physiology and anatomical
features of the muscles and ligaments of the jaw and their participation
in various mandibular movements. For the orthodontist, such matters as
knowledge of the load sharing by these muscles, the forces developed at
the temporamandibular joints during the process of mastication and
general kinesiology of the mandible are of prime importance. The
evolutionary aspect of the mandibular structure and its relation to the
mechanics of the jaw muscles are subjects of interest to anthropologists.

In this chapter we shall discuss the development of a
musculoskeletal model for the jaw. This will be done along the same
lines as that for the other parts of the skeletal structure.

The human jaw (or the mandible)articulates with the skull (or the
maxilla) through the temporomandibular joints. Unlike other parts of the
human musculoskeletal system (the lower and upper extremities, the head
and neck and the vertebral column) which represent a multi-segmented,
multi-muscled complex system, the jaw acted upon by the muscles of
mastication is a relatively simple system to model and analyze. There
are only four major muscles of mastication—temporalis, masseter, medial
pterygoid and lateral pterygoid which connect the mandible to the skull
through the temperamandibular joints, one on each side of the sagittal
plane. With the muscles modelled by straight lines connecting the
respective points of origin and points of insertion, the four muscles
described above can be represented by four straight lines. However, by
taking into account the independent actions of the anterior and posterior
temporalis, as revealed by EMG signals, and the superficial and deep
fibers of masseter, one arrives at a total of six muscle forces. The
maxilla can be assumed to be fixed or grounded so that the only rigid
body is the mandible, connected to the maxilla at the left and the right
temporomandibular joints. It can be therefore seen that the jaw model is
a simple system with only one rigid body acted upon by a total of twelve
muscle forces described in the following section.

DESCRIPTION OF MUSCLES OF MASTICATION

The process of mastication involves four major muscles: the masseter, the temporalis, and the two pterygoid muscles, lateral and medial. The suprahyoid muscles, extending between the jaw and the hyoid bone, may assist the masticatory process. However, since their participation in jaw movements is of secondary importance this group of muscles could be excluded from the analysis.

The **temporalis** (Fig. 1) has an extensive origin arising from the entire temporal fossa and the deep surface of the temporal fascia and converging to end in a tendon which passes medial to the zygomatic arch. It is inserted into the upper and anterior border and the medial surface of the coronoid process, and into the anterior border of the ramus of the mandible as far as the vicinity of the last molar tooth. Since the anterior fibers of the muscle run almost vertically downward, and the posterior fibers both forward and downward it is imperative to separte the muscle into two parts for the purpose of analysis of its biomechanical action. The corresponding line of action for each division is obtained by connecting the estimated centroid of area of origin to the estimated centroid of the area of insertion by a straight line. The lines of action of the anterior and posterior temporalis have been superimposed on the muscle sketch (Fig. 1). These force lines also reveal the action of the jaws for which the temporalis would be normally responsible namely, closure of the jaws, with the posterior part strongly retracting or pulling them backward.

The **masseter** (Fig. 2) is a thick, quadrilateral muscle covering the angle of the mandible. It consists of a larger superficial portion arising from the zygomatic process of the maxilla and from the anterior two-thirds of the inferior border of the zygomatic arch, and a deep smaller portion arising from the psoterior third of the inferior border and from the entire medial surface of the zygomatic arch. The superficial part extends downward and posteriorly and is inserted into the angle and lower half of the lateral surface of the ramus of the mandible. The deep part extends vertically downward is inserted into the upper half of the ramus and the lateral surface of the coronoid process of the mandible. The muscle elevates the mandible such as during closure and biting. The lines of action corresponding to the superficial part and the deep part are shown in Fig. 2.

The **medial (or internal) pterygoid** muscle takes its origin from the medial surface of the lateral pterygoid plate and from the pyramidal process of the palative bone. The fibers extend inferiorly, posteriorly and lateralward to insert on the medial surface of the angle and ramus of the mandible as high as the mandibular foramen. The muscle operates in closing the jaws and is active in protrusion and grinding movements. The medial pteryoid muscle and its mechanical line of action is shown in Fig. 3.

The **lateral (or external) pterygoid** muscle is a thick, short muscle lying horizontally between the infratemporal fossa and the mandibular condyle. It has two heads of origin; the smaller superior part which arises from the infratemporal surface of the greater wing of the sphenoid and the larger inferior part which arises from the lateral surface of the lateral pterygoid plate. The two heads are directed horizontal posteriorly and laterally, with the superior head inserting into the articular disk of the temporomandibular joint and into the superior part

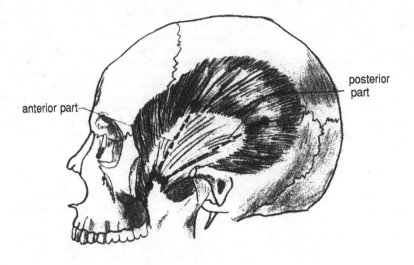

Figure 7-l.   The temporalis muscle and its model.

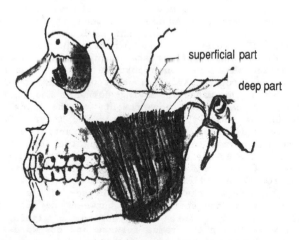

Figure 7-2. The masseler muscle and its model.

of the neck of the mandibular condyle, and the inferior head inserting into the pterygoid forea of the neck of the condyle. The lateral pterygoid muscle opens the jaws, protrudes the mandible and moves the mandible from side to side. Acting alone on one side, the muscle causes the condyle on that side to be displaced forward and downward so that the chin is thrust controlaterally. Acting alternately and separately, the lateral pterygoids move the mandible side-to-side as in grinding and chewing. The model for this muscle is depicted in Fig. 3. The muscle is modelled by a single line connecting the appropriate point of origin to the point of insertion. There is not sufficient evidence to warrant its simulation by two lines, one corresponding to each of the two heads. Electromyographic studies on thesus monkeys (see Refs. 10, 11) indicate that the inferior and the superior head may act independently to certain movements. Whether this finding could be applied to humans as well is questionable.

KINEMATICS OF MANDIBULAR MOVEMENTS

The mandible is held in position by the muscles of mastication and is attached to the cranium by the ligaments of the temporomandibular joint, namely, capusular sphenomandibular, temporomandibular and stylomandibular as shown in Fig. 4. The temporomandibular joint is a combined hinge and gliding joint. It is formed by the anterior part of the mandibular fossa of the temporal bone, the eminentia articularis of the zygomatic process of the temporal bone, and the mandibular condyle. The capsular ligament is connected to the circumference of the mandibular fossa and the articular eminence, and to the neck of the mandibular condyle. This articular capsule envelopes an articular disk or menisus interposed between the mandibular condyle and the mandibular fossa. The disk is connected to the articular capsule and to the tendon of the external pterygoid muscle (Fig. 5), and divides the joint into two synovial cavities.

Normal mandibular movements occuring through the temporomandibular articulation are the elevation and depression of the mandible, protrusion of the mandible and side-to-side movements of the mandible. Since the joint has two compartments, a superior between the menisus and the mandibular fossa and an inferior betwen the meniscus and the mandibular condyle, the type of movement of the mandible is influenced by the portion of the joint used. The first step during mouth opening involves a simple hinge action with the lower portion of the joint alone being used. The condyle head rotates around a point on the inferior surface of meniscus, the body of the mandible moving downward and backward (Fig. 6a). The second step calls both compartments of the joint into action and the condyle and miniscus glide anteriorly and inferiorly along the articular tubercle (Fig. 6b). This type of action occurs during protrusion, where the normal rotation of opening the joint is prevented by the synergetic action of the closing muscles. In lateral movements of the mandible, one disc glides forward while the other remains in place. A combination of rotary and gliding actions occurs during wider openings of mouth, where, to prevent the posterior surface of the ramus of the mandible from compressing the soft tissues between the mandible and the mastoid process caused by a pure hinge action, a gliding action is essential. This would bring the ramus forward and downward and provide greater freedom for further rotary movements (Fig. 7a).

In grinding or chewing movements, forward movement of one condyle preceeds, and in twin, causes a lateral displacement of the mandible.

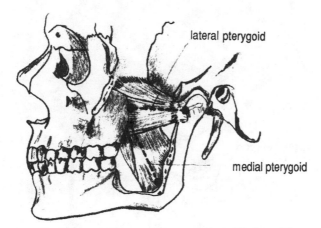

Figure 7-3. The medial and lateral pterygoids and their models.

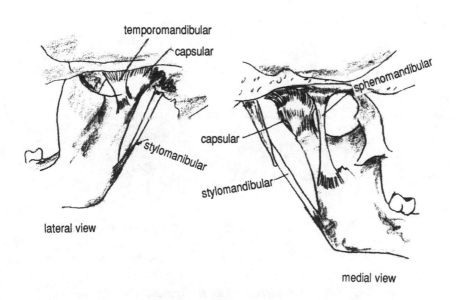

Figure 7-4. Ligaments at the temperomandibular joint.

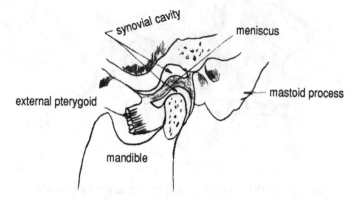

Figure 7-5.   The temperomandibular joint

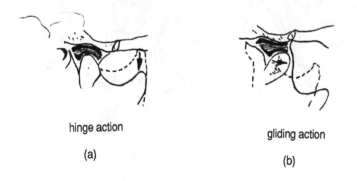

hinge action

(a)

gliding action

(b)

Figure 7-6.   Hinge and gliding action at the temperomandibular joint.

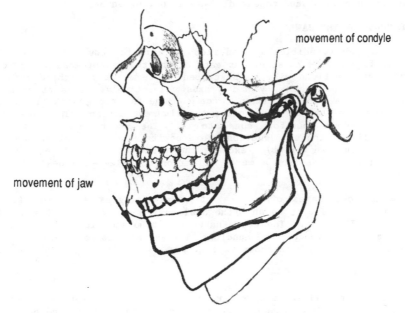

Figure 7-7a.   Kinematics of jaw rotation: sliding and rotation.

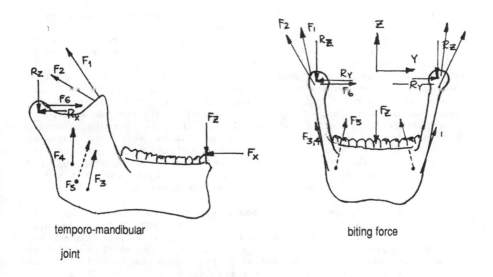

temporo-mandibular

joint

biting force

Figure 7-7b.   Free body diagram of the jaw.

373

The action of the closing muscles and the meshing of the teeth brings the mandible back into place. The condylar displacement may be alternate, or the same one displaced repeatedly as in chewing on one side.

THE MODEL OF THE JAW

The jaw (mandible) is modelled as a rigid body acted upon by six muscle forces on each side. At each of the two temporomandibular joints, the joint reactions will be described in terms of the three orthogonal components along the x, y and z axes respectively. Thus at each joint there will be three reaction force components, along with three moment components on the jaw to describe the ligament action. The total number of variables is therefore equal to 21. The jaw being a rigid body in space will require six equations to describe its equilibrium. Since the number of unknown variables exceeds the number of available equations for solution we will select a merit criterion to select the desired optimal solution.

The coordinates of points of muscle origins and insertions are selected with respect to an orthogonal system of axes with its origin at the right TM joint. The coordinates are tabulated in Table I. For the sake of convenience the mandible is assumed to be weightless. A schematic of the jaw is shown in Fig. 7b.

MODEL APPLICATIONS

During normal masticatory processes considerable forces are produced by the jaw muscles. However, whether or not the temporomandibular joint is load-bearing during such activities is a matter of controversy. Based on simplified mechanical analysis of the jaw (such as by Wilson (1)) or on anatomical and histological grounds (Scott (2), Steinhardt (3), Robinson (4)) it appears that there is no load at the joint during biting. That the joint may indeed be subjected to reactive forces has been proposed by Craddock (5), Roydhouse (6) who carried out mathematical analyses using assumed lines of muscle action. Their assumptions, however, in neglecting the horizontal components of the forces render the studies inconclusive. A recent investigation into the forces at the TM joint has been reported by Barbenel (7) who has modeled the muscles by their lines of action and obtained the values for joint force based on linear programming techniques. He used two different criteria namely minimize total muscle action and minimize joint force to obtain the results during occlusion. The analysis however treats the jaw as a plane problem rather than a rigid body in space. Also, the temporalis muscle has been modelled by only one line contrary to EMG evidence. Barbenel concludes that the TM joint is load bearing during function and that the minimum muscle force objective as suggested by some (MacConaill (8)) and which yields only one muscle (Marseter) to be active using the model, contrary to EMG evidence (Moller (9)), does not apply.

The model developed herein should be applicable to more general funcitons of the jaw since it is a three-dimensional model. We will investigate two main static activities——biting with jaws closed and biting with jaws open. Pertinent information regarding the kinematics of the mandible and the muscle activity for these studies is readily available for verfication of the model results.

CRITERION USED

The 3-D model described in terms of 21 variables and 6 equations will be solved for muscle forces and joint reactions and moments using several different criteria:

1. minimize $U = \sum_{i=1}^{6} F_i + \sum_{j=1}^{6} F_j$

    F_i = muscles on the left side

    F_j = muscles on the right side

2. minimize $U = (R_x + R_y + R_z)_{\text{right jt.}} + (R_x + R_y + R_z)_{\text{left jt.}}$

3. minimize

$$U = \sum_{i=1}^{6} F_i + \sum_{j=1}^{6} F_j + 2(R_x + R_y + R_z)_{\text{right}}$$

$$+ 2(R_x + R_y + R_z)_{\text{left}} + 4(M_x + M_y + M_z)$$

The first criterion simply implies minimization of total muscle effort while the second is based on minimization of the total force on the temporomandibular joints. The latter will be used to solve for the unknown variables for the case of biting with closed jaws, with a view to ascertain whether this action can be performed without any load on the temporomandibular joints. The third criterion extends the optimization process to include all the muscle forces, joint reactions and moments with suitable weighting factors and matches the criterion used to investigate the upper extremities. (See Chapter V). There it was discussed why the joint reaction forces should be included in the previously selected criterion (minimize $U = \Sigma F + 4M$) adopted for the lower extremities and the vertebral column.

Since the joints of the upper extremities may be subjected to tensile forces unlike those of the lower extremities which perform normally under compression, it may be desirable to include the reaction forces at the former in the objective functions. Since the temporamandibular joints are also normally subjected to tensile forces it would be only appropriate to investigate the application of this modified criterion to the jaw.

CLOSED JAWS BITING

The action studied was biting with jaws closed, with the load acting at arbitrarily selected distances from the intercondylar axis. This represents the process of occlusion with particular segments of the dental arch involved, for example, a load acting close to the intercondylar axis representing biting with posterior (molars) teeth while a load distant from the TM joint represents force exertion on the anterior (incisors) teeth. In this action, it is assumed that at any location, the corresponding teeth on both sides of the sagittal plane are employed. Accordingly the resultant occlusal load may be assumed to lie in the sagittal plane and to have no lateral components. Since the exact orientation of the biting force is not known, the action was analyzed for a wide range of sagittal inclination of the force. This inclination

375

TABLE I.   Coordinates of muscle origins and insertions.

Muscle i	Muscle Name	x^1	y^1	z^1	x^2	y^2	z^2
1	Temporalis (Anterior)	-1.5	-4.5	5.3	3.3	-3.6	-0.5
2	Temporalis (Posterior)	-4.5	-5.7	4.8	2.9	-3.6	0.3
3	Masseter (Superficial)	3.5	-5.0	-0.5	1.0	-4.5	-4.7
4	Masseter (deep)	2.0	-5.5	-0.1	1.5	-4.5	-2.7
5	Medial Pterygoid	2.3	-2.0	-0.6	0.8	-3.8	-4.7
6	Lateral Pterygoid	3.0	-3.0	-0.2	0.5	-3.9	0

1. Points located on the skull.
2. Points located on the mandible.

TABLE II.   Rotation and translation of mandible during jaw opening.

Jaw Opening (mm)	DX (cm)	DZ (cm)	ϕ (degrees)
7	0.05	-0.05	5.3
12	0.35	-0.35	8.45
17	0.55	-0.5	11.8
22	0.8	-0.7	14.95
27	1.0	-0.75	19.2

ranged from −50° to 50° with the vertical at 5-degree intervals. Similarly the load magnitude was arbitrarily selected as 5 kg; however since the final results for the joint loads are normalized with respect to the biting force, the initial value of the latter should not be of significance. That is, the resultant TM joint force is computed as $\sqrt{R_x^2 + R_y^2 + R_z^2}$, where R_x, R_y and R_z are the three respective components at either of the TM joints. Since the action simulated is symmetrical about the sagittal plane, the forces in the corresponding muscles on both sides and the forces at the TM joints should be equal.

OPEN JAWS BITING

In this study the model will be applied to the situation when the biting is done with the jaws open. The muscle load sharing will be analyzed for six different biting pressures (1.2, 2.2, 3.4, 4.6, 5.7 and 6.9 kg) at five different jaw opening (7,12,17,22,27 mm) as investigated in the experimental study by Garrett et al (12). The authors recorded the electromyographic signals from the masseter muscle during this activity in a large number of subjects with normal and abnormal occlusion, and this information will be utilised to check the theoretical results from the model.

As discussed earlier when the jaws are opened, there is both mandibular sagittal rotation about the intercondylar axis and translation of the condyles. Using antomical information on jaw opening kinematics (see Fig. 8), the mandible is rotated and translated to the required extent so that the desired jaw opening is obtained. This is the distance between the tip of the maxillary incisor and the tip of the mandibular incisor. Note that during the rest position of the mandible, the tips of the upper and lower incisors do not contact but the maxillary incisor projects over the mandibular incisor. In a horizontal direction, the two teeth are apart by a distance called 'overjet' and in the vertical direction the distance separating them is called 'overbite'. Any configuration of the jaw resulting from rotation and translation of the mandible so as to produce a given joint opening should therefore take into account the initial overjet and overbite. The selected rotations and translations in terms of X and Z components corresponding to the five jaw openings are shown in Table II. The overjet and overbite values were selected to be 1.8 mm and 2.3 mm respectively as reported by Garrett et al (12). The biting load is assumed to act at the tip of the mandibular incisor and is taken to be directed along a line joining the upper and lower incisors. Note that due to the sagittal symmetry of the biting action, the biting force lies in the sagittal plane with no lateral components.

The merit criterion used in this case is based on minimization of all muscle forces, twice the joint reaction forces and four times the joint moments.

RESULTS AND DISCUSSIONS

OPEN JAW BITING

The results for muscle forces, joint reaction forces and moments during open jaw biting with different pressures at different jaw openings are shown in Table III. Since the activity analyzed is symmetric about the sagittal plane, the results for muscle forces and joint reactions

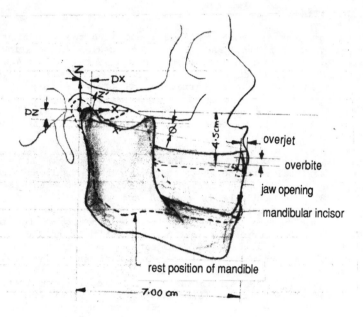

Figure 7-8.   Modelling of the jaw kinematics by translations DX, DZ, and rotation θ.

378

which are shown for the right side only in the table will be identical to those for the left side.

When plotted, the force in the masseter muscle at different jaw openings at various levels of biting intensities is as shown in Fig. 9. The corresponding electromyographic pattern for this muscle as reported by Garrett et al (12) is shown in Fig. 10. It is obvious that the theoretical results predicted from the model are in excellent agreement with those obtained by experimental studies. This correlation is emphasized in Fig. 11, where the ratio of EMG/EMG_{max} for different biting pressures at the constant opening of 17 mm and the ratio of F/F_{max} (ratio of theoretical force in masseter at any particular biting intensity to the theoretical force at maximum biting pressure) display an almost perfect linear relationship. Similar plots can be obtained for the other jaw openings.

The agreement between the experimental EMG signals (μv) vs biting pressure, and the calculated force in the masseter vs biting pressure, for each jaw opening is shown in Figs. 12a-c. At a constant jaw opening the EMG signals increase linearly with increasing biting pressure, while the theoretical force in the muscle as obtained from the model also increase linearly with increasing biting pressure. One may therefore obtain a ratio for the recorded EMG signal to the calculated muscle force from these plots, and these ratios are pltoted in Fig. 13 for different biting pressures. The scatter in the plot is surprisingly minimal and a value of 70 μv/kg appears to be the average. If one were to use this number and calculate the force in the muscle corresponding to the EMG signals (force (kg) = EMG signal (μv)/70) a plot as shown in Fig. 14 is obtained. In this plot, the force value as obtained from the model and as calculated using the mean EMG to force conversion factor (70 μv = 1 kg) are compared for different biting intensities. The model prediction agrees with the converted force value.

CLOSED JAW BITING

The variation in the TM joint force when the biting force is located at various distances from the intercondylar axis (3.0, 4.0, 5.0, 6.0 and 6.82 cm) and acts at different orientations is shown in Fig. 15. These results are based on the criterion that minimizes the total joint force ($U = \Sigma R_x + \Sigma R_y + \Sigma R_z$) and are tabulated in Table IV. The results indicate that it is possbile to have no load on the TM joint during activity in certain situations. The likelihood is greater the closer the biting pressure is applied from the TM joint and if the applied pressure orientation favors certain specific angles. For example, at a distance of 3 cm from the TM (a position which favors biting with the posterior teeth) there is no reaction force at the TM joint if the applied biting load is inclined to the vertical between $\alpha = +15°$ and $\alpha = -5°$. This implies that over $15° - (-5°) = 20°$ range there is no force at the TM joint at a point distant DX = 3 cm. As one moves away from the joint however, the range over which the TM joint is load-free decreases. For example at a point distant 4 cm from the TM joint, the reaction force at this joint is zero over a $-5° - (-15°) = 10°$ range only. For DX = 5 cm, this range is further diminished to only $-20° - (-25°) = 5°$. For DX = 6 and DX = 6.82 cm locations which would involve biting pressure on the canines and the incisors, the TM joint is always load bearing, and for no value of the load inclination is the joint force zero. The closer the biting pressure to the TM joint therefore, the greater the likelihood that the joint is free of load for

379

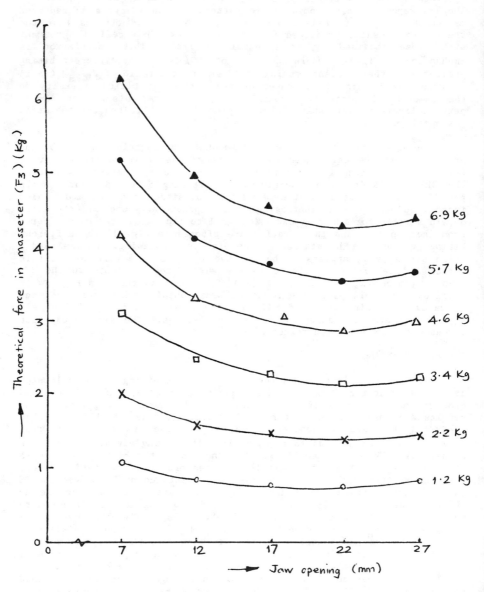

Figure 7-9. Theoretical results for masseter force (F_3)
(Criterion: $U = \Sigma F + 2R + 4M$)

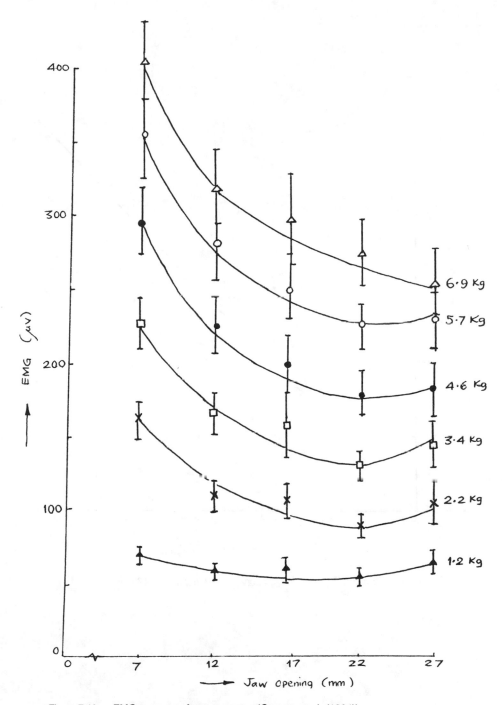

Figure 7-10.　EMG response from masseter (Garrett, et al. (1964)).

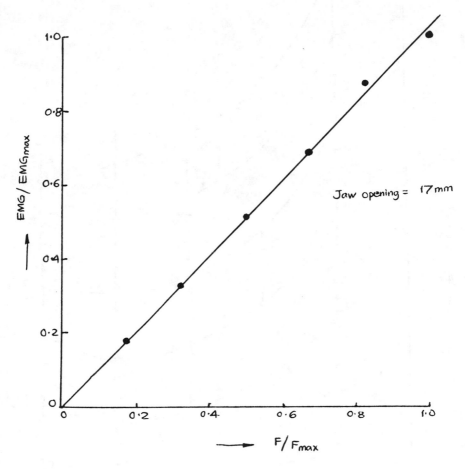

Figure 7-11. EMG/EMB$_{max}$ vs. F/F$_{max}$ curve for jaw opening of 17 mm.

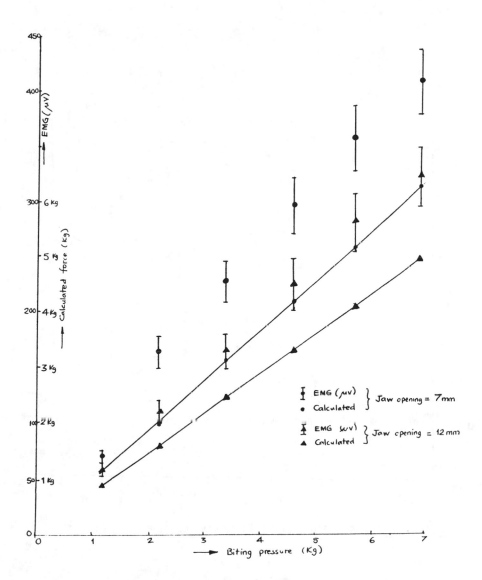

Figure 7-12a. Theoretical and experimental results for masseter at J.0 = 7 & 12 mm.

383

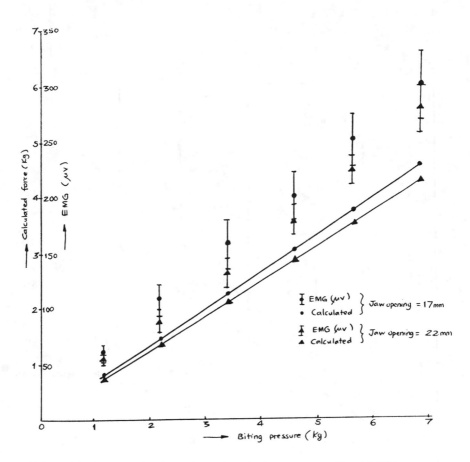

Figure 7-12b. Theoretical and experimental results for masseter at J.0 = 17 & 22 mm.

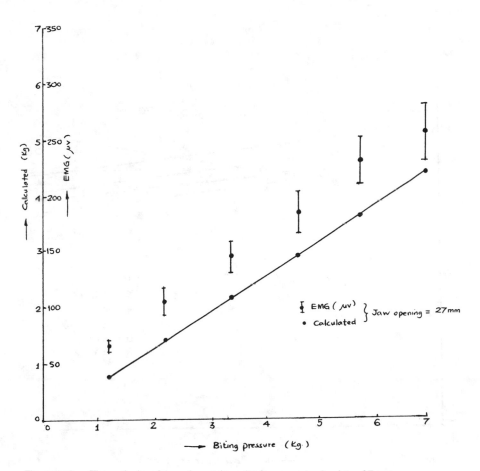

Figure 7-12c. Theoretical and experimental results for masseter for J.0 = 27 mm.

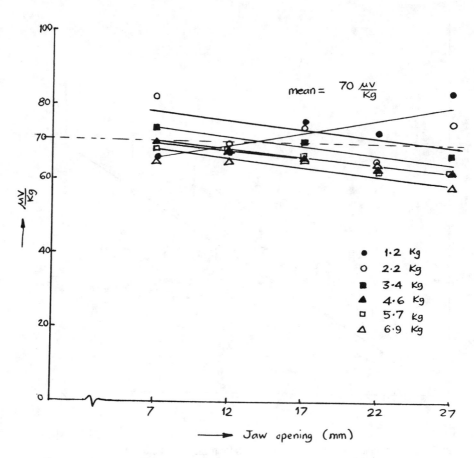

Figure 7-13.   Ratio EMG(μv)/Force in masseter (kg) for various biting pressures.

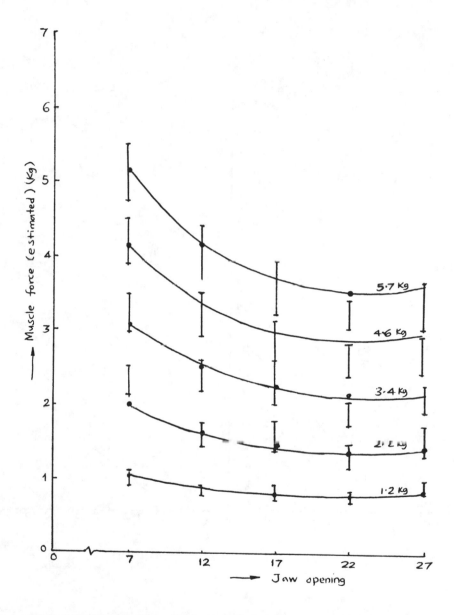

Figure 7-14. Estimated force (kg) in masseter (using 1 kg = 70 μv) for various biting pressure.

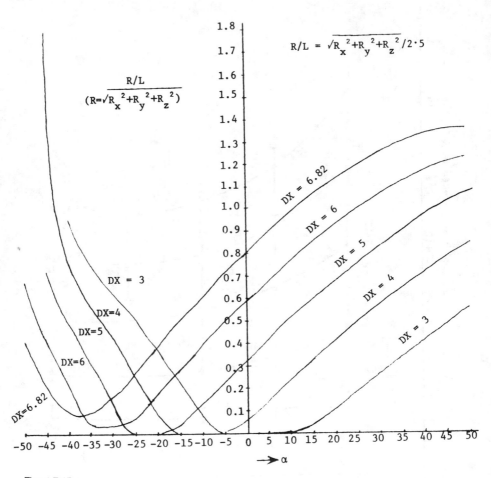

Figure 7-15.

TABLE III. Results for biting with open jaw.*

Jaw Opening	7m	12m	17m	22m	27m	Biting Pressure
F_1	0.46	0.79	0.86	0.89	0.94	
F_3	1.08	0.86	0.79	0.74	0.76	
R_x	0	0	0	0	0	1.2 Kg
R_y	0.17	0.18	0.18	0.17	0.17	
R_z	-0.96	-0.93	-0.91	-0.89	-0.94	
U	7.59	7.74	7.62	7.48	7.84	
F_1	0.84	1.46	1.57	1.63	1.72	
F_3	1.99	1.58	1.45	1.36	1.39	
R_x	0	0	0	0	0	2.2 Kg
R_y	0.31	0.33	0.32	0.31	0.31	
R_z	-1.76	-1.69	-1.66	-1.63	-1.73	
U	13.93	14.19	13.97	13.71	14.38	
F_1	1.29	2.26	2.43	2.52	2.65	
F_3	3.07	2.44	2.24	2.09	2.16	
R_x	0	0	0	0	0	3.4 Kg
R_y	0.47	0.51	0.49	0.47	0.48	
R_z	-2.73	-2.62	-2.57	-2.51	-2.67	
U	21.53	21.94	21.58	21.18	22.23	
F_1	1.75	3.06	3.29	3.41	3.59	
F_3	4.16	3.29	3.03	2.84	2.93	
R_x	0	0	0	0	0	4.6 Kg
R_y	0.64	0.69	0.67	0.64	0.65	
R_z	-3.69	-3.5	-3.47	-3.39	-3.61	
U	29.13	29.69	29.19	28.66	30.07	
F_1	2.17	3.79	4.08	4.23	4.44	
F_3	5.15	4.08	3.75	3.52	3.62	
R_x	0	0	0	0	0	5.7 Kg
R_y	0.79	0.86	0.83	0.79	0.80	
R_z	-4.57	-4.39	-4.29	-4.21	-4.48	
U	36.09	36.78	36.18	35.51	37.26	
F_1	2.63	4.59	4.94	5.12	5.38	
F_3	6.23	4.94	4.54	4.26	4.39	
R_x	0	0	0	0	0	6.9 Kg
R_y	0.96	1.04	1.01	0.96	0.97	
R_z	-5.53	-5.32	-5.21	-5.09	-5.42	
U	43.69	44.53	43.79	42.99	45.10	

*1. Criterion used, minimize $U = \Sigma R_i + 2(R_x + R_y + R_z) + 4(M_x + M_y + M_z)$.

2. Results shown for one side of the jaw only. F_1 = force in anterior temporalis. F_3 = force in masseter (superficial); forces in other muscles are zero.

certain specific orientations of the applied pressure. An individual should therefore, make greater use of the posterior teeth to minimize loads on the TM joint during mandibula activity. These findings appear significant when viewed in light of the work by other authors as described earlier, some of who claim that the TM joint is always load bearing while others suggest that the joint is load-free during mandibular activity. It is possible, as shown here, that an individual may adjust the biting pressure for both location and orientation so as to produce no reaction force at the TM joint. In situations other than these, the TM joint is always load bearing. Moreover, the technique described here which as also adopted by Barbenel (7) failed to produce similar findings. The model described by the author was a two-dimensional one.

The nature of the load on the TM joint varies from compressive to tensile over the range of locations and orientations of biting pressrues investigated. This interesting pattern is illustrated in Fig. 16 where "isobars" or equal reaction force curves are plotted as a function of distance of the biting force from the TM joint (DX) and its vertical inclination (α). The shaded area represents the situations where TM joint is load free and shows how movement away from this joint enhances the possibility of the joint being loadbearing. For negative values of α, that is, when the biting pressure acting on the teeth is directed posteroanteriorly, the vertical component R_z of the joint reaction force is mostly tensile, while for positive values of α it is mostly compressive with the transition zone lying in between.

These curves can be conveniently used as a design plot or guide for the fabrication of dentures, dental implants, etc.

The typical effect of selection of the criterion established for obtaining the solution can be observed on comparison of the results tabulated in Tables IV through VI. The muscle load sharing predicted from the model based on minimization of total joint reaction forces, minimization of total muscle effort and minimization of all muscle forces, twice the joint reactions and four time the ligament action (joint moments) are shown in Tables IV, V and VI respectively with the occlusal load located at DX = 4 cm from the TM joint. As can be seen, the objective function $U = \Sigma R_x + R_y + R_z$ yields results that are more meaningful, suggesting that at least three of the four muscles, the temporalis, the masseter and the lateral pterygoid are active for most of the entire range of inclinations of the occlusal loads investigated. However, only the third merit criterion which includes muscle forces, joint reaction forces and moments with suitable weighting factors in the objective function apears to indicate activity in the appropriate muscles— the temporalis, the masseter and the medial pterygoid for most of the range of occlusal load orientations studied. These results on criterion selection are also witnesed when the occlusal load is positioned at other locations from the TM joint. It may be recalled that this criterion was also successfully applied to the upper extremities previously (see Chapter VI) in the spinal column model, and thus, appears to be more favorable among the other formulated criteria for application to joints of the musculoskeletal structure that may normally be subjected to tensile loads.

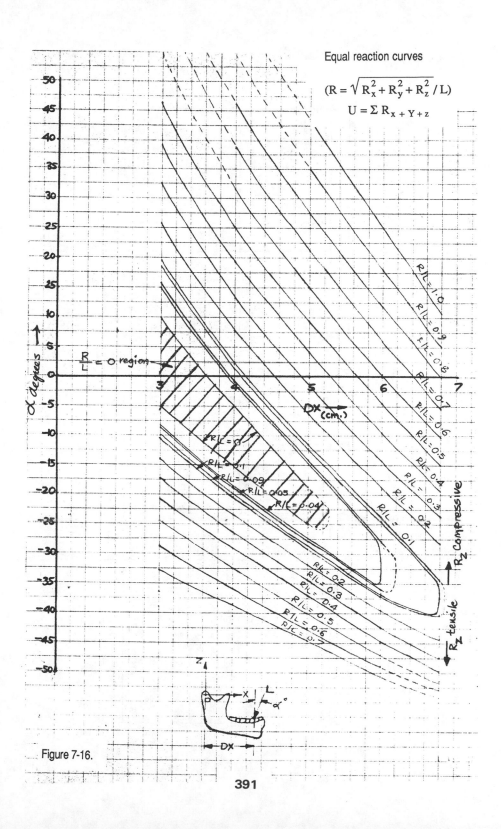

Equal reaction curves

$$\left(R = \sqrt{R_x^2 + R_y^2 + R_z^2} \,/\, L\right)$$
$$U = \Sigma\, R_{x + Y + z}$$

Figure 7-16.

TABLE IV. Results for closed jaw biting (DX=4.00cm), $U = \Sigma R_x + R_y + R_z$

Variable \ $\alpha°$	-50	-45	-40	-35	-30	-25	-20	-15	-10	-5	0	5	10	15	20	25	30	35	40	45	50
F_1	0	0	0	0	0	0	0	0	0	0	0	0	0	0	0	0	0	0	0	0	0
F_2	0	0	0.67	1.39	1.64	1.62	1.57	1.54	2.13	3.64	4.16	4.14	4.09	4.01	3.89	3.75	3.58	3.38	3.16	2.91	2.64
F_3	0	0	0	0	0	0	0	0.36	0.77	0.75	0.92	1.19	1.46	1.71	1.95	2.17	2.38	2.56	2.73	2.88	3.01
F_4	0	0	0	0	0	0	0.29	0.25	0	0	0	0	0	0	0	0	0	0	0	0	0
F_5	0	0	0	0	0.36	0.92	1.24	1.28	0.89	0.23	0	0	0	0	0	0	0	0	0	0	0
F_6	35.39	2.36	0	0	0	0	0	0	0.70	2.51	3.18	3.25	3.29	3.31	3.30	3.27	3.21	3.13	3.03	2.89	2.75
R_x	-35.12	-3.98	-1.05	-0.28	0	0	0	0	0	0	0	0	0	0	0	0	0	0	0	0	0
R_y	-11.96	-0.79	0.16	0.33	0.25	0.03	0	0	0	0	0	0	0	0	0	0	0	0	0	0	0
R_z	4.26	1.95	1.58	1.35	1.03	0.65	0.22	0	0	0	-1.49	-0.38	-0.59	-0.82	-1.03	-1.24	-1.43	-1.62	-1.79	-1.95	-2.09
M_x	0	0	0	0	0	0	0	0	0	0	0	0	0	0	0	0	0	0	0	0	0
M_y	0	0	0	0	0	0	0	0	0	0	0	0	0	0	0	0	0	0	0	0	0
M_z	0	0	0	0	0	0	0	0	0	0	0	0	0	0	0	0	0	0	0	0	0
R/L	14.94	1.79	0.76	0.57	0.42	0.26	0.09	0	0	0	0.06	0.15	0.24	0.33	0.41	0.49	0.57	0.65	0.72	0.78	0.84
U	102.68	13.43	5.57	3.90	2.55	1.37	0.43	0	0	0	0.29	0.75	1.19	1.64	2.06	2.47	2.86	3.23	3.58	3.89	4.18

TABLE V. Results for closed jaw biting (DX=4.00cm), $U = \Sigma F_i$

α° Variable	-50	-45	-40	-35	-30	-25	-20	-15	-10	-5	0	5	10	15	20	25	30	35	40	45	
F_1	0	0	0	0	0	0	0	0	0	0	0	0	0	0	0	0	0	0	0	0	0
F_2	0	0	0	0	0	0	0	0	0	0	0	0	0	0	0	0	0	0	0	0	0
F_3	0	0	0.36	0.73	1.11	1.47	1.83	2.17	2.49	2.79	3.08	3.34	3.58	3.78	3.96	4.11	4.23	4.31	4.37	4.38	4.37
F_4	0	0	0	0	0	0	0	0	0	0	0	0	0	0	0	0	0	0	0	0	0
F_5	0	0	0	0	0	0	0	0	0	0	0	0	0	0	0	0	0	0	0	0	0
F_6	35.39	2.36	0	0	0	0	0	0	0	0	0	0	0	0	0	0	0	0	0	0	0
R_x	-35.12	-3.98	-1.79	-1.81	-1.81	-1.81	-1.79	-1.75	-1.70	-1.64	-1.57	-1.48	-1.39	-1.28	-1.16	-1.04	-0.90	-0.76	-0.61	-0.46	-0.31
R_y	-11.96	-0.79	0.04	0.08	0.11	0.15	0.19	0.22	0.25	0.29	0.31	0.34	0.36	0.39	0.40	0.42	0.43	0.44	0.44	0.45	0.45
R_z	4.26	1.95	1.61	1.42	1.22	1.01	0.79	0.56	0.33	0.09	-0.13	-0.37	-0.59	-0.82	-1.04	-1.25	-1.45	-1.64	-1.82	-1.98	-2.13
M_x	0	0	0	0	0	0	0	0	0	0	0	0	0	0	0	0	0	0	0	0	0
M_y	0	0	0	0	0	0	0	0	0	0	0	0	0	0	0	0	0	0	0	0	0
M_z	0	0	0	0	0	0	0	0	0	0	0	0	0	0	0	0	0	0	0	0	0
R/L	14.94	1.79	0.96	0.92	0.88	0.83	0.78	0.74	0.70	0.67	0.64	0.63	0.62	0.63	0.64	0.67	0.70	0.74	0.79	0.83	0.88
U	70.79	4.71	0.71	1.47	2.22	2.95	3.66	4.34	4.98	5.59	6.16	6.68	7.15	7.57	7.92	8.22	8.45	8.63	8.73	8.77	8.74

TABLE VI. Results for closed jaw biting (DX=4.00cm), $U = \Sigma F + 2(R_x+R_y+R_z) + 4(M_x+M_y+M_z)$

Variable \ α°	-50	-45	-40	-35	-30	-25	-20	-15	-10	-5	0	5	10	15	20	25	30	35	40	45	50
F_1	0	0	0	0	0	0	0.60	1.31	2.22	2.13	2.03	1.92	1.79	1.65	1.49	1.33	1.15	0.97	0.77	0.58	0.37
F_2	0	0	0.67	1.39	1.64	1.62	1.19	0.67	0	0	0	0	0	0	0	0	0	0	0	0	0
F_3	0	0	0	0	0	0	0	0	0	0.25	0.59	0.92	1.25	1.57	1.87	2.16	2.44	2.69	2.93	3.14	3.33
F_4	0	0	0	0	0	0	0	0	0	0	0	0	0	0	0	0	0	0	0	0	0
F_5	0	0	0	0	0.36	0.92	0.97	0.92	0.68	0.89	0.94	0.98	1.02	1.05	1.07	1.09	1.09	1.09	1.08	1.06	1.03
F_6	0	0	0	0	0	0	0	0	0	0	0	0	0	0	0	0	0	0	0	0	0
R_x	-1.92	-1.77	-1.05	-0.28	0	0	0	0	0	0	0	0	0	0	0	0	0	0	0	0	0
R_y	0	0	0.16	0.33	0.25	0.03	0	0	0.07	0	0	0	0	0	0	0	0	0	0	0	0
R_z	1.61	1.77	1.58	1.35	1.03	0.65	0.34	0.05	-0.23	-0.51	-0.74	-0.97	-1.18	-1.39	-1.59	-1.78	-1.96	-2.11	-2.26	-2.38	-2.49
M_x	0	0	0	0	0	0	0	0	0	0	0	0	0	0	0	0	0	0	0	0	0
M_y	2.66	0.18	0	0	0	0	0	0	0	0	0	0	0	0	0	0	0	0	0	0	0
M_z	0	0	0	0	0	0	0	0	0	0	0	0	0	0	0	0	0	0	0	0	0
R/L	1.0	1.0	0.76	0.57	0.42	0.26	0.14	0.02	0.09	0.20	0.29	0.39	0.47	0.56	0.64	0.71	0.78	0.85	0.90	0.95	0.99
U	24.72	14.85	12.48	10.59	9.10	7.82	6.88	5.97	6.98	8.56	10.07	11.51	12.86	14.11	15.25	16.28	17.18	17.96	18.59	19.09	19.44

REFERENCE

1. Wilson, G.H.: A Manual of Dental Prosthetics. Henry Kimpton, London, 1920.

2. Scott, J.H.: A contribution to the study of the mandibular joint function. Br. Dent. J. 94: 345, 1955.

3. Steinhardt, G.: Die Beanspruchung der Gelk flächen bei vershiedenen Bissarten. Dt. Zahnheilk Vortr. 1: 9, 1934.

4. Robinson, M.: Theory of reflex controlled nonlever action of the mandible. J. Am. Dent. Assoc. 33: 1261, 1946.

5. Craddock, F.W.: A review of Costen's syndrome. Br. Dent. J., 91: 199,1951.

6. Roydhouse, R.: Upward force of the condyles on the cranium. J. Am. Dent. Assoc., 50: 166, 1955.

7. Barbenel, J.C.: The biomechanics of the TM Joint: A theoretical study. J. Biomech. 5: 251, 1972.

8. Mac Conaill, M.A.: The ergonomic aspects of articular mechanisms. In Studies on the Anatomy and Function of Bones and Joints (Ed. F.G. Evans) Springer-Verlag, Berlin, 1967.

9. Moller, E.: The chewing apparatus. Acta. Physiol. Scand. 69, Suppl. 280, 1966.

10. Grant, P.G.: Lateral pterygoid: 2 muscles? Am. J. Anat., 138: 1, 1973.

11. McNamara, Jr., J.A.: The independent funcitons of the two heads of the lateral pterygoid muscle. Am. J. Anat.; 138: 197, 1973.

12. Garrett, F.A., Angelone, L. and Allen, W.I.: Electrical response of masseter muscles. Am. J. Orthodont., 50: 435, 1964.

13. Basmajian, J.V.: Muscles Alive. Williams & Wilkins, Baltimore, 1974.

14. Moller, E.: The chewing apparatus. Acta. Physiol. Scand., 69, Suppl. 280, 1966.

CHAPTER VIII

The Hand

"When you represent the hand represent with it the arm as far as the elbow, and with this arm the sinews and muscles which come to move this arm away from the elbow.   And   do   the   same   in   the demonstration of the foot.

All the muscles that start at the shoulders, the shoulder-blade and the chest, serve for the movement of the arm from the shoulder to the elbow. And all the muscles that start between the shoulder and the elbow, serve for the movement of the arm between the elbow and the hand. And all the muscles that start between the elbow and the hand, serve for the movement of the hand. An all the muscles that start in the neck, serve for the movement of the head and shoulders."

Leonardo da Vinci
1452-1519

INTRODUCTION

Unlike other segments of the human musculoskeletal structure modelled so far, modelling of the hand (and the foot) offers a formidable challenge. These skeletal bodies are multi-segmented and held together by a complex architecture of numerous muscles and ligaments. The passive elements, namely, the ligaments play an active role in various functions and activities of the hand (and the foot), both from a view of kinematics as well as for synergistic participation with muscles during load sharing. Moreover, the interaction within these bodies is greatly influenced by other members of the extremities to which they are connected by extrinsic means. Therefore, in order to clearly define the functions of these bodies in terms of models, it becomes essential that they be modelled in totality and not viewed as independent parts isolated from the extremities. Comprehensive models of the hand and the foot thus will involve inclusion of the upper and the lower extremities and the interconnecting musculature in these models. Both the hand and the foot are constructed on somewhat similar lines, each having a proximal part, the carpus or the tarsus, a middle portion, the phalanges. The anatomy of the musculature—both extrinsic and intrinsic is also analogous in these segments although the overall functions may be quite different. Therefore, models of the hand and the foot will be quite similar although their applications—the hand for pulling or pinching and the foot for providing support during leaning and walking will not exactly be the same. This chapter is devoted to the development of a comprehensive

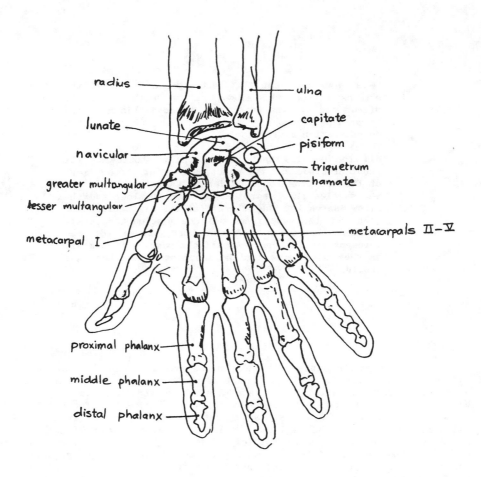

Figure 8-1. Palmar view of bones of the wrist and hand.

model of the hand including the upper extremities, and the model of the foot is deferred until the following chapter.

ANATOMY OF THE HAND: SEGMENTS AND JOINTS

The skeleton of the hand consists of three segments: the carpus or wrist bones; the vetacarpus or bones of the palm; and the phalanges or bones of the fingers. There are eight carpal bones which are arranged in two rows (see Fig. 1). The scaphoid, the lunate, the triangular and the pisiform comprise the proximal row arranged in that order from the radial to the ulnar side, while the distal row is made up of the trapezium, the trapezoid, the capitate and the lunate. There are five metacarpal bones in the palm of the hand, numbered I through V from the thumb side, which articulate at their distal extremities with the phalanges. Each finger has three phalanges—a proximal, a middle and a distal, while the thumb has only two, thus resulting in a total of fourteen. The wrist joint is a condylar articulation and is formed by the distal end of the radius and distal surface of the articular disk with the scaphoid, lunate and triangular carpal bones. The movements permitted in this joint are flexion-extension, abduction-adduction, and circumduction. These movements are combined with those of the carpals which articulate among themselves is each row and in articulations of the two rows with each other. Gliding joints are formed in the articulations of the proximal row of carpal bones and in those of the distal row, while at the midcarpal joint, between the two rows, flexion, extension and a slight amount of rotation is allowed. At the carpometacarpal joint of the thumb, the saddle-shaped articular surfaces allow considerable degree of freedom in different types of movements such as flexion and extension in the plane of the palm of the hand, abduction and adduction in a plane perpendicular to the palm, circumduction and opposition. At the carpometacarpal joints of the remaining four fingers however relatively little movement can occur, the articulations limited to slight gliding. The metacarpophalangeal joints formed by the reception of the rounded extremities of the metacarpal bones into the cavities on the proximal ends of the proximal phlanges, movements of flexion-extension, adduction-abduction and circumduction are allowed. Finally, at the interphalangeal joints which are hinge type articulations movements of flexion-extension only can occur.

MODEL OF THE HAND

For the sake of convenience the eight carpals may be lumped together into a single rigid body. With the additional five metatarsals and fourteen phalanges, a total of twenty rigid bodies will have to be included in the model. Since the muscle loadsharing in the hand is affected by other segments of the upper extremities (on which some of the extrinsic muscles originate) it becomes necessary to include all the upper extremity segments (the ulnaradius, the arm, the scapula and the clavicle) in the hand model (See Table I). The upper extremity model in which the hand was modelled as a rigid body has already been described earlier (see Chapter V) and therefore this part will not be described in this chapter. Although it will be linked to the overall hand model. Note that if this comprehensive model is applied to investigate a specific act or posture in which not all fingers come into play, one may accordingly simplify the model. For example, if the musculoskeletal model is used to analyse the pinch action between the index finger and the thumb, only these two fingers need to be considered. However, we will first attempt to develop a comprehensive model where all segments will be considered.

The extrinsic hand muscles modelled here are the following:
1. Extensor carpi radialis longus
2. Flexor carpi radialis
3. Extensor digitorum communis
4. Flexor digitorium sublimis
5. Extensor indicis proprius
6. Flexor digitorum profundus

7. Extensor pollicis longus
8. Flexor pollicis longus
9. Extensor pollicis brevis
10. Extensor carpi radialis brevis
11. Abductor pollicis longus

The intrinsic muscles of the hand that are included are the following:
1. Flexor pollicis brevis
2. Abductor pollicis brevis
3. Adductor pollicis
4. Opponens pollicis

5. Dorsal interosseous
6. Volvar interosseous
7. Lumbricales

TABLE I. Segments in the comprehensive hand model.

1. Segments of the upper extremities:

    a. Thorax
    b. Clavicle
    c. Scapula
    d. Humerus
    e. Ulna-radius

2. Segments of the hand:

    i. carpals

    ii. First Finger: a. Metacarpal I
                       b. Proximal Phalanx I
                       c. Distal Phalanx I

         Second Finger: a. Metacarpal II
                       b. Proximal Phalanx II
                       c. Middle Phalanx II
                       d. Distal Phalanx III

         Third Finger: a. Metacarpa III
                       b. Proximal Phalanx III
                       c. Middle Phalanx III
                       d. Distal Phalanx III

         Fourth Finger: a. Metacarpal IV
                       b. Proximal Phalanx IV
                       c. Middle Phalanx IV
                       d. Distal Phalanx IV

         Fifth Finger: a. Metacarpal V
                       b. Proximal Phalanx V
                       c. Middle Phalanx V
                       d. Distal Phalanx V

MUSCLE MODELS FOR THE EXTRINSIC AND THE INTRINSIC MUSCLES

The extrinsic muscles originate either on the humerus near the epicondyles or on the ulna-radius and proceed to insert on the fingers along the arm. Since the tendons of these muscles will usually wrap around the distal end of the ulna-radius (particularly in such acts as flexion-extension) an alternate location for the origins of these muscles can be selected in this vicinity. Therefore, these muscles are in general modelled by lines joining suitable origin points on the distal end of the ulna-radius to the corresponding insertion points on the hand. A muscle that originates on the epicondyle can therefore be modelled by a line connecting the origin on the epicondyle to a point on the ulnaradius and a line connecting the latter to the respective insertion point.

1. Extensor carpi radialis longus:

This muscle arises on the humerus near the lateral epicondyle and inserts into the base of the second metacarpal bone (MCII). It can be modelled by a line connecting the lateral epicondyle to the ulna-radius and by a line connecting the latter to the MCII (Fig. 2).

2. Flexor carpi radialis:

The flexor carpi radialis arises from the medial epicondyle of the humerus, passes through a groove on the trapezium bone and inserts into the base of the second metacarpal bone. The muscle is modelled by three lines:

    a.   A line connecting the medial epicondyle to the ulna-radius

    b.   A line connecting the ulna-radius to a point on the carpals representing the contact at the trapezium.

    c.   A line connecting the carpals to MCII (Fig. 3).

3. Extensor digitorum communis:

The extensor digitorum originates on the lateral epicondyle of the humerus, passes through the extensor retinaculum across the wrist and divides into four tendons which insert into the second and third phalanges of the fingers. The manner of insertion of the tendons is rather complex, forming the extensor apparatus for each finger. After crossing the metacarpophalangeal articulation the tendon is joined by the tendons, of the interossei and lumbricales, spreading into a broad aponeurosis covering the dorsal surface of the first phalanx (Fig. 4a). Opposite the first interphalangeal joint this aponeurosis divides into three bands. The central band is inserted into the base of the second phalanx, while the two collateral bands continue onward along the side of the second phalanx, unite across the distal interphalangeal joint and insert into the dorsal surface of the distal phalanx. It is aparent that when the fingers are flexed, the tendon will wrap around the digits and consequently the muscle is modelled by several lines, with appropriate reactions at assumed points of contact.

The muscle model for the extensor digitorum consists of four major parts—each part representing the part of the muscle responsible for the extension of an individual finger. Each part in turn consists of three components to represent the three bands—the central, the medial and the lateral which form the individual extensor mechanism. Hence, the central band of the extensor apparatus of each finger consists of:

    (i)   a line connecting the lateral epicondyle to the ulna-radius.

    (ii)   a line connecting the ulna-radius to a point on the dorsal surface of the metacarpal head.

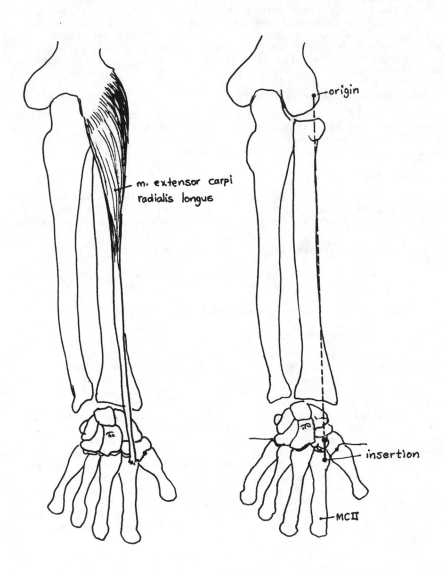

Figure 8-2. Muscle model **for** extensor carpi radialis longus.

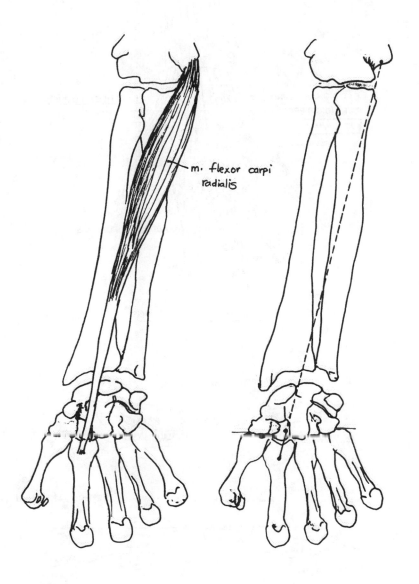

Figure 8-3. Muscle model for flexor carpi radialis.

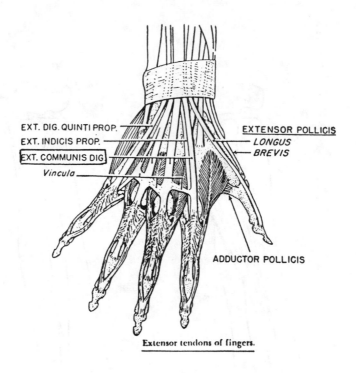

EXT. DIG. QUINTI PROP.

EXT. INDICIS PROP.

EXT. COMMUNIS DIG.

Vincula

EXTENSOR POLLICIS

LONGUS

BREVIS

ADDUCTOR POLLICIS

Extensor tendons of fingers.

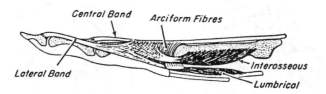

Central Band   Arciform Fibres

Lateral Band

Interosseous

Lumbrical

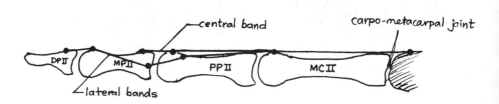

central band

carpo-metacarpal joint

DP II   MP II   PP II   MC II

lateral bands

Figure 8-4a. Model of the **ex**tensor apparatus of the finger.

(iii)   a line connecting the metacarpal head to a point on the
         dorsal surface of the proximal phalanx head.

(iv)   a line connecting the proximal head to the middle phalanx.

The collateral bands (medial and lateral) are modelled as lines
connecting the humerus to the ulna-radius and from there to the distal
phalanx through several points on the intermediate simulating the
wrapping of the tendons around these bones at the appropriate points as
shown in Fig. 4b and 4c.

4. Flexor digitorum sublimis:

This muscle arises from three heads of origin—humeral, ulnar and
radial, and divides into two planes of muscular fibers—the superficial
plane carrying tendons for insertion into the middle and ring fingers,
and the deep plane carrying tendons for insertion into the index and
little fingers.   The tendons pass beneath the flexor retinaculum, and
opposite the bases of the first phalanges each tendon divides into two
slips to allow the corresponding tendon of flexor digitorum profundus to
pass through.   The two slips reunite to form a grooved channel for the
reception of the accompanying tendon of the flexor digitorum profundus
and finally divided to insert into the sides of the second phalanx about
its middle.   The tendons, during their course of insertion, enter into
the fibrous tendon sheaths beginning proximally over the heads of the
metacarpal bones, which prevent their "bowstringing".   It is assumed that
the tendons are held to the digits by the fibrous sheaths at
approximately the middle of each segment, where, due to the change in the
direction of the muscular line of action a reaction will occur.   The
muscle therefore can be modelled by four components—each component
responsible for the flexion of an individual finger.   Each flexor slip in
turn can be modelled by a numerical origin as well as a ulna-radial
origin which then is extended to the finger in the following fashion.

a.   line connecting the ulna-radius to the carpals at a point below
      the flexor retinaculum where the tendon is assumed to pass in the
      carpal tunnel.

b.   a line connecting the carpals to a point on the volar side of the
      metacarpal in the vicinity of the head where the tendon enters
      the digital sheath.

c.   a line connecting the metacarpal to the middle of the volar side
      of the proximal phalanx.

d.   a line connecting the proximal phalanx to the middle of the
      middle phalanx on the volar side (see Fig. 5) representing the
      final insertion.

5. Flexor digitorum profundus:

The flexor digitorum profundus arises from the ulna and ends in four
tendons which pass through the openings in the tendons of the flexor
digitorum sublimis as described earlier, and insert into the bases of the
last phalanges.   The model of this muscle is therefore similar to that of
the flexor digitorum superficialis described above—four components, each
going to an individual finger.   Each component is modelled by five lines
similar to the superficial digital flexor with an additional line

405

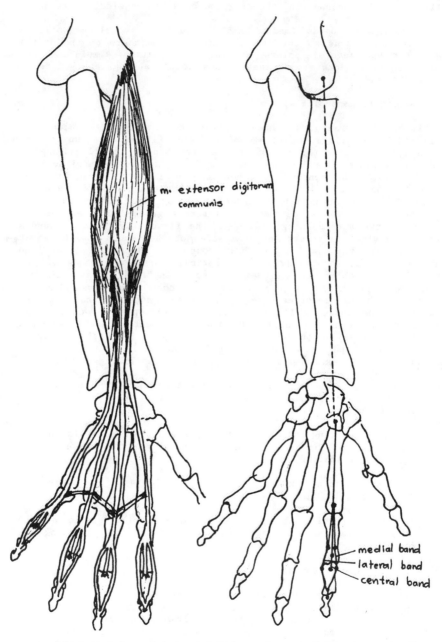

m. extensor digitorum communis

medial band
lateral band
central band

Figure 8-4b. Muscle model for extensor digitorum communis.

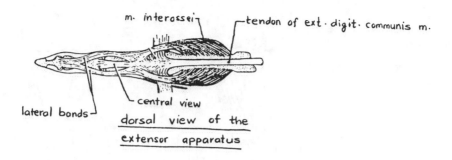

m. interossei

tendon of ext. digit. communis m.

lateral bonds

central view

dorsal view of the
extensor apparatus

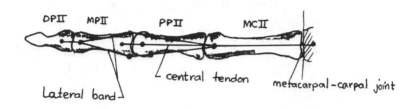

DP II   MP II   PP II   MC II

Lateral band

central tendon

metacarpal-carpal joint

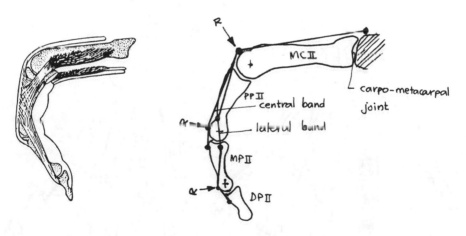

R

MC II

carpo-metacarpal
joint

PP II
central band

lateral band

MP II

DP II

lateral view of the finger in flexion
showing wrapping of the extensor
tendon.

Figure 8-4c.   Wrapping of extensor digitorum communis tendon.

407

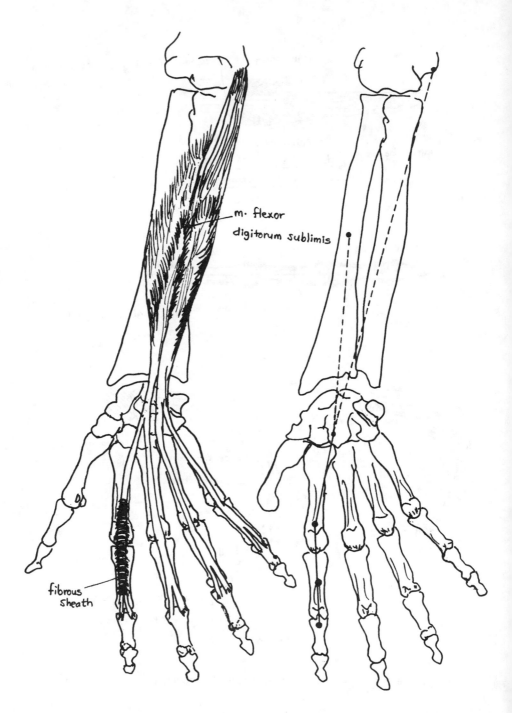

Figure 8-5. Muscle model for flexor digitorum sublimis.

connecting the middle phalanx to the terminal or digital phalanx to represent the final insertion on that segment (see Fig. 6).

6. Extensor indicis propruus:

This muscle originates on the dorsal surface of the body of the ulna and passes under the extensor retina culum along with the extensor digitorum communis. Opposite the head of the second metacarpal (MCII) it joins the ulnar side of the tendon of the extensor digitorum communis to the index finger. This muscle is modelled identical to the extensor digitorum communis, with the same three bands (medial, central and lateral) of the extensor apparatus of the index finger as described earlier (Fig. 7).

7. Extensor pollicis longus:

The extensor pollicis longus arises from the ulna and from the ulna and from the interosseous membrane, and inserts into the base of the last phalanx (DPI) of the thumb. The muscle can be modelled by:

a. A line connecting the ulnar origin to the metacarpal head (MCI).

b. A line connecting the MCI to the proximal phalanx I (PPI).

c. A line connecting the proximal phalanx I to the distal phalanx I (DPI).

The muscle is therefore assumed to wrap around the PPI and MCI contact points on its way to insertion on the distal phalanx (see Fig. 11).

8. Extensor pollicis brevis:

This muscle arises from the radius and the interosseous membrane and inserts into the base of the first phalanx of the thumb. It is therefore modelled as a line connecting the ulna-radius to a point on the metacarpal I and a line connecting the latter point to the insertion on the proximal phalanx I (Fig. 11).

9. Flexor pollicis longus:

The flexor pollicis longus has its origin on the radius, the interosseous membrane and the medial epicondyle of the humerus. Its fibers converge into a flattened tendon which passes under the flexor retinaculum, enters into an osseo-aponeurotic tunnel similar to those for the flexor tendons of the fingers and finally inserts into the base of the distal phalanx of the thumb. Its model is shown in Fig. 12. It is assumed that the restraint to bowstringing provided by the fibrous sheaths is at the center of the phalanx. The muscle is modelled by two origins—one on the ulna-radius and one on the humerus, each extending to the distal phalanx of the thumb by means of four lines connecting the ulna-radius, the carpals, the metacarpal I, the proximal phalanx I and the distal phalanx I.

10. Abductor pollicis longus:

This muscle arises from the dorsal surface of the ulna and the radius and from the interosseous membrane between them. Its tendon

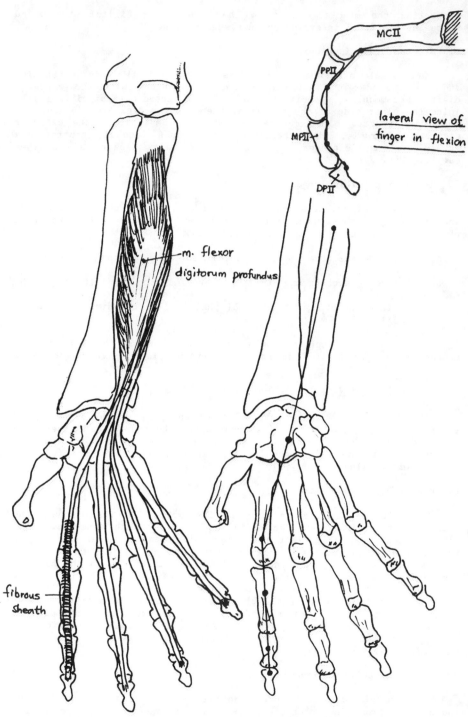

MCII

PPII

MPII

DPII

lateral view of
finger in flexion

m. flexor
digitorum profundus

fibrous
sheath

Figure 8-6. Muscle model for flexor digitorum profundus.

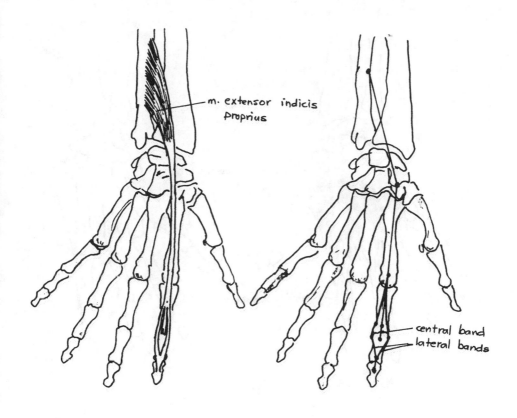

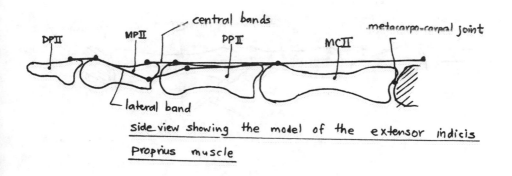

side view showing the model of the extensor indicis
proprius muscle

Figure 8-7. Model of the extensor indicis proprius muscle.

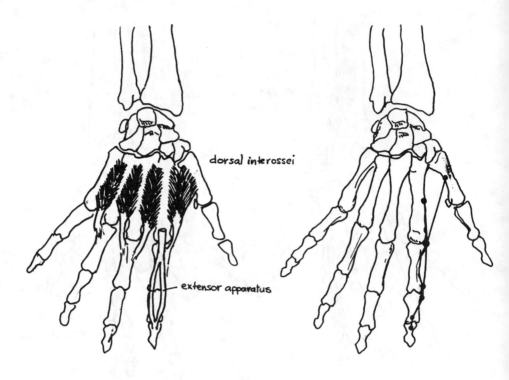

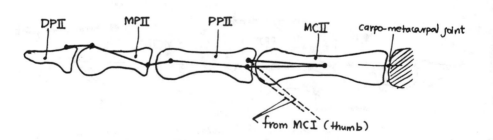

Figure 8-8. Side view of finger showing the dorsal interosseous model.

412

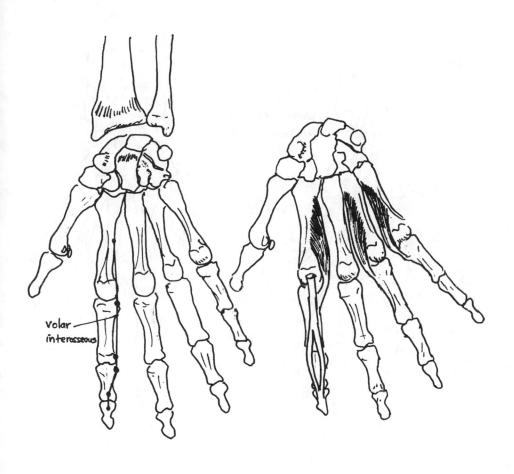

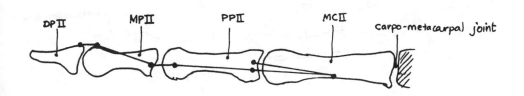

Figure 8-9. Sideview of finger showing the volar interosseous model.

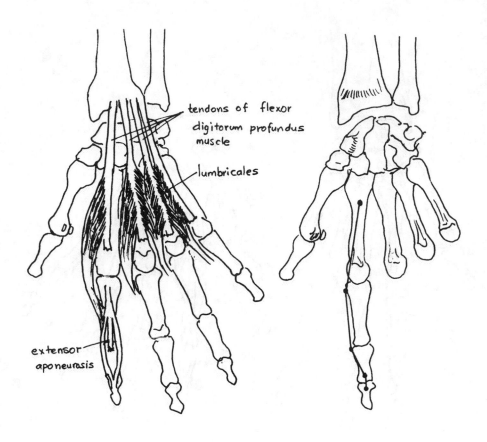

tendons of flexor
digitorum profundus
muscle

lumbricales

extensor
aponeurosis

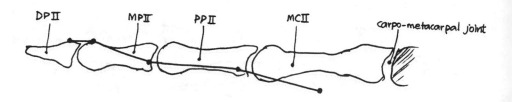

DPII MPII PPII MCII carpo-metacarpal joint

Figure 8-10. Sideview of finger showing lumbricales model.

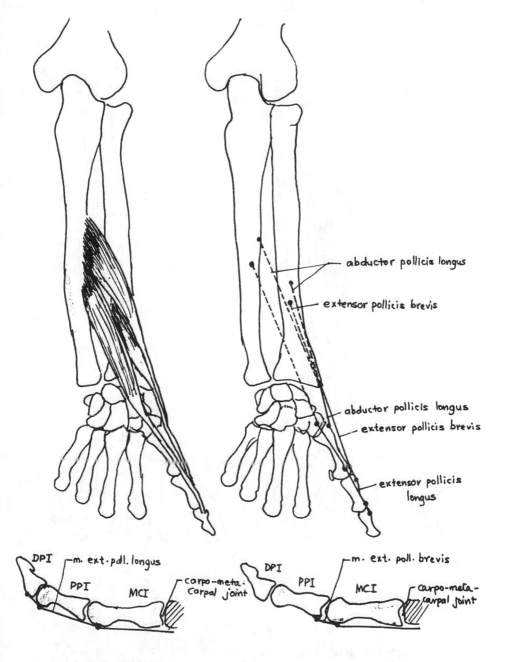

Figure 8-11. Muscle models for ext. poll. longus and brevis, and abd. poll. longus.

415

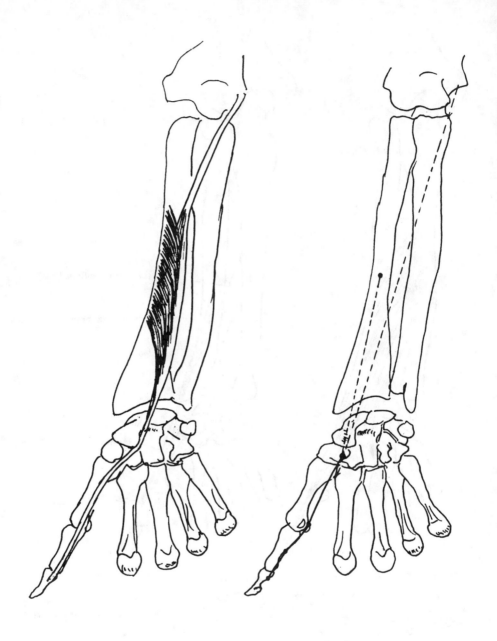

Figure 8-12. Muscle model for m. flexor pollicis longus.

passes through a groove on the lateral side of the distal end of the radius and inserts into the radial side of the base of the first metacarpal bone MCI. The muscle can be modelled by a line connecting the ulna-radius to a point on the base of the first metacarpal bone (Fig. 11).

11. Extensor carpi radialis brevis:

This muscle arises from the extensor condyle of the humerus and inserts on the dorsum of the base of the third metacarpal bone. It is modelled by a line connecting the humerus to the ulna-radius and a line connecting the latter to the MCIII.

12. Flexor pollicis brevis:

The flexor pollicis brevis arises from the flexor retinaculum and the distal part of the ridge on the trapezium, and inserts into the base of the proximal phalanx of the thumb. The muscle is modelled by two lines—a line connecting the trapezium (the carpal) to a point on the head of the first metacarpal MCI and another joining the latter to the point of insertion on PPI. The muscle assumed to wrap around the point on the head of the metacarpal, which is also the location of a sesamoid bone in the muscle tendon (Fig. 13).

13. Abductor pollicis brevis:

The abductor pollicis brevis originates on the scaphoid and the trapezium and inserts into the radial side of the base of the first phalanx of the thumb. The muscle can be modelled as a line connecting a point on the carpals to a point on the based of PPI with a reaction on the head of the first metacarpal MCI (Fig. 13).

14. Adductor pollicis:

The adductor pollicis has two main parts—an oblique part and a transverse part. The oblique head arises from the capitate bone and the bases of the second and third metacarpal bones; the transverse head arises from the palmar surface of the third metacarpal bone. Both heads converge towards the thumb to insert into the ulnar side of the base of the proximal phalanx PPI. The muscle is modelled by four lines—from the capitum, the base of MCII, the base of MCIII, and the palmar surface of MCIII respectively, all inserting into PPI (Fig. 13).

15. Opponens pollicis:

The opponens pollicis is a small, triangular muscle placed beneath the abductor pollicis brevis. It arises from the trapezium and from the flexor retinaculuma and inserts into the radial side of the metacarpal bone of the thumb. The muscle is represented by a line connecting the trapezium (the carpals) to the MCI (Fig. 13).

16. Dorsal interosseous:

The four dorsal interossei occupy the spaces between the metacarpal bones. They arise by two heads from the adjacent sides of the metacarpal bones and insert into the bases of the proximal phalanges and into the lateral bands of the aponeuroses of the tendons of the extensor digitorum. Thus, there are four dorsal interossei:

417

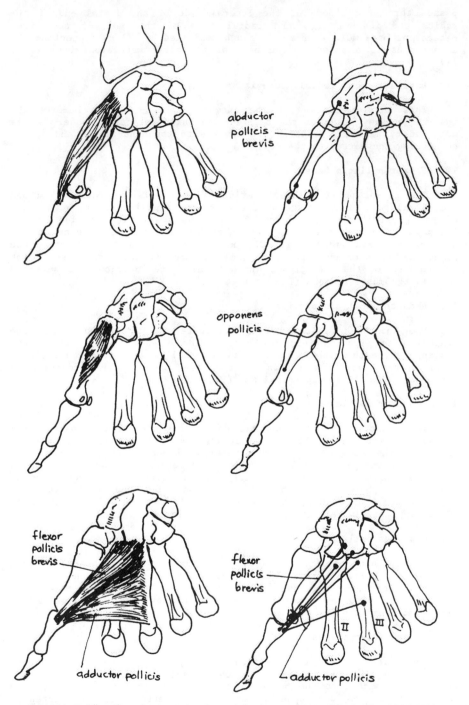

Figure 8-13. Muscle models for abductor pollicis brevis, oppenens pollicis, flexor pollicis brevis and adductor pollicis.

a.   One that is located between MCI and MCII and arises from the same two bones, and inserts into the base of PPII and into the lateral band (radial) of the extensor apparatus of the index finger.

b.   The second dorsal interosseous arises from the MCII and MCIII and inserts into the base of the PPIII and into the lateral band of the extensor apparatus of the middle finger.

c.   The third dorsal interosseous arises from the MCIII and MCIV and inserts into the base of the PPIII and into the medial band, (ulna) of the extensor apparatus of the middle finger.

d.   The fourth interosseous arises from the MCIV and MCV and inserts into the base of the PPIV and into the medial band of the extensor apparatus of the ring finger.

Each of the four dorsal interossei is modelled by two components—one that inserts directly into the proximal phalanx and the other that represents the insertion into the medial or the lateral bands of the extensor apparatus.   The latter are modelled similar to the bands described in the extensor digitorum model (see Fig. 8).

17. Volvar (or palmar) interosseous:

The three palmar interossei lie on the palmar surfaces of the metacarpal bones. They arise from the metacarpal bone of one finger and insert in to the side of the base of the first phalanx and into the lateral band of the aponeurosis of the tendon of the extensor digitorum. Each of three interossei thus can be modelled by two parts:

a.   a line connecting the metacarpal bone of the finger to the proximal phalanx of the same finger.

b.   a line connecting the point of origin on the metacarpal bone to the distal phalanx of the same finger by means of the ipsilateral band of the extensor apparatus (see Fig. 9).

18. Lumbricales:

The lumbricales are four, small fleshy muscles associated with the flexor digitorum profundus. The first lumbricale arises from the radial side and palmar surface of the flexor digitorum profundus tendon to the index finger and inserts into the lateral band of the aponeurotic expansion of the extensor digitorum tendon. The model for this muscle is identical to that for the lateral band of the extensor digitorum communis with the exception of its point of origin which is chosen in the palm on the flexor digitorum profundus tendon to the index finger (see Fig. 10). The other three lumbricales are modelled similarly.

A list of the muscle models and their points of origin and insertion is given in Table IIA.

COORDINATES OF MUSCLE ORIGINS AND INSERTIONS

The coordinates of the points of muscle origins and insertions on the various segments are identified with respect to orthogonal body-fixed axes for each segment selected with the origin at the center of the joint

TABLE II. Body fixed coordinates of muscles.

Muscle Name	Point 1				Point 2			
	Segment Name	X1	Y1	Z1	Segment Name	X2	Y2	Z2
EXPLL1	URD	-.70	-.50	-1.30	MC1	4.20	-.60	-.30
EXPLL2	MC1	4.20	-.60	-.30	PP1	3.30	-.40	0.00
EXPLL3	PP1	3.30	-.40	0.00	DP1	.70	-.40	0.00
FXPLL1	URD	-.70	-.50	1.40	CRP	-1.30	-1.50	2.30
FXPLL2	CRP	-1.30	-1.50	2.30	MC1	4.20	.60	0.00
FXPLL3	MC1	4.20	.60	0.00	PP1	1.00	.50	0.00
FXPLL4	PP1	3.30	.40	0.00	DP1	.80	.40	0.00
EXPLB1	URD	-1.00	-2.00	.70	MC1	4.20	-.60	0.00
EXPLB2	MC1	4.20	-.60	0.00	PP1	1.00	-.50	0.00
FXPLB1	CRP	-1.20	-2.00	2.40	MC1	4.20	.60	.40
FXPLB2	MC1	4.20	.60	.40	PP1	.90	.50	.40
ABPOLL	URD	-1.00	-2.00	.70	MC1	.80	0.00	.60
ABPLB1	CRP	-1.50	-2.30	2.00	MC1	4.20	.60	.40
ABPLB2	MC1	4.20	.60	.40	PP1	1.10	.50	.40
ADPOL1	CRP	-.40	-.30	1.20	PP1	.80	.40	-.80
ADPOL2	MC2	1.00	0.00	.70	PP1	.80	.40	-.80
ADPOL3	MC3	.60	0.00	.70	PP1	.80	.40	-.80
ADPOL4	MC3	3.00	0.00	.30	PP1	.80	.40	-.80
OPNPOL	CRP	-1.20	-2.20	2.50	MC1	2.50	.10	.20
DORI11	MC1	2.30	0.00	-.60	PP2	1.30	-.80	-.20
DORI12	MC1	2.30	0.00	-.60	PP2	1.30	-.80	-.20
DORI13	PP2	3.30	-.60	0.00	MP2	.60	-.60	-.20
DORI14	MP2	2.40	0.00	-.25	DP2	.40	0.00	-.30
EXCPRL	URD	0.00	-1.50	-.80	MC2	.60	-.80	-.60
FXCPR1	URD	0.00	-.80	1.20	CRP	-1.30	-1.50	2.50
FXCPR2	CRP	-1.30	-1.50	2.50	MC2	.70	-.20	.60
FXDS21	URD	-.70	1.00	1.60	CRP	-1.00	0.00	1.50
FXDS22	CRP	-1.00	0.00	1.50	MC2	6.00	0.00	.90
FXDS23	MC2	6.00	0.00	.90	PP2	4.20	0.00	.25
FXDS24	PP2	4.20	0.00	.25	MP2	1.30	0.00	.20
FXDP21	URD	-.70	1.00	1.60	CRP	-1.00	-.10	1.30
FXDP22	CRP	-1.00	-.10	1.30	MC2	6.00	0.00	.90
FXDP23	MC2	6.00	0.00	.90	PP2	4.20	0.00	.25
FXDP24	PP2	4.20	0.00	.25	MP2	2.10	0.00	.20
FXDP25	MP2	2.10	0.00	.20	DP2	.60	0.00	.20
EXIPM1	URD	-.70	.30	-1.30	MC2	6.20	0.00	-.80
EXIPM2	MC2	6.20	0.00	-.80	PP2	4.70	0.00	-.30
EXIPM3	PP2	4.70	0.00	-.30	MP2	.55	0.00	-.40
EXIPR1	URD	-.70	.30	-1.30	MC2	6.20	0.00	-.80

(continued)

TABLE II   Continued

Muscle Name	Point 1				Point 2			
	Segment Name	X1	Y1	Z1	Segment Name	X2	Y2	Z2
DORI40	PP3	1.30	0.90	-.31	MC4	2.80	-.40	-.25
DORI41	MC4	2.80	-.40	-.25	PP3	1.30	0.90	0.00
DORI42	PP3	3.50	.60·	0.00	MP3	0.70	0.70	-.20
DORI43	MP3	3.10	0.00	-.25	DP3	0.40	0.00	-.35
VOLI40	MC4	3.00	-.30	0.10	PP4	1.30	-.80	.25
VOLI41	MC4	3.00	-.30	0.10	PP4	1.30	-.80	0.00
VOLI42	PP4	3.00	-.55	0.00	MP4	0.60	-.60	-.20
VOLI43	MP4	2.50	0.00	-.20	DP4	0.35	0.00	-.30
DORI44	MC4	2.80	0.40	-.25	PP4	1.20	0.80	-.20
DORI45	MC4	2.80	0.40	-.25	PP4	1.20	0.80	-.20
DORI46	PP4	3.00	0.60	0.00	MP4	0.60	0.60	-.20
DORI47	MP4	2.50	0.00	-.20	DP4	0.35	0.00	-.30
FXDS51	URD	-.70	1.50	1.60	CRP	-.70	1.20	1.30
FXDS52	CRP	-.70	1.20	1.30	MC5	4.60	0.00	0.70
FXDS53	MC5	4.60	0.00	0.70	PP5	3.50	0.00	0.20
FXDS54	PP5	3.50	0.00	0.20	MP5	1.00	0.00	0.20
FXDP51	URD	-.70	1.80	1.50	CRP	-.70	1.00	1.10
FXDP52	CRP	-.70	1.00	1.10	MC5	4.60	0.00	0.70
FXDP53	MC5	4.60	0.00	0.70	PP5	3.50	0.00	0.20
FXDP54	PP5	3.50	0.00	0.20	MP5	1.90	0.00	0.17
FXDP55	MP5	1.90	0.00	0.17	DP5	0.60	0.00	0.20
EXCRPU	URD	-1.50	2.80	-.80	MC5	0.60	0.80	0.00
FXCRPU	URD	-1.50	3.00	1.00	CRP	-1.70	2.00	1.50
EDGM51	URD	-.70	0.90	-1.30	MC5	5.00	0.00	-.60
EDGM52	MC5	5.00	0.00	-.60	PP5	3.85	0.00	-.25
EDGM53	PP5	3.85	0.00	-.25	MP5	0.50	0.00	-.33
EDGR51	URD	-.70	0.90	-1.30	MC5	5.00	0.00	-.60
EDGR52	MC5	5.00	0.00	-.60	PP5	2.70	-.25	-.30
EDGR53	PP5	2.70	-.25	-.30	MP5	0.50	-.50	-.15
EDGR54	MP5	2.16	0.00	-.15	DP5	0.30	0.00	-.25
EDGU51	URD	-.70	0.90	-1.30	MC5	5.00	0.00	-.60
EDGU52	MC5	5.00	0.00	-.60	PP5	2.70	0.25	-.30
EDGU53	PP5	2.70	0.25	-.30	MP5	0.50	0.50	-.15
EDGU54	MP5	2.15	0.00	-.15	DP5	0.30	0.00	-.25
DORI50	MC5	1.70	-.40	-.40	PP4	1.20	0.80	-.20
DORI51	MC5	1.70	-.40	-.40	PP4	1.20	0.80	-.20
DORI52	PP4	3.00	0.60	0.00	MP4	0.60	0.60	-.20
DORI53	MP4	2.50	0.00	-.20	DP4	0.35	0.00	-.30
VOLI50	MC5	2.80	-.40	0.20	PP5	1.00	-.80	0.20

(continued)

421

TABLE II   Continued

Muscle Name	Point 1				Point 2			
	Segment Name	X1	Y1	Z1	Segment Name	X2	Y2	Z2
FXDP35	MP3	2.80	0.00	0.25	DP3	0.70	0.00	0.25
EDGM31	URD	-.70	0.60	-1.30	MC3	6.20	0.00	-.80
EDGM32	MC3	6.20	0.00	-.80	PP3	5.20	0.00	-.30
EDGM33	PP3	5.20	0.00	-.30	MP3	0.60	0.00	0.40
EDGR31	URD	-.70	0.60	-1.30	MC3	6.20	0.00	-.80
EDGR32	MC3	6.20	0.00	-.80	PP3	3.50	-.40	-.40
EDGR33	PP3	3.50	-.40	-.40	MP3	0.70	-.70	-.20
EDGR34	MP3	3.10	0.00	-.25	DP3	0.40	0.00	-.35
EDGU31	URD	-.70	0.60	-1.30	MC3	6.20	0.00	-.80
EDGU32	MC3	6.20	0.00	-.80	PP3	3.50	0.40	-.40
EDGU33	PP3	3.50	0.40	-.40	MP3	0.70	0.70	-.20
EDGU34	MP3	3.10	0.00	-.25	DP3	0.40	0.00	-.35
DORI30	MC3	3.00	-.40	-.40	PP3	1.30	-.90	-.30
DORI31	MC3	3.00	-.40	-.40	PP3	1.30	-.90	0.00
DORI32	PP3	3.50	-.60	0.00	MP3	0.70	-.70	-.20
DORI33	MP3	3.10	0.00	-.25	DP3	0.40	0.00	-.35
DORI34	MC3	3.00	0.40	-.40	PP3	1.30	1.30	-.30
DORI35	MC3	3.00	0.40	-.40	PP3	1.30	0.90	0.00
DORI36	PP3	3.50	0.60	0.00	MP3	0.70	0.70	-.20
DORI37	MP3	3.10	0.00	-.25	DP3	0.40	0.00	-.35
FXDS41	URD	-.70	1.50	1.60	CRP	-.70	1.00	1.50
FXDS42	CRP	-.70	1.00	1.50	MC4	5.00	0.00	0.70
FXDS43	MC4	5.00	0.00	0.70	PP4	4.30	0.00	0.30
FXDS44	PP4	4.30	0.00	0.30	MP4	1.30	0.00	0.25
FXDP41	URD	-.70	1.60	1.60	CRP	-.70	0.60	1.30
FXDP42	CRP	-.70	0.60	1.30	MC4	5.00	0.00	0.70
FXDP43	MC4	5.00	0.00	0.70	PP4	4.30	0.00	0.30
FXDP44	PP4	4.30	0.00	0.30	MP4	2.20	0.00	0.20
FXDP45	MP4	2.20	0.00	0.20	DP4	0.60	0.00	0.25
EDGM41	URD	-.70	0.90	-1.30	MC4	5.40	0.00	-.70
EDGM42	MC4	5.40	0.00	-.70	PP4	4.70	0.00	-.30
EDGM43	PP4	4.70	0.00	-.30	MP4	0.50	0.00	-.40
EDGR41	URD	-.70	0.90	-1.30	MC4	5.40	0.00	-.70
EDGR42	MC4	5.40	0.00	-.70	PP4	3.00	-.30	-.30
EDGR43	PP4	3.00	-.30	-.30	MP4	0.60	-.60	-.20
EDGR44	MP4	2.50	0.00	-.20	DP4	0.35	0.00	-.30
EDGU41	URD	-.70	0.90	-1.30	MC4	5.40	0.00	-.70
EDGU42	MC4	5.40	0.00	-.70	PP4	3.00	0.30	-.30
EDGU43	PP4	3.00	0.30	-.30	MP4	0.60	0.60	-.20
EDGU44	MP4	2.50	0.00	-.20	DP4	0.35	0.00	-.30

(continued)

TABLE II   Continued

Muscle Name	Point 1				Point 2			
	Segment Name	X1	Y1	Z1	Segment Name	X2	Y2	Z2
EXIPR2	MC2	6.20	0.00	-.80	PP2	3.30	-.30	-.30
EXIPR3	PP2	3.30	-.30	-.30	MP2	.60	-.60	-.20
ÉXIPR4	MP2	2.40	0.00	-.25	DP2	.40	0.00	-.30
EXIPU1	URD	-.70	.30	-1.30	MC2	6.20	0.00	-.80
EXIPU2	MC2	6.20	0.00	-.80	PP2	3.30	.30	-.30
EXIPU3	PP2	3.30	.30	-.30	MP2	.60	.60	-.20
EXIPU4	MP2	2.40	0.00	-.25	DP2	.40	0.00	-.30
EDGM21	URD	-.70	.30	-1.30	MC2	6.20	0.00	-.80
EDGM22	MC2	6.20	0.00	-.80	PP2	4.70	0.00	-.30
EDGM23	PP2	4.70	0.00	-.30	MP2	.55	0.00	-.40
EDGR21	URD	-.70	.30	-1.30	MC2	6.20	0.00	-.80
EDGR22	MC2	6.20	0.00	-.80	PP2	3.30	-.30	-.30
EDGR23	PP2	3.30	-.30	-.30	MP2	.60	-.60	-.20
EDGR24	MP2	2.40	0.00	-.25	DP2	.40	0.00	-.30
EDGU21	URD	-.70	.30	-1.30	MC2	6.20	0.00	-.80
EDGU22	MC2	6.20	0.00	-.80	PP2	3.30	.30	-.30
EDGU23	PP2	3.30	.30	-.30	MP2	.60	.60	-.20
EDGU24	MP2	2.40	.00	-.25	DP2	.40	0.00	-.30
DORI15	MC2	2.30	-.40	-.40	PP2	1.30	-.80	-.20
DORI16	MC2	2.30	-.40	-.40	PP2	1.30	-.80	-.20
DORI17	PP2	3.30	-.60	0.00	MP2	.60	-.60	-.20
DORI18	MP2	2.40	0.00	-.25	DP2	.40	0.00	-.30
DORI21	MC2	2.70	.40	-.50	PP3	1.30	-.90	-.30
DORI22	MC2	2.70	.40	-.50	PP3	1.30	-.90	-.30
DORI23	PP3	3.50	-.60	0.00	MP3	.70	-.70	-.20
DORI24	MP3	3.10	0.00	-.25	DP3	.40	0.00	.35
VOLI11	MC2	4.00	.45	0.00	PP2	1.30	.80	.35
VOLI12	MC2	4.00	.45	0.00	PP2	1.30	.80	0.00
VOLI13	PP2	3.30	.60	0.00	MP2	.60	.60	-.20
VOLI14	MP2	2.40	0.00	-.25	DP2	.40	0.00	-.30
EXCPRB	URD	0.00	-1.00	-1.00	MC3	.80	.50	-.80
FXDS31	URD	-.70	1.00	1.60	CRP	-.70	.50	1.70
FXDS32	CRP	-.70	.50	1.70	MC3	5.80	0.00	.80
FXDS33	MC3	5.80	0.00	.80	PP3	4.70	0.00	.30
FXDS34	PP3	4.70	0.00	.30	MP3	1.60	0.00	.25
FXDP31	URD	-.70	1.40	1.60	CRP	-.70	.20	1.50
FXDP32	CRP	-.70	.20	1.50	MC3	5.80	0.00	.80
FXDP33	MC3	.80	0.00	.80	PP3	4.70	0.00	.30
FXDP34	PP3	4.70	0.00	.30	MP3	2.80	0.00	.25

(continued)

TABLE II. Continued

Muscle Name	Point 1				Point 2			
	Segment Name	X1	Y1	Z1	Segment Name	X2	Y2	Z2
VOLI51	MC5	2.80	-.40	0.20	PP5	1.00	-.80	0.00
VOLI52	PP5	2.70	-.50	0.00	MP5	0.50	-.50	-.15
VOLI53	MP5	2.15	0.00	-.15	DP5	0.30	0.00	-.25
ABDGMN	CRP	-1.70	2.20	1.60	PP5	1.00	0.70	0.25
FXDGMN	CRP	-.60	1.60	2.00	PP5	1.00	0.70	0.25
ODGMN1	CRP	-.30	1.60	2.00	MC5	1.60	0.50	0.00
ODGMN2	CRP	-.30	1.60	2.00	MC5	2.90	0.40	0.20
ODGMN3	CRP	-.30	1.60	2.00	MC5	3.90	0.45	0.00
EDMNM1	URD	-.70	1.50	-1.10	MC5	5.00	0.00	-.60
EDMNM2	MC5	5.00	0.00	-.60	PP5	3.85	0.00	-.25
EDMNM3	PP5	3.85	0.00	-.25	MP5	0.50	0.00	-.35
EDMNR1	URD	-.70	1.50	-1.10	MC5	5.00	0.00	-.60
EDMNR2	MC5	5.00	0.00	-.60	PP5	2.70	-.25	-.30
EDMNR3	PP5	2.70	-.25	-.30	MP5	0.50	-.50	-.15
EDMNR4	MP5	2.15	0.00	-.15	DP5	0.30	0.00	-.25
EDMNU1	URD	-.70	1.50	-1.10	MC5	5.00	0.00	-.60
EDMNU2	MC5	5.00	0.00	-.60	PP5	2.70	0.25	-.30
EDMNU3	PP5	2.70	0.25	-.30	MP5	0.50	0.50	-.15
EDMNU4	MP5	2.15	0.00	-.15	DP5	0.30	0.00	-.25

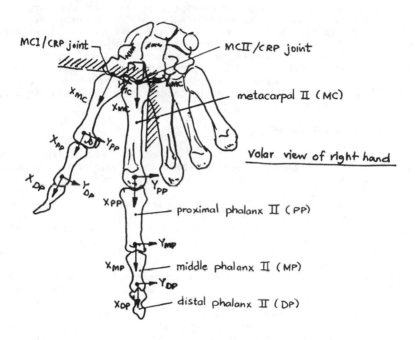

Figure 8-14. Body-fixed coordinate axes.

425

formed with the adjoining element. Thus, the coordinates of all points of interest on the metacarpals and the digits are obtained with respect to orthogonal reference frames located at the proximal carpometacarpal, or the metacarpophalangeal or the interphalangeal joints. The origin of the axes for the interphalangeal and the metacarpophalangeal joints coincides with the approximate center of the hinge joint, which is how these joints function from a mechanical point of view.

The points on the carpals are referred to axes located at the articulation between the third metacarpal and the carpals; the points on the ulna-radius are referred with respect to axes at the carpo-radial joint or the wrist joint. The reference frames for the other segments of the upper extremities and the points of interest on them have already been described in Chapter V. The coordinates of all the points were obtained by direct measurement on a skeleton and are listed in Table II. When a desired overall configuration of the hand segment is established, the coordinates of all points are referred with respect to ground-fixed axes system at the carpometacarpal joint between the carpals and the third metacarpal.

JOINT REACTIONS AND MOMENTS

The reaction force at a joint or between two articulating segments can be modelled in general, by the three orthogonal X,Y,Z components. However, the interphalangeal and the metacarpophalangeal joints which act as mechanical hinge or pivot permitting rotation about only one axes (such as during flexion and extension) it would be inappropriate to assign unconstrained, arbitrary reactions. Since the reaction force, which must be compressive, should be oriented along the longitudinal axis of the segment it was decided to model the articulating forces at these joints by two components——a component directed along the segment longitudinal axis and a component along the Y axis (medio-lateral direction). However, since the area on which the longitudinally directed reaction acts is extensive it was decided to assign two such reactions passing through the center of the joint on either side of the axis and directed normal to the joint surface as shown in Fig. 15. Thus, if the two reactions are of different magnitudes, the resultant will not necessarily lie along the segment axis, therefore permitting the resultant to select the required orientation by properly balancing the two reactions. At the carpometacarpal joints, which are gliding joints the reactions assigned should conform to the geometry of the surface on which they act. At these joints therefore, the reactions are selected to act at more than one location and partially constrained, that is, the X, Y or Z components are so selected as to produce a resultant that is directed normal to the surface (see Fig. 15). At the carpo-radial or the wrist joint, several partially constrained reactions are assigned due to the extensive joint surface over which the transmission of force between the carpals and the distal end of the radius can occur. Five locations were selected—medial and lateral to the joint center on the palmar side and at the center of the joint. At each of these locations X, Y and Z components were assigned in selected directions depending on the surface geometry.

The intermetacarpal articulations between the fourth and the fifth, the third and the fourth, and the second and the third which are gliding joints, the reaction forces are fully constrained in terms of two angles as explained in Fig. 3, Chapter IX.

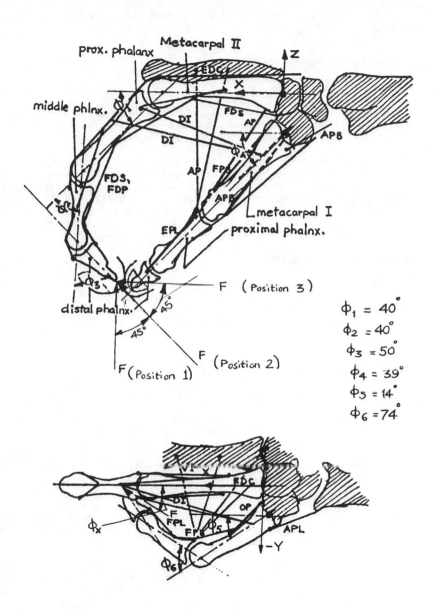

$$\phi_1 = 40°$$
$$\phi_2 = 40°$$
$$\phi_3 = 50°$$
$$\phi_4 = 39°$$
$$\phi_5 = 14°$$
$$\phi_6 = 74°$$

Figure 8-15. The pinch posture.

At all the joints the ligament action is represented by assigned moments——these are the three orthogonal X, Y and Z components. The coordinates of all the joints and the assigned reactions are described in Table III.

SUMMARY OF THE COMPREHENSIVE HAND MODEL

A. NUMBER OF EQUATIONS

The twenty segments in the hand, along with the ulna-radius and the humerus yield a total of twenty-two segments. These are rigid bodies in space and thus each one requires six equilibrium equations.

Additional equations however must also be accounted for——these described the equilibrium between the muscle forces that wrap around a segment and the consequent reaction force at that point. By treating the wrapping muscle as two unknown forces which are constrained to be equal (see Fig. 16, Chapter 9) on additional equality constraint is developed. When such constraints are included along with the equilibrium equations, the total number of equation turns out to be 278.

B. NUMBER OF UNKNOWN VARIABLES

    (i)   Number of muscle forces:
        The total number of hand and upper extremity muscle forces is equal to 248. Note that this number includes all muscle components or lines that are used to model a muscle and therefore does not necessarily mean that there are 248 individual muscles.

    (ii) Number of joint reaction forces:
        At each of the interphalangeal and the metacarpophalangeal joints three reactions were assigned——two constrained in sense of direction while the medial one (Y component) was left open to select the required direction, which effectively means there are two unknown variables for the Y reactions. Hence at the five metacarpophalangeal and the nine interphalangeal joints there are a total of 56 unknown variables. At the metacarpocarpal and the wrist joints where partially constrained reactions have been assigned the total number of unknown variable turns out to be 48.

        At the three intermetacarpal joints fully constrained reactions were assigned so that three more unknown variables are added.

        The elbow and the shoulder joint reaction forces which are assigned in terms, of X, Y and Z components, a total of 12 unknown variables must be accounted for since the components are left open as to their directions.

        The total number of unknown variables representing joint reaction forces thus amounts to 119.

    (iii) Number of joint reaction moments
        The total number of joints in the hand including the elbow and the shoulder joint is 23 so that when arbitrary three

TABLE III. Joint reactions and coordinates of joints of the hand.
A. Constrained Reactions: (constrained along the
two points defined on Segment 1 and on Segment 2).

Joint Name		Segment 1	Coordinates on Segment 1			Segment 2	Coordinates on Segment 2		
			X	Y	Z		X	Y	Z
DP1/PP1	a.	DP1	0.35	0.3	0	PP1	3.3	0	0
	b.	DP1	0.35	-0.3	0	PP1	3.3	0	0
PP1/MC1	a.	PP1	0.55	0.4	0	MC1	4.2	0	0
	b.	PP1	0.55	-0.4	0	MC1	4.2	0	0
DP2/MP2	a.	DP2	0.2	0	0.2	MP2	2.3	0	0
	b.	DP2	0.2	0	-0.2	MP2	2.3	0	0
MP2/PP2	a.	MP2	0.35	0	0.35	PP2	4.5	0	0
	b.	MP2	0.35	0	-0.35	PP2	4.5	0	0
PP2/MC2	a.	PP2	0.7	0	0.4	MC2	6.0	0	0
	b.	PP2	0.7	0	-0.4	MC2	6.0	0	0
DP3/MP3	a.	DP3	0.25	0	0.2	MP3	3.0	0	0
	b.	DP3	0.25	0	-0.2	MP3	3.0	0	0
MP3/PP3	a.	MP3	0.30	0	0.35	PP3	5.0	0	0
	b.	MP3	0.30	0	-0.35	PP3	5.0	0	0
PP3/MC3	a.	PP3	0.70	0	0.50	MC3	6.0	0	0
	b.	PP3	-0.70	0	-0.50	MC3	6.0	0	0
DP4/MP4	a.	DP4	0.2	0	0.2	MP4	2.4	0	0
	b.	DP4	0.2	0	0.2	MP4	2.4	0	0
MP4/PP4	a.	MP4	0.35	0	0.35	PP4	4.5	0	0
	b.	MP4	0.35	0	-0.35	PP4	4.5	0	0
PP4/MC4	a.	PP4	0.65	0	0.4	MC4	5.2	0	0
	b.	PP4	0.65	0	-0.4	MC4	5.2	0	0
DP5/MP5	a.	DP5	0.15	0	0.15	MP5	2.1	0	0
	b.	DP5	0.15	0	-0.15	MP5	2.1	0	0
MP5/PP5	a.	MP5	0.25	0	0.25	PP5	3.7	0	0
	b.	MP5	0.25	0	-0.25	PP5	3.7	0	0
PP5/MC5	a.	PP5	0.55	0	0.3	MC5	4.8	0	0
	b.	PP5	0.55	0	-0.3	MC5	4.8	0	0

(continued)

TABLE III. Continued

B. Partially Constrained Reactions.

Joint Name		Segment 1	Coordinates on Segment 1			Segment 2	Reaction Components on Segment 1
			X	Y	Z		
MC1/CPP	a.	MC1	-0.2	0.4	0.0	CRP	X,Y
	b.	MCL	-0.2	-0.4	0.0	CRP	X,-Y
	c.	MC1	0.0	0.0	0.0	CRP	±Z
MC2/CRP	a.	MC2	0.0	0.0	0.5	CRP	X,-Z
	b.	MC2	0.0	0.0	-0.5	CRP	X,Z
	c.	MC2	0.0	0.0	0.0	CRP	±Y
MC3/CRP	a.	MC3	0.0	0.0	0.6	CRP	X,-Z
	b.	MC3	0.0	0.0	-0.6	CRP	X,Z
	c.	MC3	0.0	0.0	0.0	CRP	Y
MC4/CRP	a.	MC4	0.0	0.0	0.4	CRP	X,-Z
	b.	MC4	0.0	0.0	-0.4	CRP	X,Z
	c.	MC4	0.0	0.0	0.0	CRP	±Y
MC5/CRP	a.	MC5	0.0	0.0	0.4	CRP	X,-Z
	b.	MC5	0.0	0.0	-0.4	CRP	X,Z
	c.	MC5	0.0	0.0	0.0	CRP	±Y
DP1/PP1		DP1	0.4	0.0	0.0	PP1	+Z
PP1/MC1		PP1	0.6	0.0	0.0	MC1	±Z
DP2/MP2		DPE	0.25	0.0	0.0	MP2	±Y
MP2/PP2		MP2	0.4	0.0	0.0	PP2	±Y
PP2/MC2		PP2	0.8	0.0	0.0	MC2	±Y
DP3/MP3		DP3	0.25	0.0	0.0	MP3	±Y
MP3/PP3		MP3	0.35	0.0	0.0	PP3	±Y
PP3/MC3		PP3	0.8	0.0	0.0	MC3	±Y
DP4/MP4		DP4	0.25	0.0	0.0	MP4	±Y
MP4/PP4		MP4	0.4	0.0	0.0	PP4	±Y
PP4/MC4		PP4	0.7	0.0	0.0	MC4	±Y
DP5/MP5		DP5	0.2	0.0	0.0	MP5	Y
MP5/PP5		MP5	0.3	0.0	0.0	PP5	Y
PP5/MC5		PP5	0.6	0.0	0.0	MC5	Y
CRP/URD	a.	CRP	0.3	1.5	0.5	URD	X,-Y,-Z
	b.	CRP	0.3	-1.5	0.5	URD	X,Y,-Z
	c.	CRP	0.3	1.5	-0.5	URD	X,-Y,Z
	d.	CRP	0.3	-1.5	-0.5	URD	X,Y,Z

(continued)

TABLE III. (continued)

C. Constrained reactions: (Constraint in terms of two angles).

Joint Name	Segment 1	Coordinates on Segment 1			ϕ_1	ϕ_2	Segment 2
		X	Y	Z			
MC4/MC5	MC4	0.4	0.3	0.10	100	-30	MC5
MC3/MC4	MC3	0.4	-0.5	0.00	70	5	MC4
MC2/MC3	MC3	0.4	-0.5	0.0	-120	-30	MC2

D. Location of External Forces on Digits (such as in pulling, pinching etc.)

Segment Name	Coordinates of Point of Action		
	X	Y	Z
MC2	1.1	0	0.7
MC3	1.2	0	0.8
MC4	1.2	0	0.8
MC5	1.0	0	0.6

431

X, Y, Z moment components are assigned at these joints, the total number of variables is 138.

The total number of available equations is 278 and the total number of unknown variables is 505. Note that this model is for one hand only——if both hands were simultaneously used or were to be modelled to represent the contributions of the left and the right upper extremities to the total body structure the problem size would be doubled.

C. MERIT CRITERION

In the upper and the lower extremity models as well as in the models developed for other parts of the human musculoskeletal system, the criterion that was proposed was minimization of the weighted sum of all muscle and ligament forces. However, in case of the upper extremities where the joints could be conceivably subjected to tensile loads it was suggested that such joint reaction forces should be included in the merit criterion in some manner. The final form of the objective function thus becomes $U = \Sigma F + 2R + 4M$. When applied to the comprehensive hand model the joints to be included in this objective function will be the elbow and the shoulder joint. At all other joints the reactions have been constrained in orientation to act normal to the joint surface and thus cannot be tensile; these are accordingly omitted from the criterion.

APPLICATION OF THE HAND MODEL TO PULLING

The comprehensive hand model was used to study the muscle load sharing during the act of pulling with the tips of the fingers. This can be readily represented in the model by means of external forces acting on the distal phalanges of the individual fingers. The configuration of the hand segments and the upper extremities in this act was "guestimated" by actually conducting a test to record the posture as well as the level of activity in the major extrinsic muscles by surface electromyography. The relative orientations of the digits can only be superficially estimated since one would require x-ray data to accurately measure these values. The established configuration of the hand in terms of three angles (ϕx, ϕy and ϕz) for each of the segments is given in Table IV. The hand and the forearm is assumed to be held horizontal during pulling, so that the forces acting on the tips of the fingers (distal phalanges) are also horizontal. The model results are calculated separately with the pull force applied successively to each finger. The results for the muscle and ligament forces and joint reaction forces are tabulated in Table V. The pull force applied on each finger was 4.53 kg (10 lb.)

As might be expected, the structural stability of the fingers is maintained by the digital flexor muscles——the flexor digitorum superficialis and profundus. However, some of the extensor muscles are also involved particularly those that insert on the carpals and metacarpals. This is because with the flexion of the fingers, the wrist joint that is, the carpals and the metacarpals will be caused to flex also. However, in order to prevent this flexion the extensor muscles and/or the interossei must be activated to counterbalance the effect of the flexor muscles.

This result is also verified by the experimental observations. In the experimental investigation the fingertips were subjected to known values of horizontal forces with the forearm held horizontal. The acting force was measured by means of a spring scale and was applied to the tip of each of the four fingers in succession. The recorded electromyographic signals (rectified, filtered and integrated) was obtained by surface electrodes from four superficial muscles—extensor

TABLE IV   Segmental Orientations of the Hand Segments (degrees)

Body	ϕ_x	ϕ_y	ϕ_z
MC1	0	-30	-35
PP1	0	-30	-35
DP1	0	-30	0
MC2	0	0	-10
PP2	0	0	0
MP2	0	-50	0
DP2	0	100	0
MC3	0	0	0
PP3	0	0	0
MP3	0	-50	0
DP3	0	-120	0
MC4	0	0	7
PP4	0	0	0
MP4	0	-50	0
DP4	0	-110	0
MC5	0	0	13
PP5	0	0	0
MP5	0	-50	0
DP5	0	-110	0
CRP	0	0	0
URD	0	-20	0
ARM	10	45	0
SCP	0	0	0
CLV	-5	0	-30

carpi radialis longus and brevis, extensor digitorum and flexor digitorum superficialis.   No attempt was made to distinguish the exact part or component of a given muscle from the signal was specifically being obtained.   The muscle activity or trend with the external loading is shown in Fig. 16b a tip force of 4 lb.   It was seen that the muscle activity is extremely sensitive to the overall orientation which varied considerably as successive fingers were loaded.

The overall trend from the electromyographic seems to agree with the theoretically calculated values for the actual force required.

APPLICATION OF THE MODEL TO PINCH ACTION

The comprehensive hand model was applied to the case of tip pinching using fingers of the hand.   This action has been analyzed by several authors (Chow et al (1976), Paul (1976), Hirsch et al (1974) etc.) who have used simplified models of the finger to reduce the problem to a statically determinate one so that the muscle forces can be explicitly evaluated.   This model was used to investigate the muscle load sharing and joint reaction forces for the tip pinch involving the middle

433

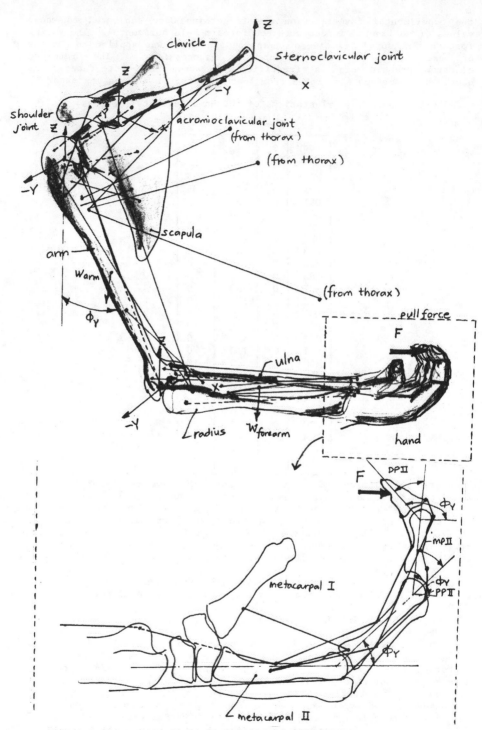

Figure 8-16a. Hand model in pulling at the fingers.

434

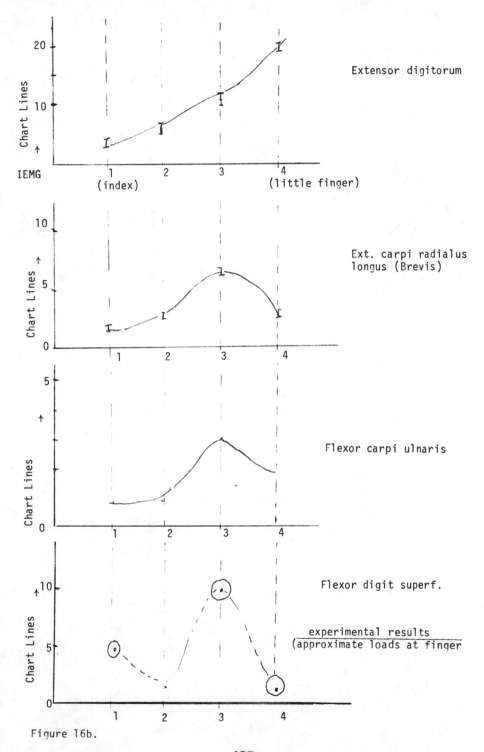

Figure 16b.

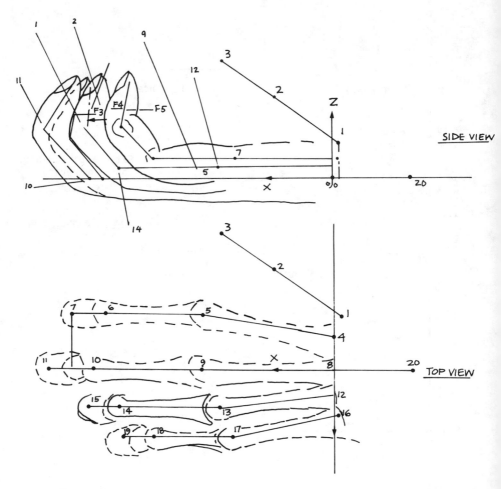

Figure 8-16c. Hand model, showing side and top views.

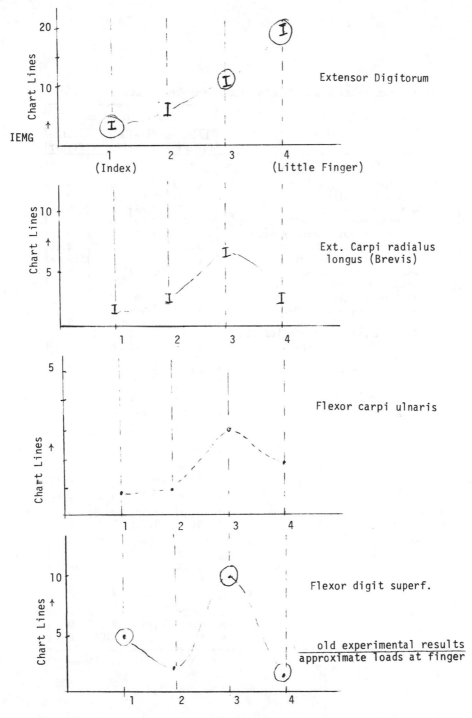

Figure 16d.

TABLE V. Muscle forces and joint reactions during pulling.

				Load applied on			
				Index Finger	Middle Finger	Ring Finger	Little Finger
I.	Muscle Forces in the Hand						
	Abductor pollicis longus			0	0	0	0
	Adductor pollicis			0.77	0.04	0	0
	Dorsal interosseous			1.52	2.88	0.49	0
	Extensor carpi radialis longus			5.29	0.57	0	0
	Extensor carpi radialis brevis			0	0.15	0	0
	Volar interosseous			4.28	0	0.59	0.11
	Extensor carpi ulnaris			0	0	0	16.19
	Flexor carpi ulnaris			0	0	0	0
	Flexor pollicis brevis			0.22		0	
	Extensor pollicis brevis			2.18	0.08	0	0
	Abductor pollicis brevis			1.19	0.02	0	0
	Flexor carpi radialis			8.56	0.02	0	0
	Flexor digitorum superficialis			1.51	7.77	7.11	7.33
	Flexor digitorum profundus			9.02	4.01	6.05	5.40
	Extensor digitorum			0	0	0	0.08
	Opponens digiti minimi			0	0	0	7.37
II.	Joint Reactions						
	DP/MP	a		9.71	5.82	6.44	5.96
		b		8.15	4.18	4.89	3.25
		y		0.52	0.23	0.03	-0.13
	MP/PP	a		3.87	3.89	3.46	3.21
		b		8.88	8.86	9.47	7.97
		y		0.52	0.23	0.02	-0.15
	PP/MC	a		3.82	4.34	3.39	1.75
		b		9.31	7.83	8.39	7.46
		y		0.65	0.58	0.22	0.08
	MC/CRP	a	x	18.07	7.89	4.33	10.71
			z	-4.67	-0.95	0	0
		b	x	1.13	0	5.19	16.46
			z	0	0	0.19	3.70
		c	y	-4.89	-0.35	2.19	8.75
	CRP/URD	a	x	4.29	5.78	7.99	6.64
			y	-7.68	-2.04	-2.24	0
			z	0	0	-4.04	0

(continued)

TABLE V (continued)

			Load applied on			
			Index Finger	Middle Finger	Ring Finger	Little Finger
	b	x	13.61	1.98	0	0
		y	0	0	0	38.72
		z	0	-5.58	-0.59	-15.98
	c	x	2.84	0	0	16.39
		y	0	0	0	-40.1
		z	0.37	1.16	5.19	22.31
	d	x	0	0	1.33	0
		y	0	0	0	0
		z	2.07	5.55	0	0
MC3/MC4			0	0	0	0
MC4/MC5			0	0	0	0
MC2/MC3			0	0	0	0
III. Joint Moments						
DP/MP		x	-0.37	-0.09	-0.02	0.07
		y	0	0	0	0
		z	-0.44	-0.27	-0.03	0.11
MP/PP		x	-1.22	-0.59	-0.07	0.77
		y	-1.64	0	0	-0.98
		z	0	0	0	0
PP/MC		x	-0.69	-0.52	-0.11	0.23
		y	0	-1.79	0	0
		z	0	0	0	0
MC/CRP		x	-4.61	-0.12	2.01	0
		y	0	0	0	0
		z	0	0	-4.66	0
CRP/URD		x	0	0	0	0
		y	0	0	0	0
		z	0	0	0	0

(continued)

TABLE V (continued)

			Load applied on			
			Index Finger	Middle Finger	Ring Finger	Little Finger
IV.	Upper Extremity Muscle Forces					
	Brachialis		5.57	7.19	8.92	10.57
	Brachioradialis		0	4.33	5.36	5.43
	Biceps Brachii		6.14	5.40	4.51	3.64
	Triceps Brachii		0	0	0	0
	Latissimus dorsi		0	0	0	0
	Pectoralis major		3.73	3.29	3.36	3.14
	Deltoid		0	1.27	2.77	3.55
	Supraspinatus		0	0	0	0
	Infraspinatus		0	0	0	0
	Subscapularis		2.00	1.24	0.42	0.06
V.	Upper Extremity Joint Reactions					
	Elbow joint	x	-19.29	-17.72	18.51	-19.09
		y	0	-1.07	-1.51	-1.77
		z	0.61	1.04	1.97	2.29
	Shoulder joint	x	0	0	0	0
		y	-4.70	-3.84	-3.13	-2.46
		z	-0.15	-1.21	-2.38	-2.72
VI.	Upper Extremity Joint Moments					
	Elbow joint	x	0	0	0	0
		y	0	0	0	0
		z	0	0	0	0
	Shoulder joint	x	0	0	0	0
		y	0	0	0	0
		z	0	0	0	0
VII.	Merit function		141.10	130.49	124.76	126.15
	$(U = \Sigma F + 2R_t + 4M)$					

finger. The phalangeal orientations and the direction of the applied tip force for ulnar, radial and tip pinch were the same as those used by Chow et al in their study (see Fig. 17a and 17b).

The muscle forces and joint reactions for three instances of pinch are tabulated in Table VI. The criterion used to obtain these results was the same as for pulling—minimize all muscle forces and four times the ligament moments.

The muscle force values correspond to 10 arbitrary units of applied force, and appear to follow the pattern indicted by Chow et al (1976). However the joint reactions are lower than those predicted by the latter. One reason why the reactions and the muscle forces are higher in these authors' model is that only the phalanges and metacarpals of the hand has been included for analysis. Each joint is analyzed separately in terms of the orientation of the tendons in the vicinity of the joint, the equilibrium of that joint is assumed to be maintained independent of the other joints of the same finger which a tendon might also cross. Moreover, both the agonist and the antagonist are active to a high extent thus resulting in a higher net joint force.

CONCLUSIONS

The modelling approach that has been used for other segments of the human body has been extended to the hand in this chapter. The hand is a multi-segmented, highly ligamentous structure and the model developed herein is a comprehensive three dimensional model of the right hand including the arm and the forearm and other connecting musculature of the upper extremities and the thorax. The criterion minimize all muscle forces and four times the joint ligments has been validated by application to the pulling and the pinch action.

441

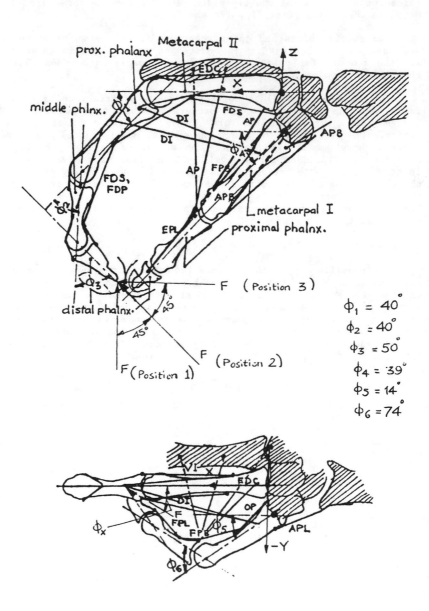

Figure 8-17a. The fingers in pinch posture (the upper extremity model is the same as in Figure 16a).

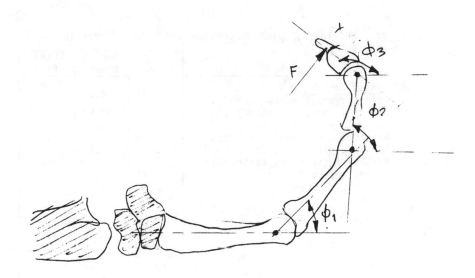

Angle	Tip pinch	Radial pinch	Ulnar pinch
ϕ_1	48	62	72
ϕ_2	98	110	87
ϕ_3	123	133	87

Figure 8-17b. Segmental orientation for the pinch posture.

443

TABLE VI. Muscle and joint forces for the middle finger in pinch action.

			Tip Pinch	Radial Pinch	Ulnar Pinch
I. Muscle forces					
Extensor carpi radialis longus			3.87	18.31	0
Dorsal interossei (ulnar)			46.34	39.94	25.98
Volar interossei			0	0	0
Extensor carpi radialis brevis			0.62	0	7.62
Flexor digitorum profundus			42.05	7.11	22.42
Flexor digitorum superficialis			0	37.39	31.93
Dorsal Interossei (radial)			3.93	17.52	14.83
Extensor digitorum			0.98	0	0
II. Joint Reactions					
DP/MP		a	34.10	17.03	41.65
		b	28.40	9.54	42.14
		y	0	-1.20	1.20
MP/PP		a	19.30	23.49	61.65
		b	35.72	46.12	77.71
		y	0	-1.43	1.11
PP/MC		a	20.02	21.66	34.47
		b	75.01	84.02	72.93
		y	1.12	2.76	2.19
MC/CRP	a	x	38.89	0	0
		z	-6.89	0	0
	b	x	0	50.58	54.99
		z	0	2.39	3.39
		y	-0.69	-1.63	-1.08
CRP/URD	a	x	56.87	0	0
		y	-15.44	-13.03	-10.70
		z	0	0	0
	b	x	32.40	3.16	0
		y	0	0	0
		z	-8.08	0	-0.17
	c	x	-51.10	0	-54.16
		y	0	0	0
		z	0	-6.40	0
	d	x	0	37.62	62.85
		y	0	0	0
		z	0.53	7.56	6.59
	e	x	0	18.04	44.75
		y	0	0	0
		z	0	0	0

(continued)

TABLE VI. (continued)

			Tip Pinch	Radial Pinch	Ulnar Pinch
III.	Joint moments				
	DP/MP	x	0	-0.09	4.71
		y	0	0	0
		z	0	-6.46	4.42
	MP/PP	x	0	0	0
		y	0	16.99	9.53
		z	0	-4.16	4.99
	PP/MC	x	-0.03	4.68	-1.63
		y	0	0	0
		z	0	0	0
	MC/CRP	x	0	-4.19	0.93
		y	0	0	0
		z	0	0	0
	CPP/URD	x	0	0	0
		y	0	0	0
		z	0	0	0
	U		105.38	294.58	208.37

CHAPTER IX

The Foot

"Mondinus says that the muscles which raise
the toes of the feet are to be found in the
outer part of the thigh; and then adds that
the back of the foot has no muscles because
nature has wished to make it light so that it
should be easy in movement, as if it had a
good deal of flesh it would be heavier; and
here experience shows that the muscles a b c d
move the second pieces of the bones of the
toes; and that the muscles of the leg r S t
move the points of the toes. Here then it is
necessary to enquire why necessity has not
made them all start in the foot or all in the
leg; or why those of the leg which move the
points of the toes should not start in the
foot instead of having to make a long journey
in order to reach these points of the toes;
and similarly those that move the second
joints of the toes should start in the leg."

Leonardo da Vinci
1452-1519

INTRODUCTION

The foot is structurally analogous to the hand in many ways. It is a
multi segmented and highly ligamentous structure, articulated at numerous
joints assisted by a meticulous arrangement of extrinsic and intrinsic
muscles to provide a diverse range of motions and functions. It consists
of a proximal part, the tarsus, a middle portion, the metatarsus and a
terminal portion, the phalanges (see Fig. 1). The proximal part is
comprised of a series of more or less cubical bones permitting limited
amount of gliding on one another and are instrumental in distributing the
forces transmitted to or from the bones of the lower leg. The middle
part is made up of slightly movable long bones which assist the tarsus in
distributing forces, while the terminal portion, the phalanges or the
toes can perform a wide range of movements, mainly those of flexion and
extension.

The foot is essential as a rigid basis of support for the body in
the erect posture. The tarsus which occupy or substantial part of the
foot, being placed at right angles to the leg greatly enhance the unique

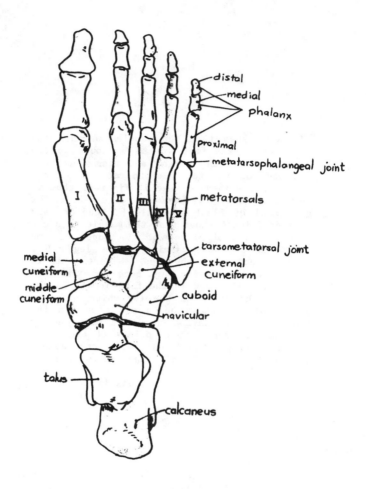

Figure 9-1. Dorsal view of the skeletal system of the right foot.

erect stature of man. The tarsus and the metatarsus are constructed in a series of arches which play a major role in the form and function of the foot. The longitudinal arch seen in the longitudinal dimension of the foot is formed by the seven tarsal and the five metatarsal bone and their binding ligaments. The arch rests posteriorly on the tuberosity of the calcaneus and anteriorly on the heads of the five metatarsals. The transverse arches are visualized in the transverse sections of the foot and are formed by the arrangement of the metatarsals and tarsals bones placed adjacently. In the mid foot the transverse arches are complete but in the mid tarsal region they are only half complete.

The human foot and the mechanics of its arches, its muscles and ligaments have long been of theoretical and practical interest to the anatomist and clinician. Numerous investigators have been studying the structure and properties of the arches, the relation of the muscles and ligaments to their support, the distribution of weight bearing in the foot, and the leverage of the foot.

Engineering concepts have been commonly used by researchers to describe the form and function of the human foot. The foot was initially presumed to act as a tripod in weight bearing, the three points of support being the heel and the heads of the first and the fifth metatarsals. Later Morton (1) demonstrated that the four lateral metatarsal heads all participate in weight bearing and that the foot did not distribute its weight like a tripod. Elftman (2) measured the pressure distribution in the human foot during locomotion by using cinematography and demonstrated that the whole area of the ball of the foot participates in weight bearing. Jones (3) applied both varying loads on the vertical tibia of a cadaver leg and tensions to the tendons duplicating the action of the muscles and measured the pressure and movement produced either by the application of the load on the tibia or tensions on the tendons or both. He observed the pressure partition between the first metatarsal and the lateral four metatarsals in the cadaver foot to be in the ratio of 1:1, while in the relaxed living foot it was found to be 1:1.67. In the living foot, in the standing position however, the ratio was 1:2.50. Morton (loc. cit.) reported a lower value of 1:2 for the ratio in the standing position. The load distribution between the various parts of the plantar surface of the foot during dynamic activities such as walking has been extensively investigated. Schwartz et al (4) used the oscillographic method to record the time, the amount of force, and the sequence and distribution of force with respect to six functionally significant areas on the sole of the foot during walking for male and female subjects. Carlsöö (5) reported the influence of frontal and dorsal loads on the activity of major postural muscles and the distribution of body weight between the forefoot and heel in standing and walking. Recent investigations of the force distribution on the foot during walking including those by Grundy et al (6), Stott et al (7).

The types of joints prevalent among the various bones of the foot and the location of the joint axes has been of interest to any researchers (for example, Barnett and Napier (8), Manter (9), Hicks (10), etc.). Hicks performed experiments on a mechanically normal amputated leg to show all movements in the foot and rotations occurring at joints. It was demonstrated that in natural conditions rotations occur at the ankle (two axes), the talo-calcano-vavicular, the mid-tarsal (two axes), and at the five ray joints. The movements at these joints were then summarized as follows: the whole foot undergoes flexion-extension upon the fixed leg (the flexion is accompanied by slight supination and the extension by slight pronation); the part distal to the talus in addition undergoes pronation-abduction-extension supination-adduction-flexion; the part distal to the mid-tarsal joint in addition undergoes (approximately) pronation supination; the forepart of the foot in addition undergoes a pronation-supination twist.

The mechanism of arch support in the foot remains controversial despite years of research. According to some the arches are maintained by the contraction of muscles (Keith (11)); while according to the others, by the strength of passive tissues (bones and ligaments) (Morton (1)). Also, according to a third theory both muscles and passive structures are important for the arch function. According to Morton (1) static strains upon the foot are relatively low in intensity, falling well within the capabilities of the ligaments, and only during strong dynamic activities such as during the take-off phase of walking the muscles come into action. Jones (3) by direct observation on cadavers and by palpation in the living, believes that of the total tension stress of the longitudinal arch not more than 15-20% is borne by the deep posterior tibial and peroveal muscles, much being borne by the planter ligaments, with support from the short plantar muscles. He suggests that by reason of their positions and relationship they are better adapted to support the longitudinal arch than the long muscles. Later works by Harris and Beath (12), Jones (13) etc. suggests that both ligaments and muscles are responsible for a normal arch. In recent times, this theory draws support from the works of Basmajian and Stecko (14) who used simultaneous electromyography of six muscles in the leg and foot in twenty subjects to reveal that only heavy loading elicits muscle activity. Loads of 100 to 200 pounds on one foot are borne easily by passive structures (ligaments and bones) that support the arches. With 400 pounds, the muscles do come into play, but even then many remain inactive. The authors conclude that the first live of defense of the arches is ligamentous, the muscles forming a dynamic reserve, called upon reflexly by excessive loads, including the take-off phase in walking. These findings are supported by Mann and Inman (15) who investigated the phasic activity of intrinsic muscles of the foot and concluded that muscle activity is not necessary to support the arches of the fully loaded foot at rest. During walking, the actions of the intrinsic muscles are related to the axes of the subtalar and transverse tarsal joints of the foot and are required to stabilize the joints.

Hicks (16) suggests that there are three types of weight bearing mechanisms namely, the beam, the arch or truss and the muscle mechanism, each capable of supporting body weight by itself. However, it seems likely that the three mechanism may take turns at supporting weight, sometimes two together, sometimes alone, depending on the magnitude of stresses induced and the position of the foot.

The foot so far has been treated as a single rigid body in the model for the lower extremities (Chapter V). This model will predict the forces in the extrinsic muscles of the foot only, that is, in the muscles connecting the upper or lower leg to the foot, and the magnitude of loads transmitted at the ankle joint. Many practical situations, however, involve interactions which necessitate the integration of the detailed model of the foot in the lower extremities model. This chapter will be devoted to the development of a comprehensive musculoskeletal model of the lower extremities including the foot. This will incorporate all the segments of the foot; the extrinsic and the intrinsic muscles; the major ligaments which are vital for maintaining the structural stability of the foot and which function along with the muscles in weight bearing and in transmission of forces between the segments. The model will be used to investigate the muscle and ligament load sharing, forces between the foot segments and the nature of distribution of the supportive ground-to-foot forces among the various sites on the side of the foot.

Before the model of the foot can be developed however, certain modifications are required in the previously described model for the lower extremities.

1. Some of the extrinsic muscles should be remodelled to take into account their manner of insertions into the different segments of the foot. These muscles have been modelled by straight lines connecting the leg to the foot, and their portions on the foot on their course to their final points of insertion were omitted. They would be primarily responsible for the segmental movements such as the flexion or extension and therefore when the foot was considered as a rigid body, these specific portions of the muscles were duly excluded. The extrinsic muscles are now remodelled in terms of their components going to different segments of the foot. These will be described in greater detail in the following.

2. In the vicinity of the ankle joint, the extensor and flexor retinacula play an important role in preventing the extrinsic muscles from bowstringing. These are fibrous bands attached to the bones, binding down the tendons situated anterior and posterior to the ankle joint in their passage to the insertions on the foot (Fig. 2). There are three such major bands near the ankle:

    (a) superior and inferior extensor retinacula
    (b) flexor retinaculum
    (c) superior and peroneal retinacula

    The extensor retinacula constrains the muscles anterior to the ankle joint, muscles which are extensors of the toes. The flexor retinaculum is located posteriorly on the leg and guides the toe flexor muscles as they pass behind the ankle joint and underneath the foot on their course to insertions on the toes. The superior peroneal retinaculum channels the peroneal muscles (longus and brevis) passing behind the lateral malleolus of the fibula to their terminations on the foot. It appears that inferior extensor retinaculum which binds down the tendons of extensor digitorum longus, extensor hallucis longus, peroneus tertius and tibialis anterior as they descend anterior to the tibia and the fibula out to be included in the model. It is modelled as a membrane attached to the tibia and the calcaneus, both of which are treated as two independent rigid bodies. The superior extensor retinaculum is attached laterally to the fibula and medially to the tibia, but since both tibia and fibula are together modelled as one rigid body, this band may be excluded. In the case of the flexor and the peroneal retinacula, the tendons that they bind pass behind the medial and the lateral malleolei, and hence they may be treated as

451

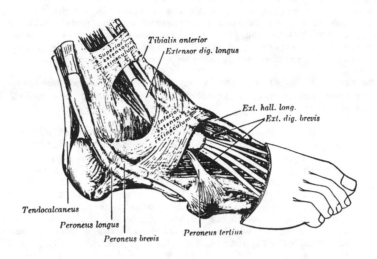

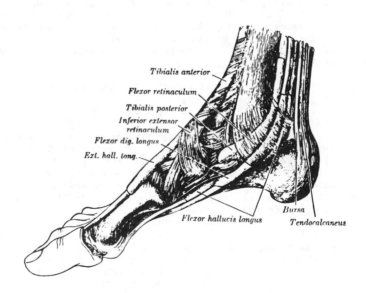

Figure 9-2. (a & b): Synovial sheaths of the tendons passing behind the extensor retinaculum and the flexor retinaculum.

wrapping around these bony segments. These retinacular therefore can also be omitted.

3. As mentioned previously, the bones of the foot are connected by numerous ligaments. Some of the ligaments may directly relieve the muscles of their loadsharing and thus should be treated as muscles. For example, the plantar aponeurosis on the sole of the foot stability (13,16). This will be therefore, included in the foot model as a muscle represented by a line or lines connecting the point of origin to the point of insertion. Similarly, the long and the short plantar ligaments and the calcaneo-navicular ligament (spring ligament) which have been shown to be important in sustaining the longitudinal arch of the foot should also be included in the model. Other ligaments will be accorded the usual treatment, that is, their action is simulated by inclusion of moments at the joints between the bony segments.

THE COMPREHENSIVE FOOT MODEL

A. SEGMENTS AND JOINTS OF THE FOOT:

The foot is comprised of seven tarsals, five metatarsals and fourteen phalanges. Since the phalanges are rather small in dimensions, the five toes may be treated as one single unit, detached at the five metatarso phalangeal joints. Thus a total of thirteen bones require to be included in the model (see Table I).

TABLE I. Bones of the foot.

1. Talus
2. Calcaneus
3. Navicular
4. Cuboid
5. Cuneiform I   (medial)
6. Cuneiform II   (middle)
7. Cuneiform III (lateral)
8. Metatarsal I
9. Metatarsal II
10. Metatarsal III
11. Metatarsal IV
12. Metatarsal V
13. Toes

The talus of the foot articulates four bones—the tibia, the fibula, the navicular and the calcaneus. The contact between the talus and the calcaneus occurs at three sites—the anterior, middle and posterior articular facets on the calcaneus. The calcaneous is the largest of the tarsal bones and is situated at the posterior part of the foot, forming the heel.

The talocalcaneonavicular articulation is a gliding joint formed by the rounded head of the talus, the posterior surface of the navicular and the anterior articular surface of the calcaneus. The calcaneus also articulates with the cuboid bone (see Fig. 1) which in turn articulates with the lateral cuneiform and the fourth and fifth metatarsals. The navicular is connected to the three cuneiform (medial, middle and lateral) which glide with each other and with the first three metatarsals. The five metatarsals articulate with the proximal phalanges

453

TABLE II. Joints of the Foot and Reaction Constraint Angles (Right foot).
A. Constrained Reactions

No.	Joint abbreviation	Segments involving the joint	Constraint angles (degrees) ϕ_1	ϕ_2	Reaction on Segment
1	CAL/TAL(a)	calcaneus/talus	-40	-50	calcaneus
2	TAL/NAL(a)	talus/navicular	15	22	talus
3	CAL/CBD	calcaneus/cuboid	0	0	calcaneus
4	NAV/CFI	navicular/cuneiform I	10	30	navicular
5	NAV/CF2	navicular/cuneiform II	-20	0	navicular
6	NAV/CF3	navicular/cuneiform III	-50	20	navicular
7	CBD/CF3	cuboid/cuneiform III	45	-30	cuboid
8	CBD/MT4	cuboid/metatarsal IV	-21	34	cuboid
9	CBD/MT5	cuboid/metatarsal V	0	0	cuboid
10	CF2/CF1	cuneiform II/cuneiform I	60	5	cuneiform II
11	CF3/CF2	cuneiform III/cuneiform II	75	-25	cuneiform III
12	CF1/MT1	cuneiform I/metatarsal I	10	20	cuneiform I
13	CF2/MT2	cuneiform II/metatarsal II	-22.5	23	cuneiform II
14	CF3/MT2	cuneiform III/metatarsal II	60	0	cuneiform III
15	MT3/MT2	metatarsal III/metatarsal II	90	-20	metatarsal III
16	CF3/MT3	cuneiform III/metatarsal III	-33	36	cuneiform III
17	MT4/MT3	metatarsal IV/metatarsal III	90	-30	metatarsal IV
18	MT5/MT4	metatarsal V/metatarsal IV	90	-40	metatarsal V
19	MT1/LFT	metatarsal I/toes	10	0	metatarsal I
20	MT2/LFT	metatarsal II/toes	0	0	metatarsal II
21	MTS/TLFT	metatarsal III/toes	0	0	metatarsal III
22	MT4/LFT	metatarsal IV/toes	0	0	metatarsal IV
23	MT5/LFT	metatarsal V/toes	0	0	metatarsal V
24	CAL/TAL(b)	calcaneus/talus	-40	-90	calcaneus
25	CAL/TAL(c)	calcaneus/talus	0	-70	calcaneus
26	CAL/TAL(d)	calcaneus/talus	35	-45	calcaneus
27	TAL/NAV(b)	talus/navicular	0	0	talus
28	TAL/TIB1	talus/tibia	0	-90	talus
29	TAL/TIB2	talus/tibia	0	-90	talus
30	TAL/TIB3	talus/tibia	90	0	talus

of the toes forming biaxial joints. The intermetatarsal articulations
between relatively planar surfaces allow only gliding motion and are
formed at the proximal ends of the metatarsal bones.

For modelling purposes each of the thirteen segments can be treated
as a rigid body in space requiring six equation of equilibrium—three for
force balance along the three orthogonal axes and three for moment
balance about the same three axes. In the past we have usually modelled
the joint reaction forces in terms of three orthogonal x, y and z
components. However, the tarsal and metatarsal joints occur between
surfaces that are either planar or curved and permit only limited gliding
or sliding motion. The forces transmitted at such joints therefore,
should in some fashion, conform to the geometrical nature of the
adjoining surfaces and be directed along paths that such articulations
permit. The joint forces are therefore constrained to act along specific
directions as defined by the orientations of the planar surfaces of
adjoining bodies. For those joints where the articulating surfaces are
non-planar or where the contact occurs over a widely distributed area,
more than one reaction force will be assigned. The orthogonal components

454

of such a force are individually assigned to act in a particular direction depending on the spatial inclination of the surface at that location. For example, the joint force at the metatarsal I/cuneiform I interface is defined by the two angles ϕ_1 and ϕ_2 (see Fig. 3). The joint forces between the talus and the navicular (Fig. 4) are assigned at three locations: at the medial locations the reaction force consists of $-x$, y and z components are allowed and at the lateral location $-x$, y and z components are allowed and at the lateral location $-x$, y and z components act as shown. Note that these components have been selected depending on whether the spatial orientation of the surface at that location can permit a compressive force with such components. The constraint angles used for the various tarsal and metatarsal joints and the reaction components for other joints are listed in Table II.

TABLE II. Joint of the Foot and Reaction Constraint Angles (Right foot).

B. Partially Constrained Reactions

No.	Joint abbreviation	Segments involving the joint	Reaction components constrained along	Reaction acting on segment
1	TAL/NV1	Talus/navicular	$-X,-Y,Z$	Talus
2	TAL/NV2	Talus/navicular	$-X,Z$	Talus
3	TAL/NV3	Talus/navicular	$-X,Y,Z$	Talus
4	CAL/CBD	Calcaneus/cuboid	$-X,-Y$	Calcaneus
5	CAL/TAL1	Calcaneus/talus	$-X,-Y,-Z$	Calcaneus
6	CAL/TAL2	Calcaneus/talus	$-X,-Y,-Z$	Calcaneus
7	CAL/TAL3	Calcaneus/talus	$-X,Y,-Z$	Calcaneus
8	CAL/TAL4	Calcaneus/talus	$-X,-Y,-Z$	Calcaneus
9	CAL/TAL5	Calcaneus/talus	$-X,Y,-Z$	Calcaneus
10	CAL/TAL6	Calcaneus/talus	$X,-Y,-Z$	Calcaneus

The ligament action at the various joints will be, as usual, simulated by the residual joint moments. These are assigned as the three respective x, y, z components between the two segments which constitute the joint.

B. MUSCLES OF THE FOOT

There are 11 major intrinsic muscles of the foot which will be modelled by lines representing their functional lines of action. The extrinsic muscles connecting the lower leg to the foot have been modelled in the lower extremities model. However, as reported earlier some of the extrinsic muscles will require remodelling as the foot is no longer treated as a single body. These muscles are described in the following.

1. Tibialis Anterior

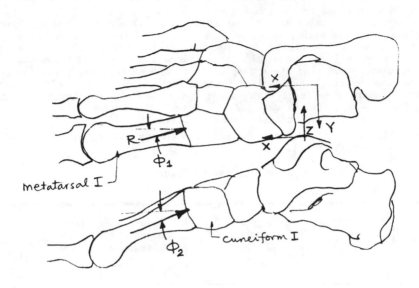

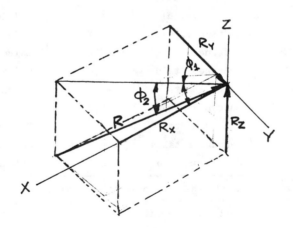

$$R_x = R \cos\phi_2 \cdot \cos\phi_1$$
$$R_Y = R \cos\phi_2 \cdot \sin\phi_1$$
$$R_Z = R \sin\phi_2$$

Figure 9-3. Constrained reaction at the metatarsal I/cuneiform I joint defined by angles θ_1, θ_2.

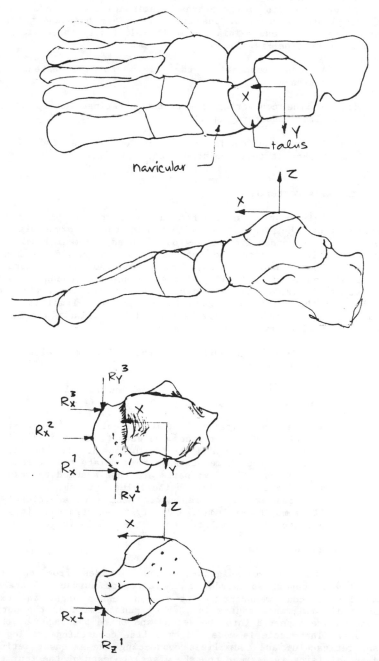

Figure 9-4. Reaction forces at the tabonavicular articulation at three locations.

The tibialis anterior arises from the lateral condyle and upper half of the lateral surface of the tibial shaft. It runs vertically down obliquely across the ankle and after passing through the most medial compartments of the extensor retinaculum is inserted into the medial and plantar surface of the first cuneiform bone and the base of the first metatarsal bone. This muscle is modelled by three lines in the following fashion (see Fig. 5).

(a) A line connecting the tibia to the extensor retinaculum.

(b) A line connecting the extensor retinaculum to the first cuneiform bone.

(c) A line representing the direction of the tendinous fibers at their point of insertion into the base of the first metatarsal bone.

2. Tibialis Posterior:

This muscle is situated in the posterior part of the leg and arises from the tibia, the fibula and the intermuscular septum between them. Its tendon lies in a groove behind the medial malleolus of the tibia and inserts into the tuberosity of the navicular bone, and gives off fibrous expansions to the three cuneiforms and the bases of the second, third and fourth metatarsal bones. Since the tendon wraps around the medial malleolus, which is a part of the tibia, the portion of the muscle between the origin on the tibia and the medial malleolus may be omitted. The muscle therefore may be assumed to originate alternately on the medial malleolus, and modelled by three main parts:

(a) a straight line connecting the medial malleolus to the navicular.

(b) a straight line connecting the medial malleolus to the first cuneiform bone.

(c) a line connecting the medial malleolus to the talus and five additional lines—one each to the second and third cuneiforms and to the bases of the second, third and fourth metatarsals to simulate the fibrous expansions. The five parts are thus assumed to wrap around a common point on the talus. If the foot is analyzed as one rigid body, these five additional lines may be excluded, and only three straight lines are needed to simulate the entire muscle (Fig. 6).

3. Extensor digitorum

The extensor digitorum longus arises from the lateral tibial condyle, upper two-thirds of the anterior surface of the fibula and the interosseous membrane. Its tendon passes under the extensor retinacula dividing into four slips, which run forward on the dorsum of the foot and are inserted into the second and third phalanges of the four lesser toes. The muscle is modelled by a line connecting the leg to the extensor retinaculum and four lines connecting the extensor retinaculum (where the division is assumed to take place) to each of the four toes at

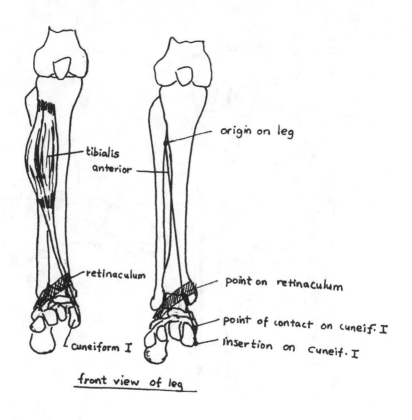

front view of leg

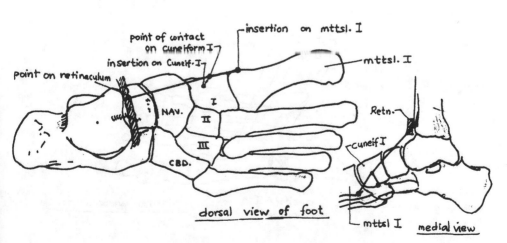

dorsal view of foot

medial view

Figure 9-5. Muscle model for tibialis anterior.

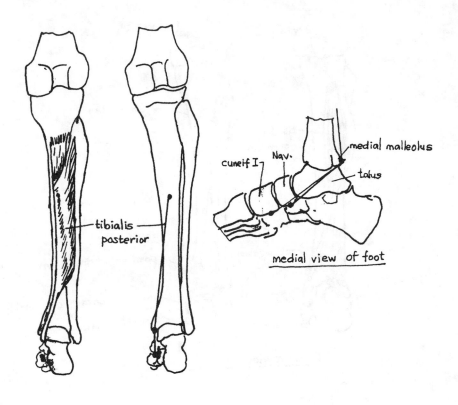

medial view of foot

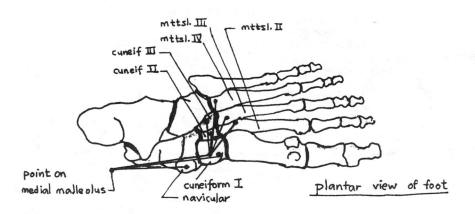

plantar view of foot

Figure 9-6. Muscle model for tibialis posterior.

the head of the corresponding metatarsal. Continuations of the tendons to the phalanges are excluded when the foot is modelled as one rigid body (Fig. 7).

4.   Extensor hallucis longus:

This muscle is modelled in a manner similar to the extensor digitorum longus described above. The extensor hallucis longus arises from the middle half of the anterior surface of the fibula and the interosseous membrane, passes through a compartment in the inferior extensor retinaculum and is inserted into the base of the distal phalax of the great toe. The muscle can therefore be modelled as two lines—one line joining the leg to the extensor retinaculum and another joining the extensor retinaculum to the great toe of the foot (Fig. 6).

5.   Flexor digitorum longus:

The flexor digitorum longus is situated in the posterior of the leg. It arises from the posterior surface of the tibia, wraps around the medial malleolus of the tibia and passes into the sole of the foot where it divides into four tendons which are inserted into the bases of the distal phalanges of the lateral four toes. As explained earlier in the case of the tibialis posterior, the portion of the muscle between the origin on the tibia and the medial malleolus of the tibia may be omitted. The muscle is modelled as a line connecting the medial malleolus to the talus, about which it is assumed to wrap. Each of the four slips to the four toes is modelled as a line connecting the metatarsal head to the talus where the division of the main tendon is assumed to occur (Fig. 8) and a line connecting the metatarsal head to the corresponding toe.

6.   Flexor hallucis longus:

The flexor hallucis longus arises from the lower two-thirds of the posterior surface of the fibula and the intermuscular septum. Its tendon lies in a groove which crosses the posterior surface of the distal and of the tibia, the posterior surface of the talus, and crossing under the surface of the sustentaculum tali of the calcaneus runs obliquely across the sole of the foot to be inserted into the base of the terminal phalanx of the great toe. The muscle is modelled by four lines:

(a)   A line connecting the posterior surface of the distal end of the tibia to the talus.

(b)   A line connecting the talus to a point under the sustentaculum tail on the calcaneus

(c)   A line connecting the sustentaculum tali to the head of the first metatarsal (Fig. 8).

(d)   A line connecting the head of the first metatarsal to the great toe.

Therefore, reactions from wrapping will be included on the talus and the sustentaculum tali.

7.   Peroneus longus:

461

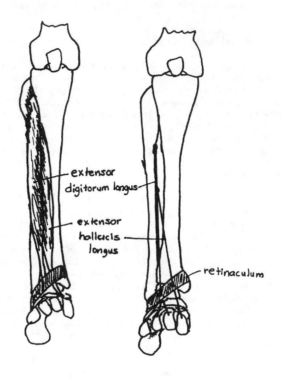

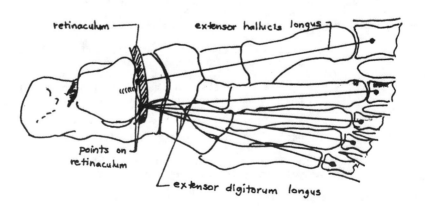

Figure 9-7. Muscle model for ext. digit long & ext. hall. long.

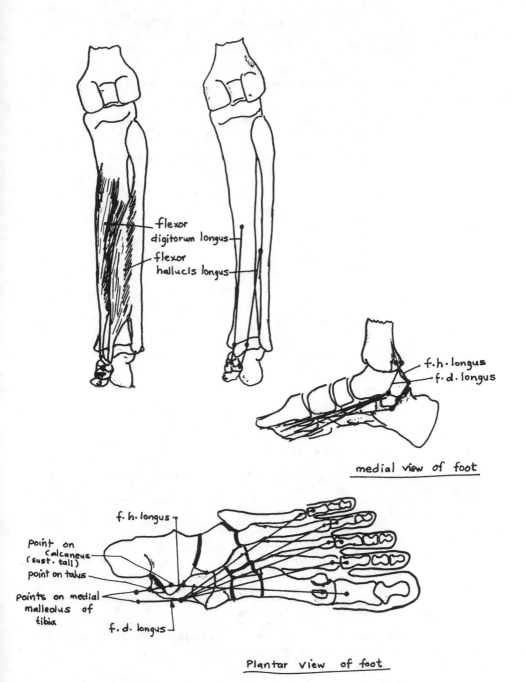

flexor digitorum longus

flexor hallucis longus

f.h. longus
f.d. longus

medial view of foot

f. h. longus

point on calcaneus (sust. tali)

point on talus

points on medial malleolus of tibia

f. d. longus

Plantar view of foot

Figure 9-8. Muscle model for flex. digit. long. & flex. hall. long.

The peroneus longus muscle arises from the lateral tibial condyle, and the upper two-thirds of the lateral surface of the fibula. Its tendon wraps around the lateral malleolus, extends obliquely forward across the lateral side of the calcaneus turning around the trochlear process, crosses under the cuboid into the sole of the foot and is inserted into the lateral side of the base of the first metatarsal and the lateral side of the medial cuneiform. The muscle is modelled by four lines:

(a) A line connecting the lateral malleolus to the trochlear process of the calcaneus.

(b) A line connecting the trochlear process of the calcaneus to the cuboid.

(c) A line connecting the cuboid to the first cuneiform to simulate the insertion into that bone.

(d) A line connecting the cuboid to the base of the first metatarsal to simulate the insertion into that bone.

The muscle therefore wraps around two points——a point on the trochlear process of the calcaneus and a point on the plantar surface of the cuboid. The latter is also the point where the two lines (c) and (d) diverge from on their way to their respective insertions (Fig. 9). Note that the lines (b), (c) and (d) can be excluded if the foot is considered to be one rigid segment.

8. Peroneus Brevis:

The peroneus brevis fibers originate on the distal two-thirds of the lateral surface of the fibula, and run vertically downward to end in a tendon which runs behind the lateral malleolus, passes through the superior peroneal retinaculum proximal to the trochlear process, inserting into the tuberosity at the base of the fifth metatarsal. The muscle is modelled by two lines—a line connecting the lateral malleolus to the calcaneus and a line connecting the calcaneus to the fifth metatarsal, with a reaction due to this wrapping on the calcaneus (Fig. 9).

9. Peroneus Tertius:

The peroneus tertius arises from the distal third of the anterior surface of the fibula and the interosseous membrane and passes under the extensor retinacula in the same canal as the extensor digitorum longus and inserts into the dorsal surface of the base of the fifth metatarsal. The model for this muscle is shown in Fig. 9, the muscle modelled by two parts—an upper part connecting the extensor retinaculum to the point of insertion on the fifth metatarsals. The muscles is thus assumed to wrap around the extensor retinaculum.

MUSCLES OF THE FOOT

The 11 major intrinsic muscles of the foot are as follows:

464

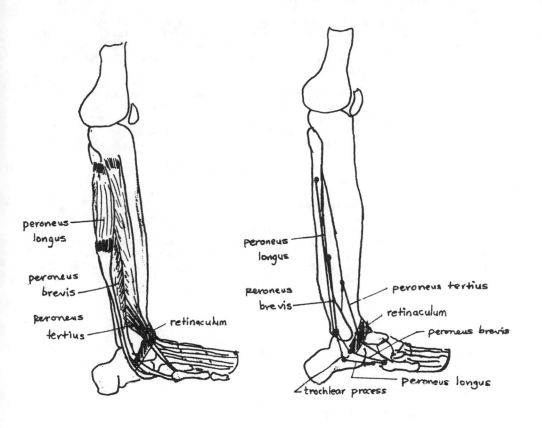

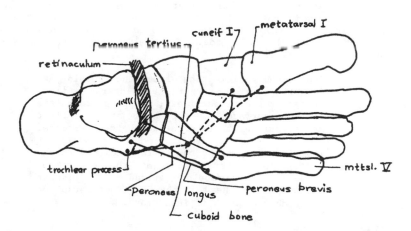

Figure 9-9. Muscle model for perenoei longus, brevis and tertius.

1. **Extensor Digitorum Brevis:**

This muscle arises from the distal and lateral surfaces of the calcaneus, passes obliquely across the dorsum of the foot, and ends in three tendons which are inserted into the tendons of the extensor digitorum longus of the second, third and fourth toes. The muscle is modelled by three lines (Fig. 10):

   (a) A line connecting the calcaneus to the second toe with a reaction on the cuneiform III.

   (b) A line directly connecting the calcaneus to the third toe.

   (c) A line directly connecting the calcaneus to the fourth toe.

2. **Extensor Hallucis Brevis:**

The extensor hallucis brevis has the same origin as the extensor digitorum brevis and inserts into the dorsal surface of the base of the proximal phalanx of the great toe. This muscle is modelled as a line connecting the calcaneus to the first toe, with a reaction on the second metatarsal bone.

3. **Flexor Digitorum Brevis:**

This muscle arises from the medial process of the calcaneal tuberosity and divides into four tendons, one for each of the four lesser toes. The muscle is modelled by four lines, each originating from the calcaneus and inserting into each of the four toes. The tendons are assumed to wrap around the plantar surfaces of the heads of the four metatarsals as shown in Fig. 11.

4. **Flexor Hallucis Brevis:**

The flexor hallucis brevis arises from the plantar surface of the cuboid and from the third cuneiform bone and divides into two portions which are inserted into the medial and lateral sides of the base of the first phalanx of the great toe. The muscle is modelled by four lines of action (Fig. 12):

   (a) Two lines originating from the cuboid and inserting into the medial and lateral side of the great toe respectively, wrapping around the plantar surface of the head of the first metatarsal on their course to their insertions.

   (b) Two lines connecting the third cuneiform to the medial and lateral side of the great toe respectively, wrapping around the plantar surface of the head of the first metatarsal on their course to their insertions.

5. **Adductor Hallucis:**

The adductor hallucis has a transverse and an oblique head of origin. Since the minor transverse head arises from ligaments rather

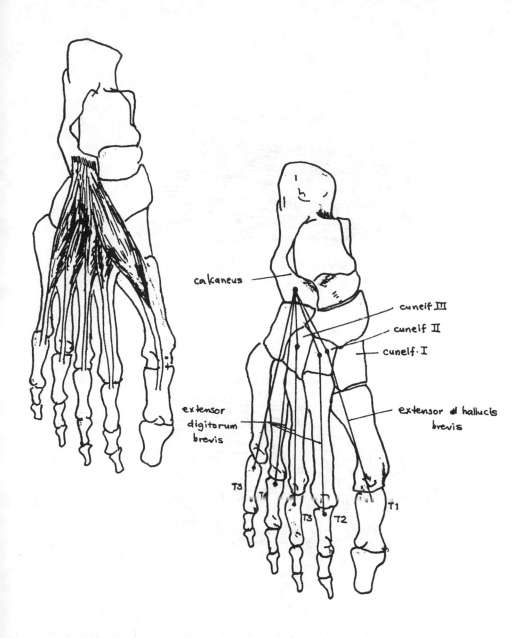

Figure 9-10. Muscle model for ext. digit. brevis & ext. hall. brevis.

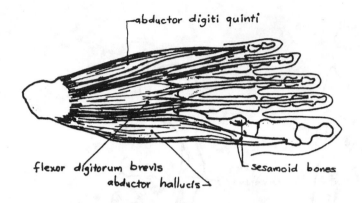

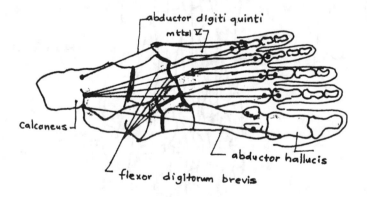

Figure 9-11. Muscle model for abductor digiti quinti, flexor digitorum brevis, and abductor hallucis.

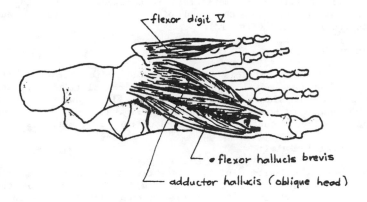

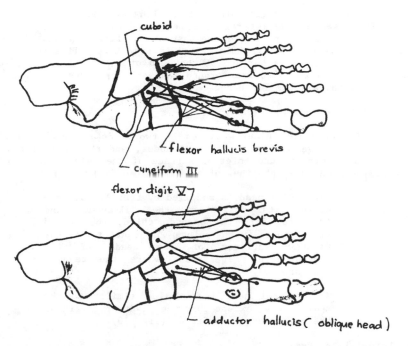

Figure 9-12. Muscle model for flexor digit V, flexor hallucis brevis and adductor hallucis (oblique head).

than any defined bones, only the oblique head is considered here. It arises from the bases of the second, third and fourth metatarsal bones, and is inserted into the lateral side of the base of the proximal phalanx of the great toe. The muscle is represented by three lines originating on the second, third and fourth metatarsal respectively, converging to wrap around a common point on the head of the first metatarsal and inserting into the first toe (Fig. 12).

6. Abductor Hallucis:

The abductor hallucis lies along the medial border of the foot. It arises from the medial process of the calcaneal tuberosity and inserts into the tibial side of the base of the first phalanx of the great toe. The muscle is modelled as shown in (Fig. 11). It is assumed to wrap around the plantar surface of the head of the first metatarsal.

7. Flexor Digiti Minimi:

The flexor digiti minimi, arises from the base of the fifth metatarsal bone and inserts into the lateral side of the base of the first phalanx of the fifth toe. The muscle is conveniently represented by a straight line connecting the fifth metatarsal to the fifth toe (Fig. 12).

8. Abductor Digiti minimi:

The abductor digiti minimi lies along the lateral border of the foot. It arises on the lateral process of the calcaneal tuberosity and inserts into the fibular side of the base of the first phalanx of the little toe. The muscle model is illustrated in Fig. 11. A line connects the calcaneus to the plantar surface of the head of the fifth metatarsal where the tendon is assumed to wrap around, and another, extends between the fifth metatarsal and the fifth toe.

9. Quadratus Plantae:

The quadratus plantae arises by two heads—the medial head from the medial concave surface of the calcaneus, and the lateral head from the lateral border of the inferior surface of the calcaneus. This muscle is represented by eight lines of action in the following manner:

(a) The medial head is modelled by four lines, each having its origin on the calcaneus, and inserting into the second, third, fourth and fifth toes (with points of contact at the plantar surfaces of the heads of the corresponding metatarsals where the tendons wrap around).

(b) The lateral head is modelled similar to the medial head, except that the point of origin on the calcaneus is different (Fig. 13).

10. Dorsal Interossei:

The dorsal interossei are four bipenniform muscles, each arising by two heads from the adjacent sides of the metatarsal bones between which it is placed. Their tendons are inserted into the bases of

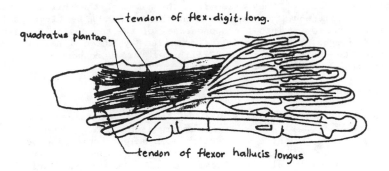

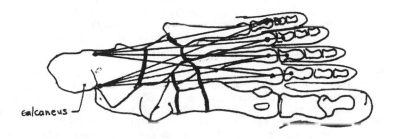

Figure 9-13. Muscle model for quadratus plantae.

the first phalanges, and inot the aponeurosis of the tendons of the extensor digitorum longus. The muscles are modelled by straight lines connecting the metatarsals to the toes as illustrated in Fig. 14.

11. Plantar Interossei:

The plantar interossei are three in number and lie beneath the metatarsal bones. They arise from the bases of the third, fourth and fifth metatarsal bones and insert into the bases of the first phalanges of the same toes, and into the aponeuroses of the tendons of the extensor digitorum longus. The muscle is modelled by three lines of action, connecting the third, fourth and fifth metatarsal respectively to their corresponding toes (Fig. 14).

MAJOR FOOT LIGAMENTS

The structural stability of the foot is maintained not only by musculature but also by several major ligaments. Such ligaments, therefore, which can serve or function as muscles should also be included in the foot model, the important ones being the plantar aponeuoirs, the long and the short plantar ligaments and the calcaneonavicular (or spring) ligament, which are important for maintaining the arches of the foot. Other ligaments connecting the various tarsal and metatarsal joints will not be included directly—they will be assumed to be represented by the residual moments at these joints as previously done.

1. The Plantar Aponeurosis:

The plantar aponeurosis a strong, thick fibrous membrane on the sole of the foot and consists of three portions—central, lateral and medial. The central portion is attached to the medial process of the calcaneal tuberosity. It is narrow and thick proximally but broadens, becoming thinner, and divides near the heads of the metatarsals bones into five processes, one for each of the toes. Each process in turn diverges near the metatarso-phalangeal joint into two layers—a superficial layer which inserts into the skin of the transverse sulcus which separates the toes from the sole, and a deep layer which blends with the sheaths of the flexor tendons of the toes. The lateral portion covers the under surface of the abductor digiti minimi and is continuous medially with the central portion, and laterally with the dorsal fascia. Likewise, the medial portion covers the under surface of the abductor hallucis and is continuous medially with the dorsal fascia and laterally with the central portion of the plantar aponeurosis. The plantar aponeurosis is modelled by three divisions corresponding to the three portions, each division consisting of five lines connecting the proper site on the calcaneus to each of the five toes, with reactions assumed to occur at the points of contact on the heads of the metatarsals. The model is thus similar to that for the quadratus pantae muscle (see Fig. 15).

2. The Long and The Short Plantar Ligaments:

The long plantar ligament arises on the calcaneal tuberosity and inserts into the bases of the third, fourth and fifth metatarsal and the cuboid bones. Accordingly, the ligament can be modelled by four separate lines connecting the calcaneus to each of these four bones. The short plantar ligaments stretches between the plantar surfaces of the calcaneus and the cuboid bone and is modelled by two

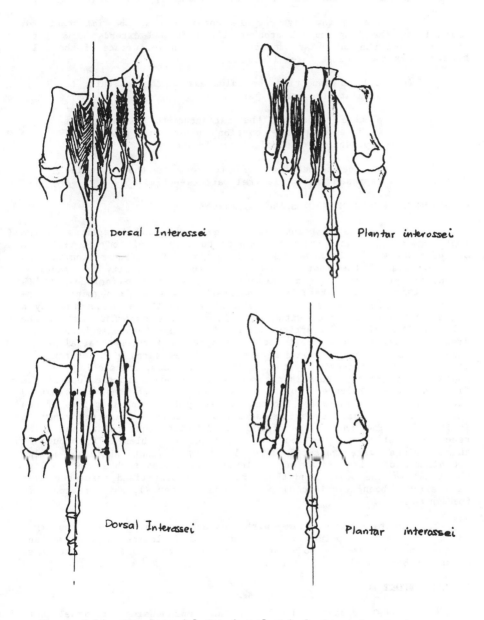

Figure 9-14. Muscle model for dorsal and plantar interossei

separate lines connecting the calcaneus and the cuboid.

3. The Calcano-navicular (Spring) Ligament:

The spring ligament is considered to be important for maintaining the longitudinal arch of the foot and extends between the calcaneus and the navicular supporting the plantar surface of the talus head. It is modelled as three lines:

(a) a line connecting the calcaneus to the navicular to simulate the medial part.

(b) a line connecting the calcaneus to the navicular with a wrapping reaction under the talus head representing the medial portion.

(c) a line connecting the calcaneus to the navicular representing the lateral part (see Fig. 15).

COORDINATES OF MUSCLE ORIGINS AND INSERTIONS

The coordinates of points of muscle origins and insertions on the foot segments are measured with respect to orthogonal body-fixed frames of reference with origins at the joint centers. Since the motion between the segments of the foot is negligible in the majority of practical situations (compared to other members of the lower extremities) it is assumed that the seven tarsals can be treated as a single segment for the purpose of kinematics, and the points of interest can be obtained by a segment-fixed axis system with its origin at the superior part of the talus at the tolo-tibial articulation. The five metatarsals however are treated as individually mobile at the respective tarso-metatarsal joints and any point of interest on a metarsal can be measured with reference to its own coordinate-axis system. The five toes, consisting of the phalanges of the five toes is assumed to articulate with the foot at the first metatarso-phalangeal joint, and all points on the toes are measured with a reference axis system located at this point. The locations of the body-fixed axis for the various segments of the foot are described in Table IIIA. The coordinates of all muscle attachment points and joint reaction locations on the foot are tabulated in Tables IIIA through F; these values were measured directly from a foot skeleton. The coordinates of other points on the lower extremities such as on the leg, the thigh or the pelvis and their respective body-fixed reference axes have already been described elsewhere (see Chapter V), but repeated here for convenience.

The general ground frame with respect to which the transformed coordinates of all points will be computed following segment motions (consequent to a desired body posture or configuration) is taken to be at the talotibial or the ankle joint.

GROUND SUPPORT REACTIONS

One of the purposes of developing the comprehensive foot model is to be able to predict the ground-to-foot pressure distribution on the sole of the foot. In particular, one would be interested in knowing how the given posture whether static (such as leaning forward or backward) or dynamic (such as during walking) influences the load sharing among the various support points, typically, the heads of the metatarsals and the

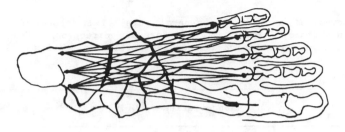

(a) the plantar aponeurosis model

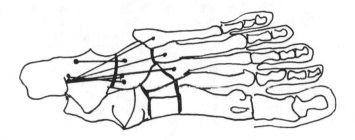

(b) the long and the short plantar ligament model

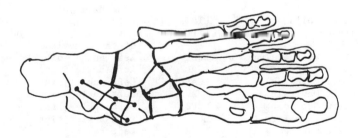

(c) the calcaneo-navicular ligament model

Figure 15. The major ligaments of the foot.

TABLE IIIA. Locations of body-fixed axes.

The body-fixed axes are in general parallel to those of the ground
reference frame: X-axis in the fore-aft direction.   F-axis in the
medio-lateral direction, Z-axis in the vertical direction.

No.	Body	Abbreviation	Origin of body-fixed axes
1	Pelvis	PELV	Right hip joint
2	Right thigh	RTHI	Right knee joint at the center of the condyles
3	Right leg	RLEG	Right ankle joint (TIBTAL)
4	Right Retinaculum	RRET	Right ankle joint
5	Right Talus	RTAL	Right ankle joint
6	Right Calcaneus	RCAL	Right ankle joint
7	Right Navicle	RNAV	Right ankle joint
8	Right Cuboid	RCVB	Right ankle joint
9	Right Cuneiform I	RCF1	Right ankle joint
10	Right Cuneiform II	RCF2	Right ankle joint
11	Right Cuneiform III	RCF3	Right ankle joint
12	Right Metatarsal I	RMT1	Right Metatarsal I/Cuneform I joint
13	Right Metatarsal II	RMT2	Right Metatarsal II/Cuneform II joint
14	Right Metatarsal III	RMT3	Right Metatarsal III/Cuneform III joint
15	Right Metatarsal IV	RMT4	Right Metatarsal IV/Cuboid joint
16	Right Metatarsal V	RMT5	Right Metatarsal V/Cuboid joint
17	Right toes	RLFT	Right Metatarsal I/Proximal Phalanx I joint
18	Left thigh	LTHI	Left hip joint
19	Left leg	LLEG	Left knee joint at the center of the condyles
20	Left Retinaculum	LRET	Left ankle joint
21	Left Talus	LTAL	Left ankle joint
22	Left Calcaneus	LCAL	Left ankle joint
23	Left Navicale	LNAV	Left ankle joint
24	Left Cuboid	LCUB	Left ankle joint
25	Left Cuneiform I	LCF1	Left ankle joint
26	Left Cuneiform II	LCF2	Left ankle joint
27	Left Cuneiform III	LCF3	Left ankle joint
28	Left Metatarsal I	LMT1	Left Metatarsal I/Cuneiform I joint
29	Left Metatarsal II	LMT2	Left Metatarsal II/Cuneiform II joint
30	Left Metatarsal III	LMT3	Left Metatarsal III/ Cuneiform III joint
31	Left Metatarsal IV	LMT4	Left Metatarsal IV/Cuboid joint
32	Left Metatarsal V	LMT5	Left Metatarsal V/Cuboid joint
33	Left Toes	LLFT	Left Metatarsal I/Proximal phalanx I joint

TABLE IIIB Muscles of the Lower Extremities.

No.	Muscle	Muscle Name*	Body 1	Body 2
1.	Gracilis	GRACLS	Pelvis	Thigh
2.	Adductor longus	ADLONG	Pelvis	Thigh
3.	Adductor magnus:			
	1. adductor part	ADMAGA	Pelvis	Thigh
	2. extensor part	ADMAGS	Pelvis	Thigh
4.	Adductor brevis	ADBREV	Pelvis	Thigh
5.	Semitendinosus	SEMITS	Pelvis	Leg
6.	Semimembranosus	SEMIMB	Pelvis	Leg
7.	Biceps Femoris:			
	1. long head	BICFLH	Pelvis	Leg
	2. short head	BICFSH	Thigh	Leg
8.	Sartorius	SARTOR	Pelvis	Leg
9.	Gluteus maximus:			
	1. Insertion into femur	GLMAX1	Pelvis	Thigh
	2. Insertion into ilio-	GMAXT1	Pelvis	Thigh
	tibial tract	GMAXT2	Thigh	Leg
10.	Tensor Fascial Latal	TFASL1	Pelvis	Thigh
		TFASL2	Thigh	Leg
11.	Iliopsoas	ILPSOS	Pelvis	Thigh
12.	Glutens Medius	GLUMED	Pelvis	Thigh
13.	Glutens Minimus	GLUMIN	Pelvis	Thigh
14.	Quadriceps Femoris:			
	1. Rectus Femoris	RECFEM	Pelvis	Patella (Thigh)
	2. Vastus Medialis	VASMED	Thigh	Patella (Thigh)
	3. Vastus Intermedius	VASINT	Thigh	Patella (Thigh)
	4. Vastus Lateralis	VASLAT	Thigh	Patella (Thigh)
	5. Patellar Tendon	PATELL	Thigh	Leg
15.	Gastrocnemius:			
	1. Medial head	GASTMH	Thigh	Calcaneus
	2. Lateral head	GASTLH	Thigh	Calcaneus
16.	Soleus:			
	1. Tibial part	SOLEST	Leg	Calcaneus
	2. Fibular part	SOLESF	Leg	Calcaneus
17.	Tibialis Anterior:			
	1. Part inserting on	TBANTI	Leg	Retinaculum
	Cuneif. I	TANCF1	Retinaculum	Metatarsal I

*abbreviation   (continued)

477

No.	Muscle	Muscle Name*	Body 1	Body 2
18.	Tibialis Posterior:			
	1. Part to Cuneiform I	TPOCF1	Leg	Cuneiform I
	2. Part to Navicular	TPONAV	Leg	Navicular
	3. Part to Cuneiform II	TPOST1	Leg	Talus
		TPOCF2	Talus	Cuneiform II
	4. Part to Cuneiform III	TPOST2	Leg	Talus
		TPOCF3	Talus	Cuneiform III
	5. Part to Metatarsal II	TPOST3	Leg	Talus
		TPOMT2	Talus	Metatarsal II
	6. Part to Metatarsal III	TPOST4	Leg	Talus
		TPOMT3	Talus	Metatarsal III
	7. Part to Metatarsal IV	TPOST5	Leg	Talus
		TPOMT4	Talus	Metatarsal IV
19.	Extensor Digitimum Longus:			
	1. Part to second toe	EXDL21	Leg	Retinaculum
		EXDL22	Retinaculum	Toes
	2. Part to third toe	EXDL31	Leg	Retinaculum
		EXDL32	Retinaculum	Toes
	3. Part to fourth toe	EXDL41	Leg	Retinaculum
		EXDL42	Retinaculum	Toes
	4. Part to fifth toe	EXDL51	Leg	Retinaculum
		EXDL52	Retinaculum	Toes
20.	Extensor Hallucis Longus	EXHL11	Leg	Retinaculum
		EXHL12	Retinaculum	Toes
21.	Flexor Digitorum Longus:			
	1. Part to second toe	FXDL21	Leg	Talus
		FXDL22	Talus	Metatarsal II
		FXDL23	Metatarsal II	Toes
	2. Part to third toe	FXDL31	Leg	Talus
		FXDL32	Talus	Metatarsal III
		FXDL33	Metatarsal III	Toes
	3. Part to fourth toe	FXDL41	Leg	Talus
		FXDL42	Talus	Metatarsal IV
		FXDL43	Metatarsal IV	Toes
	4. Part to fifth toe	FXDL51	Leg	Talus
		FXDL52	Talus	Metatarsal V
		FXDL53	Metatarsal V	Toes
22.	Flexor Hallucis Longus	FXHL11	Leg	Talus
		FXHL12	Talus	Calcaneus
		FXHL13	Calcaneus	Metatarsal I
		FXHL14	Metatarsal I	Toes

*abbreviation

(continued)

478

TABLE IIIB (continued)

No.	Muscle	Muscle Name*	Body 1	Body 2
23.	Peroneus Longus:			
	1. Part To Cuneiform I	PRLNG1	Leg	Calcaneus
		PRLNG2	Calcaneus	Cuboid
		PRLCF1	Cuboid	Cuneiform I
	2. Part to Metatarsal I	PRLNG3	Leg	Calcaneus
		PRLNG4	Calcaneus	Cuboid
		PRLMT1	Cuboid	Metatarsal I
24.	Peroneus Brevis	PRBRV1	Leg	Calcaneus
		PRBRV2	Calcaneus	Metatarsal V
25.	Peroneus Tertius	PRTRT1	Leg	Retinaculum
		PRTRT2	Retinaculum	Metatarsal V
26.	Extensor Digitorum Brevis:			
	1. Part to second toe	EDGB21	Calcaneus	Cuneiform III
		EDGB22	Cuneiform III	Toes
	2. Part to third toe	EDGB31	Calcaneus	Toes
	3. Part to fourth toe	EDGB41	Calcaneus	Toes
27.	Extensor Hallucis Brevis	EHLB11	Calcaneus	Metatarsal II
		EHLB12	Metatarsal II	Toes
28.	Flexor Digitorum Brevis:			
	1. Part to second toe	FXDB21	Calcaneus	Metatarsal II
		FXDB22	Metatarsal II	Toes
	2. Part to third toe	FXDB31	Calcaneus	Metatarsal III
		FXDB32	Metatarsal III	Toes
	3. Part to fourth toe	FXDB41	Calcaneus	Metatarsal IV
		FXDB42	Metatarsal IV	Toes
	4. Part to fifth toe	FXDB51	Calcaneus	Metatarsal V
		FXDB52	Metatarsal V	Toes
29.	Flexor Hallucis Brevis:			
	1. Part from Cuboid	FXHLB1	Cuboid	Metatarsal I
		FXHLB2	Metatarsal I	Toes
	2. Part from Cuneiform III	FXHLB3	Cuneiform III	Metatarsal I
		FXHLB4	Metatarsel I	Toes
	3. Part from Cuboid	FXHLB5	Cuboid	Metatarsal I
		FXHLB6	Metatarsal I	Toes
	4. Part from Cuneiform III	FXHLB7	Cuneiform III	Metatarsal I
		FXHLB8	Metatarsal I	Toes
30.	Flexor digiti minimi	FXDIG5	Metatarsal V	Toe
31.	Abductor Hallucis	ABDHL1	Calcaneus	Metatarsal I
		ABDHL2	Metatarsal I	Toes

*abbreviation

(continued)

TABLE IIIB (continued)

No.	Muscle	Muscle Name*	Body 1	Body 2
32.	Adductor Hallucis:			
	1. Part from Metatarsal II	ADHAL1	Metatarsal II	Metatarsal I
		ADHAL2	Metatarsal I	Toes
	2. Part from Metatarsal III	ADHAL3	Metatarsal III	Metatarsal I
		ADHAL4	Metatarsal I	Toes
	3. Part from Metatarsal IV	ADHAL5	Metatarsal IV	Metatarsal I
		ADHAL6	Metatarsal I	Toes
33.	Abductor digiti minimi:			
		ABDG51	Calcaneus	Metatarsal V
		ABDG52	Metatarsal V	Toes
34.	Dorsal interossei:			
	1. Into medial side of second toe	DORI11	Metatarsal I	Toes
	2. Into medial side of second toe	DORI21	Metatarsal II	Toes
	3. Into lateral side of second toe	DORI22	Metatarsal II	Toes
	4. Into lateral side of second toe	DORI31	Metatarsal III	Toes
	5. Into lateral side of third toe	DORI32	Metatarsal III	Toes
	6. Into lateral side of third toe	DORI41	Metatarsal IV	Toes
	7. Into lateral side of fourth toe	DORI42	Metatarsal IV	Toes
	8. Into lateral side of fourth toe	DORI51	Metatarsal V	Toes
35.	Volar (Plantar) Interossei:			
	1. Into medial side of third toe	VOLI31	Metatarsal III	Toes
	2. Into medial side of fourth toe	VOLI41	Metatarsal IV	Toes
	3. Into medial side of fifth toe	VOLI51	Metatarsal V	Toes
36.	Quadratus Plantae:			
	1. Medial Head to second toe	QDPL21	Calcaneus	Metatarsal II
		QDPL22	Metatarsal II	Toes
	2. Medial head to third toe	QDPL31	Calcaneus	Metatarsal III

*abbreviation

(continued)

No.	Muscle	Muscle Name*	Body 1	Body 2
	3. Medial head to fourth toe	QDPK41 QDPL42	Calcaneus Metatarsal IV	Metatarsal IV Toes
	4. Medial head to fifth toe	QDPL51 QDPL52	Calcaneus Metatarsal V	Metatarsal V Toes
	5. Lateral head to second toe	QDPL23 QDPL24	Calcaneus Metatarsal II	Metatarsal II Toes
	6. Lateral head to third toe	QDPL33 QDPL34	Calcaneus Metatarsal III	Metatarsal III Toes
	7. Lateral head to fourth toe	QDPL43 QDPL44	Calcaneus Metatarsal IV	Metatarsal IV Toes
	8. Lateral head to fifth toe	QDPL53 QDPL54	Calcaneus Metatarsal V	Metatarsal V Toes
37.	Plantar Aponeurosis: 1. Middle band to first toe	PLNT10 PLNT11	Calcaneus Metatarsal I	Metatarsal I Toes
	2. Middle band to second toe	PLNT12 PLNT13	Calcaneus Metatarsal II	Metatarsal II Toes
	3. Middle band to third toe	PLNT14 PLNT15	Calcaneus Metatarsal III	Metatarsal III Toes
	4. Middle band to fourth toe	PLNT16 PLNT17	Calcaneus Metatarsal IV	Metartarsal IV Toes
	5. Middle band to fifth toe	PLNT18 PLNT19	Calcaneus Metatarsal V	Metatarsal V Toes
	6. Medial band to first toe	PLNT20 PLNT21	Calcaneus Metatarsal I	Metatarsal I Toes
	7. Medial band to second toe	PLNT22 PLNT23	Calcaneus Metatarsal II	Metatarsal II Toes
	8. Medial band to third toe	PLNT24 PLNT25	Calcaneus Metatarsal III	Metatarsal III Toes

*abbreviation

(continued)

No.	Muscle	Muscle Name*	Body 1	Body 2
	9. Medial head to fourth toe	PLNT26	Calcaneus	Metatarsal IV
		PLNT27	Metatarsal IV	Toes
	10. Medial head to fifth toe	PLNT28	Calcaneus	Metatarsal V
		PLNT29	Metatarsal V	Toes
	11. Lateral head to first toe	PLNT30	Calcaneus	Metatarsal II
		PLNT31	Metatarsal II	Toes
	12. Lateral head to second toe	PLNT32	Calcaneus	Metatarsal III
		PLNT33	Metatarsal III	Toes
	13. Lateral band to fourth toe	PLNT34	Calcaneus	Metatarsal IV
		PLNT35	Metatarsal IV	Toes
	14. Lateral band to fourth toe	PLNT36	Calcaneus	Metatarsal V
		PLNT37	Metatarsal V	Toes
	15. Lateral head to fifth toe	PLNT38	Calcaneus	Metatarsal V
		PLNT39	Metatarsal V	Toes
38.	Long Plantar Ligament:			
	1. Part to third metatarsal	LIGLP1	Calcaneus	Metatarsal III
	2. Part to fourth metatarsal	LIGLP2	Calcaneus	Metatarsal IV
	3. Part to fifth metatarsal	L1GLP3	Calcaneus	Metatarsal V
	4. Part to cuboid	L1GLP4	Calcaneus	Cuboid
39.	Short Plantar Ligament			
	1. Medial part	L1GSP1	Calcaneus	Cuboid
	2. Lateral part	L1GSP2	Calcaneus	Cuboid
40.	Calcaneo-navicular ligament			
	1. Medial part	L1GCN1	Calcaneus	Navicular
	2. Middle part	L1GCN2	Calcaneus	Talus
		L1GCN3	Talus	Navicular
	3. Lateral part	L1GCN4	Calcaneus	Navicular

*abbreviation

TABLE IIIC Coordinates of Joints of the Lower Extremities and the Foot (right side).

Joint Abbreviation	with respect to body	coordinates (cm) x	y	z
HIPPJT	Thigh	0.0	−0.5	41.5
KNEEJT	Leg	0.0	0.0	40.0
TIB/TAL	Leg	0.0	0.0	0.0
RET/LEG	Leg	3.0	2.0	0.0
RET/CAL	Calcaneus	1.5	−2.2	−3.3
CAL/TAL(a)	Calcaneus	−0.8	−1.7	−2.0
TAL/NAV(a)	Talus	3.0	0.8	−2.3
CAL/CBD	Calcaneus	2.0	−1.7	−3.5
NAV/CF1	Navicular	4.2	1.1	−2.8
NAV/CF2	Navicular	4.0	−0.1	−1.9
NAV/CF3	Navicular	3.5	−1.0	−2.5
CBD/CF3	Cuboid	3.5	−1.7	−2.8
CBD/MT4	Cuboid	4.7	−3.0	−3.4
CBD/MT5	Cuboid	4.0	−3.7	−4.0
CF2/CF1	Cuneiform II	5.0	0.3	−2.0
CF3/CF2	Cuneiform III	4.3	−0.8	−2.1
CF1/MT1	Cuneiform I	6.4	0.5	−3.0
CF2/MT2	Cuneiform II	5.6	−0.6	−2.2
CF3/MT2	Cuneiform III	5.3	−1.3	−2.3
MT3/MT2	Metatarsal III	0.4	0.6	0.2
CF3/MT3	Cuneiform III	5.3	−2.0	−2.6
MT4/MT3	Metatarsal IV	0.8	0.6	0.4
MT5/MT4	Metatarsal V	0.8	0.4	0.4
MT2/CF1	Metatarsal II	0.4	0.4	0.5
MT1/LFT	Metatarsal I	5.6	−0.1	−2.0
MT2/LFT	Metatarsal II	6.4	−1.4	−2.9
MT3/LFT	Metatarsal III	6.0	−1.0	−2.5
T4/LFT	Metatarsal IV	6.0	−1.2	−1.8
MT5/LFT	Metatarsal V	5.6	−1.5	−1.4
CAL/TAL(b)	Calcaneus	−0.5	0.3	−2.0
CAL/TAL(c)	Calcaneus	−1.7	−0.5	−1.7
CAL/TAL(d)	Calcaneus	0.5	0	0.7
TAL/NAV(b)	Talus	2.8	−0.2	−2.0
TAL/TIB1	Talus	0	−1.0	0
TAL/TIB2	Talus	0	1.0	0
TAL/TIB3	Talus	1.1	1.3	−0.7
TAL/NV1	Talus	2.7	1.5	−2.3
TAL/NV2	Talus	3.0	0.8	−2.3
TAL/NV3	Talus	2.7	−0.4	−2.3
CAL/CBD	Calcaneus	2.0	−1.8	−3.0
CAL/TAL1	Calcaneus	1.6	−0.7	−2.7
CAL/TAL2	Calcaneus	0.4	1.0	−2.4
CAL/TAL3	Calcaneus	−0.5	−1.8	−2.5
CAL/TAL4	Calcaneus	−1.8	−0.5	−1.5
CAL/TAL5	Calcaneus	−0.4	0	−2.3
CAL/TAL6	Calcaneus	−1.0	0.4	−2.5

TABLE IIID   Coordinates of Ground-to-Foot Support Reactions on the Foot.

Location	with respect to body	coordinates (cm)		
		x	y	z
Head of metatarsal I	Metatarsal I	4.6	-0.1	-3.0
Head of metatarsal II	Metatarsal II	5.5	-1.4	-3.8
Head of metatarsal III	Metatarsal III	5.1	-1.0	-3.4
Head of metatarsal IV	Metatarsal IV	5.2	-1.2	-2.6
Head of metatarsal V	Metatarsal V	4.8	-1.5	-2.0
Heel (lateral)	Calcaneus	-3.8	-1.3	-6.0
Heel (medial)	Calcaneus	-3.8	0.5	-6.0
Toes	Toes	2.0	-3.4	-1.0

TABLE IIIE   Weights and Centers of Gravity of Body Segment.

Location	with respect to body	coordinates (cm)			Weight (kg)
		x	y	z	
Leg	Leg	-1.0	0	25.2	3.8
Thigh	Thigh	2.0	-2.0	23.6	7.2
Upper Torso (including pelvis)	Pelvis	0	8.0	26.5	48.3

TABLE IIIF Coordinates of Muscles of the Right Side

Muscle Name	Body 1	Coordinates (cm) x	Coordinates (cm) y	Coordinates (cm) z	Body 2	Coordinates (cm) x	Coordinates (cm) y	Coordinates (cm) z
GRACLS	PELV	3.50	4.50	-3.50	RLEG	1.00	2.50	33.00
ADLONG	PELV	5.00	4.70	-1.50	RTHI	1.50	-1.00	18.00
ADMAGA	PELV	-2.00	1.00	-5.50	RTHI	0.00	4.00	2.00
ADMAGE	PELV	0.00	2.00	-5.50	RTHI	1.00	-1.00	22.02
ADBREV	PELV	4.00	4.00	-2.50	RTHI	0.50	-2.00	28.00
SEMITS	PELV	-4.00	1.00	-2.50	RLEG	0.00	3.00	33.00
SEMIMB	PELV	-3.80	0.00	-3.00	RLEG	-1.00	3.00	37.00
BICFLH	PELV	-3.80	0.00	-2.50	RLEG	-2.50	-3.50	36.00
BICFSH	PELV	0.50	-1.00	13.00	RLEG	-2.50	-3.50	36.00
SARTOR	PELV	4.50	-3.00	-8.50	RLEG	1.00	2.50	35.00
GLMAX1	PELV	-7.00	2.00	6.50	RTHI	-1.50	-4.00	33.00
GMAXT1	PELV	-7.50	11.00	10.00	RTHI	0.00	-8.00	38.50
GMAXT2	RTHI	0.00	-8.00	38.50	RLEG	3.00	-3.50	38.00
TFASL1	PELV	4.00	-3.00	10.00	RTHI	0.00	-8.00	38.50
TFASL2	RTHI	0.00	-8.00	38.50	RTHI	3.00	-3.50	38.00
ILPSOS	PELV	3.00	0.00	1.00	RTHI	-1.00	-2.70	34.00
GLUMED	PELV	-3.00	-3.00	11.00	RTHI	-1.50	-6.00	40.00
GLUMIN	PELV	2.00	-2.00	8.50	RTHI	0.00	-6.50	40.00
RECFEM	PELV	2.50	-2.00	2.50	RTHI	5.00	-2.00	0.00
VASMED	RTHI	2.50	-1.00	18.00	RTHI	5.00	-2.00	0.00
VASINT	RTHI	3.00	-2.00	20.00	RTHI	5.00	-2.00	0.00
VASLAT	RTHI	2.50	-3.00	18.00	RTHI	5.00	-2.00	0.00
PATELL	RLEG	4.00	-1.00	35.00	RTHI	5.00	-2.00	0.00
GASTMH	RTHI	-2.00	1.50	0.00	RCAL	-4.50	-1.00	-3.70
GASTLH	RTHI	-2.00	-2.50	0.00	RCAL	-4.50	-1.00	-3.70
SOLEST	RLEG	-0.50	1.00	29.00	RCAL	-4.50	-1.00	-3.70
SOLESF	RLEG	-3.00	-3.00	31.00	RCAL	-4.50	-1.00	-3.70
TBANTI	RLEG	2.00	0.00	25.00	RRET	2.50	1.00	0.70
TANCF1	RRET	2.50	1.00	0.00	RCF1	5.50	1.30	-3.70
TBANT2	RLEG	2.00	0.00	25.00	RRET	2.50	1.00	0.00
TANMT1	RRET	2.50	1.00	0.00	RMT1	-0.20	0.50	-1.10
TPDCF1	RLEG	-2.50	2.00	0.00	RCF1	4.20	1.20	-4.30
TPDNAV	RLEG	-2.50	2.00	0.00	RNAV	3.00	2.00	-3.60
TPOST1	RLEG	-2.50	2.00	0.00	RTAL	1.80	0.50	-3.50
TPOCF2	RTAL	1.80	0.50	-3.50	RCF2	4.80	-0.50	-2.70
TPOST2	RLEG	-2.50	2.00	0.00	RTAL	1.80	-0.50	-3.50
TPOCF3	RTAL	1.80	0.50	-3.50	RCF3	4.00	-1.20	-3.30
TPOST3	RLEG	-2.50	2.00	0.00	RTAL	1.80	-0.50	-3.50
TPOMT2	RTAL	1.80	0.50	-3.50	RMT2	0.30	-0.20	-0.80

(continued)

Muscle Name	Body 1	Coordinates (cm)			Body 2	Coordinates (cm)		
		x	y	z		x	y	z
TPOST4	RLEG	−2.50	2.00	0.00	RTAL	1.80	0.50	−3.50
TPOMT3	RTAL	1.80	0.50	−3.50	RMT3	0.50	0.50	−0.90
TPOST5	RLEG	−2.50	2.00	0.00	RTAL	1.80	0.50	−3.50
TPOMT4	RTAL	1.80	0.50	−3.50	RMT4	0.30	0.50	−0.70
EXDL21	RLEG	0.00	−2.50	15.00	RRET	2.00	−1.00	−1.00
EXDL22	RRET	2.00	−1.00	−1.00	RLFT	0.00	−2.40	0.80
EXDL31	RLEG	0.00	−2.50	15.00	RRET	2.00	−1.00	−1.00
EXDL32	RRET	2.00	−1.00	−1.00	RLFT	−0.70	−3.40	0.80
EXDL41	RLEG	0.00	−2.50	15.00	RRET	2.00	−1.00	−1.00
EXDL42	RRET	2.00	−1.00	−1.00	RLFT	−1.30	−4.60	0.60
EXDL51	RLEG	0.00	−2.50	15.00	RRET	2.00	−1.00	−1.00
EXDL52	RRET	2.00	−1.00	−1.00	RTFT	−2.40	−5.60	0.20
EXHL11	RLEG	0.00	−2.00	16.00	RRET	2.50	0.40	0.00
EXHL12	RRET	2.50	0.40	0.00	RLFT	0.00	0.00	1.00
FXDL21	RLEG	−2.50	0.90	0.00	RTAL	1.50	1.50	−3.40
FXDL22	RTAL	1.50	1.50	−3.40	RMT2	5.50	−1.40	−3.80
FXDL23	RMT2	5.50	−1.40	−3.80	RLFT	0.00	−2.40	−1.00
FXDL31	RLEG	−2.50	0.90	0.00	RTAL	1.50	1.50	−3.40
FXDL32	RTAL	1.50	1.50	−3.40	RMT3	5.10	−1.00	−3.40
FXDL33	RMT3	5.10	−1.00	−3.40	RLFT	−0.70	−3.40	−1.00
FXDL41	RLEG	−2.50	0.90	0.00	RTAL	1.50	1.50	−3.40
FXDL42	RTAL	1.50	1.50	−3.40	RMT4	5.20	−1.20	−2.60
FXDL43	RMT4	5.20	−1.20	−2.60	RLFT	−1.30	−4.60	−1.00
FXDL51	RLEG	−2.50	0.90	0.00	RTAL	1.50	1.50	−3.40
FXDL52	RTAL	1.50	1.50	−3.40	RMT5	4.80	−1.50	−2.00
FXDL53	RLT5	4.80	−1.50	−2.00	RLFT	−2.40	−5.60	−1.00
FXHL11	RLEG	−1.80	−0.50	0.50	RTAL	−2.00	0.50	−1.70
FXHL12	RTAL	−2.00	0.50	−1.70	RCAL	0.00	1.30	−3.20
FXHL13	RCAL	0.00	1.30	−3.20	RMT1	4.60	−0.10	−3.00
FXHL14	RMT1	4.60	−0.10	−3.00	RLFT	0.00	0.00	−1.00
PRLNG1	RLEG	−1.50	−2.00	−1.80	RCAL	−1.00	−2.50	−4.00
PRLNG2	RCAL	−1.00	−2.50	−4.00	RCUB	3.50	−3.00	−4.60
PRLCF1	RCUB	3.50	−3.00	−4.60	RCF1	5.50	−0.50	−3.80
PRLNG3	RLEG	−1.50	−2.80	−1.80	RCAL	−1.00	−2.50	−4.00
PRLNG4	RCAL	−1.00	−2.50	−4.00	RCUB	3.50	−3.00	−4.60
PRLMT1	RCUB	3.50	−3.00	−4.60	RMT1	−0.20	−1.00	−0.80
PRBRV1	RLEG	−1.50	−2.80	−1.80	RCAL	0.00	−2.50	−3.30
PRBRV2	RCAL	0.00	−2.50	−3.30	RMT5	−0.30	−0.30	−1.00
PRTRT1	RLEG	−1.00	−2.00	10.00	RRET	2.00	−1.00	−1.00
PRTRT2	RRET	2.00	−1.00	−1.00	RMT5	0.50	−0.10	−0.20
EDGB21	RCAL	1.70	−2.20	−2.50	RCF3	5.00	−1.80	2.10
EDGB22	RCF3	5.00	−1.80	−2.10	RLFT	0.00	−2.40	−0.80
EDGB31	RCAL	1.70	−2.20	−2.50	RLFT	−0.70	−3.40	0.80
EDGB41	RCAL	1.70	−2.20	−2.50	RLFT	−1.30	−4.60	0.60
EHLB11	RCAL	1.70	−2.20	−2.50	RMT2	9.40	−0.70	0.40
EHLB12	RMT2	0.40	0.70	.40	RLFT	0.00	−0.00	1.00

(continued)

486

Muscle Name	Body 1	Coordinates (cm) x	y	z	Body 2	Coordinates (cm) x	y	z
FXDB21	RCAL	-2.50	0.00	-5.50	RMT2	5.50	-1.40	-3.80
FXDB22	RMT2	5.50	-1.40	-3.80	RLFT	0.00	-2.40	-1.00
FXDB31	RCAL	-2.50	0.00	-5.50	RMT3	5.10	-1.00	-3.40
FXDB32	RMT3	5.10	-1.00	-3.40	RLFT	-0.70	-3.40	-1.00
FXDB41	RCAL	-2.50	0.00	5.50	RMT4	5.20	-1.20	-2.60
FXDB42	RMT4	5.20	-1.20	-2.60	RLFT	-1.30	-4.60	-1.00
FXDB51	RCAL	-2.50	0.00	5.50	RMT5	4.80	-1.50	-2.00
FXDB52	RMT5	4.80	-1.50	-2.00	RLFT	-2.40	-5.60	-1.00
FXHLB1	RCUB	3.50	-1.20	3.70	RMT1	4.60	0.50	-3.00
FXHLB2	RMT1	4.60	0.50	-3.00	RLFT	0.00	0.60	-1.00
FXHLB3	RCF3	4.00	-1.00	3.20	RMT1	4.60	0.50	-3.00
FXHLB4	RMT1	4.60	0.50	-3.00	RLFT	0.00	0.60	-1.00
FXHLB5	RCUB	3.50	-1.20	3.70	RMT1	4.60	-1.00	-2.80
FXHLB6	RMT1	4.60	-1.00	2.80	RLFT	0.00	-0.90	-0.80
FXHLB7	RCF3	4.00	-1.00	3.20	RMT1	4.60	-1.00	-2.80
FXHLB8	RMT1	4.60	-1.00	-2.80	RLFT	0.00	-0.90	-0.80
FXDIG5	RMT5	0.00	-0.10	-1.30	RLFT	-2.40	-5.90	-0.80
ABDHL1	RCAL	-3.00	0.50	5.00	RMT1	4.60	0.55	-3.00
ABDHL2	RMT1	4.60	0.50	-3.00	RLFT	0.00	0.60	-1.00
ADHAL1	RMT2	1.00	-0.20	-0.90	RMT1	4.60	-1.00	-2.80
ADHAL2	RMT1	4.60	-1.00	2.80	RLFT	0.00	-0.90	-0.80
ADHAL3	RMT3	0.90	0.50	-1.00	RMT1	4.60	-1.00	-2.80
ADHAL4	RMT1	4.60	-1.00	-2.80	RLFT	0.00	-0.90	-0.80
ADHAL5	RMT4	0.90	0.50	0.70	RMT1	4.60	-1.00	-2.80
ADHAL6	RMT1	4.60	-1.00	-2.80	RLFT	0.00	-0.90	-0.80
ABDG51	RCAL	-3.80	-1.00	-5.80	RMT5	-0.30	-0.30	-1.50
ABDG52	RMT5	-0.30	-0.30	-1.50	RLFT	-2.40	-5.90	-0.80
DORI11	RMT1	2.40	-0.90	0.00	RLFT	0.00	-2.00	-0.70
DORI21	RMT2	3.00	-0.20	-0.80	RLFT	0.00	-2.00	-0.70
DORI22	RMT2	3.00	-1.00	-0.80	RLFT	0.00	-2.80	-0.70
DORI31	RMT3	2.70	0.00	-0.90	RLFT	0.00	-2.80	-0.70
DORI32	RMT3	2.70	-0.70	-0.90	RLFT	-0.70	-3.70	-0.70
DORI41	RMT4	2.90	-0.10	-0.60	RLFT	0.70	3.70	0.70
DORI42	RMT4	-2.90	-0.80	-0.60	RLFT	-1.30	-4.90	-0.80
DORI51	RMT5	-3.00	-0.60	-0.50	RLFT	-1.30	-4.90	-0.80
VOLI31	RMT3	2.20	-0.30	-1.20	RLFT	-0.70	-3.10	-0.70
VOLI41	RMT4	-2.30	-0.50	-0.90	RLFT	-1.30	-4.30	-0.80
VOLI51	RMT5	2.50	-0.50	-0.80	RLFT	-2.40	-5.30	-0.80
QDPL21	RCAL	-1.20	0.00	-4.00	RMT2	5.50	-1.40	-3.80
QDPL22	RMT2	5.50	-1.40	-3.80	RLFT	0.00	-2.40	-1.00
QDPL31	RCAL	-1.20	0.00	-4.00	RMT3	5.10	-1.00	-3.40
QDPL32	RMT3	5.10	-1.00	-3.40	RLFT	-0.70	-3.40	-1.00
QDPL41	RCAL	-1.20	0.00	-4.00	RMT4	5.20	-1.20	-2.60
QDPL42	RMT4	5.20	-1.20	-2.60	RLFT	-1.30	-4.60	-1.00

(continued)

Muscle Name	Body 1	Coordinates (cm)			Body 2	Coordinates (cm)		
		x	y	z		x	y	z
QDPL51	RCAL	-1.20	0.00	0.00	RTAL	1.80	0.50	-3.50
QDPL52	RMT5	4.80	-1.50	-3.50	RMT3	0.50	0.50	-0.90
QDPL23	RCAL	-3.00	-2.00	0.00	RTAL	1.80	0.50	-3.50
QDPL24	RMT2	5.50	-1.40	-3.50	RMT4	0.30	0.50	-0.70
QDPL33	RCAL	-3.00	-2.00	15.00	RRET	2.00	-1.00	-1.00
QDPL34	RMT3	5.10	-1.00	-1.00	RLFT	0.00	-2.40	0.80
QDPL43	RCAL	-3.00	-2.00	15.00	RRET	2.00	-1.00	-1.00
QDPL44	RMT4	5.20	-1.20	-1.00	RLFT	-0.70	-3.40	0.80
QDPL53	RCAL	-3.00	-2.00	15.00	RRET	2.00	-1.00	-1.00
QDPL54	RMT5	4.80	-1.50	-1.00	RLFT	-1.30	-4.60	0.60
PLNT10	RCAL	-3.00	0.50	15.00	RRET	2.00	-1.00	-1.00
PLNT11	RMT1	4.60	0.50	-1.00	RTFT	-2.40	-5.60	0.20
PLNT12	RCAL	-3.00	0.50	0.00	RLFT	0.00	0.00	1.00
PLNT13	RMT2	5.50	-1.40	0.00	RTAL	1.50	1.50	-3.40
PLNT14	RCAL	-3.00	0.50	-3.40	RMT2	5.50	-1.40	-3.80
PLNT15	RMT3	5.10	-1.00	-3.80	RLFT	0.00	-2.40	-1.00
PLNT16	RCAL	-3.00	0.50	0.00	RTAL	1.50	1.50	-3.40
PLNT17	RMT4	5.20	-1.20	-3.40	RMT3	5.10	-1.00	-3.40
PLNT18	RCAL	-3.00	0.50	-3.40	RLFT	-0.70	-3.40	-1.00
PLNT19	RMT5	4.80	-1.50	0.00	RTAL	1.50	1.50	-3.40
PLNT20	RCAL	-3.00	-1.20	-3.40	RMT4	5.20	-1.20	-2.60
PLNT21	RMT1	4.60	0.50	-2.60	RLFT	-1.30	-4.60	-1.00
PLNT22	RCAL	-3.00	-1.20	0.00	RTAL	1.50	1.50	-3.40
PLNT23	RMT2	5.50	-1.40	-3.40	RMT5	4.80	-1.50	-2.00
PLNT24	RCAL	-3.00	-1.20	-2.00	RLFT	-2.40	-5.60	-1.00
PLNT25	RMT3	5.10	-1.00	0.50	RTAL	-2.00	0.50	-1.70
PLNT26	RCAL	-3.00	-1.20	-1.70	RCAL	0.00	1.30	-3.20
PLNT27	RMT4	5.20	-1.20	-3.20	RMT1	4.60	-0.10	-3.00
PLNT28	RCAL	-3.00	-1.20	-3.00	RLFT	0.00	0.00	-1.00
PLNT29	RMT5	4.80	-1.50	-3.40	RLFT	-0.70	-3.40	-1.00
PLNT30	RCAL	-3.00	-2.20	0.00	RTAL	1.50	1.50	-3.40
PLNT31	RMT1	4.60	0.50	-3.40	RMT4	5.20	-1.20	-2.60
PLNT32	RCAL	-3.00	-2.20	-2.60	RLFT	-1.30	-4.60	-1.00
PLNT33	RMT2	5.50	-1.40	-3.80	RLFT	0.00	-2.40	-1.00
PLNT34	RCAL	-3.00	2.20	-6.00	RMT3	5.10	-1.00	-3.40
PLNT35	RMT3	5.10	1.50	-3.40	RMT5	4.80	-1.50	-2.00
PLNT36	RCAL	-3.00	-2.20	-6.00	RMT4	5.20	-1.20	-2.60
PLNT37	RMT4	5.20	-1.20	2.60	RLFT	-1.30	4.60	-1.00
PLNT38	RCAL	-3.00	-2.20	-6.00	RMT5	4.80	-1.50	-2.00
PLNT39	RMT5	4.80	-1.50	-2.00	RLFT	-2.40	-5.60	-1.00
LIGLP1	RCAL	-1.30	-2.00	-4.80	RMT3	0.70	0.40	-1.00
LIGLP2	RCAL	-1.30	-2.00	-4.80	RMT4	0.60	0.50	-0.80
LIGLP3	RCAL	-1.30	-2.00	-4.80	RMT5	0.50	0.00	-0.80
LIGLP4	RCAL	-1.30	-2.00	-4.80	RCUB	3.50	-2.40	-4.50
LIGSP1	RCAL	1.00	-1.20	-4.60	RCUB	3.30	-1.20	-4.00
LIGSP2	RCAL	1.00	-2.10	-4.60	RCUB	3.50	-2.00	-4.20
LIGCN1	RCAL	0.00	1.00	-2.70	RNAV	2.20	-1.50	-2.90
LIGCN2	RCAL	1.00	0.20	-3.10	RTAL	1.80	0.50	-3.50
LIGCN3	RTAL	1.80	0.50	-3.50	RNAV	2.40	-0.40	-3.00

heel. This is modelled by three force components along x (fore and aft), along y (medio-lateral) and z (vertical) axes respectively at seven locations under the foot—under each of the five metatarsals at their heads, at two locations (medial and lateral) under the calcaneous because of the extensive heel area and at the center of the segment representing the "toes". The coordinates of each of the seven points are listed in Table III-D. A schematic diagram of the foot model is shown in Fig. 15A.

SUMMARY OF THE FOOT MODEL

A. NUMBER OF EQUATIONS

The thirteen segments in each foot (Table I) together with the pelvis, the left and the right femur, the left and the right leg and the left and the right extensor retinaculum yield a total of thirty three segments. These are rigid bodies in space and thus each one requires six equilibrium equations.

Additional equations however, must also be accounted for—these describe the equilibrium between the muscle forces that wrap around a segment and the consequent reaction force at that point. There are two ways in which the wrapping can be modelled:

(a) By considering the equilibrium of the point of wrapping as a particle in space, that is, three equations for force balance along the three axes can be written. These equations will include the muscle force F and the three components of the wrapping reaction R_x, R_y and R_z. This analysis results in three additional equations and three additional unknown variables; an example where this is used in the model is the quadriceps muscle group where the four members of the group wrap at the patella on the femur.

(b) By treating the wrapping muscle as two unknown forces which are constrained to be equal (see Fig. 16). In this way the reactions at the point of contact are excluded, reducing the number of unknown variables as well as the number of additional equations. This approach is followed for all the muscles that wrap.

When the equations to account for wrapping are considered, for both left and right sides, the total number of equations including all the segmental equilibrium equations, turns out to be 344.

B. NUMBER OF UNKNOWN VARIABLES

i. Number of muscle and ligament forces
The total number of unknown muscle and ligament forces for the foot and the lower extremities on each side is 193— which would be doubled if both sides were to be included.

ii. Number of joint reaction forces
As explained earlier, at some joints the reaction force was left open so that it could act in either direction for each of the three coordinate axes. This yields six unknown variables at a joint. At a joint where the reaction is fully

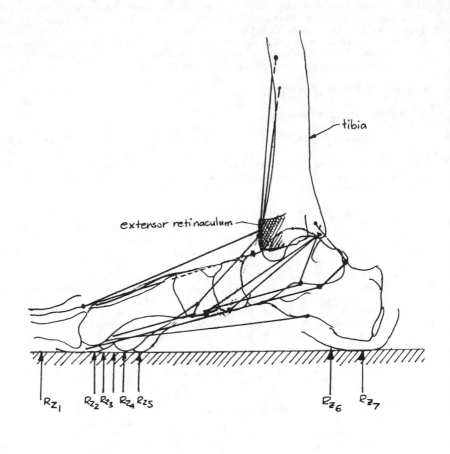

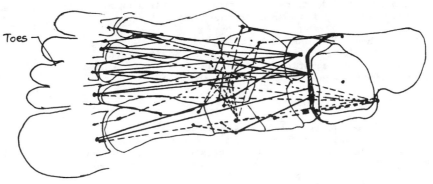

Figure 9-15A. The foot model.

constrained in terms of the two pre-defined angles (Fig. 3) of course one would need only one variable, the magnitude of the reaction force. At other joints where the reactions are "partially" constrained, that is, where the respective reaction force components are known beforehand to act along a given direction due to geometry, no more than three unknown variables need to be defined.

The total number of unknown variables that define joint reaction forces for the foot and the lower extremities on each side is 91—which would be doubled if both sides were to be included.

iii. Number of joint reaction moments

The joint reaction moment at any joint are modelled in terms of the respective bi-directional components along x, y and z, that is, a total of six unknown variables at a joint. The total number variables of this type is 174 for each side of the lower extremities and the foot, or 348 for both sides.

iv. Other unknown variables

As explained above, the quadriceps group is modelled to wrap at the patella and will thus result in three bi-directional x, y and z components—or six variables for each side.

The ground reaction forces at eight locations on the sole of the foot (five metarsal heads, the "toes", medial and lateral heel) are modelled by five reaction components: bi-directional x and y components and a vertically upward z component since this can act only in this direction and not bind or pull the heel to the ground. Therefore, ground-to-foot support reactions will number 40 for each foot.

To sum up, the total number of unknown variables is equal to 1008 while the total number of equations is 344.

Merit criterion

In order to be consistent with the merit criterion that has been previously established for other parts of the body, it was decided to use the same criterion, minimize $U = \Sigma F + 4\Sigma M$ where F and M have the usual connotations. The ligament forces however should also be included in this criterion as they have been explicitly modelled as forces. The criterion is therefore modified to minimize $U = \Sigma F + 0.5 \ \Sigma F_L + 4 \ \Sigma F_L$ where ΣF_L is the ligament force summation.

APPLICATION OF THE FOOT MODEL

The foot model was applied to the act of leaning forward as well as walking acts which were earlier investigated by the lower extremities model with the foot as a rigid segment. However, with the comprehensive foot model one can now predict not only the muscle loadsharing among the muscles of the foot but the pressure distribution on the foot as well. Moreover, the pressure distribution will be experimentally verified by means of a suitable transducer assembly that measures the vertical supportive forces at the points of interest on the foot.

A. LEANING STUDY:

APPLICATION OF THE MODEL TO LEANING AND RESULTS

The static leaning posture is simulated by pelvic tilt at the hip joint and represents a posture that is symmetric in the sagittal plane. The angles for the various segments that define the configuration to be studied are listed in Table IV-A.

The results for the various muscle and ligament forces, joint reaction forces and the ground-to-foot supportive reaction forces for the forward leaning positive are tabulated in Table IV-B. They were obtained by using the merit criterion described in the preceding section. The corresponding pressure distribution normalized with respect to the body weight is plotted in Fig. 17. Of the total ground force on each foot (half of body weight due to symmetry) roughly 21.5% is borne by the heel while the remainder is distributed among the metatarsal heads. This sharing of course depends on the leaning posture—the larger the angle of forward leaning the lesser will be the force on the heel and greater the force on the toes. The distribution of load on the metatarsals, on the other hand, seems to be influenced by the tarsal–metatarsal geometry. For the set of angles that were used in this model (the reaction constraint angles at the metatarso-tarsal joints, see Table V), presumably in a normal individual, the sharing among the metatarsals is not necessarily equal. The first metatarsal bear a major part of the total force on the forefoot. The distribution among the metatarsals however is significantly affected by the geometry. To illustrate this effect, the tarso-metatarsal reaction constraint angles were arbitrarily changed ±5 and ±10 percent approximately from the "normal" values (Table IV-C), and the corresponding results for the same leaning posture are shown (Table IV-B). The pressure distribution on the foot is plotted in Fig. 17 for these geometries, normalized with respect to body weight. As can be seen, increasing the angles by 5 and 10 percent, results in a shift of the pressure to the medial side of the foot—with a 10 percent change, the loads on the first and the second metatarsals in fact are greater than those on the remaining three. There is very little change in the force on the heel as should be expected since the sharing between the forefoot and the hindfoot is governed totally by the extent of leaning which has been kept the same. Decreasing the angles by 5 and 10 percent on the other hand, exhibits the reverse effect—with a 10 percent change, the loads shift laterally with the greatest share being borne by the third and the fourth metatarsals while the first metatarsal contribution to loadsharing is almost negligible. As before there is very little change in the force carried by the hindfoot or the heel.

The pressure distribution or the loadsharing among the metatarsals is therefore quite sensitive to the orientations of the bones of the foot and the corresponding directions along which the intermetatarsal and tarso-metatarsal reaction forces are directed.

EXPERIMENTAL VERIFICATION OF FOOT PRESSURE DISTRIBUTION DURING LEANING

A special strain-gage transducer assembly was built to measure the vertical forces on the foot during leaning studies to check the theoretically obtained pressure distribution. A sketch of the assembly is shown in Fig. 18. The force transducer consists of a ring instrumented with a strain gage placed on the inside surface at the

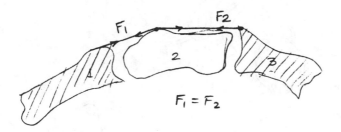

Figure 9-16. Modelling wrapping of a muscle.

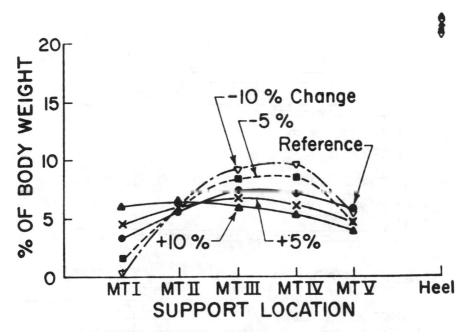

Figure 9-17. Predicted load distribution under the foot for the leaning stance showing sensitivity to metatarso-tarsal joint force orientation.

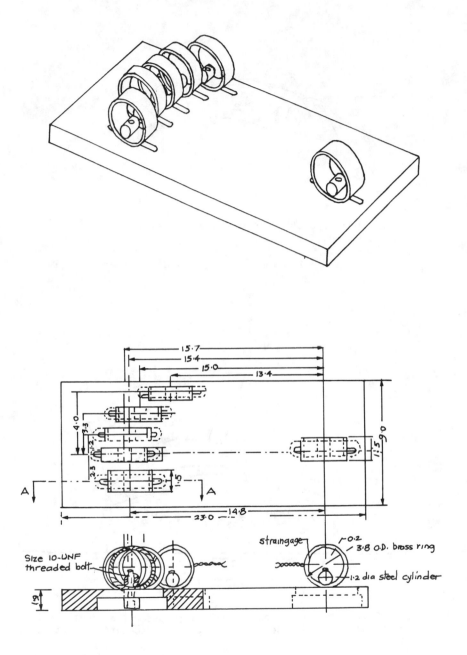

Figure 9-18. Transducer assembly for measuring foot pressure distribution.

TABLE IVA Segmental Configuration for Forward Leaning

Segment	ϕ_x (degrees)	ϕ_y (degrees)	ϕ_z (degrees
Right leg	2.2	-3.0	0
Right thigh	2.2	-3.0	0
Pelvis	0	25.0	0
Retinaculum	0	-3.0	0
Right tarsals	0	-3.0	0
Right metatarsal I	0	0	0
Right metatarsal II	0	0	0
Right metatarsal III	0	0	0
Right metatarsal IV	0	0	0
Right metatarsal V	0	0	0

Note: 1. The tarsals include the talus, calcaneus, navicular, cuboid, the three cuneiforms all treated as
a segment for kinematics purpose.

2. The angles for the left extremity are identical to
those for the right side due to symmetry.

TABLE IVB Muscle and Joint Forces for 25° Forward Leaning.

Variable → ↓		tarso-metatarsal joint angle constraint cases				
		Case 1	Case 2	Case 3	Case 4	Case 5
Muscle forces:						
Seminembranosus		23.42	23.02	22.61	23.78	24.01
Biceps femoris		28.56	29.02	29.51	28.14	27.86
Gluteus maximus		13.16	1.306	12.96	13.25	13.30
Glastrocnemius		54.42	54.06	53.68	54.74	54.96
Tibialis posterior		30.48	30.38	31.31	31.08	31.93
Peroneus longus		3.14	3.29	3.79	2.93	2.94
Plantar Aponeurosis		3.73	0	0.83	4.36	5.87
Plantar ligament		11.7	9.89	9.56	12.01	11.78
Calcanonavicular ligament		2.95	3.54	3.68	2.73	2.74
Joint moments:						
TAL/CAL	x	0	0	0	0	0
	y	0	0	0	0	0
	z	0	0	0	0	0
TAL/NAV	x	0	0	0	0	0
	y	0	0	0	0	0
	z	0	0	0	0	0
CBD/CAL	x	0	0	0	0	0
	y	0	0	0	0	0
	z	0	0	0	0	0
NAV/CF1	x	0	0	-0.29	0	-0.36
	y	0	0	0	0	0
	z	-3.46	-5.37	-8.18	-2.36	-1.58
NAV/CF2	x	0	0	0	0	0
	y	0	0	0	0	0
	z	0	0	0	0	0
NAV/CF3	x	0	0	0	0	0
	y	0	0	0	0	0
	z	0	0	0	0	0
CBD/CF3	x	0.45	0.31	0.77	0.39	0.27
	y	0	0	0	0	0
	z	12.42	10.61	9.49	14.33	15.56
CBD/MT4	x	0.86	0.35	0	1.5	1.84
	y	0	-0.54	-3.38	0	0
	z	0	0	0	0	0
CBD/MT5	x	3.73	3.17	1.85	3.14	2.72
	y	0	0	0	0	0
	z	6.77	5.32	3.78	5.77	4.97
CF1/CF2	x	-2.61	-3.33	-3.73	-3.23	-2.16
	y	-6.38	-7.86	-8.77	-5.45	-4.85
	z	0	0	0	0	0

(continued)

TABLE IVB continued

Variable ↓ →		tarso-metatarsal joint angle constraint cases				
		Case 1	Case 2	Case 3	Case 4	Case 5
Joint moments:						
CF2/CF3	x	0	0	0	0	0
	y	0	0	0	0	0
	z	−0.32	−0.69	−0.97	−0.13	−0.01
MT1/CF1	x	0.89	0.24	0.68	1.04	0.78
	y	0	0	0	0	0
	z	2.01	0.49	0	2.27	2.5
MT2/CF1	x	−0.01	−0.65	0	0	−0.05
	y	0	0	0	0	0
	z	−1.02	−2.34	−1.52	−1.04	−1.23
MT2/CF2	x	0	0	0	0	0
	y	0	−0.45	−0.66	−0.12	−0.28
	z	0	−5.37	0	0	0
MT2/CF3	x	0	0	0	0	0
	y	0	0	0	0	0
	z	0	0	0	0	0
MT2/MT3	x	0	0	0	0	0
	y	0	0	0	0	0
	z	0	0	0	0	0
MT3/CF3	x	0	0	−0.14	0	0
	y	0	0	−1.33	0	0
	z	0	−0.11	0	0	0
MT3/MT4	x	0	0	0	0	0
	y	0	0	0	0	0
	z	0	0	0	0	0
MT4/MT5	x	0	0	−0.09	0	0
	y	−0.11	−0.69	0	0	0
	z	0	0	0	0	0
MT1/LFT	x	0	0	0	0	0
	y	0	0	0	0	0
	z	0	0	0	0	0
MT2/LFT	x	0	0	0	0	0
	y	0	0	0	0	0
	z	0	0	0	0	0
MT3/LFT	x	−0.02	−0.20	0	−0.12	−0.18
	y	0	0	0.67	0	0
	z	0	0	0	0	0
MT4/LFT	x	−0.05	−0.03	0	0	0
	y	0	0.90	0	0	0
	z	0	0	0	0	0
MT5/LFT	x	0	0	−0.12	0	0
	y	0	0	0	0	0
	z	0	0	0	0	0

(continued)

Variable ↓ →		tarso-metatarsal joint angle constraint cases				
		Case 1	Case 2	Case 3	Case 4	Case 5
Unconstrained or partially constrained joint reactions						
TIB/TAL	x	−1.23	−1.43	−2.18	−1.32	−1.63
	y	−81.13	−81.07	−82.55	−81.78	−82.85
	z	0	0	0	0	0
KNEEJT	x	5.49	5.49	5.49	5.49	5.49
	y	1.36	1.39	1.34	1.34	1.33
	z	−137.25	−136.96	−137.51	−137.51	−137.68
HIPPJT	x	7.65	7.63	7.67	7.67	7.68
	y	−1.57	−1.53	−1.62	−1.62	−1.64
	z	−88.04	−88.02	−88.06	−88.06	−88.07
RET/LEG	x	−21.98	−21.75	−22.28	−22.28	−22.59
	y	−61.59	−60.96	−62.44	−62.44	−63.34
	z	−51.91	−51.38	−52.63	−52.63	−53.39
RET/CAL	x	21.98	21.75	22.28	22.28	22.59
	y	61.59	60.96	62.44	62.44	63.34
	z	51.91	51.38	−52.63	52.63	53.39
TAL/NV1	x	−7.74	−8.64	−8.52	−7.45	−7.46
	y	0	0	0	0	0
	z	0	0	0	0	0
TAL/NV2	x	0	0	0	0	0
	y	0	0	0	0	0
	z	1.23	2.16	2.91	0.61	0
TAL/NV3	x	0	0	0	0	0
	y	−25.23	−23.55	−22.24	−27.02	−28.26
	z	12.91	11.75	11.17	13.98	15.02
CAL/CBD	x	−6.31	−5.75	−11.65	−8.52	−10.19
	y	0	0	0	0	0
	z	0	0	0	0	0
CAL/TAL1	x	0	0	0	0	0
	y	43.43	40.47	42.30	0	42.04
	z	−11.44	0	0	−9.9	0
CAL/TAL2	x	0	0	0	0	0
	y	0	0	0	49.52	0
	z	0	0	0	0	−8.08
CAL/TAL3	x	−22.04	−30.29	−31.34	−37.68	−24.34
	y	0	0	0	0	0
	z	−77.29	−80.79	−78.62	−75.16	−92.44
CAL/TAL4	x	0	0	0	0	0
	y	6.41	9.68	9.21	0	7.69
	z	0	0	0	0	0
CAL/TAL5	x	0	0	0	0	0
	y	0	0	0	0	0
	z	0	−20.29	−21.49	0	0
CAL/TAL6	x	0	0	0	18.28	0
	y	0	0	0	0	0
	z	0	0	0	0	0

(continued)

TABLE IVB continued

Variable → ↓	tarso-metatarsal joint angle constraint cases				
	Case 1	Case 2	Case 3	Case 4	Case 5
Constrained joint reactions:					
CAL/TAL(a)	0	0	0	0	0
TAL/NAV(a)	0	0	0	0	0
CAL/CBD	20.09	24.07	14.57	27.71	26.02
NAV/CF1	7.76	9.85	11.81	6.59	6.06
NAV/CF2	14.07	17.87	19.95	12.40	11.53
NAV/CF3	29.99	26.28	23.88	33.02	35.04
CBD/CF3	9.79	8.46	8.27	10.60	11.25
CBD/MT4	11.98	9.52	8.22	13.13	13.69
CBD/MT5	8.23	6.95	5.83	8.30	8.31
CF2/CF1	0	0	0	0.13	0.21
CF3/CF2	0.01	0.23	0.39	0	0
CF1/MT1	7.02	8.89	10.43	5.95	5.36
CF2/MT2	5.70	6.93	7.48	5.13	4.87
CF3/MT2	0.60	0	0	0.65	0.67
MT3/MT2	0	0.37	0.46	0	0
CF2/MT3	12.46	10.78	8.76	13.96	15.17
MT4/MT3	4.44	3.95	3.14	4.83	4.96
MT5/MT4	6.22	5.14	3.99	5.84	5.53
MT1/LFT	0.42	1.31	0.68	0.67	1.01
MT2/LFT	0	0	0	0	0
MT3/LFT	0	0	0	0	0
MT4/LFT	0	0	0	0	0
MT5/LFT	0	0	0	0	0
CAL/TAL(b)	0	0	0	0	0
CAL/TAL(c)	40.72	26.52	27.55	45.27	29.52
CAL/TAL(d)	0	0	0	0	0
TAL/NAV(b)	33.15	35.98	38.41	32.50	32.45
TAL/TIB1	100.04	128.34	126.08	134.34	136.51
TAL/TIB2	0	33.77	36.86	29.82	29.09
TAL/TIB3	150.5	149.61	151.37	152.22	154.47
Ground reactions (vertical):					
MT1	2.34	3.21	4.32	1.07	0.13
MT2	3.66	4.28	4.64	3.35	3.23
MT3	5.11	4.79	4.22	5.85	6.48
MT4	4.92	4.27	3.95	6.08	6.76
MT5	3.99	3.30	2.57	3.76	3.56
LFT	0	0	0	0	0
Heel	15.13	15.30	15.46	15.05	15.00
Merit Function:					
U = F + 4M	657.49	672.01	696.22	659.74	663.01

TABLE IVC   Tarso-metatarsal, λ Joint Constraint Angles.

Joint	Case 1*		Case 2		Case 3		Case 4		Case 5	
	ϕ_1	ϕ_2	ϕ_1	ϕ_2	ϕ_1	ϕ_2	ϕ_1	ϕ_2	ϕ_1	ϕ_2
MT1/CF1	10	20	10.5	21	11	22	9.5	19	9	18
MT2/CF2	-22.5	23	-21	24	-20	25	-24	22	-25	21
MT3/CF3	-33	36	-31	38	-30	40	-35	34	-36	32
MT4/CBD	-21	34	-20	36	-19	37	-22	32	-23	31
MT5/CBD	0	0	0	0	0	0	0	0	0	0

Note: 1.   Case 1 refers to "normal" values for the tarsometatarsal
joint angles.   Case 2 and 3 are obtained by +5 and +10
percent change in these values respectivley;   Case 4 and 5
are obtained by -5 and -10 percent change in these values
respectively.

2.   The joint reaction orientation is defined by the two
angles ϕ_1 and ϕ_2.   The x,y and z components of the
reaction 'R' will then be given by

$$R_x = -R \cdot \cos \phi_2 \cdot \cos \phi_1$$

$$R_y = -R \cdot \cos \phi_2 \cdot \sin \phi_1$$

$$R_z = R \cdot \sin \phi_2$$

horizontal diameter. When placed in this location, the strains recorded are due entirely to the vertical force acting on the ring and effects of any horizontal forces are eliminated. The strain gage is cemented to the surface and the two leads are connected to one of the arms of an electrical balancing bridge integral with a carrier-amplifier. The amplifier voltage output corresponding to the change in the strain gage resistance resulting from the force induced strains can be recorded on a strip-chart recorder.

The transducer assembly consists of six transducers mounted on and fastened on a steel plate. The location of the rings correspond to the five metatarsal heads and the heel. The rings used to measure the supportive forces under the heel and under thye first metatarsal are wider than those for the other metatarsals since the pressure at these locations is distributed over a larger surface area. The transducers are mounted in slots so that they can be conveniently adjusted to conform to different feet size (see Fig. 18). To record the foot pressure distribution, each of the transducer channels is first balanced and then calibrated by applying a static load of 25 lb. For leaning studies the subject is first made to stand erect with the right foot on the transducer and the other foot on a scale with the same height as the transducer assembly so that both feet are level. The subject is then asked to lean forward by tilting the upper torso on the thigh and the angle of leaning as well as the overall configuration is checked by an assistant (Fig. 19) subject at the same time is also asked to visually ensure that his total body weight is evenly divided between the two feet by means of the scale. It was observed that this was a task somewhat difficult to accomplish since the subjects tended to continually oscillate to where the load on the right feet was seen to fluctuate by about five pounds with respect to the desired load on that leg. The resulting pressure distribution is then recorded and the subject asked to dismount from the platform. The process is repeated for several trials, and the actual value of the force on each of the transducer rings, that is, the metatarsal head forces and the heel force are then computed.

The experimental pressure distributions for leaning are tabulated in Table V-A. These results were obtained from tests on a 130 lb weight male and correspond to several trials made on the subject. The fluctuations in the location of the overall center of support under the foot during the various trials is shown in Figs. 20 and 21 and does not appear to be very significant.

The pressure distribution normalized with respect to the body weight is illustrated in Fig. 22. The scatter in the result could be attributed to the difficulty in duplicating the leaning posture particularly where the foot is concerned. As shown in the theoretical analysis, the pressure distribution is sensitive to the orientations of the bones and conceivably there will be some variation in the geometry of the foot segments for different postures. The overall pattern of load sharing among the metatarsels, however, seems to agree quite well with the calculated distribution using the comprehensive foot model. Of the total load on the forefoot, the first metatarsel share is considerably less than the remaining four.

B. APPLICATION OF THE MODEL TO WALKING

In earlier investigations of the quasi-static walking act the foot was modelled as a rigid body and the ground-to-foot support forces were

501

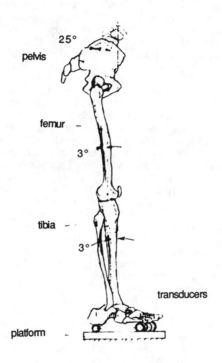

Figure 9-19. The forward leaning stance and
the foot placement on the transducer.

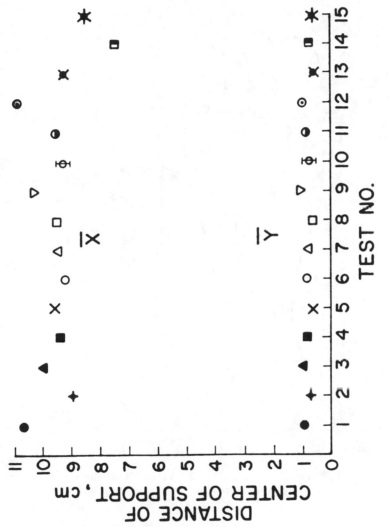

Figure 9-20. $\overline{X}$, $\overline{Y}$ coordinates of overall center of support for various tests with respect to the heel.

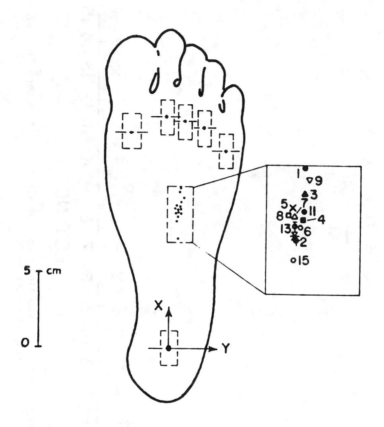

Figure 9-21. Scaled illustration of location of center of support in the foot for the various tests.

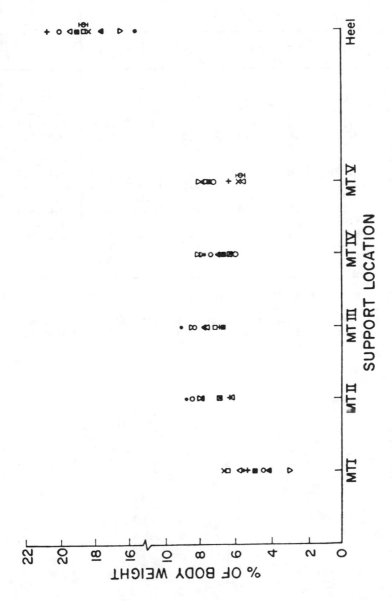

Figure 9-22. Experimental distribution of total support force at different locations under the foot (MTI: first metatarsal head; MTII: second metatarsal head, etc.). Notation corresponding to the different test numbers are the same as in Figure 3.

TABLE VA Experimental Load Distribution Results for Leaning (lbs).

Test No.	Location of Transducer						$\bar{x},\bar{y}$**
	MTI*	MTII	MTIII	MTIV	MTV	Heel	
1	7.1	11.4	11.7	9.9	9.8	20.2	10.61,0.86
2	7.0	8.1	8.9	8.5	8.0	27.0	8.93,0.69
3	5.5	10.3	10.1	9.0	10.0	22.7	9.88,0.90
4	6.5	8.2	8.7	8.7	9.5	24.5	9.33,0.82
5	8.7	8.8	9.0	8.3	7.5	23.7	9.55,0.60
6	6.0	8.8	8.8	8.7	9.1	25.8	9.15,0.799
7	7.5	8.0	10.1	9.8	7.0	25.1	9.38,0.68
8	8.7	8.8	8.9	8.0	7.0	24.0	9.45,0.57
9	4.0	10.4	11.0	10.5	10.6	21.2	10.21,1.06
10	6.5	8.0	9.2	8.0	7.5	24.0	9.24,0.71
11	5.5	10.2	10.0	8.0	10.0	24.8	9.48,0.85
12	6.0	11.0	11.2	10.2	10.0	18.4	10.79,0.96
13	8.0	8.8	9.1	7.0	7.5	25.0	9.21,0.59
14	4.0	5.2	9.2	6.8	6.8	32.5	7.37,0.71
15	7.6	7.0	10.0	8.5	7.6	30.5	8.51,0.63

*MTI, MTII, etc. represent the first, second metatarsal head respectively

**The overall center of support is expressed with respect to the heel transducer, $\bar{x}$ measured longitudinally and $\bar{y}$ measured mediolaterally.

TABLE VB Experimental Load Distribution (Normalized with Respect to Body Weight = 130 lbs.)

Test No.	Location of Transducer					
	MTI*	MTII	MTIII	MTIV	MTV	Heel
1	5.46	8.77	8.99	7.62	7.54	15.54
2	5.23	6.23	6.85	6.54	6.15	20.77
3	4.23	7.92	7.77	6.92	7.69	17.46
4	4.99	6.31	6.69	6.69	7.31	18.85
5	6.69	6.77	6.92	6.39	5.77	18.23
6	4.62	6.77	6.77	7.54	6.99	19.85
7	5.77	6.15	7.77	7.54	5.39	19.31
8	6.69	6.77	6.85	6.15	5.39	18.46
9	3.08	7.99	8.46	8.08	8.15	16.31
10	4.99	6.15	7.08	6.15	5.77	18.46
11	4.23	7.85	7.69	6.15	7.69	19.08
12	4.62	8.46	8.62	7.85	7.69	14.15
13	6.15	6.77	6.99	5.39	5.77	19.23
14	3.08	3.99	7.08	5.23	5.23	24.99
15	5.85	5.39	7.69	6.54	5.85	23.46

assumed concentrated at a single point on the foot. In the comprehensive model now, the support forces will be distributed among the metatarsals and the heels.

In the walking act there is considerable motion between the segments of the foot; for example, towards the termination of the single-legged stance phase as the body weight is gradually shifted to the forefoot the heel is no longer in contact with the ground and the foot is flexed at the metatarso-phalangeal joints. Due to lack of necessary information about the segmental configuration during this part of the walking cycle and about the sharing of the body weight between the two feet at the instant of heel strike, the new model will be used only for the single-legged stance phase when the entire foot is on the ground. Three instances from this part of the walking cycle were arbitrarily selected for investigating the muscle load sharing and the foot pressure pattern— these correspond to the 15, 27 and 36% of the walking cycle with the right extremity in the stance phase and the left, in the swing phase. The segmental configuration for these positions is given in Table VI-A.

The calculated muscle and ligament forces and joint reactions are tabulated in Table VI-B. The corresponding foot pressure distribution is also shown in the same table and compared in Fig. 23.

At 15% of the walking cycle which represents the instant of the beginning of the right leg single legged stance or the instant of left toe off, the body weight is being transferred to the right limb. Since the total body center of gravity falls in the area of the ankle joint or slightly posterior to it, one would expect the heel to carry a major part of the load. This is evident from the results—the total force on the heel is much greater than all the forces on the metatarsals combined. Also, the forefoot load is distributed more or less uniformly among the five metatarsals. At 27% of the walking cycle, the overall center of support has moved forward to around midfoot and one would expect the hindfoot and the forefoot forces to be roughly equal or nearly so. This is reflected in the calculated pressure distribution. Also, among the metatarsals, the first two carry a much greater part of the total load on the forefoot while the third and the fourth contribute very little to this load, and the fifth one does not carry any load at all.

At 36% of walking cycle, the load on the foot has shifted more forward and this is seen in the calculated pressure distribution—the first three metatarsals carry almost the entire body weight while the pressure on the heel is insignificant. This is to be expected since this instant of walking cycle represents the limiting condition when the foot is still flat and the entire sole is in contact with the ground. At the next instant of walking cycle the heel will be off the ground and the total body weight will be supported by the forefoot alone.

C. APPLICATION OF THE MODEL TO PATHOLOGICAL FEET

The model can be applied to situations where the foot geometry is abnormal due to pathological reasons. This can be quite conveniently analyzed through the model by merely altering the segmental configuration to represent the pathological case under question. Three pathological cases will be investigated:

(a) **Pes cavus** where the tarso-metatarsal arch as viewed in the sagittal plane is higher than the normal. The foot geometry in this case

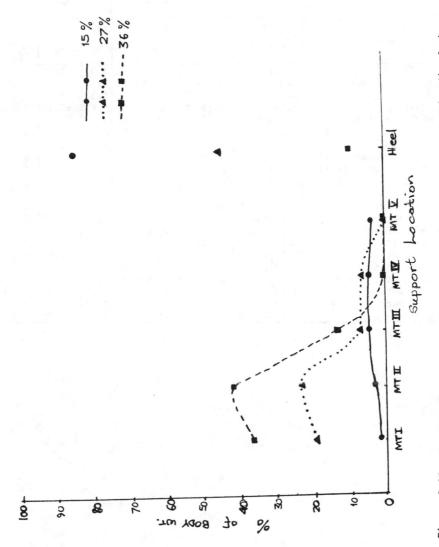

Figure 9-23. Ground-to-foot force distribution among various sites under the foot during walking.

TABLE VIA Segmental Configuration for Walking.

Segment	ϕ_x	ϕ_y	ϕ_z	ϕ_x	ϕ_y	ϕ_z	ϕ_x	ϕ_y	ϕ_z
	15%			27%			35%		
Right leg	5.12	4.79	0	5.35	5.10	0	5.62	7.02	0
Right retinaculum	5.12	4.79	0	5.35	5.10	0	5.62	7.02	0
Right thigh	5.12	-9.20	0	5.35	-0.89	0	5.62	5.02	0
Pelvis	5.12	5.79	0	5.35	0.10	0	5.62	-2.97	0
Right tarsals	0	0	0	0	0	0	0	0	0
Right metatarsal I	0	0	0	0	0	0	0	0	0
Right metatarsal II	0	0	0	0	0	0	0	0	0
Right metatarsal III	0	0	0	0	0	0	0	0	0
Right metatarsal IV	0	0	0	0	0	0	0	0	0
Right metatarsal V	0	0	0	0	0	0	0	0	0
Right toes	0	0	0	0	0	0	0	0	0
Left thigh	5.12	5.79	0	0	0	0	0	0	0
Left leg	5.12	55.79	0	5.35	-22.39	0	5.62	-32.97	0
Left retinaculum	5.12	55.79	0	5.35	42.60	0	5.62	11.02	0
Left tarsals	5.12	74.79	0	5.35	42.60	0	5.62	11.02	0
Left metatarsal I	5.12	74.79	0	5.35	42.60	0	5.62	9.02	0
Left metatarsal II	5.12	74.79	0	5.35	42.60	0	5.62	9.02	0
Left metatarsal III	5.12	74.79	0	5.35	42.60	0	5.62	9.02	0
Left metatarsal IV	5.12	74.79	0	5.35	42.60	0	5.62	9.02	0
Left metatarsal V	5.12	74.79	0	5.35	42.60	0	5.62	9.02	0
Left toes	5.12	74.79	0	5.35	42.60	0	5.62	9.02	0

TABLE VIB Muscle and Joint Forces for Walking.

Variable	Instant of walking cycle					
	15%		27%		35%	
	Right Leg	Left Leg	Right Leg	Left Leg	Right Leg	Left Leg
Muscle forces:						
Sartorius	0	0	0	10.95	0	16.34
Semembranosus	49.38	10.58	0	0	0	0
Gluteus Medius	64.52	3.24	61.63	0	0	0
Gluteus minimus	22.18	0	12.44	0	63.79	0
Tensor fasciae latae	0	4.77	23.69	2.41	19.13	2.55
Biceps femoris	0	3.94	31.59	3.45	76.09	4.97
Glutens maximus	0	3.47	3.05	0.33	26.34	0.22
Gastrocnemius	67.13	1.36	32.11	0	149.25	0
Tibialis posterior	17.27	0	73.01	0	132.58	0
Flexor digitorum longus	0	0	0	0	29.55	0
Peroneus longus	1.78	0	11.19	0	20.45	0
Preoneus Brevis	0	0	1.43	0	2.81	0
Peroneus Tertius	0	0	5.03	0	10.81	0
Flexor hallucis brevis	1.18	0	0	0	0	0
Plantar aponeurosis	0.93	0	2.50	0	8.87	0
Plantar ligament	6.67	0	0	0	0.68	0
Quadriceps	125.4	0	0	3.31	0	7.84
Calcanonavicular lig.	1.66	0	0	0	0.89	0
Joint moments:						
NAV/CF1 x	0	0	−3.95	0	−9	0
y	0	0	0	0	0	0
z	−1.97	0	−30.29	0	−56.28	0
CBD/CF3 x	0.25	0	0	0	0	0
y	0	0	0	0	0	0
z	7.04	0	15.42	0	28.24	0
CBD/MT4 x	0.49	0	0	0	0	0
y	0	0	0	0	0	0
z	0	0	0	0	0	0
RET/CAL x	0	0	0	0	0	0
y	0	0	3.78	0	0	0
z	0	0	0	0	0	0
RET/LEG x	0	0	0	0	8.7	0
y	0	0	0	0	0	0
z	0	0	0	0	9.13	0
CBD/CAL x	0	0	4.99	0	0	0
y	0	0	0	0	0	0
z	0	0	0	0	0	0

(continued)

TABLE VIB continued

Variable		Instant of walking cycle					
		15%		27%		35%	
		Right Leg	Left Leg	Right Leg	Left Leg	Right Leg	Left Leg
CBD/MT5	x	2.11	0	0	0	0	0
	y	0	0	0	0	0	0
	z	3.84	0	0	0	0	0
CF1/CF2	x	-1.48	0	-10.99	0	-20.03	0
	y	-3.59	0	-22.62	0	-36.19	0
	z	0	0	0	0	0	0
CF2/CF3	x	0	0	0	0	0	0
	y	0	0	0	0	0	0
	z	-0.18	0	-1.35	0	-2.47	0
MT1/CF1	x	0.51	0	1.59	0	1.54	0
	y	0	0	8.14	0	20.94	0
	z	1.14	0	0	0	0	0
MT2/CFL	x	-0.01	0	0	0	0	0
	y	0	0	0	0	0	0
	z	-0.58	0	0	0	0	0
MT4/MT5	x	0	0	0.06	0	0	0
	y	-0.06	0	0	0	0	0
	z	0	0	0	0	0	0
MT3/LFT	x	-0.01	0	0	0	0	0
	y	0	0	0	0	0	0
	z	0	0	0	0	0	0
MT4/LFT	x	-0.03	0	-0.47	0	0	0
	y	0	0	2.67	0	0	0
	z	0	0	0	0	0	0
NAV/CF2	x	0	0	0	0	0	0
	y	0	0	2.39	0	0	0
	z	0	0	0	0	0	0
MT2/CF2	x	0	0	0	0	0	0
	y	0	0	-1.78	0	-4.42	0
	z	0	0	0	0	0	0
MT2/CF3	x	0	0	-3.66	0	-7.09	0
	y	0	0	0	0	0	0
	z	0	0	0	0	0	0

(continued)

TABLE VIB continued

Variable		Instant of walking cycle					
		15%		27%		35%	
		Right Leg	Left Leg	Right Leg	Left Leg	Right Leg	Left Leg
Unconstrained Joint reactions:							
TIB/TAL	x	57.09	0	-62.47	0	-80.25	0
	y	-239.65	2.74	-18.39	0	-192.9	0
	z	182.09	-0.34	60.71	0	166.82	0
KNEEJT	x	-21.69	-3.66	-7.81	0.26	-41.51	6.68
	y	41.23	0.79	6.48	0.69	-12.68	0.99
	z	-300.78	-15.92	-155.51	-15.72	-332.28	-26.47
HIPPJT	x	23.45	-2.74	0.29	2.54	-9.02	10.88
	y	-1.96	5.38	-20.15	1.42	-36.18	2.63
	z	-264.09	-13.39	-155.89	-5.18	-175.27	-15.25
RET/LEG	x	-61.12	-2.95	9.21	0	-20.44	0
	y	172.95	2.74	23.29	0	-64.75	0
	z	132.49	0.36	20.41	0	-42.37	0
RET/CAL	x	61.12	2.95	-7.28	0	25.03	0
	y	172.95	-2.74	-27.34	0	55.96	0
	z	132.49	-0.36	-18.28	0	47.14	0
TAL/NV1	x	-4.38	0	-21.93	0	-39.92	0
	y	0	0	0	0	0	0
	z	0	0	0	0	0	0
TAL/NV2	x	0	0	0	0	0	0
	y	0	0	0	0	0	0
	z	0.70	0	8.82	0	0	0
TAL/NV3	x	0	0	0	0	0	0
	y	14.29	0	45.71	0	83.32	0
	z	7.31	0	23.83	0	45.43	0
CAL/EBD	x	-3.58	0	-21.42	0	-39.13	0
	y	0	0	0	0	0	0
	z	0	0	0	0	0	0
CAL/TAL1	x	0	0	0	0	0	0
	y	158.67	0	0	0	0	0
	z	35.71	0	0	0	-20.78	0
CAL/TAL2	x	0	0	-34.47	0	-7.09	0
	y	0	0.62	0	0	0	0
	z	0	-3.06	0	0	0	0
CAL/TAL3	x	-95.12	0	0	0	-149.20	0
	y	0	0	0	0	106.43	0
	z	-99.24	-0.26	0	0	0	0
CAL/TAL4	x	17.57	0	0	0	0	0
	y	0	0	184.9	0	-149.0	0
	z	-125.71	0	0	0	0	0
CAL/TAL5	x	0	0	0	0	0	0
	y	0	0	184.9	0	0	0
	z	0	0	0	0	0	0
CAL/TAL6	x	0	8.02	0	0	120.94	0
	y	0	0	141.70	0	0	0
	z	0	0	0	0	0	0

(continued)

TABLE VIB continued

| | Instant of walking cycle | | | | | |
| | 15% | | 27% | | 35% | |
Variable	Right Leg	Left Leg	Right Leg	Left Leg	Right Leg	Left Leg
Constrained Joint reactions						
CAL/TAL(a)	0	0	0	0	0	0
CLA/NAV	0	0	0	0	0	0
CAL/CBD	16.49	0	4.65	0	8.54	0
NAV/CF1	4.40	0	35.96	0	66.79	0
NAV/CF2	7.94	0	59.44	0	107.79	0
CBD/CF3	5.55	0	19.06	0	34.84	0
CBD/MT4	6.79	0	10.16	0	18.58	0
CBD/MT5	4.67	0	3.53	0	6.48	0
CF2/CF1	0	0	0	0	0.10	0
CF3/CF2	0.01	0	0.06	0	61.56	0
CF1/MT1	3.98	0	33.14	0	43.68	0
CF2/MT2	3.22	0	23.97	0	6.07	0
CF3/MT2	0.34	0	3.33	0	0	0
MT3/MT2	0	0	0	0	22.26	0
CF3/MT3	7.06	0	12.16	0	7.63	0
MT4/MT3	2.51	0	4.32	0	10.65	0
MT5/MT4	3.52	0	5.14	0	0	0
MT1/LFT	0.24	0	2.71	0	0	0
MT2/LFT	0	0	0	0	0	0
MT3/LFT	0	0	0	0	0	0
MT4/LFT	0	0	0	0	0	0
MT5/LFT	0	0	0	0	0	0
CAL/TAL(b)	0	0	0	0	0	0
CAL/TAL(c)	0	7.25	8.25	0	209.69	0
CAL/TAL(d)	0	5.01	0	0	0	0
TAL/NAV	18.77	0	90.22	0	166.65	0
TAL/TIB1	99.74	3.35	84.96		232.72	0
TAL/TIB2	0	0.96	0	.	0	0
TAL/TIB3	409.07	5.44	8.93		277.82	0

(continued)

TABLE VIB   continued

| Variable | Instant of walking cycle | | | | | |
| | 15% | | 27% | | 35% | |
	Right Leg	Left Leg	Right Leg	Left Leg	Right Leg	Left Leg
Ground Reaction (vertical)						
MT1/TAL(a)	0	0	0	0	0	0
NT2/NAV	0	0	0	0	0	0
MT3/CBD	16.49	0	4.65	0	8.54	0
MT4/CF1	4.40	0	35.96	0	66.79	0
MT5/CF2	7.94	0	59.44	0	107.79	0
LFT/CF3	5.55	0	19.06	0	34.84	0
Heel	6.79	0	10.16	0	18.58	0
Merit Function						
$U = \Sigma F + 4m + \Sigma F_L$	474.46		733.12		1497.29	

is as defined in Table VII-A. The high arch is simulated by tilting the metatarsals and the toes for both feet +10 degrees and the tarsals for both feet -10 degrees about the medio-lateral axis (Y-axis). It will be assumed that the tarso-metatarsal joint reaction constraint angles remain unchanged from their normal values.

(b) **Hallus valgus** - a deformity resulting from excessive abduction of the first metatarsal with respect to the foot center line. The segmental orientation for this condition is shown in Table VII-A and is represented by abduction of the MIT (that is, rotation about the Z axis) through an angle of 8 degrees and adduction of the toes through an angle of 5 degrees. Both feet are assumed to be under the hallus valgus condition to an identical degree. In this case however, there will be a change in the reaction constraint angles for the MTI/CFI and the MTI/Toes joints. The constraint angles for MTI/CFI are changed to new values of $\phi_1 = 18°$, $\phi_2 = 20°$ (normal: $\phi_1 = 10°$, $\phi_2 = 20°$) and the angles for MTI/Toes are changed to $\phi_1 = -5°$, $\phi_2 = 0°$ (normal: $\phi_1 = 10°$, $\phi_2 = 0°$).

The results for pressure distribution along with muscle and joint forces for the three pathological cases are tabulated in Table VII-B. The same results for a normal foot are also listed in that table for comparison.

The pressure distribution under the foot for the normal case and the three pathological cases are plotted in Fig. 24. From that figure it can be seen that the pressure distributions for the normal foot and the hallus valgus foot are not significantly different. This might be expected since for the latter condition the only change in the foot geometry is the first metatarsal while the remaining foot geometry is identical to that in the normal foot.

The most significant variation from the normal pressure distribution occurs for the pes cavus and the pes planus conditions. In the pes cavus or high arched foot, the pressure distribution is shifted to the lateral side of the foot with the fourth metatarsal (MTIV) carrying a major portion of the total support force. In other words, the "flow of force" in the foot tends to favor the lateral side. Exactly the opposite occurs in the pes planus condition where the flow of force tends to favor the medial side and consequently, the major portion of the support force is carried by the second metatarsal (MTII). Note that for both of these cases the tarso-metatarsal joint reaction constraint angles were kept unchanged from their normal values. It is conceivable that in the pes cavus and the pes planus feet these constraint angles do not remain the same as in the normal foot. The deformed conditions are primarily a result of rotations and rearrangement of the foot segments thus affecting the orientations of the articulating segments drastically. As demonstrated earlier, the orientations of the tarsal and metatarsal articulating surfaces which control the transmission of joint forces greatly influence the overall pressure distribution under the foot. In view of this, for accurate representation of the pathological feet in the model considerably more information regarding the segmental orientation is required.

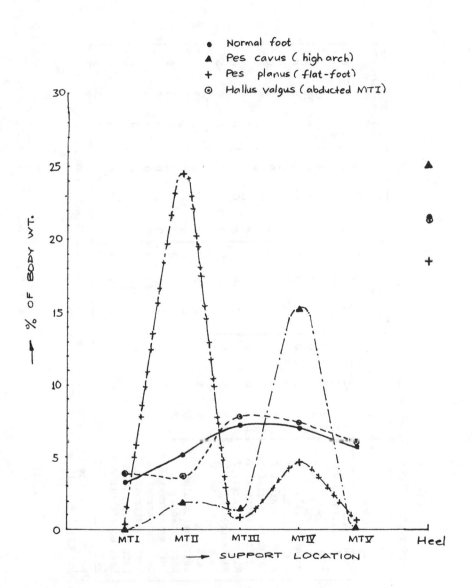

Figure 9-24. Pressure distribution for pathological feet.

TABLE VIIA. Segmental configuration for pathological foot.

Segment	Normal			Pes Cavis			Pes Planus			Hallus Valgus*		
	ϕ_x	ϕ_y	ϕ_z	ϕ_x	ϕ_y	ϕ_z	ϕ_x	ϕ_y	ϕ_z	ϕ_x	ϕ_y	ϕ_z
Right leg	2.2	-3	0	2.2	-3	0	0	-3	0	2.2	-3	0
Right retinaculum	2.2	-3	0	2.2	-3	0	0	-3	0	2.2	-3	0
Right thigh	2.2	-3	0	2.2	-3	0	0	-3	0	2.2	-3	0
Pelvis	0	25	0	0	25	0	0	25	0	0	25	0
Right tarsals	0	0	0	0	-10	0	0	-10	0	0	0	0
Right metatarsal I	0	0	0	0	10	0	0	-10	0	0	0	0
Right metatarsal II	0	0	0	0	10	0	0	-10	0	0	0	0
Right metatarsal III	0	0	0	0	10	0	0	-10	0	0	0	0
Right metatarsal IV	0	0	0	0	10	0	0	-10	0	0	0	0
Right metatarsal V	0	0	0	0	10	0	0	-10	0	0	0	0
Right toes	-2.2	0	0	0	10	0	0	-10	0	0	0	-5
Left thigh	-2.2	-3	0	0	-3	0	0	-3	0	-3	-3	0
Left leg	-2.2	-3	0	0	-3	0	0	-3	0	-3	-3	0
Left retinaculum	0	-3	0	0	-3	0	0	-3	0	-3	-3	0
Left tarsals	0	0	0	0	-10	0	0	10	0	0	0	0
Left metatarsal I	0	0	0	0	10	0	0	-10	0	0	0	-8
Left metatarsal II	0	0	0	0	10	0	0	-10	0	0	0	0
Left metatarsal III	0	0	0	0	10	0	0	-10	0	0	0	0
Left metatarsal IV	0	0	0	0	10	0	0	-10	0	0	0	0
Left metatarsal V	0	0	0	0	10	0	0	-10	0	0	0	0
Toes	0	0	0	0	10	0	0	-10	0	0	0	5

*For this case the tarsometatarsal joint constraint angles were also changed:
MTL/CFl: $\phi_1 = 18$, $\phi_2 = 20$; MTl/LFT: $\phi_1 = -5$, $\phi_2 = 0$

TABLE VIIB. Muscle and joint forces for pathological foot.

Variable ↓	Normal	Pes Cavus (high arch)	Pes Planus (flat-foot)	Hallus Valgus
Muscle Forces:				
Semimembranosus	23.42	24.19	46.63	23.36
Biceps femoris	28.56	28.43	16.80	28.63
Glutens maximus	13.16	11.72	0	13.14
Gastrocnemius	54.42	54.92	64.43	54.37
Tibialis Posterior	30.48	34.65	57.79	31.68
Peroneus longus	3.14	2.44	0.91	3.18
Plantar aponeurosis	3.73	13.74	0	0
Plantar ligament	11.7	13.2	5.83	13.29
Calcaneo navicular lig.	2.95	0.07	14.42	1.71
Adductor hallucis	0	5.96	3.92	0
Joint moments:				
TAL/CAL x	0	0	0	0
y	0	0	0	0
z	0	0	0	0
TLA/NAV x	0	0	0	0
y	0	0	0	0
z	0	0	0	0
CBD/CAL x	0	0	0	-0.52
y	0	0	0	0
z	0	0	0	0
NAV/CF1 x	0	0	0	-0.59
y	0	0	0	0
z	-3.46	-4.78	0	-5.99
NAV/CF2 x	0	0	5.24	0
y	0	0	0	0
z	0	0	0	0
NAV/CF3 x	0	0	0	0
y	0	0	0	0
z	0	0	0	0
CBD/CF3 x	0.45	0	1.11	0.26
y	0	3.26	0	0
z	12.42	16.54	0	14.10
CBD/MT4 x	0.86	1.99	0	1.01
y	0	0	-5.72	0
z	0	0	0	0
CBD/MT5 x	3.73	0.03	0	3.69
y	0	0	0	0
z	6.77	0	0.41	6.77

(continued)

TABLE VIIB (continued)

Variable		Normal	Pes Cavus (high arch)	Pes Planus (flat-foot)	Hallus Valgus
Joint moments:					
CF4/CF2	x	-2.61	0	-3.09	-1.59
	y	-6.38	-1.59	-0.78	-1.45
	z	0	-0.03	0	0
CF2/CF3	x	0	0.66	0	0
	y	0	0	0	0
	z	-0.32	0	-1.07	-0.19
MT1/CF1	x	0.89	1.37	0	0
	y	0	0	0	0
	z	2.01	-0.59	0	0
MT2/CF1	x	-0.01	0	0	-0.56
	y	0	0	0	0
	z	-1.02	-1.04	-4.17	-2.60
MT2/CF2	x	0	0	-5.78	0
	y	0	0	-4.93	0
	z	0	0	3.25	0
MT2/CF3	x	0	0.01	0	0
	y	0	0	0	0
	z	0	0	0	0
MT2/MT3	x	0	0	0	0
	y	0	0	0	0
	z	0	0	0	0
MT3/CF3	x	0	0.89	0	0
	y	0	0	0	0
	z	0	0	0	0
MT3/MT4	x	0	0	0	0
	y	0	0	0	0
	z	0	0	0	0
MT4/MT5	x	0	0	-0.46	0
	y	-0.11	0	0	0
	z	0	0	0	0
MT1/LFT	x	0	0	0	0
	y	0	0	0.65	0.95
	z	0	0	0	0.43
MT2/LFT	x	0	0	0	0
	y	0	0	0	0
	z	0	0	0	0
MT3/LFT	x	-0.02	0	-0.43	-0.02
	y	0	0	0	0
	z	0	0	0	0

(continued)

TABLE VIIB (continued)

Variable		Normal	Pes Cavus (high arch)	Pes Planus (flat foot)	Hallus Valgus
Joint moments:					
MT4/LFT	x	-0.05	0	0	0
	y	0	0	3.28	0
	z	0	0	0	0
MT5/LFT	x	0	0	-0.60	0
	y	0	0	0	0
	z	0	0	0	0
Unconstrained or partially constrained joint reactions:					
TIB/TAL	x	-1.23	7.8	0	2.33
	y	-81.13	-88.73	-117.63	-84.97
	z	0	0	150.95	0
KNEEJT	x	5.49	6.42	6.34	5.49
	y	1.36	1.39	7.06	1.37
	z	-137.25	-138.38	-158.05	-137.20
HIPPJT	x	7.65	7.54	7.60	7.64
	y	-1.57	-1.11	5.53	-1.57
	z	-88.04	-87.33	-86.51	-88.04
RET/LEG	x	-21.98	-11.57	-57.55	-22.12
	y	-61.59	-51.29	-105.32	-61.99
	z	-51.91	-39.44	-94.06	-52.25
RET/CAL	x	21.98	11.57	57.55	22.12
	y	61.59	51.29	105.32	61.99
	z	51.91	39.44	94.06	52.25
TAL/NVL	x	-7.74	-3.69	-27.25	-6.96
	y	0	0	0	0
	z	0	0	0	0
TAL/NV2	x	0	0	0	0
	y	0	0	0	0
	z	1.23	0	6.13	3.61
TAL/NV3	x	0	0	0	0
	y	-25.23	26.09	30.14	24.23
	z	12.91	15.43	3.85	13.34
CAL/CBD	x	-6.31	-18.01	-6.39	-5.82
	y	0	0	0	0
	z	0	0	0	0
CAL/TAL1	x	0	-14.05	0	0
	y	43.43	-45.49	-54.02	0
	z	-11.44	0	0	-16.37

(continuous)

TABLE VIIB   (continued)

Variable		Normal	Pes Cavus (high arch	Pes Planus (flat-foot)	Hallus Valgus
Unconstrained or partially constrained joint reactions:					
CAL/TAL2	x	0	0	0	0
	y	0	0	0	0
	z	0	0	39.83	0
CAL/TAL3	x	-22.04	0	-94.84	-98.91
	y	0	0	0	0
	z	-77.29	-69.83	-112.15	-34.28
CAL/TAL4	x	0	0	0	0
	y	6.41	0	-36.49	-50.93
	z	0	-45.06	-18.26	-76.76
CAL/TAL5	x	0	0	0	0
	y	0	3.81	0	0
	z	0	0	0	0
CAL/TAL6	x	0	0	0	65.06
	y	0	0	0	0
	z	0	0	0	0
Constrained joint reactions:					
CAL/TAL(a)		0	0	0	0
TAL/NAV(a)		0	0	0	0
CAL/CBD		20.09	13.64	3.84	31.61
NAV/CF1		7.76	7.02	6.28	11.97
NAV/CF2		14.01	3.52	78.82	8.60
NAV/CF3		29.99	36.03	5.73	32.08
CBD/CF3		9.79	12.56	2.29	9.96
CBD/MT4		11.98	16.61	2.09	12.48
CBD/MT5		8.23	0	1.12	9.11
CF2/CF1		0	0	0	0
CF3/CP2		0.01	0	0.08	0.01
CF4/MT1		7.02	4.61	3.14	7.61
CF2/MT2		5.70	1.43	31.94	3.49
CF3/MT2		0.60	0	1.19	0
MT3/MT2		0	0	3.44	0
CF2/MT3		12.46	12.81	2.23	13.29
MT4/MT3		4.44	3.99	4.34	4.73
MT5/MT4		6.22	0.02	1.77	6.59
MT1/LFT		0.42	7.54	0	1.22
MT2/LFT		0	0	0	0

(continued)

TABLE VIIB (continued)

Variable	Normal	Pes Cavus (high arch)	Pes Planus (flat-foot)	Hallus Valgus
Constrained joint reactions:				
MT3/LFT	0	12.17	0	0
MT4/LFT	0	0	0	0
MT5/LFT	0	0	0	0
CAL/TAL(b)	0	0	0	0
CAL/TAL(c)	40.72	1.32	0	0
CAL/TAL(d)	0	0	0	0
TAL/NAV(b)	33.15	30.17	68.74	32.41
TAL/TIB1	100.04	124.15	84.39	132.08
TAL/TIB2	0	27.52	0	31.59
TAL/TIB3	150.5	148.37	235.64	154.35
Ground reactions (vertical):				
MT1	2.34	0	0.24	2.72
MT2	3.66	1.38	17.26	2.58
MT3	5.11	1.03	0.58	5.45
MT4	4.92	10.67	3.29	5.11
MT5	3.99	0.06	0.46	4.23
LFT	0	4.34	0	0
Heel	15.13	17.66	13.0	15.06
Merit function: $U = F + 4M$	657.49	619.96	948.01	672.44

CONCLUSIONS AND DISCUSSIONS

The comprehensive foot model which includes all tarsal and metatarsal segments of the foot can be used to predict the foot pressure distribution in static activities. The predicted results seem to agree very well with experimental findings and seem to be significantly affected by the orientations of the segments within the foot itself and by the directions along which intersegmental articulating forces are transmitted.

REFERENCES

1. Morton, D.J.: The human foot. Columbia Univ. Press, New York, 1935.

2. Elftman, H.: A cinematic study of the distribution of pressure in the human foot. Anat. Rec. 59: 481, 1934.

3. Jones, R.L.: The human foot: An experimental study of its mechanics and the role of its muscles and ligaments in the support of the arches. Am. J. Anat., 68: 1, 1941.

4. Schwartz et al.: A quantitative analysis of recorded variables in the walking pattern of normal adults. J.B. Jt. Surg., 46A: 324, 1964.

5. Carlsoo, S.: Influence of frontal and dorsal loads on muscle activity and on the weight distribution in the feet. Acta Orthop. Scand., 34: 249, 1964.

6. Grundy et al.: An investigation of the centers of pressures under the foot while walking. J.B. Jt. Surg. 57B: 98, 1975.

7. Stott et al.: Forces under the foot. J.B. Jt. Surg., 55B: 335, 1973.

8. Barnett, C.H. and Napier, R.: The axis of rotation of the ankle joint in man. J. Anat. (London) 86:1, 1952.

9. Manter, J.T.: Movements of the subtalar and transverse tarsal joints. Anat. Res., 80: 397, 1941.

10. Hicks, J.H.: The mechanics of the foot. I. The Joints. J. Anat. (London) 87:345, 1953.

11. Keith, A.: The history of the human foot and its bearing on orthopaedic practices. J.B. Jt. Surg., 11: 10, 1929.

12. Harris, R.J. and Beath, T.: Hypermobile flat-foot with short tenco achillis. J.B. Jt. Surg., 30A: 116, 1948.

13. Jones, F.W.: "Structure and function as seen in the foot. Bailliere, Tindall and Cox, London, 1949.

14. Basmajian, J.V. and Stecko, G.A.: The role of muscles in the arch support of the foot. J.B. Jt. Surg. 45A: 1963.

15. Mann, R. and Inman, V.T.: Phasic activity of intrinsic muscles of the foot, J.B. Jt. Surg.: 66A: 469, 1964.

16. Hicks, J.H.: The three weightbearing mechanisms of the foot. **In Biomechanical Studies of the Musculoskelatal System.** (Ed. F.G. Evans), C.G. Thomas, Springfield, Ill. (pp. 161-191) 1961.

17. Basmajian, J.B.: Muscles Alive: Their functions revealed by electromyography. Williams & Wilkins. Baltimore. 1974.

18. Gresczyk, E.G.: EMG study of the effect of leg muscles on the arches of the normal and flat foot. MS Thesis, University of Vermont, 1965.

CHAPTER X

Biomechanical Consideration in Robotics

"I fear however that the whole of this subject is of so
dark a nature as to be more usefully investigated by
experiment than by reasoning, and in absence of any
conclusive evidence from either, the only way that
presents itself is to copy nature; accordingly I shall
instance the spindles of the trout and woodcock."

Sir George Cayley
1773-1857

INTRODUCTION

Since recorded history analysis of the structure and operating
principles of living mechanisms have provided impetus for the development
of mechanical devices to extend human reach, augment human functions and
compensate for human disabilities. Nature has endowed biological systems
with great capabilities which attained perfection in their specialized
functions through long evolutionary processes. Many of these systems are
possible to emulate by "intelligent" devices which can be utilized for
the benefit of mankind. Consequently Biomechanics is recently attaining
great importance in the development of many specialized robots for
commercial, scientific and medical purposes. This chapter discusses
several aspects of robotics where biomechanical concepts can be adapted
to aid in their development. The cited examples represent developments
at the University of Wisconsin which include walking machines, prosthetic
and orthotic devices, aids for improving human manual dexterity and
sensory ability; as well as systems for safe extension of the human
domain into unreachable and hazardous environments.

LEGGED LOCOMOTION

The possibility of attaining the advantages of animal locomotion
with walking mechanisms has intrigued innovators for centuries. The
earliest known patent for a walking device was issued to Gompertz[1] in
1814. The terrain encountered in the earth-moving industry has provided
the incentive for development of several walking machines[2-3]. One such
machine is the 0.2 mile/hr. Walking Dragline developed by Bucyrus-
Erie. It is a huge excavating machine with two long feet moved by
rotating cams. When at work the machine rests on its central base plate.

Walking tractors and horses have also been developed as a result of
the difficult terrain in some farming applications involving loose soil,
mud, steep inclines, or rough roads. These include a steam tractor with
feet invented by Boydell[3], an elaborate, cam-controlled four-legged
tractor[4], a hydraulically powered, three-legged walking tractor [5], and
"walking horses"[6-7]. The latter attempted to duplicate the form of the
animal's leg motions with mechanical linkages.

Military interest in walking machines include a British proposal in 1940, to construct a 1000 ton walking tank[8], and many investigations in the United States[9-10].

The moon exploration program precipitated interest in walking machines in the 1960's[11-12-13] and [14].

An important aspect of the development of walking machines is the determination of possible gait patterns and the formulation of general schemes to control them. Examples of such control schemes for leg positioning and actuation systems are reported in Refs. 8, 15-29. Townsend and Seireg[30-32] developed an optimization scheme for synthesizing not only forces for control but also the patterns of displacement and rotation of body segments for optimal stability and energy expenditue. In all the above systems the control of motion is accomplished by applying appropriate moments at the hips.

LEGGED LOCOMOTION ON COMPLIANT AND ROUGH TERRAIN

A recent study[33] deals with the optimum design of structures and walking machines which are elastically supported and moving on rough and compliant terrain (Fig. 1). Any arbitrary number of legs and redundant actuators can be considered. A design algorithm is developed for determining the optimum gait configuration, actuation scheme and spring support which maximize stability and equalize the support reactions under the feet with minimum actuation effort.

An efficient piece-wise linear optimization procedure based on linear programming is developed which guarantees a unique solution and by-passes the computational difficulties of non-linear algorithms. Besides legged machines the procedure can also be readily applied to ocean platforms and stationary structures subjected to earthquakes. An extensive program at Ohio State University deals with the development of a six-legged vehicle for movement over obstacles and on rought terrain[34].

MECHANICAL ARMS

A comprehensive model of the human arm discussed in previous chapters (Ref. [35]) is shown in Figure 2. The model incorporates the clavicle, the scapula, the arm, the forearm and the hand which are connected together by 34 muscles. The articulation of the wrist is simulated in the model but the hand is treated as a single rigid body.

Although the scapula, the clavicle and the thorax have been included in the model, they are treated as ground and are used only to establish the locations of the arm and forearm muscles that originate on these segments. The overall configuration is established by defining 13 angles which describe the mostions of all the segments.

An antropomorphic robotic arm was developed by Seireg and Rodriguez[36] based on a study of the minimum number of degrees of freedom and the necessary actuators for duplicating the actions of the human arm. The actuators are placed in the torso and produce the forces and movements by a system of cables and pulleys placed at the different joints.

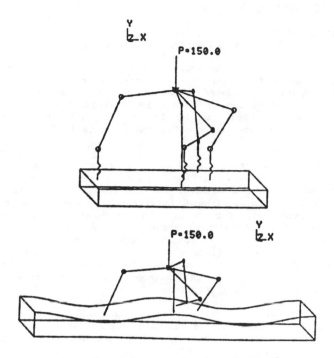

Figure 10-1. Legged locomotion on rough and compliant terrain.

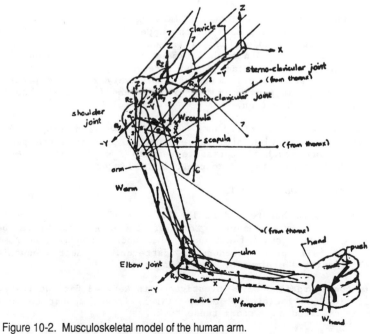

Figure 10-2. Musculoskeletal model of the human arm.

MECHANICAL HANDS

The human hand is an intricate and complex system capable of a multitude of sensory and actuation functions.

A comprehensive three-dimensional model for the hand is described in previous chapters[37]. Except for the thumb, which has only two phalanges, all the fingers are modeled by three phalangers and the entire hand is modeled as a collection of twenty-one individual segments. The eight carpal or wrist bones can be lumped together as a single segment articulating with the five metacarpals distally and the ulna-radius proximally.

All major muscles of the fingers are considered and are modelled by their individual lines of action. Muscles such as the digital flexors and extensors or the interossei require special treatment to take into account their wrapping around intervening segments and their complex pattern of insertion. For example, the extensor digitorum tendon for the index finger splits into three bands on the dorsal side of the proximal phalanx and is connected with the interossei tendons to form the extensor apparatus. The digital flexors pass through fibrosseous tunnels on the volar side of the fingers and are modelled by series of lines connecting the middle parts of adjacent segments. The coordinates of muscle attachments are estimated from Braus[38].

Since many of the extrinsic muscles controlling the finger originate either on the humerus or on the ulna-radius or on both, these bones should also be included in any comprehensive model of the hand.

The reactions at the joints are simulated by their 3-dimensional force and moment components along XYZ coordinate axes. The configuration for the pinch posture is illustrated in Figure 3.

The Japanese Society of Biomechanism work on the artificial hand during the last decade covers work on the upper limb, wrist to the tips of fingers. Aproximately 20 degrees of freedom for the hand alone and 7 for the arm and wrist[39]. Since the primary mechanical function of the hand is its ability to grasp and pinch different types of objects, not all the degrees of freedom are necessary for every robot or prothetic hand. Consequently, the minimum number of degrees of freedom should be determined for each application.

The new trend in robotic hands is to allow each finger to be controlled independently with miniature actuators placed in the palm of the hand. Miniature touch and pressure sensors 3 mm in size have also been used to identify objects and to control the finger grip.

HANDS FOR MATERIAL HANDLING ROBOTS

Considerable work has been done in the area of multiple prehension manipulator systems development. Most of these efforts have been attempts to duplicate certain functins of the hand. According to[40], the human hand has six basic prehensile patterns. Their mechanical equivalents[41] require a minimum of three fingers[42].

Various configurations were investigated in Ref. 43 for the tasks of grasping, lifting and rotating objects. Three fingers appear to be the minimum number required to perform these tasks.

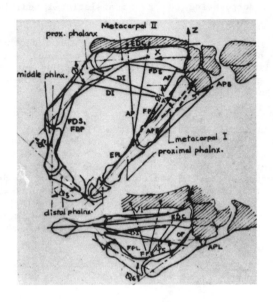

Figure 10-3 Comprehensive model of the human hand in the pinch posture.

The selected configuration for the hand is shown in Figure 4. It consists of three "two link" fingers. For all three fingers, the joint connecting the links is a simple revolute joint. The joint connecting F3 to the palm is also a simple revolute joint. The joints connecting the two adjacent fingers to the palm consist of a revolute and translational joint combination while that connecting the third finger to the palm is a simple revolute. The workspace of these fingers is increased substantially by incorporating the two translational additional degrees of freedom.

The actual dimensions of the hand will depend upon the application. The palm, must be large enough to accommodate the laser and photoelectric cell matrix. For applications requiring very small hands a separate arm may be required to house the laser and photoelectric matrix and direct the other to the proper location.

A computer based procedure is developed for use in locating, identifying and handling of objects placed on a moving conveyor. A reflective surface with known geometry is attached to the object at a pedetermined position to simplify the search. The hand moves in a peprogrammed pattern to allow the laser beam to traverse the search space. The reflected signal is monitored and utilized to determine the exact spacial position of the reflective surface (and consequently the object attached to it). It is found that an optimal relationship exists between the dimensions of the reflective surface and the radius of search for minimizing total task time which includes the location, identification, and handling of the object.

The procedure is readily applicable to tracking and handling objects moving at random on a known reference surface.

HUMAN POWER AMPLIFICATION

Examples of mechanisms for this purpose include human-augmentation exoskeleton named Handi-Man[15] developed by General Electric Company to enable a man to pick up to 1500 lb. load from the floor to a 6 ft. height and walk up a ladder with it. Hydraulic amplifiers reinforce the movement of the operator while a force feedback system insures the proper action control. The force feedback system consists of hydraulic servos which are sensitive to position, velocity and force.

Power amplification exoskeletons can have considerable impact on extending the load carrying capabilities of the operator. It can also contribute significantly to the reduction of injury to the human structure such as lower back pain in the work place.

Power amplification concepts can also be effectively used to aid the elderly and the handicapped. An example of such devices has been developed by the author and his students at the University of Wisconsin which enables a quadraplegic child with only a very low level activity in his left arm to operate a go-cart. The movement of this arm activates the servo mechanisms producing the motion and control of the cart. Besides allowing mobility and recreation it provides an incentive for the patient to exercise his arm with graded resistance by periodically changing a special adjustment as needed.

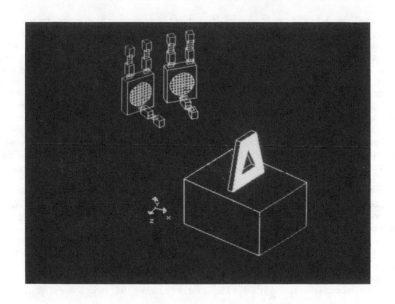

Figure 10-4. Three-finger robotic hand with laser system for search, identification, and handling of objects.

     (a) - search and identification
     (b). - handling of object

IMPROVEMENT OF MANUAL DEXTERITY AND PERFORMANCE ACCURACY

Mechanisms with displacement and tactile sensors can be incorporated in antropomorphic robots to attenuate the motion of the human hand. This consequently can improve the dexterity of the human operator in the situations requiring delicate positioning and small scale motions such as in special surgery, dentistry, defusing explosive devices and other delicate operations. Such intelligent systems can be constructed to copy or attenuate the hand motions of the skilled operator to any desired scale, and with constraints on the maximum excursions force and velocity for the entire operation or any particular segment of the motion. The system is programmed to disregard unintentional or unsteady movements. Robots with tactile sensors would be capable of providing controlled forces as needed and continually feed-back information on the response of the system to the executed action.

A continuous real time three-dimensional display of the performed motion can be automatically superimposed on a graphic display of the preplanned operation with a quantitative evaluation of differences for the operator's information.

Also, a computer can be adjusted to monitor the actions of the skilled operator over several months and to record the successful actions while ignoring the unsuccessful ones. The computer can then provide valuable information and data during the training of new operators.

Programs for such antropomorphic robots are now under development by the author and his students. The robots are teleoperators with artificial intelligence. The general dimensions and field of vision are automatically scaled on an interactive terminal to correspond to those of the operator.

ASSISTIVE DEVICES FOR THE PHYSICALLY HANDICAPPED

Because of the advances in robotics, considerable attention has been given to developing aids for the handicapped. The activities in this field are too numerous to cite and illustrative examples of some of the many orthotic and prosthetic devices which utlize robot technology are given in Ref. [44] to [48].

Serious consideration has been given to the construction of exoskeletal walking machines for providing locomotion to paraplegics during the past two decades. This includes work in Yugoslavia[45], Japan[46] and the U.S.A.[44]. The later (Figure 5) has preprogrammed mutli-task capabilities and patient control over its operation is accomplished by activating the appropriate switches for the desired motion.

AUGMENTATION OF PERFORMANCE IN CRITICAL OR HAZARDOUS ENVIRONMENTS

There are many environments requiring the human skill and judgement which are potentially hazardous to the human. Outer space, underwater, underground, difficult terrain and fire are examples of the numerous ssituations where the human operator function can be made safer and more effective. There are currently space robots[49] mining machinery[50] and deep water submersibles for this purpose[51]. Most of these utilize remote manipulators controlled by a human operator who can remain in a safe environment.

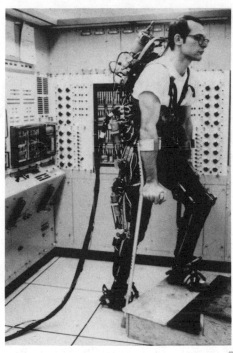

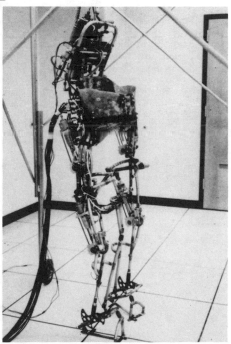

Figure 10-5. Computer controlled exoskeleton.

UNDERWATER SYSTEMS

Recent interest in utilization of ocean resources led to the development of human controlled manipulators for deep ocean work and assistive mechanisms to augment the diver's capabilities when working in relatively shallow waters.

Examples of assistive devices for the diver developed at the University of Wisconsin include intelligent controllers for depth and orientation, life support monitoring systems with automatic surfacing capability, and an untethered underwater platform capable of supporting the working diver [52] to [57].

Biomechanical investigations of the movements of the fish and birds aided in the early stages of development of the airplane and submarine. Such studies have been instrumental in developing the divers glider[57] and the concept of a automated underwater cargo vessel utilizing variable buoyancy, gravity for gliding propulsion (Fig. 6). The propulsion system emulates that used by certain species of fish where a bladder is inflated and deflated to induce forward propulsion by changing the angle of attack. A procedure for optimum design and automatic control of the vessel is described in Ref. [58]. The concept can be used in a variety of underwater winged robotic devices as well as automatic submarines capable of transporting liquid cargo over long distances under ice covered oceans.

CONCLUSIONS

The examples discussed illustrate the great potential for biomechanical applications in the robotics field. Thomas Jefferson once said "no knowledge can be more satisfactory to man that that of his own frame, its parts, their functions and actions." Beside the satisfaction and the challenge such knowledge can have profound impact on the development of useful and efficient robots which augment, enhance and expand the human capabilities.

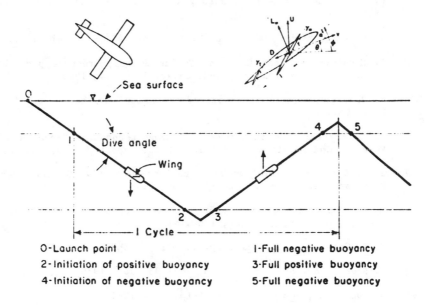

O-Launch point
2-Initiation of positive buoyancy
4-Initiation of negative buoyancy

1-Full negative buoyancy
3-Full positive buoyancy
5-Full negative buoyancy

Figure 10-6. Underwater cargo vessel which emulates the gliding fish.

REFERENCES

1. L. Gompertz, Sundry Improvements in Carriages (and Substitues for Wheel Carriages) and Other Machines, British Patent 3,804, 1814.

2. Ransomes and Rapier, Ltd., Improvements in, or Relating to Excavating Machines, British Patent 546,700, 1944.

3. M. Mann, Pop. Sci., pp. 51-54, 216, July 1960.

4. W. E. Urschel, Walking Tractor, U.S. Patent 2,491,064, Dec. 13, 1949.

5. P. E. Corson, Walking Tractor, U. S. Patent 2,833,878, Feb. 11, 1958.

6. L. A. Rygg, Mechanical Horse, U. S. Patent 491,927, Feb. 14, 1893.

7. S. Maratori, Project Chin Ma (Golden Horse), Brochure published by the author, Piazza Cavour, 4-3, Chiavari, Italy, 1960.

8. A. C. Hutchinson, "The Chartered Mechanical Engineer," pp. 480-484, Nov. 1967.

9. R. K. Bernhard, Ordinance Corpos, Land Locomotion Research Branch, Research and Development Division, OTAC, Report No. 43, July 1958.

10. J. E. Shigley, U. S. Army Ordinace Tank-Automotive Command, Detroit, Michigan, Report No. RR, LL-71, Sept. 1960.

11. M. G. Bekkder, Off-the-Road Locomotion, University of Michigan Press, Ann Arbor, Michigan, 1960.

12. W. S. Griswald, Pop. Sci., pp. 214-217, 996-999, Mar. 1962.

13. D. Scott, Pop. Sci., pp. 61-63, 192-193, Jan. 1970.

14. W. von Braun, Pop. Sci., pp. 62-64, Mar. 1971.

15. R. S. Mosher, SAE Trans., 76, pp. 558-597, 1968.

16. R. S. Mosher, U.S. Army Tank-Automoive Center, Warren, Michigan, 1966.

17. R. A. Liston and R. S. Mosher, ASME Publication 67-Tran-34, 1967.

18. R. J. Williams and A. Seireg, "Interactive Modeling and Analysis of Open or Closed Loop Dynamic Systems with Redundant Actuators," Journal of Mechanical Design, ASME, Vol. 101, pp. 407-416, July 1979.

19. I Kato, S. Onteru, H. Kabayashi, K. Shirai, and A. Uchiyma, Paper presented at the 1st CISM-IFTOMM International Symposium on Theory and Practice of Robots and Manipulators, Udine, Italy, 1973.

20. R. B. McGee, Simulation 3, 135, 1967.

21. R. Tomovic, Cybernetica 4, 1961.

22. A. A. Frank, J. Terramechanics 8, 1971.

23. A. Seireg and R. Peterson, Proceedings of the 4th World Congress International Federation of Theory Machines and Mechanisms, Newcastle, England, Sept. 1975.

24. A. A. Frank and R. B. McGee, J. Terramechanics 6, 23, 1969.
25. Proceedings of the 1st CISM-IFTOMM Symposium on Theory and Practice of Robots and Manipulators, Udine, Italy, 1973.

26. A. Morecki and K. Kedzior, (eds.), "On Theory and Practice of Robots and Manipulators," 2nd CISM-IFTOMM Symposium, Warsaw, Poland, 1976.

27. K. Tagushi, K. Ikrda and S. Matsumoto, Ref. 26, pp.172-181.

28. A. P. Bessonov and N. W. Umnov, Ref. 26, pp. 67-73.

29. V. V. Kalinin, Mechanics of Solids, Vol. 15, No. 2, 1980, pp. 37-42.

30. M. A. Townsend and A. Seireg, ASME, Trans., Ser. B, pp. 472-482, May 1972.

31. M. A. Townsend and A. Seireg, J. Biomechanics 5, 71, 1972.

32. M. A. Townsend and A. Seireg, IEEE Trans., J. Biol. Eng., Sept. 1973.

33. A. Seireg and N. Dabir-Ebrahimi, "Optimum Design of Legged Structures on Complian and Rough Terrain," University of Wisconsin Report, M.E., 1983.

34. K. Waldron, R. McGee, et al., Several reports and publications, Ohio State University.

35. R. Arvikar and A. Seireg, "Evaluation of Upper Extremity Joint Forces During Exercise," Advances in Engineering, ASME, NY, 1978.

36. A. Seireg and L. Rodriguez, University of Wisconsin, M.E. Report, 1983.

37. R. Arvikar and A. Seireg, "A Musculoskeletal Analysis of the Pinch Action," Proceedings of the 30th ACEMB, Los Angeles, California, p. 181, Nov. 1977.

38. H. Braus, -Anatomic Des Menschen, J. Springer, Berlin, 1954.

39. K. Sadamoto, Editor, "Mechanical Hands Illustrated," Hemisphere Pub., NY, 1986.

40. Taylor, C. L. and Schwarz, R. J., "The Anatomy and Mechanics of the Human Hand," Artificial Limbs: Vol. 2, pp. 22-35, May 1955.

41. Skinner, F., "Designing a Multiple Prehension Manipulator," Mechanical Engineering, pp. 30-37, September 1975.

42. Van Der Loos, H.F.M., "Design of Three-Fingered Gripper," The Industrial Robot, Dec. 1978.

43. D. Foral and A. Seireg, "An Algorithm for Fast Location, Identification and Handling of Objects."

44. A. Seireg and J. Grundman, "Design of a Multitask Exoskeletal Walking Device," Biomechanics of Medical Devices, edited by D. N. Ghista, Marcel Dekker Inc., NY, Chapter 13, pp. 569-639, 1981.

45. M. Vukobratovic, D. Hristic, and Z. Stojiljkovic, Med. Biol. Eng., pp. 66-80, Jan. 1974.

46. I. Kato, Wasada University, Tokyo, Japan, Several unpublished reports, 1971-1973.

47. T. O. Kautz and A. Seireg, "Feasibility Study of a Computer Controlled Hydraulic Above-Knee Prosthetic Limb," Proceedings of ASME Century II International Medical Devices and Sports Equipment Conference, pp. 81-88, Aug. 1980.

48. Research Programs to Aid the Handicapped, Hearings before the Committee on Science and Technology, U.S. House of Representatives [104], Sept. 22-23, 1976.

49. A. K. Bejczy, "Effect of Hand-Based Sensors on Manipulator Control Performance," Mechanisms and Machine Theory, Vol. 12, No. 5, 1977.

50. E. R. Palowitch and P. H. Broussard, Some Opportunities in Teleoperated Mining, Mechanisms and Machine Theory, Vol. 12, No. 5, 1977.

51. J. Charles and J. Vertut, "Cable Controlled Deep Submergence Teleoperator System," Mechanisms and Machine Theory, Vol. 12, No. 5, 1977.

52. A. Seireg and A. Baz, "Optimum Design of an Orientation Control device for Submersibles," ASME Trans. Series B, Vol. 96, 1974.

53. A. Baz and A. Seireg, "Optimum Design of Automatic Depth Control System for Underwater Divers," ASME (75-DET-96), Trans. Series B, 1976.

54. A. Baz and A. Seireg, "A Diver Life Support Monitoring and Warning System with Automatic Surfacing Capabilities," Proceedings of Century II International Conference on Medical Devices and Sporets Equipment, ASME, Aug. 1980.

55. A. Baz and A. Seireg, "Equipment for the Underwater Diver," Trans., ASME Journal of Mechanical Design, Vol. 102, pp. 663-671, Oct. 1980.

56. A. Seireg and E. Kassem, "A Supporting Platform for the Working Diver," Ocean 77, Proceedings, 1977.

57. A. Baz and A. Seireg, "Optimum Design and Control of Underwater Gliders," ASME Trans., Series B, (73-DET-13), 1974.

58. Z. K. Qi and A. Seireg, "Underwater cargo Vessel Utilizing Variable Buoyancy System for Gliding Propulsion, Trans. of ASME, Journal of Energy Resources Technology, 1982.

APPENDIX A

Coordinate
Transformation Matrices

In a system of connected rigid bodies, each capable of independent motion it is often necessary to describe the location of any desired point on any one of the bodies with respect to a coordinate axis system fixed in space, commonly referred to as the ground or earth. This is accomplished through coordinate transformation matrices which relate the coordinates of any point in the system with respect to the ground-fixed axes for any selected configuration. For example, the coordinates of muscle origins and insertions on a skeletal segment can be obtained with respect to the ground axes through this process if the skeletal geometry is pre-established for a particular body posture.

The technique of obtaining the coordinates of a point on a segment consists of locating the origin of segment-fixed axes with respect to the ground axes which in effect translates the latter to the segment. The desired Eularian rotations of the segment can now be performed with respect to these translated axes. The Euler angle system used here is successive rotation about the body-fixed z,y,x axes, in that order, each rotation taking place about the present location as defined by the previous rotation. The transformed coordinates of a point on a body undergoing the three rotations ϕ_x, ϕ_y, and ϕ_z about the x,y,z axes respectively can be

$$\begin{bmatrix} 1 \\ X_G \\ Y_G \\ Z_G \end{bmatrix} = [T_{GB}][T_{\phi_x}] \ [T_{\phi_y}][T_{\phi_z}] \ \begin{bmatrix} 1 \\ X_B \\ Y_B \\ Z_B \end{bmatrix}$$

where:

X_G, Y_G, Z_G = coordinates of the point with respect to ground axes

$X_\beta, Y_\beta, Z_\beta$ = coordinates of the point with respect to its own body-fixed axes

541

$$[T_{\phi_x}] = \begin{bmatrix} 1 & 0 & 0 & 0 \\ 0 & 1 & 0 & 0 \\ 0 & 0 & \cos \phi_x & -\sin \phi_x \\ 0 & 0 & \sin \phi_x & \cos \phi_x \end{bmatrix}$$

$$[T_{\phi_y}] = \begin{bmatrix} 1 & 0 & 0 & 0 \\ 0 & \cos \phi_y & 0 & \sin \phi_y \\ 0 & 0 & 1 & 0 \\ 0 & -\sin \phi_y & 0 & \cos \phi_y \end{bmatrix}$$

$$[T_{\phi_z}] = \begin{bmatrix} 1 & 0 & 0 & 0 \\ 0 & \cos \phi_s & -\sin \phi_z & 0 \\ 0 & \sin \phi_z & \cos \phi_z & 0 \\ 0 & 0 & 0 & 1 \end{bmatrix}$$

$[T_{GB}]$ = matrix of coordinates of the origin of the body-fixed axes with respect to ground axes.

$$= \begin{bmatrix} 1 & 0 & 0 & 0 \\ X_{GB} & 1 & 0 & 0 \\ Y_{GB} & 0 & 1 & 0 \\ Z_{GB} & 0 & 0 & 1 \end{bmatrix}$$

Once all the quantities, namely the coordinates of the point with respect to its own body-fixed axes, the rotations ϕ_x, ϕ_y, ϕ_z and the coordinates of the origin of the body-fixed axes with respect to the ground axes are known, through successive multiplication of the above matrices the final coordinates of the point as referred to the ground axes can be completely established. If a body undergoes only one rotation such as in planar motion, say $\phi_x = \phi_z = 0$, the translated coordinates then can be obtained simply by the equations:

$$X_G = X_B \cos \phi_y + Z_B \sin \phi_y + X_{GB}$$

$$Y_G = Y_{GB}$$

$$Z_G = -X_B \sin \phi_y + Z_B \cos \phi_y + Z_{GB}$$

Note that for a linkage of multiple bodies, as in the human skeletal system, one may have to perform numerous successive matrix multiplications, moving from the desired point on a body i to the joint (usually the same as the origin of the body-fixed axes) it makes with the adjoining element (i-1) from there to the joint the element (i-1) makes with its adjoining element (i-2) and so on till one reaches the origin of the ground axes or the joint between body 1 and ground. This is illustrated in Figure 1. The coordinates of a point **P** on the body i are thus obtained by the following expression:

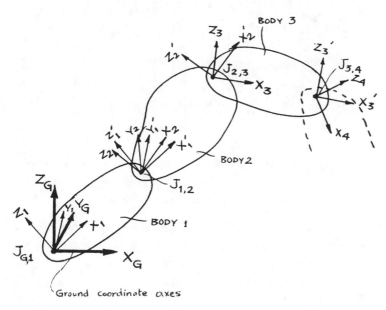

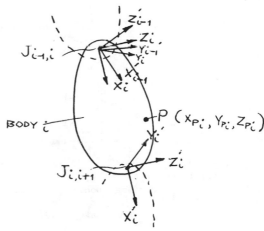

Figure A-1. Application of coordinate transformation matrices to multi-segmented system.

$$
\begin{bmatrix} 1 \\ X_{PG} \\ Y_{PG} \\ Z_{PG} \end{bmatrix} = \begin{array}{l} [T_{\phi_x}]_1 \, [T_{\phi_y}]_1 \, [T_{\phi_z}]_1 \, [T_{J_{1,2}}][T_{\phi_x}]_2 \, [T_{\phi_y}]_2[T_{\phi_z}]_2[T_{J2,3}] \,_2 \\[2mm] [T_{\phi_x}]_3 \, [T_{\phi_y}]_3 \, [T_{\phi_z}]_3 \, [Y_{J3,4}]_3 \cdots \cdot [T_{\phi_x}]_{i-1}[T_{\phi_y}]_{i-1}[T_{\phi_z}]_{i-1} \end{array}
$$

$$
[T_{J_{i-1,i}}]_{i-1}[T_{\phi_x}]_i[T_{\phi_y}]_i[T_{\phi_z}]_i \begin{bmatrix} 1 \\ X_{P_i} \\ Y_{P_i} \\ Z_{P_i} \end{bmatrix}
$$

where $[T_{\phi_x}]_j$, $[T_{\phi_y}]_j$, $[T_{\phi_z}]_j$ (for j=1,2,...,i etc.) imply the matrices

$[T_{\phi_x}]$, $[T_{\phi_y}]$, $[T_{\phi_z}]$ evaluated with the corresponding clockwise

individual rotations ϕ_x, ϕ_y, ϕ_z for body j performed with respect to body j-1. For example the angles ϕ_x, ϕ_y, ϕ_z for body 2 would be the corresponding angles between the (x_2, y_2, z_2) and the (X_1', Y_2', Z_1') coordinates systems as shown in Figure 1. The latter set of axes is assumed fixed with reference to body 1 at the joint $J_{1,2}$ between bodies 1 and 2.

Similarly, $[T_{J_{1,2}}]_1$, $[T_{J_{2,3}}]_2$, $[T_{J_{3,4}}]_3$,...,$[T_{J_{i-1,i}}]_{i-1}$ are matrices of the form:

$$
[T_{J_{i-1,i}}]_{i-1} = \begin{bmatrix} 1 & 0 & 0 & 0 \\ X_{J_{i-1,i}} & 1 & 0 & 0 \\ Y_{J_{i-1,i}} & 0 & 1 & 0 \\ Z_{J_{i-1,i}} & 0 & 0 & 1 \end{bmatrix}_{i-1}
$$

where $X_{J_{i-1,i}}$, $X_{J_{i-1,i}}$, $Z_{J_{i-1,i}}$ are the respective xyz coordinates of the joint $J_{i-1,i}$ (between bodies i-1 and i) with respect to body-fixed X_{i-1}, Y_{i-1}, Z_{i-1} system (on body i-1).

$(X_{P_G}, Y_{P_G}, Z_{P_G})$ and $(X_{P_i}, Y_{P_i}, Z_{P_i})$ are the coordinates of point P on body i with respect to the ground axes and with respect to body-fixed axes respectively.

Example: Consider a two-body planar configuration as shown in Figures 2(a) and (b). The original configuration shown in 2(a) is transformed into the final position shown in 2(b). It is required to find the coordinates of point P with respect to ground axes prior to and after the

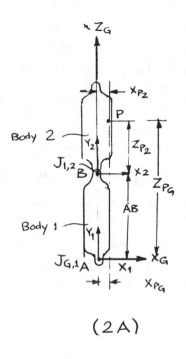

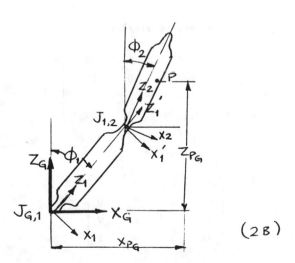

Figure A-2a & 2b. Point P with respect to ground before and after segmental rotations.

segmental rotations. In the erect posture $\phi x = \phi y = \phi z = 0$. for both bodies, so that matrices

$[T_{\phi x}]$, $[T_{\phi y}]$, $[T_{\phi z}]$ become unit matrices. Consequently,

$$
\begin{bmatrix} 1 \\ X_{PG} \\ Y_{PG} \\ Z_{PG} \end{bmatrix} = [T_{J_{1,2}}] \begin{bmatrix} 1 \\ X_{P2} \\ Y_{P2} \\ Z_{P2} \end{bmatrix} = \begin{bmatrix} 1 & 0 & 0 & 0 \\ 0 & 1 & 0 & 0 \\ 0 & 0 & 1 & 0 \\ AB & 0 & 0 & 1 \end{bmatrix} \begin{bmatrix} 1 \\ X_{P2} \\ Y_{P2} \\ Z_{P2} \end{bmatrix}
$$

which yields $X_{PG} = X_{P2}$, $Y_{PG} = Y_{P2}$, and $Z_{PG} = AB + Z_{P2}$

where X_{PG}, Z_{PG} =   coordinates of point P with respect to ground

X_{P2}, Z_{P2} =   coordinates of point P with respect to body-fixed axes (on body 2)

These results can also be directly obtained from the Figure 2(a).

In the tilted posture, using the coordinate transformation technique, and noting that $\phi_x = \phi_z = 0$ so that the matrices $[T_{\phi x}]$ and $[T_{\phi z}]$ become unit matrices,

$$
\begin{bmatrix} 1 \\ X_{PG} \\ Y_{PG} \\ Z_{PG} \end{bmatrix} = [T_{\phi y}]_1 \ [T_{J_{1,2}}]_1 [T_{\phi y}]_2 \begin{bmatrix} 1 \\ X_{P2} \\ Y_{P2} \\ Z_{P2} \end{bmatrix}
$$

Substitution of the known parameters yields:

$$
[T_{\phi y}]_1 = \begin{bmatrix} 1 & 0 & 0 & 0 \\ 0 & \cos \phi y & 0 & \sin \phi y \\ 0 & 0 & 1 & 0 \\ 0 & -\sin \phi y & 0 & \cos \phi y \end{bmatrix}_{\phi_y = \phi_1}
$$

$$
[T_{J_{1,2}}]_1 = \begin{bmatrix} 1 & 0 & 0 & 0 \\ 0 & 1 & 0 & 0 \\ 0 & 0 & 1 & 0 \\ AB & 0 & 0 & 1 \end{bmatrix}
$$

$$[T_{\phi_y}]_2 = \begin{bmatrix} 1 & 0 & 0 & 0 \\ 0 & \cos \phi_y & 0 & \sin \phi_y \\ 0 & 0 & 1 & 0 \\ 0 & -\sin \phi_y & 0 & \cos \phi_y \end{bmatrix} \quad \phi_y = -\phi_2$$

Note that while $\phi_y = \phi_1$ for the first body, $\phi_y = -\phi_2$ for the second body; since rotation of the latter with respect to the former through angle ϕ_2 is anteclockwise and hence negative. On the other hand, rotation of the first body with respect to the ground is clockwise and hence ϕ_1 is positive in keeping with the sign convention.

Successive multiplication of the matrices, pairing from right, can now be performed as follows

$$\begin{bmatrix} 1 \\ X_{PG} \\ Y_{PG} \\ Z_{PG} \end{bmatrix} = [T_{\phi_y}]_1 [T_{J_{1,2}}]_1 \begin{bmatrix} 1 & 0 & 0 & 0 \\ 0 & \cos \phi_2 & 0 & \sin \phi_2 \\ 0 & 0 & 1 & 0 \\ 0 & \sin \phi_2 & 0 & \cos \phi_2 \end{bmatrix} \begin{bmatrix} 1 \\ X_{P2} \\ Y_{P2} \\ Z_{P2} \end{bmatrix}$$

$$= [T_{\phi_y}]_1 [T_{J_{1,2}}]_1 \begin{bmatrix} 1 \\ X_{P2} \cos \phi_2 - Z_{P2} \sin \phi_2 \\ Y_{P2} \\ X_{P2} \sin \phi_2 + Z_{P2} \cos \phi_2 \end{bmatrix}$$

$$= [T_{\phi_y}] \begin{bmatrix} 1 & 0 & 0 & 0 \\ 0 & 1 & 0 & 0 \\ 0 & 0 & 1 & 0 \\ AB & 0 & 0 & 1 \end{bmatrix} \begin{bmatrix} 1 \\ X_{P2} \cos \phi_2 - Z_{P2} \sin \phi_2 \\ Y_{P2} \\ X_P \sin \phi_2 + Z_P \cos \phi_2 \end{bmatrix}$$

$$= [T_{\phi_y}] \begin{bmatrix} 1 \\ X_{P2} \cos \phi_2 - Z_{P2} \sin \phi_2 \\ X_{P2} \\ AB + X_{P2} \sin \phi_2 + Z_{P2} \cos \phi_2 \end{bmatrix}$$

547

$$= \begin{bmatrix} 1 & 0 & 0 & 0 \\ 0 & \cos\phi_1 & 0 & \sin\phi_1 \\ 0 & 0 & 1 & 0 \\ 0 & -\sin\phi_1 & 0 & \cos\phi_1 \end{bmatrix} \begin{bmatrix} 1 \\ X_{P2}\cos\phi - Z_{P2}\sin\phi_1 \\ X_{P2} \\ AB + X_{P2}\sin\phi_2 + Z_{P2}\cos\phi_2 \end{bmatrix}$$

$$= \begin{bmatrix} \cos\phi_1(X_{P2}\cos\phi_2 - Z_{P2}\sin\phi_2) + \sin\phi_1(AB + X_{P2}\sin\phi_2 + Z_{P2}\cos\phi_2) \\ -\sin\phi_1(X_{P2}\cos\phi_2 - Z_{P2}\sin\phi_2) + \cos\phi_1(AB + X_{P2}\sin\phi_2 + Z_{P2}\cos\phi_2) \end{bmatrix}$$

so that,

$$X_{PG} = \cos\phi_1(X_{P2}\cos\phi_1 - Z_{P2}\sin\phi_2) + \sin\phi_1(AB + X_{P2}\sin\phi_2 + Z_{P2}\cos\phi_2)$$

or $X_{PG} = X_{P2}\cos(\phi_1 - \phi_2) + Z_{P2}\sin(\phi_1 - \phi_2) + AB\sin\phi_1$

$$X_{PG} = Y_{P2}$$

$$Z_{PG} = -\sin\phi_1(X_{P2}\cos\phi_2 - Z_{P2}\sin\phi_2) + \cos\phi_1(AB + X_{P_2}\sin\phi_2 + Z_{P_2}\cos\phi_2)$$

or

$$Z_{PG} = -X_{P2}\sin(\phi_1 - \phi_2) + Z_{P2}\cos(\phi_1 - \phi_2) + AB\cos\phi_1$$

This may be easily verified by simple algebra as shown in Figure 3.

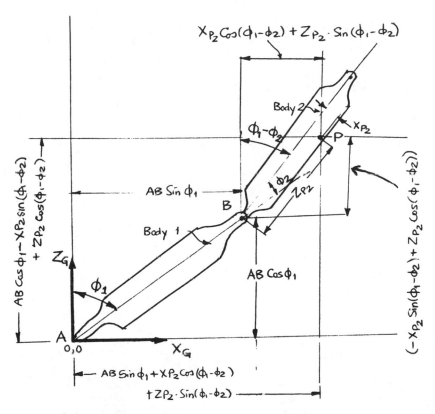

Figure A-3. Algebraic determination of coordinates of P after segmental rotations ϕ_1, ϕ_2.

APPENDIX B

Moment of Inertia

The analysis of engineering problems such as for the evaluation of stresses in beams or rotation of rigid bodies frequently involves certain integrals known as moment of inertia, or second moment of areas and masses. Consider for example a planar mass as shown in Figure 1, with rectangular coordinate axes x,y passing through a point 0 in (or outside) the body.

By definition, the mass moment of inertia of a body about any desired axis is the integral of the product of the elementary mass dm times the square of the distance from that axis. Thus, the

moment of inertia about the x axis is:

$$I_x \int y^2 \, dm$$

and, the mass moment of inertia about the y axis is written as:

$$I_y \int x^2 \, dm$$

The product of inertia is similarly defined as:

$$I_{xy} = \int xy \, dm$$

Note that while I_x and I_y are always positive, I_{xy} may be of any sign. Further, if either of the axes is one of symmetry, the produce of inertia vanishes.

The quantities I_x and I_y which represent moments of inertia taken with respect to axes lying in the plane xy are also called rectangular moments of inertia. This differentiates them from what is known as the polar moment of inertia, which is the moment of inertia about the axis perpendicular to the plane xy. Thus

$$J = \int r^2 \, dm = \int (y^2 + x^2) dm$$

$$= \int y^2 \, dm + \int x^2 \, dm = I_x + I_y$$

Hence, the sum of the two retangular moments of inertia always equals the polar moment of inertia for any set of coordinate axes if two of them lie in the plane of the given area.

Frequently it is assumed that the mass of a body is concentrated at an alternate location from the axes about which its moment of inertia is known. This distance is known as the radius of gyration and implies concentration of total body mass at this distance to produce the same moment of inertia. Thus the radius of gyration with respect to the x axis would be.

$$K_x = \sqrt{\frac{\overline{I_y}}{\int dm}}$$

and, with respect to the x axis,

$$K_y = \sqrt{\frac{\overline{I_x}}{\int dm}}$$

Sismilarly the equivalent distance of the concentrated mass from the point 0 to produce the same polar moment of inertia would be

$$K_r = \sqrt{\frac{\overline{J}}{\int dm}} = \sqrt{\frac{\overline{I_x + I_y}}{\int dm}}$$

The moments of inertia of any object of known geometry can be readily computed using the above expressions. However one may encounter composite objects of different geometric shapes or may be required to obtain the moments of inertia for the body with respect to axes displaced parallel to those about which its inertial parameters are known. This can be accomplished by the parallel-axis theorem. By this procedure, the moment of inertia of a body about an axis parallel to the centroidal axis is equal to the moment of inertia about the centroidal axis plus the product of mass of the body and the square of the distance between the two axes. Referring to Figure 2, consider a planar body whre x and y axes pass through the centroid 0_G and the mass moments of inertia about these axes are respectively $\overline{I}_x$ and $\overline{I}_y$. Let us suppose the moment of inertia about the axes x',y' parallel to the centroidal axes and distant dy and dx respectively are required. If M abe the mass of the body then using the parallel-axis theorem,

$$I_{x'} = \overline{I}_x + M\, dy^2$$

$$I_{y'} = \overline{I}_y + M\, dx^2$$

$$I_{x'y'} = \overline{I}_{xy} + M\, dx\, dy$$

$$J' = \overline{J} + M\, r'^2 = \overline{J} + M(dx^2 + dy^2)$$

$$J' = \overline{J} + M\, r'^2 = \overline{J} + M(dx^2 + dy^2)$$

$$= \overline{J}_x + \overline{I}_y + M(dx^2 + dy^2)$$

$$= I_{x'} + I_{y'}$$

The parallel-axis theorem can be conveniently extended to compute the moments of inertia of composite objects. If an object of mass M is made up of n smaller objects of masses $m_i (i=1,2,\ldots,n)$, then the moment

552

of inertia of the total object about an axis is obtained by scalar addition of the moment of inertia of each individual mass about the same axis: Furthermore, if the locations of the centroids of the individual masses with respect to the desired axis are known, and if the moments of inertia of the individual masses with respect to their own centroidal axes are also known, then the moment of inertia of the complete body about the axis, say the x axis can be readily expressed as:

$$I_x = \sum_{i=1}^{i=n} (\overline{I}_{x_i} + m_i \, d^2_{y_i})$$

Similar expressions can be formulated to compute the moments of inertia about the other axis or the product of inertia or the polar moment of inertia and so on.

It may be desired frequently to obtain the moments of inertia of a given planar mass with reference to a set of axes that is inclined to that about which the moments of inertia are known (see Fig. 3). Let us suppose the inertial quantities are known with respect to the xy axis system and it is required to obtain the same in terms of an inclined x'y' system rotated through an angle θ. This can be achieved through the inertia transformation equations.

Since the coordinates of an elementary mass dm in the old system (x,y) are related to the coordinates (x',y') in the new system as follows:

$$x' = x \cos \theta + y \sin \theta$$

$$y' = -x \sin \theta + y \cos \theta,$$

it can be shown that

and
$$I_x' = I_x \cos^2 \theta + I_y \sin^2 \theta - I_{xy} \sin^2 \theta$$
$$I_{y'} = I_x \sin^2 0 + I_y \cos^2 0 + I_{xy} \sin^2 \theta$$

Similarly,
$$I_{x'y'} = \frac{I_x - I_y}{2} \sin^2 \theta + I_{xy} \cos^2 \theta$$

It also follows that

$$I_{x'} + I_{y'} = I_x + I_y = J$$

which imply that for any inclination θ, the polar moment of inertia is an invariant. The above equations transform the moments of inertia with respect to given set of axes to another inclined at θ. One may further show through simple mathematical manipulation of the expressions for $I_{x'}$ and $I_{y'}$, that when,

$$\tan^2 \theta = \pm \frac{2I_{xy}}{I_x - I_y}$$

the quantities $I_{x'}$ and $I_{y'}$ are the maximum and minimum moments of inertia, or the principal moments of inertia, and the product of inertia are then given by:

$$I_{max \atop min} = \frac{I_x + I_y}{2} \pm \frac{1}{2} \left[(I_x - I_y)^2 + 4P^2_{xy} \right]^{1/2}$$

For the sake of simplicity, the mass was assumed to be planar so far. However, all the above expressions can be easily extended for a three-dimensional space. In this case, one requires nine quantities to completely define the "inertia tensor". As shown in Figure 4, it follows that these quantities are

$$I_{xx} = \int (y^2 + z^2) dm$$

$$I_{yy} = \int (z^2 + x^2) dm$$

$$I_{zz} = \int (x^2 + y^2) dm$$

and

$$I_{xy} = I_{yx} = \int xy \ dm$$

$$I_{yz} = I_{zy} = \int y_z \ dm$$

$$I_{zx} = I_{xz} = \int z_x \ dm$$

The corresponding inertia tensor is then expressed by

$$[I] = \begin{vmatrix} I_{xx} & -I_{xy} & -I_{xz} \\ -I_{yx} & -I_{yy} & -I_y \\ -I_{zx} & -I_{zy} & I_{zz} \end{vmatrix}$$

Additional formulae may be similarly derived from the parallel-axis theorem for a composite three-dimensional mass or for transformation of the inertia tensor with respect to another set of inclined cartesian axes, and the location of principal axes. The algebra is somewhat complicated and interested readers may refer to any standard textbook on Engineering Mechanics.

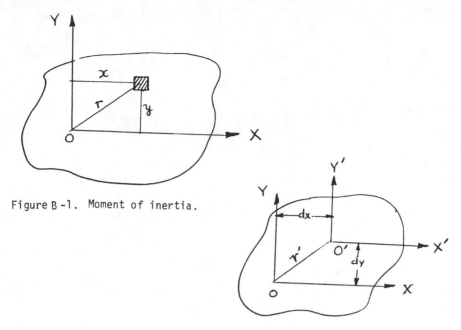

Figure B-1. Moment of inertia.

Figure B-2. Parallel-axis theorem.

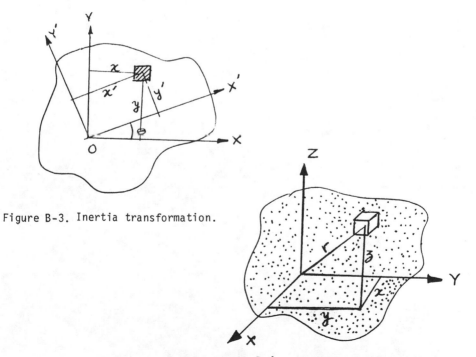

Figure B-3. Inertia transformation.

Figure B-4. Moment of inertia in 3-D.

Linear Programming and the Simplex Method

As emphasized in the main text, the equations of equilibrium for the skeletal segments are solved for muscle load sharing using linear programming techniques. In this appendix we shall describe briefly what is meant by a linear programming problem and illustrate its solution by the 'Simplex Technique'. This will be further elucidated by its application for the analysis of human body structure, such as, for the investigation of muscle load shring in the human jaw during biting using a revised Simplex algorithm. Because of certain advantages this algorithm ha been widely adopted for computer applications. For example, the "SIMPLX" subroutine package available on the UNIVAC 1110 computer at University of Wisconsin Computing Center is based on this method and was ued for all the optimization problems described in the foregoing chapters. The procedure of setting up of a problem so as to facilitate applications of this package is also described later.

LINEAR PROGRAMMING PROBLEM

The linear programming problem consts of finding a set of n variables x_j (j=1,2,...,n) so that a given linear objective function $U = \sum_{j=1}^{n} c_j x_j$ is optimum (either maximum or minimum) with respect to a set of m linear constraints of the form:

$$\sum_{j=1}^{n} a_{ij} x_j q_i b_i \qquad (i=1,2,\ldots,m) \qquad (1)$$

and where

$$X_j > 0 \qquad (j=1,2,\ldots,n)$$

where q_i stands for the sign of equality or inequality: =, > or <.

The elements C_j (j=1,2,...,n) are known as the cost functions associated with variable X_j, and the elements b_i are known as right hand side elements. By introducing the slack variables $X_{n+1},\ldots,X_{n+m}$ and introducing constraint types $t_1, t_2,\ldots,t_m$ the above m constraints are transformed into pure equations with equality constraints only. The problem can thus be restated as:

$$\text{minimize or maximize} \quad U = \sum_{j=1}^{n+m} C_j X_j, \quad \text{subject to}$$

$$\sum_{j=1}^{n} A_{ij} X_j + t_i X_{n+i} = b_i \qquad (i=1,2,\ldots,m) \quad (2)$$

with

$$X_j > 0 \qquad\qquad j = 1,2,\ldots,(n+m)$$

The constraint type t_i can be one of the following:

$t_i = 0$, when q_i is a sign of equality $=$,

$t_i = -1$, when q_i is a sign on inequality $>$,

$t_i = 1$, when q_i is a sign of inequality $<$,

for $i = 1,2,\ldots,m$)

The equation (2) is a more general form of (1) because costs C_{n+1}, $C_{n+2},\ldots,C_{n+m}$ are assigned to the slack variables X_{n+1}, $X_{n+2},\ldots$, X_{n+m}.

GEOMETRIC INTERPRETATION OF THE LINEAR PROGRAMMING PROBLEM

Consider a simple two-variable problem as follows:

$$\text{maximize} \quad U = 3x_1 + 4x_2$$

subject to the following linear conditions:

$$-x_1 + x_2 < 3$$

$$x_2 < 5$$

$$x_1 + 2x_2 < 14$$

$$4x_1 - x_2 < 20$$

Consider a set of coordinate axes x_1, x_2 as shown in Figure 1. The conditions $x_1 > 0$ and $x_2 > 0$ imply that the optimization region is to be limited to the positive sides (and inclusive of) of x_1 and x_2 only. The remaining constraints which are obviously equations of straight lines in the $x_1 x_2$ plane further constrain the optimization region to the closed polygon OABCDE over which the function $U = 3x_1 + 4x_2$ is to be maximized. The optimal solution, that is, a set of x_1 and x_2 which will make the function U maximum must therefore lie inside this region or on the boundaries of the polygon. Note that the expression $3x_1 + 4x_2 = C$ represents a family of parallel straight lines as shown in Fig. 1. These are isomerit curves of different values of U and as is obvious from the figure, the merit value increases as one moves away from the origin 0, (where U=0) to a maximum value of $U = 34$ at the corner point D of the

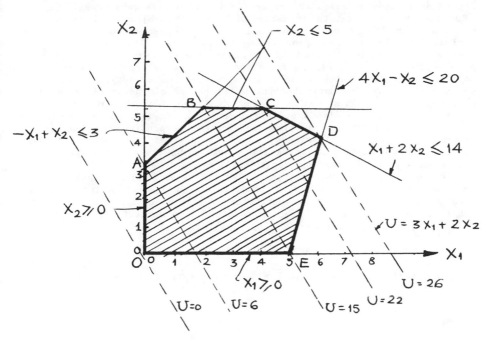

Figure C-1. Region defined by linear constraints.

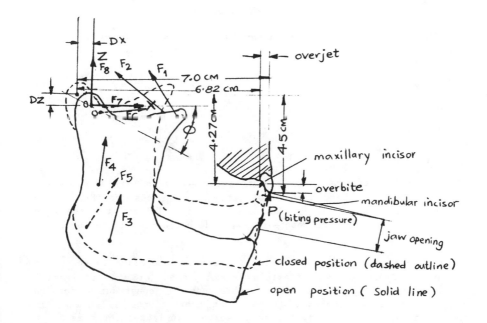

Figure C-2. Kinematics of jaw opening.

559

polygon. It can be verified that at any other point inside the shaded figure or on it boundaries the objective function is always lower. Since a point D all constraints are satisfied, and the objective function is maximum, this is the desired optimal point. It can be shown that the optimal point for a linear programming problem always lies at a corner or vertex of the region defined by the constraints. For a n dimensional problem thus, the constraints would establish a polytrope over which the objective function (which would then be represented by a hyperplane), is to be optimized. The optimal point would be one of the vertices of the polytrope. One could evaluate the objective function at all such possible vertices and find the desired optimum. Needless to say such an exhaustive process would b time-conuming and expensive and therefore alternate methods or algorithms have been devised to search for the optimum. These algorithms guide the search process very efficiently in the direction of increasing merit till the optimal point is reached eliminating excessive computations. One of several such algorithms is the simplex technique, which i illustrated in the following section through application to the two parameter problem discussed above. For a proof of the technique, the reader is referred to any suitable textbooks on the subject (see reference (1) for example).

THE SIMPLEX ALGORITHM

The problem is to maximize $U = 3x_1 + 4x_2$ subject to

$$-x_1 + x_2 \leqslant 3$$

$$x_2 \leqslant 5$$

$$x_1 + 2x_2 \leqslant 14$$

$$4x_1 - x_2 \leqslant 20$$

The first step is to transform the inequality constraints into equations by introducing the slack variables x_3, x_4, x_5 and x_6. Accordingly,

$$-x_1 + x_2 + x_3 \qquad\qquad\qquad = 3$$

$$0x_1 + x_2 \qquad + x_4 \qquad\qquad = 5$$

$$x_1 + 2x_2 \qquad\qquad + x_5 \qquad = 14$$

$$4x_1 - x_2 \qquad\qquad\qquad + x6 \quad = 2$$

The coefficient (a_{ij} and b_i) of these equations are then used to construct the right hand section of Table I. The cost coefficients c_j are also tabulated. Notice that the cost coefficients for the slack variables are all equal to zero since they do not appear in the objective function.

A convenient starting point for the simplex procedure is the origin ($x_1 = x_2 = 0$). Since it is a feasible extreme point (or a basic feasible point). Accordingly, the basis table can be constructed for the remainder of the variables and then cost coefficients as shown. The constraint equations can now be easily solved singly for the remained of

TABLE I

Basis		$\stackrel{\displaystyle j}{i}$	1	2	3	4	5	6		
C_i	U_i		x_1	x_2	x_3	x_4	x_5	x_6	b_i	$\dfrac{b_i}{(a_{ij})_e}$
0	x_3	1	-1	1	1	0	0	0	3	3 ←
0	x_4	2	0	1	0	1	0	0	5	5
0	x_5	3	1	2	0	0	1	0	14	7
0	x_6	4	4	-1	0	0	0	1	20	-20

Cost
Coefficients C_j

			3	4	0	0	0	0		

Solution of
constraint equa-
tions at selected
point
  $(x_1=0, x_2=0)$

| 0 | 0 | 3 | 5 | 14 | 20 |

$\Delta_j = C_j - \sum\limits_i C_i A_{ij}$

| 3 | 4 | 0 | 0 | 0 | 0 |

↑
enter

the variables since two of the six variables of the four equations are taken equal to zero. The result is shown in the second line from the bottom of the table and give the values for the slack variables corresponding to point 0. The values of Δj for $j = 1,7,\ldots,6$ which are tabulated in the bottom line of the table are calculated from:

$$\Delta j = Cj = \sum_i c_i a_{ij}$$

for the given table, these values are:

$$\Delta_1 = 3 - [0 \times -1 \quad +0 \times 0 \quad +0 \times 1 \quad +0 \times 4] = 3$$
$$\Delta_2 = 4 - [0 \times 1 \quad +0 \times 1 \quad +0 \times 2 \quad +0 \times -1] = 4$$
$$\Delta_3 = 0 - [0 \times 1 \quad +0 \times 0 \quad +0 \times 0 \quad +0 \times 0] = 0$$
$$\Delta_4 = 0 - [0 \times 0 \quad +0 \times 1 \quad +0 \times 0 \quad +0 \times 0] = 0$$
$$\Delta_5 = 0 - [0 \times 0 \quad +0 \times 0 \quad +0 \times 1 \quad +0 \times 0] = 0$$
$$\Delta_6 = 0 - [0 \times 0 \quad +0 \times 0 \quad +0 \times 0 \quad +0 \times 1] = 0$$

The variable with the <u>largest</u> positive value of Δ_j (x_2 in this example) is selected to enter the basis. The right hand column is then obtained by dividing the b_i values by the a_{ij} coefficient corresponding to the variable selected to enter the basis. The variable to be dropped from the basis (x_3 in this case) is the one corresponding to the <u>lowest</u> positive value of $b_i/(a_{ij})_e$.

A new table can then be constructed where x_2 replaces x_3. This is shown in Table II. Notice that the cost value corresponding to x_2 also replaces that corresponding to x_3 in the basis. The a_{ij} coefficients for x_2 are replaced by those for x_3 and the constraint equations are consequently changd to conform with this change. The first equation needs no change in this case since a_1 in Table I is already equal to 1 which is the new value for this coefficient. In the second equation, however, should be changed to render $a_{22} = 0$. This can be accomplished by subracting equation 1 from equation 2 in Table I which produces the new equation 2 in Table II. Similarly in the third row $a_{32} = 2$ in Table I which should be made to $a_{32} = 0$ ($= a_{22}$). This is done by multiplying the first equation by 2 and subtracting from the third equation in Table I. Finally, in the fourth row $a_{32} = -1$ in Table I which should be changed to read $a_{32} = 0$ ($= a_{32}$). This can be done by simply adding the first and the fourth row from Table I and the corresponding result is shown in Table II.

The solution of these new equations for the new condition ($x_1 = 0$, $x_3 = 0$) as shown in Table II gives a value for $x_2 = 3$ and $x_1 = 0$ which corresponds to point A in Fig. 1. Calculating Δj for this condition indicates that x_1 (which has the largest value for Δj) should enter the basis to replace x_4 which has the lowest value of $b_i/(a_{ij})_e$ ratio.

Table III can now be constructed by following the previous

TABLE II

Basis		j \ i	1	2	3	4	5	6		
c_i	u_i		x_1	x_2	x_3	x_4	x_5	x_6	b_i	$\dfrac{b_i}{(a_{ij})_e}$
4	x_2	1	-1	1	1	0	0	0	3	-3
0	x_4	2	1	0	-1	1	0	0	2	2
0	x_5	3	3	0	-2	0	1	0	8	8/3
0	x_6	4	3	0	1	0	0	1	23	23/3

← drop

Cost Coefficients C_j

3	4	0	0	0	0

Solution of constraint equations at selected point $(x_1=0, x_3=0)$

0	3	0	2	8	23

$\Delta_j = c_j - \sum_i c_i A_{ij}$

7	0	-4	0	0	0

↑
enter

TABLE III

c_i	U_i	j / i	1 x_1	2 x_2	3 x_3	4 x_4	5 x_5	6 x_6	b_i	$\dfrac{b_i}{(a_{ij})_e}$	
4	x_2	1	0	1	0	1	0	0	5	∞	
3	x_1	2	1	0	-1	1	0	0	2	-2	
0	x_5	3	0	0	1	-3	1	0	2	2	← drop
0	x_6	4	0	0	4	-3	0	1	17	17/4	
Cost Coefficients c_j			3	4	0	0	0	0			

Solution of constraint
equations at selected   2   5   0   0   2   17
point $(x_3=0, x_4=0)$

$\Delta_j = C_j - \sum_i C_i A_{ij}$ 0   0   3   -7   0   0

                              ↑

                        enter

TABLE IV

c_i	U_i	i \ j	1 x_1	2 x_2	3 x_3	4 x_4	5 x_5	6 x_6	b_i	$\dfrac{b_i}{(a_{ij})_e}$	
4	x_2	1	0	1	0	1	0	0	5	5	
3	x_1	2	1	0	0	-2	1	0	4	-2	
0	x_3	3	0	0	1	-3	1	0	2	-2/3	
0	x_6	4	0	0	0	9	-4	1	9	1	← drop
Cost Coefficients c_j			3	4	0	0	0	0			

Solution of constraint
equations at selected   4   5   2   0   0   9
point $(x_4=0, x_5=0)$

$\Delta_j = C_j - \sum\limits_i C_i A_{ij}$ 0   0   0   2   -3   0

                                                    ↑
                                                  enter

TABLE V

c_i	u_i	j \\ i	1 x_1	2 x_2	3 x_3	4 x_4	5 x_5	6 x_6	b_i	$\dfrac{b_i}{(a_{ij})_e}$
4	x_2	1	0	1	0	0	$\dfrac{4}{9}$	$-\dfrac{1}{9}$	4	
3	x_1	2	1	0	0	0	$\dfrac{1}{9}$	$\dfrac{2}{9}$	6	
0	x_3	3	0	0	1	0	$-\dfrac{1}{3}$	$\dfrac{1}{3}$	5	
0	x_4	4	0	0	0	1	$-\dfrac{4}{9}$	$\dfrac{1}{9}$	1	

Cost Coefficients c_j	3	4	0	0	0	0

Solution of constraint equations at selected point $(x_5=0, x_6=0)$ 6   4   5   1   0   0

$$\Delta_j = c_j - \sum_i c_i A_{ij} \qquad 0 \quad 0 \quad 0 \quad 0 \quad -\dfrac{19}{9} \quad -\dfrac{2}{9}$$

procedure. In this case x_1 replaces x_4 in the basis. The equations in Table II are also changed to produce $a_{11} = a_{13} = a_{14} = 0$, and $a_{12} = 1$ which are the same as the coefficients of the fourth column.

The solution of the new set of equations for the new conditions ($x_3=0$, $x_4=0$) produces $x_1 = 2$ and $x_2 = 5$ which corresponds to point B in Figure 1. The calculation of Δj indicates that x_3 should enter the basis to replace x_5 which has the lowest positive value of $b_i/(a_{ij})_e$. Table IV can accordingly be constructed and the solution of the constraint equations for the new condition ($x_4=0,x_5=0$) produces $x_1 = 4$ and $x_2 = 5$ which corresponds to point C in Figure 1. However, this still is not an optimal solution as the calculation of new Δj indicates that variable x_4 should enter the basis to replace x_6 which has the lowest $b_i/(a_{ij})_e$ ratio. This transformation is diaplayed in Table V and the solution of the contraint equations for this condition ($x_5=x_6=0$) yield $x_1 = 6$ and x_4 which corresponds to point D in Figure 1.

The optimality condition is now satisfied since all the Δj values at this point are either zero or negative. The search can now be stopped and the point $x_1 = 6$, $x_2 = 4$ gives the optimum solution. Since this represents the interseciton of the $U = 3x_1 + 4x_2$ line farthest from the origin at the extreme point D with ($U=34$) this quick geometric check indicates that it is the optimum solution (Fig. 1).

APPLICATION OF SIMPLEX TECHNIQUE TO OPEN JAW BITING

We will further illustrate the application of the simplex technique for the biomechanical investigation of a segment of the human body. We will consider the case of the human jaw during open jaw biting and obtain a solution for muscle load sharing using the revised simplex algorithm which is widely used for computer applications.

As mentioned earlier (see (Chapter 9)) opening of the jaws involves not only rotation of the mandible but also a displacement of the condyles along a curved path. As shown in Figure 2, to attain a particular jaw opening, the center of the condyles is displaced forward through a distance DX horizontally and downward through a distance DZ vertically. On this condylon movement one must superimpose the rotation of the mandible through an angle ϕ about a medio-lateral axis passing through the centers of the two condyles. The jaw opening is then the distance between the tips of the maxuillary and mandibular incisors. Note that this should also take into account the initial overjet and overbite (distances through which the maxillary incisor projects over the mandibular incisor in the horizontal and the vertical directions respectively) when the mandible is in the rest position. Let us consider as an example the case when the jaw opening equals 7 mm. This is achieved when DX = 0.05 cm, DZ = -0.05 cm and $\phi = 5.3°$. The overjet and overbite were taken as 1.8 mm and 2.3 mm respectively, and the biting pressure were used by Garrett, et al., (4) in their study dealing with electromyographic investigation of the masseter muscle during biting with several different pressure intensities for a wide range of jaw openings.

The mandible is a rigid body in space and will require six equations to describe its equilibrium. However, since the act in question implies only sagittal forces on the teeth, it follows that the muscle forces on the right side of the jaw will be identical to their counterparts on the left side. The problem thus becomes one of planar equilibrium and therefore one need write three equations only: sum of forces along X, sum of forces along Z and sum of moments about Y (axis perpendicular to the XZ plane).

There are six major masticatory muscles on each side (see Chapter 9). Their coordinates of origin and insertion, which will define the individual directions along which they act, are tabulated in Table VI(a). These coordinates are measured in the rest position of the mandible with respect to a set of cartesian axes fixed with its origin at the center of the right condyle. With a jaw opening of 7 mm, the coordinates of the points on the mandible (X,Z) should be transformed to $(x'y'z')$ reflect this motion according to the following equations:

$$x' = X \cdot \cos \phi + Z \cdot \sin \phi + DX$$
$$y' = Y$$
$$z' = -X \cdot \sin \phi + Z \cdot \cos \phi + DZ$$

These new coordindates with respect to the ground axes at the right TM joint are listed in Table VI(b). Note that the y coordinates are not affected because the rotation is purely sagittal. Also the coordinates of the points on the skull remain unchanged since the maxilla remains fixed.

The next step after coordinate transformation involves calculation of the direction cosines ℓ_i, m_i, n_i along the respective axes and moment components M_{x_i}, M_{y_i}, and M_{z_u} about the three axes respectively as follows:

$$L_i \sqrt{(x'_{2_i} - x_{1_i})^2 + (y'_{2_i} - y_{1_i})^2 + (z'_{2_i} - z_{1_i})^2}$$

$$\ell_i = (x_{1_i} - x'_{2_i})/L_i$$

$$m_i = (y_{1_i} - y'_{2_i})/L_i$$

$$n_i = (z_{1_i} - z'_{2_i})/L_i$$

$$M_{x_i} = n_i \cdot y'_{2_i} - m_i \cdot z'_{2_i}$$

$$M_{y_i} = \ell_i \cdot z'_{2_i} - n_i \cdot x'_{2_i}$$

$$M_{z_i} = m_i \cdot x'_{2_i} - \ell_i \cdot y'_{2_i}$$

Since the loading is symmetrical about the sagittal plane, the three equations of equilibrium can now be written.

TABLE VI(a). Coordinates of the muscles of the jaw.

CLOSED JAW POSITION

Muscle No. i	Name	x	y	z	x^2	y^2	z^2
		(Points on the skull) cm			(Points on the mandible) cm		
1	Anterior Temporalis	1.5	-4.5	5.3	3.3	-3.6	-0.5
2	Posterior Temporalis	-4.5	-5.7	4.8	2.9	-3.6	0.3
3	Masseter (Superficial)	3.5	-5.0	-0.5	1.0	-4.5	-4.7
4	Masseter (deep)	2.0	-5.5	-0.1	1.5	-4.5	-2.7
5	Medial Pterygoid	2.3	-2.0	-0.6	0.8	-3.8	-4.7
6	Lateral Pterygoid	3.0	-3.0	-0.2	0.5	-3.9	0.0

TABLE VI(b). Coordinates of the muscles of the jaw.

OPEN JAW POSITION (Jaw opening = 7 mm)

Muscle no.	Points on the skull (cm)			Points on the mandible (cm)			Direction Cosines			Moment components		
	x^1	y^1	z^1	$x^{2\prime}$	$y^{2\prime}$	$z^{2\prime}$	DCX	DCY	DCZ	XM	YM	ZM
1	1.5	-4.5	5.3	3.2897	-3.6	-0.8527	-0.27659	-0.13909	0.95087	-3.5417	-2.8922	-1.4533
2	-4.5	-5.7	4.8	2.9653	-3.6	-0.0192	-0.81763	-0.23	0.52781	-1.9045	-1.5495	-3.6255
3	3.5	-5.0	-0.5	0.6116	-4.5	-4.8223	0.55307	-0.095739	0.82762	-4.186	-3.1732	2.4302
4	2.0	-5.5	-0.1	1.2942	-4.5	-2.8770	0.23257	-0.32951	0.91506	-5.0658	-1.8534	0.62013
5	2.3	-2.0	-0.6	0.4124	-3.8	-4.8038	0.38154	0.36384	0.84973	-1.4812	-2.1833	1.5
6	3.0	-3.0	-0.2	0.5479	-3.9	-0.0962	0.93803	0.34428	-0.039713	0.18799	-0.068467	3.8469

Sum of forces along x:

$$\Sigma F_x = 0$$

Due to symmetry,

$$2 \cdot \sum_{i=1}^{6} F_i \cdot \ell_i + 2 \cdot F_7' + P \cdot \ell_p = 0$$

where F_i = force in muscle i (kg)

P = biting pressure (kg)

ℓ_p = direction cosine along x of load P

F_7 = x-component of the TM joint reaction force at each joint

Substitution of the values for ℓ_i, P and ℓ_p yields:

$$2\{-0.26659 \cdot F_1 - 0.81763 \cdot F_2 + 0.55307 \cdot F_3 + 0.23257 \cdot F_4$$
$$+ 0.38154 \cdot F_5 + 0.93803 \cdot F_6\} + 2 \cdot F_7'$$
$$- 0.94627 = 0 \qquad\qquad (1)$$

Sum of forces along z:

$$\Sigma F_z = 0$$

Due to symmetry,

$$2 \cdot \sum_{i=1}^{6} F_i \cdot n_i + 2 \cdot F_8' + P \cdot n_p = 0$$

where
n_p = direction cosines along z of load P

F_8' = z-component of the TM joint reaction force at each joint

Substituting the values for all the known parameters in the above equations we obtain:

$$2\{0.95087 F_1 + 0.52781 F_2 + 0.82762 F_3 + 0.91506 F_4$$
$$+ 0.84973 F_5 - 0.039713 F_6\} + 2 \cdot F_8'$$
$$- 0.73795 = 0 \qquad\qquad (2)$$

Sum of moments about y:

$$\Sigma M_y = 0$$

or

$$2 \cdot \sum_{i=1}^{6} F_i \cdot M_{y_i} + 2(F_7' \cdot Z_{TM} - F_8' \cdot x_{TM}) + F_9' + P \cdot M_{y_P} = 0$$

where M_{y_P} = moment about the Y axis due to load P

X_{TM}, Z_{TM} = x,z coordinates respectively of the TM joint w.r.t. ground axes

F_8' = moment about Y axis at the TM joint, representing ligament action at this joint.

Substitution of the known variables yields:

$$2\{-2.8922\ F_1 - 1.5495\ F_2 - 3.1732\ F_3 - 1.8534\ F_4$$
$$- 2.1833\ F_3 - 0.068467\ F_6\} + 2Z(F_7'\ (-0.05)$$
$$- F_8'\ (0.05)\} + 9.4239 + F_9' = 0 \qquad (3)$$

The variables F_7', F_8' which are x and z reactions respectively at the TM joint and F_9', which is y-moment at the TM joint are arbitrarily assumed to act as shown. They can be therefore either positive or negative depending on their true directions which will be indicated from the final solution.

For a linear programming problem however, all variables must be $\geqslant 0$. These variables (and any such variables when desired) therefore are each broken into two parts both constrained to be $\geqslant 0$, but the net value being the difference of two positive parts, can be either +ve or −ve. Thus

$$F_7' = F_7 - F_{10}$$
$$F_8' = F_8 - F_{11} \qquad \text{where } F_7, F_8, \ldots, F_{12} \geqslant 0$$
$$F_9' = F_9 - F_{12}$$

This process yields a total of twelve variables. The objective function for this example is selected to be minimization of all muscle forces, twice the reaction forces at the TM joint and four time the moments at the TM joint. This can be mathematically expressed as:

$$\text{minimize } U = 2 \sum_{i=1}^{6} F_i + 2(F_7 + F_8 + F_{10} + F_{11}) + 4(F_9 + F_{12})$$

Note that by manipulation of the cost coefficients of the variables in U, the objective function can be conveniently modified to represent any desired objective. The form of the function should however be always linear.

The optimization problem for the jaw therefore takes the form of:

minimize: U

572

where

$$U = \sum_{j=1}^{12} C_j F_j = 2 \sum_{j=1}^{6} + 2(F_7+F_8+F_9+F_{10}+F_{11}) + 4(F_9+F_{12})$$

subject to:

$$-0.55318\ F_1 - 1.63526\ F_2 + 1.10614\ F_3 +).46514\ F_4 + 0.76308\ F_5$$
$$+ 1.87606\ F_6 + 2\ F_7 - 2\ F_{10} = 0.94627 \qquad (4)$$

$$1.90174\ F_1 + 1.05562\ F_2 + 1.65524\ F_3 + 1.83012\ F_4 + 1.69946\ F_5$$
$$- 0.079426\ F_6 + 2\ F_8 - 2\ F_{11} = 0.73795 \qquad (5)$$

$$5.7844\ F_1 + 3.099\ F_2 + 6.3464\ F_3 + 3.76068\ F_4 + 4.3666\ F_5$$
$$+ 0.136934\ F_6 + 0.1\ F_7 + 0.1\ F_8 - F_9 - 0.1\ F_{10}$$
$$- 0.1\ F_{11} + F_{12} = 9.4239$$
$$\qquad (6)$$

and where $\quad F_j > 0 \qquad$ for j =1,2,...,12

For the revised simplex procedure with a full artificial basis (see reference) we rewirte the equations to read:

$$\text{maximize } F_{16}$$

Subject to:

$-0.55318F_1\ -1.63526F_2\ +1.10614F_3\ +0.46514F_4\ +0.76308F_5\ +1.87606F_6+2\ F_7\ +0\ F_8\ +0F_9\ +2\ F_{10}\ +0\ F_{11}\ +0F_{12}\ +F_{13} = 0.94627$

$1.90174F_1\ +\ 1.05562F_2\ +1.65524F_3\ +\ 1.83012F_4\ +\ 1.69946F_5\ -0.079426F_6+0\ F_7+2\ F_8+0F_9+0\ F_{10}-2\ F_{11}+0F_{12}+F_{14} = 0.73795$

$5.7844F_1\ +3.099\ F_2\ +6.464F_3\ +3.7068F_4\ +4.3666F_5\ +0.136934F_6\ +0.1F_8\ -\ F_9-0.1F_{10}-0.1F_{11} + F_{12} + F_{15} = 9.4239$

$2f_1+\quad 2F_2\quad 2F_3+\quad 2F_4\quad 2F_5\quad 2F_6+2\quad F_7+2\quad F_8+4F_9+2\quad F_{10}-2\ F_{11}+4F_{12}+F_{16} = 0$

$-7.13296F_1\ -2.51936F_2\ -9.10778F_3\ -6.00206F_4\ -6.82914F_5\ -1.933568F_6\ +2.1F_7\ -2.1F_8\ +F_9+2.1F_{10}\ +2.1F_{11}\ -F_{12}+F_{17} = -11.10812$

573

The first three rows are the old equations (4), (5) and (6) as modified by adding artificial variables F_{13}, F_{14} and F_{15} respectively to their left hand sides. The coefficients of the first 12 variables for the fourth row are equal to the corresponding coefficients C_j as written in the objective function to be minimized. The coefficients of the variables F_1, F_2,...,F_{12} and the right hand side of the fifth equation are obtained by adding the coefficients of the variables F_1, F_2,...,F_{12} and the right hand side elements respectively in the first three rows and reversing the sign.

We next define a matrix $\overline{A}$ formed by the coefficients of variables F_1, F_2,...,F_{12} in the above five equations, and a matrix U formed by the coefficients of artificial variables F_{13}, F_{14},...,F_{17} in the same five equations so that,

		COLUMN				
ROW	1	2	3	4	5	6
1	−0.55318	−1.63526	1.10614	0.46514	0.76308	1.87606
2	1.90174	1.05562	1.65524	1.83012	1.69946	−0.079426
$\overline{A}=$ 3	5.7844	3.099	6.3464	3.7068	4.3666	0.136934
4	2.0	2.0	2.0	2.0	2.0	2.0
5	−7.13296	−2.51936	−9.10778	−6.00206	−6.82914	−1.933568

	7	8	9	10	11	12
1	2.0	0.	0.	−2.0	0.	0.
2	0	2.0	0.	0.	−2.0	0.
3	0.1	0.1	−1.0	−0.1	−0.1	1.0
4	2.0	2.0	4.0	2.0	2.0	4.0
5	−2.1	−2.1	1.0	2.1	2.1	−1.0

and

$$U = \begin{vmatrix} 1 & 0 & 0 & 0 & 0 \\ 0 & 1 & 0 & 0 & 0 \\ 0 & 0 & 1 & 0 & 0 \\ 0 & 0 & 0 & 1 & 0 \\ 0 & 0 & 0 & 0 & 1 \end{vmatrix}$$

If m is the number of rows in the ooriginal problem ($m-3$ in this case) and n is the number of variables ($n=12$ in this case), then the size of $\overline{A}$ is $(m+2) \times n$ and the size of U is $(m+2) \times (m+2)$.

The starting tableau for the problem and the sequence of iterations is illustrated below. The problem is solved in two phases--Phase I consists of finding a basic feasible solution and Phase II consists of optimizing.

PHASE I

$$\delta_k = \delta_3 = U_{m+2} \cdot \overline{A}_3$$

$$= -9.10778$$

$$\theta_0 = \frac{F_{14}}{F_{14,3}} = 0.4458$$

In Phase I, which indicate there are artificial variables in the solution with positive values, the first step is to compute:

$$\delta_j = U_{m+2} \cdot \overline{A}_j$$

$$= U_{m+2,1}\, a_{1j} + U_{m+2,2}\, a_{2j} + \ldots + U_{m+2,m+2}\, a_{m+2,j} \qquad (j=1,2,\ldots,n)$$

What this implies is finding out the product of the $(m+2)$ row of the U matrix in the tableau with each of the n columns of $\overline{A}$ matrix. If $F_{m+n+2} = 0$, an exit to the step 1 of Phase II is made at this point. If all $\delta_j > 0$, then F_{m+n+2} is at its maximum and hence no feasible solution exists for the problem. If at least one $\delta_j < 0$, then the variable to be introduced into the solution, F_k corresponds to

$$\delta_k = \min \delta_j \qquad\qquad j = 1,2,\ldots,n.$$

In our case

$$\delta_1 = U_{m+2} \cdot \overline{A}_1$$

or

$$\delta_1 = [0 \quad 0 \quad 0 \quad 0 \quad 1] \begin{vmatrix} -0.55318 \\ 1.90174 \\ 5.7844 \\ 2.0 \\ -7.13296 \end{vmatrix}$$

$$= -7.13296$$

$$= [0 \quad 0 \quad 0 \quad 0 \quad 1] \begin{vmatrix} -1.63526 \\ 1.05562 \\ 3.099 \\ 2.0 \\ -2.51936 \end{vmatrix}$$

$$= -2.51936$$

Similarly

$$\delta_3 = -9.210778$$
$$\delta_4 = -6.00206$$
$$\delta_5 = -6.82914$$
$$\delta_6 = -1.933568$$
$$\delta_7 = -2.1$$
$$\delta_8 = -2.1$$
$$\delta_9 = 1.0$$
$$\delta_{10} = 2.1$$

$$\delta_{11} = 2.1$$
$$\delta_{12} = -1.0$$

So that

$$\delta_k = \min \quad \delta_j \text{ for } j=1,2,\ldots,12 \text{ is}$$

$$\delta_3 = -0.10778$$

The third step is to compute

$$F_{ik} = U_i \overline{A}_k$$
$$= u_{i1} a_{1k} + u_{i2} \cdot a_{2k} + \ldots + U_{i,m+2} a_{m+2,k}$$
$$\text{for } i=1,2,\ldots,m,(m+1),(m+2)$$

The variable F_ℓ to be eliminated from the solution corresponds to the ratio

$$\theta_0 = \min \frac{F_i}{F_{ik}} = \frac{F_\ell}{F_{\ell k}} \qquad i=1,2,\ldots,m$$

where the ratio is formed only for those $F_{ik} > 0$

In our case we found k = 3. So that,

$$F_{1,3} = U_1 \overline{A}_3 = [1 \quad 0 \quad 0 \quad 0 \quad 0] \begin{vmatrix} 1.10614 \\ 1.65524 \\ 6.3464 \\ 2.0 \\ -9.10778 \end{vmatrix}$$

$$= 1.10614$$

$$F_{2,3} = U\,\overline{A}_3 = [0 \quad 1 \quad 0 \quad 0 \quad 0] \begin{vmatrix} 1.110614 \\ 1.65524 \\ 6.3464 \\ 2.0 \\ -9.10778 \end{vmatrix}$$

$$= 1.65524$$

Similarly,

$$F_{3,3} = U_3\overline{A}_3 = 6.3464$$

$$F_{4,3} = U_4\overline{A}_3 = 2.0$$

$$F_{5,3} = U_5\overline{A}_3 = -9.10778$$

The computed column F_{ik} is shown in the starting tableau. To find the variable F_ℓ to be eliminated we compare the ratios

$$\frac{0.94627}{1.10614}, \quad \frac{0.73795}{1.65524} \quad \text{and} \quad \frac{9.4239}{6.3464}$$

The minimum ratio corresponds to $\dfrac{0.73795}{1.65524}$ which means the pivot element corresponds to $\theta_0 = \dfrac{F_{14}}{R_{14,3}}$ or F_{14} should be eliminated from the tableau and replaced by F_3. This is circled in Table VII.

In the fourth step, the new values of the variables in the basic solution are obtained by the formulae:

$$F'_i = F_i - \frac{F_\ell}{F_{\ell k}} F_{ik} \qquad \text{for } i \neq k$$

$$F'_k = \frac{F_\ell}{F_{\ell k}}$$

and the new elements of the matrix U are transformed by

$$U'_{ij} = U_{ij} - \frac{U_{\ell j}}{F_{\ell k}} F_{ik} \qquad \text{for } i \neq \ell$$

$$U'_{\ell j} = \frac{U_{\ell j}}{F_{\ell k}}$$

Under this transformation the (m+1)st and (m+2)nd columns of U will never change. Using the transformed rows of U the steps of Phase I are repeated (that is compute δ_j, compute θ_0 and obtain new U_{ij}, etc.) until it is determined that $F_{m+n+2} = 0$ or that no feasible solutions exits.

The second iteration is shown in Table VIII. Note that F_{14} has been replaced by F_3 in the solution. To reflect this change the new values of the variables and elements of the matrix U are obtained using the above formulae. For example,

$$F_1' = 0.94627 - \frac{0.73795}{1.65524} \times 1.10614 = 0.45312$$

$$F_2' = \frac{0.73795}{1.65524} = 0.44583$$

$$F_3' = 9.4239 - \frac{0.73795}{1.65524} \times 6.3464 = 6.59449$$

$$F_4' = 0 - \frac{0.73795}{1.65524} \times 2.0 = -0.89166$$

$$F_5' = -11.10812 - \frac{0.73795}{1.65524} \times (-9.10778) = -7.04759$$

Similarly,

$$U_1' = 1 - (1.10614) \frac{1}{1.65524} = 1$$

$$U_{12}' = 0 - 1.10614 \times \frac{1}{1.65524} = -0.66826$$

$$U_{13}' = 0 - 1.10614 \times \frac{0}{1.65524} = 0$$

$$U_{21}' = \frac{0}{1.65524} = 0$$

$$U_{22}' = \frac{1}{1.65524} = 0.60414$$

$$U_{32}' = 0 - 1.0 \times \frac{6.3464}{1.65524} = -3.83411$$

$$U_{42}' = 0 - 2.0 \times \frac{1.0}{1.65524} = -1.20828$$

$$U_{52}' = 0 - (-9.10778) \times \frac{1.0}{1.65524} = 5.50237 \quad \text{and so on.}$$

Note that in the pivot row, that is, the row corresponding to the minimum ratio θ_0 if an element is zero then the columns in U matrix (or the F_i column) in which this element lies will remain unchanged after the transformation. The new tableau shown in Table VIII indicates that F_{m+n+2} (=F_{17}) is negative, and

$$\delta_j = U_{m+2} \overline{A}_j$$

$$= [0 \quad 5.50237 \quad 0 \quad 0 \quad 1] \overline{A}_j \quad (j=1,2,\ldots,2)$$

are not all positive so we perform the third iteration. This results in a tableau as shown in Table IX. The following fourth iteration is shown

578

TABLE VII

Row index of tableau	Index of variables in solution	Value of variables	Matrix U					F_{ik}
1. STARTING TABLEAU								
1	13	0.94627	1	0	0	0	0	1.10614
2	14	0.73795	0	1	0	0 · 0		1.65524
3	15	9.4239	0	0	1	0	0	6.3464
4	16	0	0	0	0	1	0	2.0
5	17	-11.10812	0	0	0	0	1	-9.10778

PHASE I

$$\delta_k = \delta_3 = U_{m+2} \cdot \bar{A}_3$$
$$= -9.10778$$
$$\theta_0 = \frac{F_{14}}{F_{14,3}} = 0.4458$$

TABLE VIII.   Second Iteration

Phase I

i		F_i	U					F_{ik}
1	13	0.45312	1	-0.66826	0	0	0	1.33652
2	3	0.44583	0	0.60414	0	0	0	-1.20828
3	15	6.59449	0	-3.83411	1	0	0	7.56822
4	16	-0.89166	0	-1.20828	0	1	0	4.41656
5	17	-7.04759	0	5.50237	0	0	1	-8.90474

$$\delta_k = \delta_{11} = U_{m+2} \cdot \bar{A}$$
$$= -8.90474$$
$$\theta_0 = \frac{F_{13}}{F_{13,11}}$$
$$= 0.33903$$

579

in Table X.

At this stage (Table X) since $F_{m+n+2}(=F_{17})$ is ≈ 0 the problem enters Phase II. The first step is to compute

$$\nu_j = U_{m+1}\bar{A}_j$$
$$= U_{m+1,1}A_{1j} + U_{m+1,2}A_{2j} + \ldots + U_{m+1,m+2}A_{m+2,j} \qquad j=1,2,\ldots,n$$

If all $\nu_j \geqslant 0$, then F_{m+n+1} is at its maximum value and the corresponding basic feasible solution is an optimum solution. The negative value of F_{m+n+1} is the true value of the objective function to be minimized. If at least one $\nu_j < 0$, then the variable to be introducted into the solution, F_k corresponds to $\nu_k = \min \nu_j$.

In this example,

$$\nu_1 = [+.03831 \quad 1.03831 \quad - \quad -0.76698 \quad 1.0 \quad 0]\begin{vmatrix} -0.55318 \\ 1.90174 \\ 5.7844 \\ 2.0 \\ -7.13296 \end{vmatrix}$$

$$= -1.03629$$

$$\nu_2 = [1.03831 \quad 1.03831 \quad -0.76698 \quad 1.0 \quad 0]\begin{vmatrix} -1.63526 \\ 1.05562 \\ 3.099 \\ 2.0 \\ -2.51936 \end{vmatrix}$$

$$= -0.97872$$

Similarly, $\nu_3 = -0.0039$, $\nu_4 = 1.564015$, $\nu_5 = 1.20779$, $\nu_6 = 1.76044$, $\nu_7 = 3.99992$, $\nu_8 = 3.99992$, $\nu_9 = 4.76698$, $\nu_{10} = 0.000078$, $\nu_{11} = 0.000078$, $\nu_{12} = 3.23302$. Obviously this is not an optimal solution.

So that
$$\nu_k = \min \nu_j = \nu_1 = -1.03629$$

we then compute
$$F_{ik} = U_i\bar{A}_k$$
$$= U_{i1}a_{1k} + U_{i2}a_{2k} + \ldots + u_{i,m+2}a_{m+2,k}$$
$$\text{for } i=1,2,\ldots,m,m+1,m+2$$

TABLE IX   Third Iteration

Phase I

i	F_i		U					F_{ik}
1	11	0.33903	0.748212	-0.5	0	0	0	-1.49642
2	3	0.85547	0.90405	0	0	0	0	-1.80810
3	15	4.02864	-5.66263	-0.05	1	0	0	11.22526
4	16	-2.38901	-3.30452	1.0	0	1	0	8.60904
5	17	-4.02862	6.66263	1.05	0	0	1	-11.22526

$$\delta_k = \delta_{10} = U_{m+2} \cdot \bar{A}_{10}$$

$$= -11.22526$$

$$\theta_0 = \frac{F_{15}}{F_{15,10}} = 0.35889$$

TABLE X   Fourth Iteration

Phase II

i	F_i		U					F_{ik}
1	11	0.87608	-0.00666	-0.50666	0.13332	0	0	-0.18868
2	3	1.50438	-0.00805	-0.00805	0.16108	0	0	0.92089
3	10	0.35889	-0.50445	-0.00445	0.08909	0	0	0.78592
4	16	-5.47871	1.03831	1.03831	-0.76698	1	0	-1.03629
5	17	0.00001	1.00005	1.00005	1.00006	0	1	0.00042

$$F_{17} = 0$$

All $\nu_i \geq 0$

$$\nu_k = \nu_1 = -1.03629$$

$$\theta_0 = \frac{F_{10}}{F_{10,1}}$$

The variable F_ℓ to be eliminated from the solution is determined by finding the ratio

$$\phi_0 = \min_i \frac{F_i}{F_{ik}} = \frac{F_\ell}{F_{\ell k}} \qquad i=1,2,\ldots,m$$

for those $F_{ik} > 0$. If all $F_{ik} < 0$, then the problem has a solution for objective function whose value is unbounded, that is, can be made arbitrarily large. If θ_0 can be found, however, for $F_{ik} > 0$ then the variable F_k enters the basis in place of F_ℓ and the tableau is transformed to correspond to this alteration by the same expressions as used in Phase I.

Here

$$\theta_0 \min_i \frac{F_i}{F_{ik}} = \frac{F_{10}}{F_{10,1}} = 0.45665$$

The transformed tableau for the fifth iteration (Phase II) is shown in Table XI. The steps in Phase II are repeated until an optimal solution is obtained or it is determined that the solution is unbounded.

After the fifth iteration we compute $\nu_j = U_{m+1}\overline{A}_j$ which yields

$$\nu_1 = 0.000001 \qquad \nu_2 = 0.46683 \qquad \nu_3 = -0.0003$$
$$\nu_4 = 2.00444 \qquad \nu_5 = 1.20321 \qquad \nu_6 = 2.52013$$
$$\nu_7 = 2.68137 \qquad \nu_8 = 3.99995 \qquad \nu_9 = 4.64951$$
$$\nu_{10} = 1.31863 \qquad \nu_{11} = 0.00005 \qquad \nu_{12} = 3.35049$$

Since all $\nu_i \geqslant 0$ (the small negative value of ν_3 may be regarded as ≈ 0 within the extent of manual computational accuracy) the solution is optimal and the values of the variables in the optimal basis are:

$$F_1 = 0.45665 \text{ kg}$$
$$F_3 = 1.08386 \text{ kg}$$
$$F_{11} = 0.96224 \text{ kg}$$

and the value of the objective function is $U = -F_{16} = 5.00549$.

Solution using compuer

All linear programming problems involving muscle load sharing as exemplified in the course of this book were solved using the 'SIMPLX' subroutine package on the UNIVAC 1110 computer at the University of Wisconsin-Madison. This subroutine is based on the revised simplex algorithm as outlined in the previous section. The problem is stated in the Fortran language in matrix form as follows:

Subject to:

$$\sum_{J=1}^{N} A(I,J) \cdot Z(J)T(I)RHS(I)$$

$$\text{for } I=1,2,\ldots,M$$

and $\quad Z(J) \geqslant 0 \quad \text{for } I=1,2,\ldots,(N+M)$

TABLE XI   Fifth Iteration

i		F_i	U					F_{ik}
1	11	0.96224	-0.12777	-0.50773	0.15471	0	0	
2	3	1.08386	0.58303	-0.00284	0.05669	0	0	
3	1	0.45665	-0.64186	-0.00566	0.11336	0	0	
4	16	-5.00549	0.37316	1.03245	-0.64951	1	0	
5	17	-0.00018	1.00032	1.000052	1.0000	0	1	

Optimal Tableau

see $\cup_j \not\geq 0$

where   COST = cost vector (size N+M)

                     A = two-dimensional constraint matrix
                         (size MxN)

                     T = alphanumeric type vector indicating
                         constraint type (+ means <, - means >
                         and (blank) means =)

                   RHS = right hand side vector (size TM)

                     Z = basic solution vector specified by using the
                         X and JX arrays

The variables numbered N+1, N+2,...,N+M are slack variables.

    The system input thus are the A matrix, T vector, RHS vector and the
COST vector.   If a solution to the problem is known beforehand one may
provide a partial or full basis to initiate the iterations.   The
advantages of using a computer package of this kind are many.   It can
quickly find solutions to any other modified versions of the objective
function or of the right hand side vector or both.   The right hand side
vector in our case is primarily dependent on external forces and moments
acting on the musculoskeletal structure and if solutions are desired with
different values of some external force or moment this change can be
conveniently incorporated into the program.   For example, during the
analysis of muscle loadsharing for the situation when the arm is flexed
at the elbow and the hand is weighted one may readily compute the effect
of carrying different magnitudes of weights by changing only a few
elements in the RHS vector for the hand and proceeding from the solution
already known for a different magnitude of the weight.   Similarly if the
solution for a problem is known with a preselected objective function one
may readily obtain the solution to the same problem but with a different
objective function by making necessary changes to the COST vector to
conform to the desired alterations.

General outline of programming:

    In this section a general flow-chart approach to the adaptation of
the muscle load sharing problem for application the SIMDX subroutine is
given.   The main part of the program usually consists in creating the A
matrix, the coefficients of which relate to the direction cosines and
moment components of the various muscles, and the coefficients of joint
reaction forces and moments.

General outline of programming--flow chart.

```
┌─────────────────────────┐
│   Identify muscles;     │
│ their coordinates of    │
│ points of attachment.   │
└─────────────────────────┘
            │
            ▼
┌─────────────────────────┐
│  Input orientation of   │
│ segments and compute the│
│ transformed coordinates │
│ of muscle attachment    │
│ points joints, C.G.'s   │
│ etc.                    │
└─────────────────────────┘
            │
            ▼
┌─────────────────────────┐
│ Compute direction cosines│
│  and moment components  │
│ of muscles with respect │
│ to a suitable ground    │
│ axis system             │
└─────────────────────────┘
            │
            ▼
┌─────────────────────────┐
│  Create A matrix using  │
│ the direction cosines   │
│ and moment components   │
│ computed above. These   │
│ coefficients are equal  │
│ and opposite between    │
│ two bodies.             │
└─────────────────────────┘
            │
            ▼
┌──────────────────────────────┐
│ Set A matrix coefficients    │
│ corresponding to joint       │
│ reaction forces and moments  │
│ at joints between bodies.    │
│ These will also be equal and │
│ opposite between two bodies. │
└──────────────────────────────┘
            │
            ▼
┌──────────────────────────────┐
│ Break variables (such as     │
│ joint reactions and moments) │
│ which may be either +ve or   │
│ -ve into two parts each, the │
│ net value of the variable    │
│ being the difference of two  │
│ components                   │
└──────────────────────────────┘
            │
            ▼
```

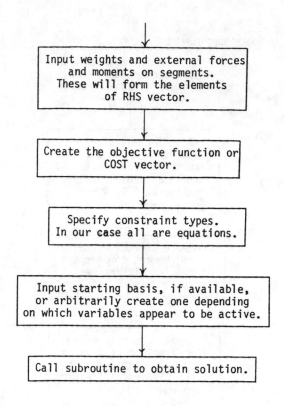

Input weights and external forces
and moments on segments.
These will form the elements
of RHS vector.

Create the objective function or
COST vector.

Specify constraint types.
In our case all are equations.

Input starting basis, if available,
or arbitrarily create one depending
on which variables appear to be active.

Call subroutine to obtain solution.

REFERENCES

1. Dantzig, G.: <u>Linear Programming</u>. McGraw Hill, New York, 1968.

2. Grass, S. I.: <u>Linear Programming</u>. McGraw Hill, New York, 1958, (Chapter 6, pp. 32-101).

3. University of Wisconsin. SIMPLX/SIMPDX Linear Programming Subroutines Reference Manual for 1110. Operations Research Series. Madison, Wisconsin.

4. Garrett, F. A., Angelone, L. and Allen, W. I.: "Electrical response of masseter muscle," <u>Am. J. Orthodant.</u>, <u>50</u>:435, 1964.

Clinical Applications
of the Musculoskeletal Model

In the foregoing chapters the models developed for the various parts of the body were applied to the investigation of normal activities or non-clinical cases. The models, however, are indispensable as diagnostic tools and can be readily used for the understanding of clinical or pathological cases involving the musculoskeletal structure. Several applications of the lower and the upper extremity model to clinical cases are described here. Based on the mode of computing operation these can be broadly classified into two groups.

i. Analysis using the batch mode of operation.

In case of the lower extremities this implies using the computer program listed in Appendix E. Examples of utilizing the batch mode program will include the analysis of joint forces during walking under varus-valgus (bowlegged-knockneed) conditions and the analysis of knee-joint forces in a patient walking with and without auxillary crutches.

ii. Analysis using the interactive mode of operation.

This approach utilizes the interactive computing and graphics program (see Appendix F) to model the musculoskeletal system and analyze the muscle load sharing. The use of this program will be demonstrated for calculating the hip joint reaction force during walking for a person with a degenerative joint and for determining the optimal wedge angle for corrective osteotomy using interactive graphical display.

A. EXAMPLES UTILIZING THE BATCH MODE OPERATION

I. EFFECT OF VALGUS-VARUS ON JOINT FORCES DURING WALKING

The lower extremities model was used to study the influence of valgus (knock-kneed) and varus (bow-legged) deformities on the distribution of knee joint forces during walking or leaning forward. The knee joint reaction force was modelled by the X (fore-aft) and Y(medio-lateral) components acting at the center of the knee joint, and two vertical Z reactions to simulate the pressures on the medial and lateral plateaus. The valgus (knock-kneed) configuration is simulated by tilting medially the longitudinal axis of the tibia (line joining the center of the knee joint to the ankle joint) in the frontal plane through an angle θ, and then tilting laterally the longitudinal axis of the femur (line joining the center of the femoral head to the center of the condyles at the knee joint) in the same plane through the angle θ. The varus (bowlegged) configuration is achieved in a similar manner, except that the tibia is tilted laterally and the femur, medially in the frontal plane.

The muscle load sharing and joint reaction forces were studied at 15%, 27% and 39% of the walking cycle. These instants represent the one-legged stance condition when the right extremity is in contact with the ground and the left extremity is swinging freely. Moreover these locations also happen to be the occurrences of maximum joint forces during walking as shown by previous analyses. The valgus-varus angles in the present study were arbitrarily selected to be 0°, ±5°, ±10° and ±15°. It was assumed that both the tibia and the femur are tilted through the same angle. Thus, for example, in the bowlegged condition if the right tibia was tilted in the frontal plane laterally through θ = 5 degrees, then the right femur was tilted medially through the same θ = 5 degrees (see Figure 1). It was also assumed for convenience that the left leg is "normal" or has no valgus or varus deformity, and the vertical reaction at the left joint is modelled by a single force acting at the center of the knee joint.

As may be expected, while the sagittal notations of the segments remain the same the resultant ground-to-foot center of support to shift from the position where they would normally lie. However, by adjusting the overall body tilts in the sagittal and in the frontal plane, the resulting center of support is brought back to their abnormal positions with no valgus-varus deviations.

RESULTS

The results for the 0, 5, 10 and 15 degrees of valgus or varus deformities for 15, 27 and 39 percent of walking cycle are shown in Tables I, II and III respectively. Results for the right side extremity only, considered to be abnormal are shown in the tables. The muscle activity on the left side extremity which is passing through the swing phase, is usually less significant compared to the right side one. The overall center of support on the right foot is in the same position as that for the case of normal walking with a normal right knee.

The results for the normal case, that is, when no valgus or varus deformity is considered are also shown in the tables for comparison. Note that in this case, the right knee joint vertical pressure was simulated by a single component at the center of the knee joint rather than two vertical components on the medial and lateral plateaus.

The criterion used was the same as before, that is minimize U = Σf + 4M, M being the moments at the ankle joint. However, the moments in the merit function now also include the moments at the knee joint.

DISCUSSIONS

A.   15% of Walking Cycle:

TABLE I. Muscle forces and joint reactions for right leg (15% of walking cycle).

Muscle or Muscle Group	Varus ←			0	→ Valgus			With single z-reaction at the knee joint (0-0)
	-15%	-10°	-5°	0°	5°	10°	15°	
Hamstrings:								
1. Semitendinosus	4.7	124.8	0	0	0	0	0	0
2. Semimembranosus	105.2	0	157.0	98.6	39.0	0	0	156.2
3. Biceps femoris (LH)	0	0	0	0	0	0	0	0
4. Biceps femoris (SH)		0	0	0	0	0	0	0
Quadriceps Group:								
1. Rectus femoris	0	0	0	133.9	76.3	18.3	0	20.7
2. Vastus Medialis	0	0	0	0	0	0	0	0
3. Vastus Intermedius	55.4	0	0	0	0	0	114.5	0
4. Vastus Lateralis	0	79.5	72.8	0	33.5	73.7	0	57.5
Abductor Group:								
1. Tensor fasciae latae	156.6	130.3	151.0	0	0	1.8	7.4	138.9
2. Gluteus Medius	0	30.0	1.9	47.0	61.9	73.9	77.9	14.1
3. Gluteus Minimus	5.5	1.4	2.4	25.5	33.7	40.2	42.9	1.9
Adductor Magnus	62.3	0	0	0				
Sartorius	0	0	0	15.0	0	0	4.5	0
Gluteus Maximus	0	0	0	0	10	0	0	0
Gastrocnemius (calf)	0	0	0	0	0	0	56.8	0
Tibialis Anterior	34.1	41.3	45.9	49.3	52.2	33.4	152.1	49.3
Extensor digitorum longus	104.9	75.0	55.1	40.8	29.9	23.0	43.3	40.8
Extensor hallucis longus	0	0	0	0	0	18.8	5.6	0
Joint Reactions and Moments:								
Hip Joint: R_x	-52.18	-39.43	-47.38	-44.65	-19.4	0.34	2.83	-50.8
R_y	35.39	10.92	-14.7	-14.9	-5.5	11.53	10.12	-35.7
R_z	386.64	342.41	367.27	370.45	263.44	190.57	190.07	384.5
Knee Joint: R_x	-29.5	-25.42	-35.14	3.57	13.24	18.6	32.73	-31.9
R_y	3.98	-20.53	-35.25	-45.7	-24.48	-6.81	7.62	-52.6
(medial) R_{z_1}	380.34	354.98	287.38	220.48	89.09	2.22	0	431.2
(lateral) R_{z_2}	0	38.36	152.38	85.64	122.12	156.89	245.83	
Ankle Joint: R_x	3.85	3.78	3.63	3.44	3.22	0.43	15.91	3.44
R_y	-59.68	-39.21	-24.65	-13.58	-4.73	0.07	34.85	-13.6
R_z	196.65	180.59	169.05	160.19	153.1	146.2	325.63	160.2
M_z	0	-64.33	-45.73	-28.24	-11.4	0	0	-28.24

TABLE II. Muscle forces and joint reactions for right leg (27% of walking cycle).

Muscle or Muscle Group	Varus +			0	→ Valgus			With single z-reaction at the knee joint (θ=0)
	-15°	-10°	-5°	0°	5°	10°	15°	
Hamstrings:								
1. Semitendinosus	0	0	0	0	0	0	0	0
2. Semimenbranosus	46.9	46.9	27.0	0	0	5.2	62.2	0
3. Biceps femoris (LH)	30.1	3.3	0	0	0	0	0	20.6
4. Biceps femoris (SH)	0	0	0	0	0	0	0	0
Quadriceps Group:								
1. Rectus femoris	0	0	0	0	11.6	0.9	0	0
2. Vastus medialis	0	0	0	0	0	0	0	3.3
3. Vastus intermedius	36.7	26.7	5.2	3.9	0.7	0	0	0
4. Vastus lateralis	0	0	0	0	0	0	0	0
Abductor Group:								
1. Tensor fasciae latae	96.7	66.2	41.2	9.8	0	0	21.9	31.2
2. Glutens medius	0	33.5	45.8	56.5	59.1	61.2	61.3	47.0
3. Glutens minimus	24.4	20.7	31.5	47.2	52.1	54.9	43.9	44.4
Sartorius	0	0	0	0	0	12.7	54.2	0
Glutens Maximus	11.0	0	0	0	0	0	0	0
Gastrocnemius (calf)	131.5	111.7	53.8	30.0	36.1	11.5	11.8	28.5
Soleus (calf)	0	0	0	0	0	27.9	29.4	0
Tibialis posterior	0	0	0	0	0	22.3	19.0	0
Extensor hallucis longus	154.3	117.3	33.9	0	0	0	0	0
Flexor digitorum longus	0	0.3	43.2	54.8	28.3	0	0	60.9
Peroneus tertius	9.8	0	45.5	50.5	6.2	0	0	59.7
Joint Reactions and Moments:								
Hip joint: R_x	-2.98	-3.69	-1.64	1.84	1.48	2.79	-0.21	-1.7
R_y	33.19	16.46	15.06	22.06	19.55	13.33	-36.88	17.5
R_z	263.69	227.82	201.9	169.42	179.06	191.01	292.00	199.4
Knee joint: R_x	25.44	20.96	8.27	5.95	8.96	2.62	1.19	4.8
R_y	-29.54	-26.94	-10.67	-3.77	-1.92	-4.79	-47.57	-6.5
(medial) R_{z1}	398.11	316.53	193.15	91.94	47.96	0	0	150.3
(lateral) R_{z2}	0	0	0	18.61	67.09	96.77	207.69	
Ankle joint: R_x	7.9	9.55	7.23	7.87	11.03	12.78	12.38	7.3
R_y	-119.56	-73.97	-40.25	-20.07	-4.78	-4.45	1.45	-22.9
R_z	334.97	283.44	239.27	202.21	140.22	130.78	129.51	215.4
M_z	0	0	0	0	0	0	-26.3	0

TABLE III. Muscle forces and joint reactions for right leg (39% of walking cycle).

Muscle or Muscle Group	Varus ←			0°	→ Valgus			With single z-reaction at the knee joint (θ=0)
	-15°	-10°	-5°		5°	10°	15°	
Hamstrings:								
1. Semitendinosus	0	0	0	0	0	0	0	0
2. Semimembranosus	0	0	0	0	0	0	13.6	0
3. Biceps femoris (LH))	9.6	0	0	10.8	69.7	3.3	0	82.3
4. Biceps femoris (SH)	0	0	0	37.6	0	0	0	34.6
Quadriceps group:								
1. Rectus femoris	0	0	0	0	0	0	0	0
2. Vastus medialis	0	0	0	0	0	0	0	0
3. Vastus intermedius	0	0	0	0	0	0	0	0
4. Vastus lateralis	75.1	35.3	0	0	0	0	0	0
Abductor group:								
1. Tensor fasciae latae	18.9	16.1	31.0	0	0	0	0	46.5
2. Glutens medius	4.9	7.6	5.9	49.2	59.4	57.7	62.6	28.6
3. Glutens minimus	75.8	78.4	74.1	62.2	108.9	60.0	57.0	43.6
Sartorius	0	0	0	0	0	0	18.5	49.1
Glutens maximus	28.3	32.1	36.5	0	0	0	0	0
Adductor longus	0	0	0	0	42.4	0	0	0
Gastrocnemius (calf)	369.2	299.6	249.7	165.2	126.7	208.1	214.0	172.5
Tibialis anterior	0	0	0	0	15.9	151.9	157.2	0
Tibialis posterior	0	0	0	0	45.4	21.9	17.3	0
Extensor hellucis longus	428.5	305.9	213.9	81.6	0	0	0	90.5
Flexor digitorum longus	0	17.0	11.8	43.8	0	0	0	35.4
Peroneus tertius	41.2	35.9	0	10.9	0	0	0	0
Joint Reactions and Moments:								
Hip joint: R_x	5.13	2.63	2.24	3.09	26.81	1.87	4.89	22.21
R_y	47.2	43.45	37.54	26.21	32.97	19.03	3.28	2.22
R_z	185.45	183.24	197.98	177.31	331.49	177.82	206.76	204.6
Knee joint: R_x	96.48	71.39	53.65	44.75	32.62	42.64	47.53	64.3
R_y	-123.91	-67.91	-28.87	-5.09	-2.2	13.07	14.28	-20.7
(medial) R_{z_1}	509.86	403.44	307.84	197.89	106.25	56.77	0	
(lateral) R_{z_2}	0	0	33.46	78.28	154.36	217.26	304.85	445.3
Ankle joint: R_x	44.57	42.23	41.43	37.53	37.9	59.63	60.46	38.04
R_y	-340.32	-212.19	-112.45	-42.08	-12.3	11.56	41.16	-41.8
R_z	818.08	679.58	522.63	362.73	252.17	446.22	451.08	359.7
M_z	0	0	0	0	-2.74	-51.72	-93.81	0

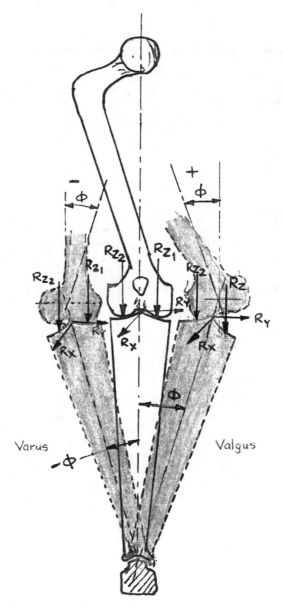

Figure D-1. Simulation of valous-varus.

The major active muscle groups (the ham strings, the quadriceps and the abductors) for the different varus-valgus angles are plotted in Figure 2. The hip and the ankle, and the knee joint reaction forces are plotted in Figure 3 and 4 respectively. Only the vertical Z forces, which are significantly higher than the lateral X and Y components are plotted.

It can be seen from these figures that the valgus-varus configuration has an influence on the muscle load sharing and the joint reaction forces. With increasing valgus, both quadriceps and abductors show increase while the hamstring force decreases to zero at and beyond 10 degrees of valgus. Corresponding to these muscle forces, the ankle joint reaction increases beyond 10 degrees of valgus, and also with increasing varus (Figure 3). The hip joint force is high during the varus condition but decreases with increasing valgus.

The knee joint pressures, medial and lateral, on the other hand show continual decrease or increase with valgus or varus. The medial part decreases with increasing valgus while the lateral part increases with increasing valgus. On the other hand, with increase in varus the lateral part diminishes to zero. This suggests that while for low degrees of valgus or varus, both the medial and lateral plateaus share in knee joint force bearing with extreme situations of valgus or varus, only one is load bearing (Figure 4). With zero degrees of valgus or varus, the medial plateau pressure is almost twice that on the lateral plateau. This distribution is anatomically compatible since the medial plateau is usually larger than the lateral.

This shifting of load from medial to lateral as the knee condition changes from varus to valgus should be normally expected from a mechanical point of view. When the knee is in varus, the resultant center of gravity lies closer to the medial plateau than it is to the lateral plateau. With the knee in valgus, the reverse is true, the center of gravity becoming closer to the lateral plateau.

B.   27% of Walking Cycle:

The activity of the various muscle groups with changing valgus-varus is depicted in Figure 5. The resultant knee joint pressure distribution between the medial and lateral plateaus is shown in Figure 6, and the pressures on the hip and the ankle joints in Figure 7.

It can be seen from Figure 5 that, the abductor group is not affected by the knee deformity. While the distribution of this muscle group force among its components does indeed vary with valgus or varus, the total value does not seem to change substantially. The other muscle group forces, namely, the hamstrings, the quadriceps and the calf however show significant increase at extreme values of both valgus and varus. This behavior is also illustrated by the hip joint force. On the other hand, the ankle joint force decreases continually from extreme varus to extreme valgus (Figure 6).

The knee joint force behavior with changing valgus or varus is similar to that for the case of 15% of walking. The medial plateau pressure is high at high values of varus and decreases to zero at high valgus. The lateral plateau pressure behaves exactly the opposite, from high values at extreme conditions of valgus to zero at extreme values of

593

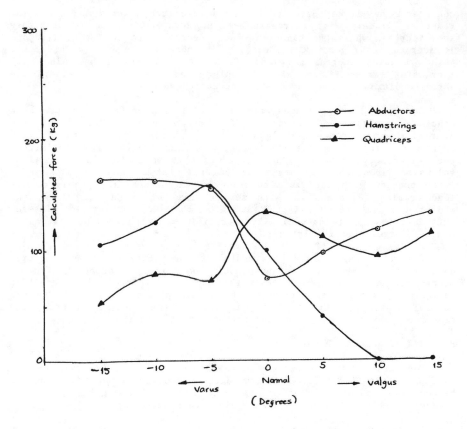

Figure D-2.Muscle force variation with valgus-varus
(15% walking cycle).

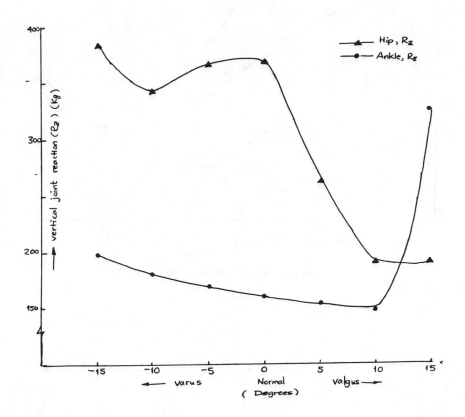

Figure D-3. Hip and knee joint reaction at 15% of walking cycle.

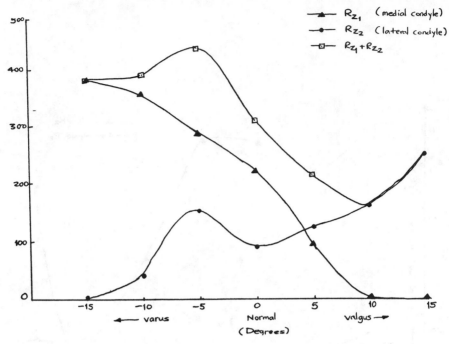

Figure D-4. Knee joint reaction at 15% walking cycle as a function of valgus-varus.

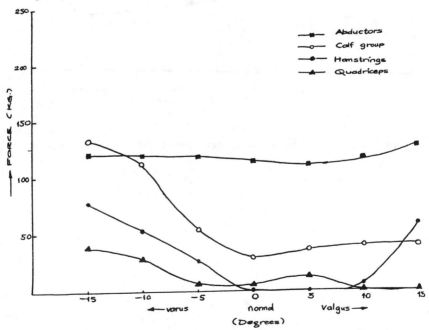

Figure D-5.   Muscle forces at 27% walking cycle.

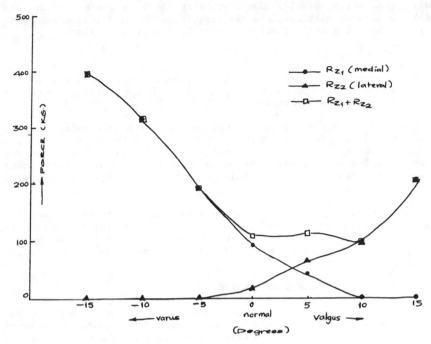

Figure D-6. Knee joint reaction at 27% of walking cycle.

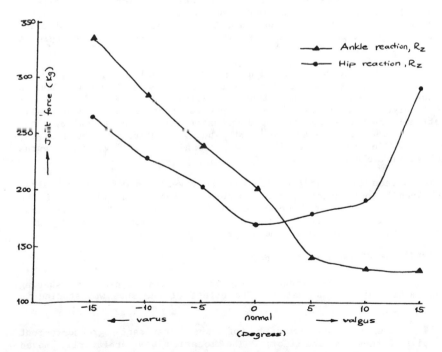

Figure D-7. Hip and ankle joint reactions (R_z) at 27% of walking cycle.

varus. Thus while normally the knee joint force is shared by both the medial and lateral plateaus, the resultant pressure shifts from medial to lateral as the knee configuration changes from varus to valgus.

C. 39% of Walking Cycle:

The muscle group forces, knee joint pressure and the hip and ankle joint reaction forces corresponding to 39% of the walking cycle are illustrated in Figures 8 through 10.

The calf group shows a very large increase with increasing varus or valgus condition. The other major active groups such as the abductors and the hamstrings on the other hand show large activity only at low values of varus or valgus. The high calf muscle forces and other muscles in the proximity of the ankle joint reaction (Figure 9) for large varus angles.

The knee joint pressure distribution pattern is similar to those for 15 and 27 percent of the walking cycle. The knee joint pressure is normally borne by both the medial and lateral plateaus but the weight-bearing shifts from medial to lateral as the knee joint configuration goes from varus to valgus.

II. EFFECT OF VALGUS-VARUS ON LEANING

In this study the influence of valgus-varus on muscle activity and joint forces during leaning was investigated. The forward or backward leaning is performed by a pelvic tilt in the sagittal plane about the hip joint. For illustration the 20° forward leaning posture was investigated with 5 degrees of valgus or varus. As discussed earlier for the case of walking, the valgus or arus configurations of the right knee are obtained by tilting the right tibia in the coronal plane about the ankle joint and then tilting the right femur in the same plane through the same angle at the knee joint.

The reaction force at the knee joint is modelled by its x and y components acting at the center of the condyles and two vertical z components acting on the tibial medial and lateral plateaus respectively. Since the leaning forward posture is bi-sagittally symmetrical, only one side of the lower extremities need be analyzed, the results for the contralateral extremity being identical to those on the other side. The results for 20° forward leaning with 5° valgus or varus are shown in Table I.

It is seen that when the knee goes from varus (bowlegged) to valgus (knock-kneed) configuration, the pressure on the medial plateau of the knee decreases, while the pressure on the lateral plateau increases. Similar results are seen for the muscle forces, some increasing with varus to valgus change while some decrease.

III. EFFECT OF FOOTWEAR ON WALKING

In this study the effect of footwear on the muscle load sharing during walking was investigated. The effect of the shoe was studied in two ways:

1. Changing the location of the resultant ground-to-foot supportive forces on the foot: The location was arbitrarily moved

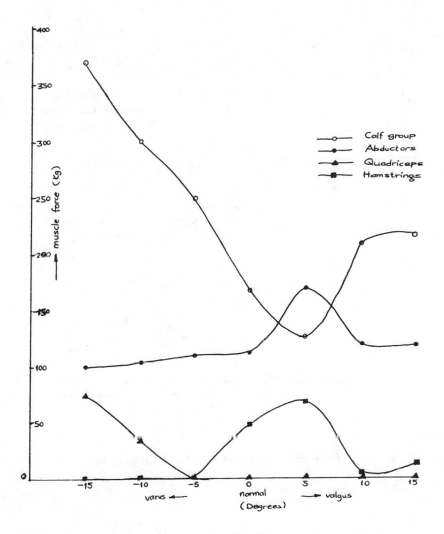

Figure D-8. Muscle forces at 39% of walking cycle.

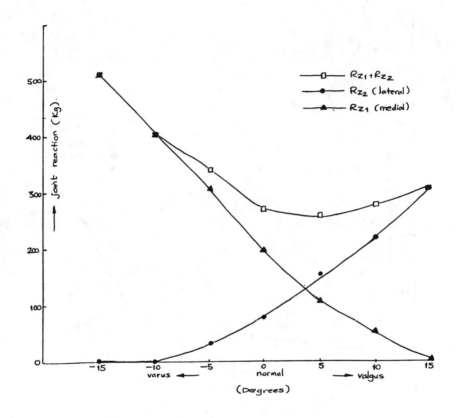

Figure D-9. Knee joint reaction at 39% of walking cycle.

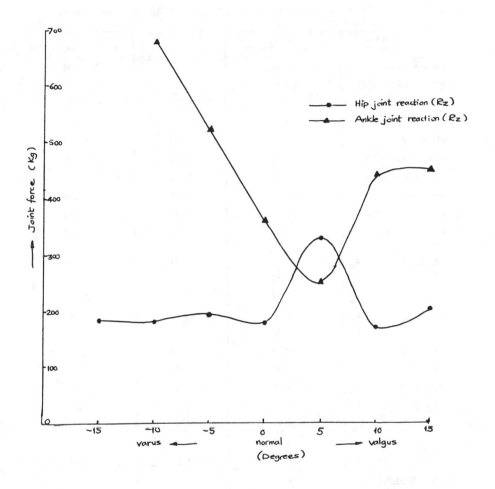

Figure D-10. Hip and knee joint reactions at 39% of walking cycle.

TABLE IV. Muscle forces and joint reactions for 20° forward leaning.

Muscle force or joint reaction		varus Bowlegged, 0 = -5°	Normal, C = 0	valgus Knock-kneed, 0 = 5°
Semimembranosus		53.41	23.61	18.41
Biceps Femoris (LH)		1.58	27.26	50.79
Sartorius		0	0	14.98
Semitendinosus		6.22	0	0
Iliopsoas		0	5.49	0.26
Biceps femoris (SH)		70.88	26.51	8.91
Grastrocnemius (MH)		6.68	17.78	0
Grastrocnemius (LH)		0	13.46	31.54
Tibialis posterior		0	8.28	7.68
Adductor longus		16.39	0	0
Hip joint:	R_x	-2.93	-0.44	-1.81
	R_y	4.87	1.56	-5.05
	R_z	100.17	79.21	107.83
Knee joint:	R_x	10.32	3.97	2.42
	R_y	10.18	3.8	-2.94
(medial)	R_{z1}	97.47	42.59	0
(lateral)	R_{z2}	68.69	96.65	155.19
Ankle joint:	R_x	4.55	3.91	3.91
Ankle joint:	R_y	-3.17	-2.83	-0.49
	R_z	39.79	73.97	73.83
	M_x	33.69	0	0
	M_z	0	0	-5.74
Ground-to-foot	R_x	0	0	0
	R_y	-1.99	-0.59	-0.16
	R_z	36.15	36.15	36.15
$(U = \Sigma F + 4M) = U$		289.95	122.49	155.53

forward, backward and outward by 1 cm from the position that it would normally occur at a given instant of walking cycle. For example, at 15% of walking cycle, when the left leg is off the ground the location of the resultant ground force is GX = 2.61 and GY = -0.14 cm (w.r.t. the tip of the heel or the rear edge of the footprint). The location was arbitrarily moved to GX = 3.61, GY = -0.14; GX = 1.61, GY = -0.14 and GX = 2.61, GY = -1.14, representing forward, backward and lateral shift respectively. This is achieved by keeping the relative segmental orientation the same as that for the normal situation (GX=2.61, GY=-0.14) but changing the overall tilts of the body about the right ankle joint, in the sagittal and in the coronal plane to yield the desired location of the center of support.

2. Applying external inertial forces at the upper torso: Inertial forces along x and along y were applied at the center of gravity of the upper torso. The magnitudes of these forces are selected so as to yield the same center of support as for the normal case without any inertia forces. Thus at 15% of cycle when a force F_x = 9.508 Kg is applied at the C.G. of the upper torso and the body is tilted about the right ankle joint in the coronal plane, the support falls at GX = 2.61, GY = -0.14. Similarly, when a force F_y = -4.96 Kg is applied at the C.G. of the upper torso and the body is tilted in the sagittal plane, the center of support is GX = 2.61 and -0.14.

The results for the above five cases are shown in Table V. All five cases seem to affect the muscle load sharing and distribution of joint forces. Of all cases however, applying a Y-inertia force on the torso (while keeping the resultant center to support unchanged from the normal position) seems to result in lower muscle and joint forces. The hip joint force is reduced from 385 kg to 315 kg while the knee joint force is reduced from 431 kg to 355 kg. Conceptually the Y inertia forces can be created by motion of the body segments in the coronal plane. Conceivably the so-called "flared" shoe where the lateral side of the sole is thicker compared to the medial side alter the motion pattern in this plane and produce Y-inertia forces. Such shoes therefore can be beneficial in athletic activities such as running, jogging, etc., where the joint forces are expected to be even higher than those during normal walking.

IV. EFFECT OF FEMORAL ANTEVERSION ON WALKING

In this study the influence of abnormal femoral anteversion on the muscle load sharing during walking was investigated. Anteversion implies the angle through which the neck of the femur is twisted forward in a horizontal plane. The effect of excessive anteversion can be simulated by twisting the pelvis forward about a vertical axis at the right hip joint. The pelvic twist angle was arbitrarily selected as 15°. Due to this twist the posture at a given instant of walking is altered as compared to normal, resulting in a shift of the center of support. The resultant ground-to-foot center of support however, can be brought back to its normal position by modifying the overall body tilt in the sagittal and the coronal plane about the right ankle joint.

Results for muscle load sharing and joint reactions were computed for the entire walking cycle (except for the double-legged support) with the pelvic twist included.

TABLE V. Muscle forces and joint reactions at 15% of walking cycle.

Muscle Force Number	Normal		Move support back		Move support forward		Move support lateral		X Inertia force		Y Inertia force	
	Right Leg	Left Leg	Right Leg	Left Leg	Right Leg	Left Leg	Right Leg	Left Leg	Right Leg	Left Leg	Right Leg	Left Leg
1	0	0	0	0	0	0	0	0	0	0	0	0
2	0	0	0	0	0	0	0	0	0	0	0	0
3	0	0	0	0	0	0	0	0	0	0	0	8.8
4	0	0	0	0	0	0	0	0	0	0	0	0
5	0	0	0	0	0	0	0	0	1.4	0	0	0
6	0	0	0	0	0	0	0	0	0	0	0	0
7	156.2	20.3	172.7	20.1	139.6	20.5	159.2	20.8	206.9	12.1	109.8	14.2
8	0	15.8	0	15.6	0	15.9	0	16.1	0	5.8	0	12.0
9	20.7	0	33.8	0	7.7	0	41.5	0	0	0	18.2	0
10	0	0	0	0	0	0	0	0	0	0	0	0
11	138.9	0	145.3	0	132.4	0	127.8	0	173.0	0	101.1	0
12	0	1.0	0	0	0	2.4	0	0	0	0	0	5.7
13	0	0	0	0	0	0	0	0	0	0	0	0
14	14.1	2.2	5.4	3.0	22.7	1.2	13.2	3.1	0	2.0	29.2	0
15	1.9	10.9	3.3	10.1	0.6	12.0	2.7	11.1	3.6	9.4	0.6	6.2
16	0	0	0	0	0	0	0	0	3.2	6.5	0	0
17	57.5	0	56.3	0	58.7	0	44.2	0	72.9	1.3	0	0
18	0	9.9	0	9.7	0	9.9	0	10.7	0	0	63.3	0
19	0	0	0	0	0	0	0	0	0	0	0	0
20	0	0.4	0	0.4	0	0.4	0	0.4	0	0	0	0.2
21	0	0	0	0.1	0	0	0	0.1	0	0.7	0	0.1
22	0	0	0	0	0	0	0	0	0	0	0	0
23	49.3	0	67.5	0	30.4	0	61.6	0	54.5	0	47.0	0
24	0	0	0	0	0	0	0	0	0	0	0	0
25	40.8	0	58.5	0	22.4	0	2.6	0	46.3	0	46.1	0
26	0	0.3	0	0.4	0	0.3	0	0.5	0	1.0	0	0
27	0	0	0	0	0	0	0	0	0	0	0	0
28	0	0	0	0	0	0	0	0	0	0	0	0
29	0	0	0	0	0	0	0	0	0	0	0	0
30	0	0	0	0	0	0	0	0	0	0	0	0
31	0	0.2	0	0.2	0	0.2	0	0.3	0	0.5	0	0

Muscle Force Number		Normal		Move support back		Move support forward		Move support lateral		X Inertia force		Y Inertia force	
		Right Leg	Left Leg	Right Leg	Left Leg	Right Leg	Left Leg	Right Leg	Left Leg	Right Leg	Left Leg	Right Leg	Left Leg
Hip:	R_x	-50.8	7.53	-60.73	6.97	-41.49	8.1	-54.38	7.9	-132.43	0.03	-37.9	4.29
	R_y	-35.69	-9.43	-40.66	-8.81	-30.71	-10.24	-40.56	-9.91	-44.82	-6.4	4.36	-3.74
	R_z	384.51	35.5	411.66	34.26	357.18	36.76	396.1	36.15	420.28	15.56	314.67	33.17
Knee:	R_x	-31.92	12.99	-38.23	12.31	-26.35	13.5	-30.07	13.78	-131.77	2.98	-17.88	3.79
	R_y	-52.64	-2.72	-58.2	-2.7	-47.07	-2.72	-57.04	-3.22	-61.72	-2.75	-12.27	.01
	R_z	431.22	37.18	464.61	36.72	397.71	37.26	429.8	38.4	497.58	20.41	354.85	21.31
Ankle:	R_x	3.44	0.72	3.2	0.81	2.71	0.73	2.79	0.97	-25.2	1.37	3.5	0.28
	R_y	-13.58	-0.02	-19.0	-0.01	-7.96	-0.02	-10.41	-0.02	-15.17	0.05	0.43	.01
	R_z	160.19	-0.56	195.77	-0.46	123.39	-0.57	134.48	-0.35	169.59	0.51	164.04	-0.96
	M_z	-28.24	0	-39.46	0	-17.03	0	-32.93	0	-56.95	0	-9.96	0
C. of Support:													
	GX	2.61		1.61		3.61		2.61		2.61		2.61	
	GY	-0.14		-0.14		-0.14		-1.14		-0.14		-0.14	
Ground forces:													
	FX	0		0		0		0		-9.51		0	
	FY	0		0		0		0		0		4.96	
	FZ	72.3		72.3		72.3		72.3		72.3		72.3	
	TZ	0								-19.4		10.67	
Objective:													
	U	653.46		760.20		545.45		647.47		828.79		502.42	

The results for the hip, knee and ankle joint reactions, with and without the effect of anteversion are shown in Figures 11 through 12 respectively. It is evident that while for some part of the walking cycle the inclusion of anteversion increases the joint reaction, for some, the effect is to decrease the joint force.

V. EFFECT OF CANE SUPPORT AND GAIT VARIATION ON KNEE JOINT FORCES

In this study the lower extremities model was used to calculate the knee joint force in a person walking with the aid of auxillary crutches. The subject as a 51-year old male patient with left knee pain, weighing 187 lb (84.88 kg). The experimental kinesiological gait data recorded consisted of the following:

i.   Segmental orientation as viewed from the front as well as from the side at three instants in the left walking cycle--left heel strike, beginning right heel rise and single limb support with the right foot crossing the left. A typical configuration for the single limb support instant is shown in Fig. 14.

ii. The ground reaction forces and the center of pressure as shown in Figs. 15 and 16.

iii. The cane force in the left and the right canes for the entire gait cycle as shown in Fig. 17.

iv. The electromyograms from five major muscle groups (hamstrings, vastus lateralis, rectus femoris, gastrocnemius, anterior tibialis) for the left and right walking cycles. Typical results from several trials for the left walking cycle for the muscles monitored are shown in Fig. 18.

The configuration for the segments was read off the front and side view and is listed in Table VI for the three instants. Note that although the patient had a bad left knee, in the model the orientation is switched around so that the right extremity becomes the imparied side and the information for the left side is used for the right side. This is because the reference frame in the model was selected to be at the right ankle joint and so it is easier to input configuration data for the right side.

The results for the three instants of the walking cycle are shown in Table VII. In the left heel strike and the right heel rising positions, both feet are on the ground. For proper analysis one must known how the overall support force is shared by the two feet. Since this was not readily available (ground support forces and center of pressure being measured for one side only) only the extremity on which the ground support forces are known was analyzed. The rest of the body was treated as ground. Thus the analyzed segments would involve the right foot, the right leg, the right thigh while the other segments as well as the canes would be excluded. The ground support forces on the right foot and its location form the input to the model.

For the single limb support instant since one of the extremity is off the ground, the entire ground-to-foot support force acts on the other extremity. In this case therefore all the segments of the lower extremities can be analyzed, the input being the cane forces, and the known inertia forces. The output ground support reactions can then be compared to the measured value. In the first trial only the cane forces

606

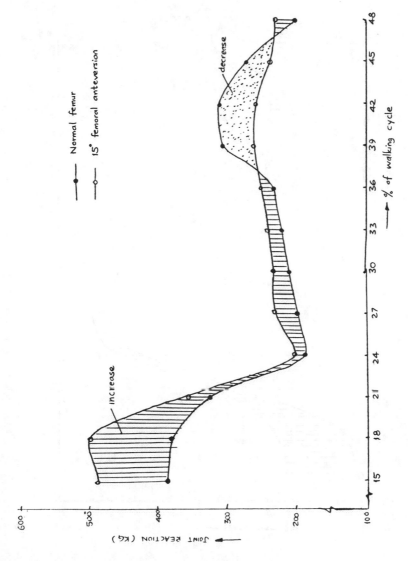

Figure D-11. Hip joint reaction (R_z) with normal and anteverted femur.

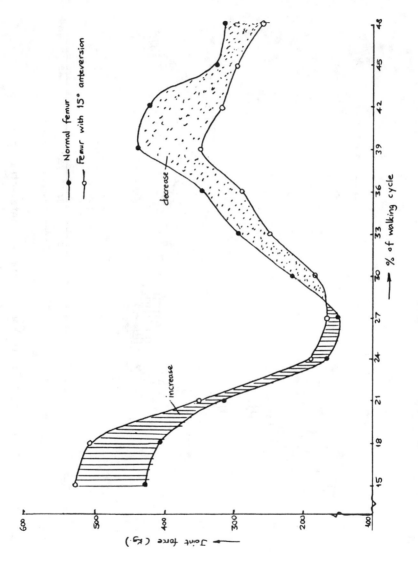

Figure D-12. Knee joint reaction (R_z) with normal and anteverted femur.

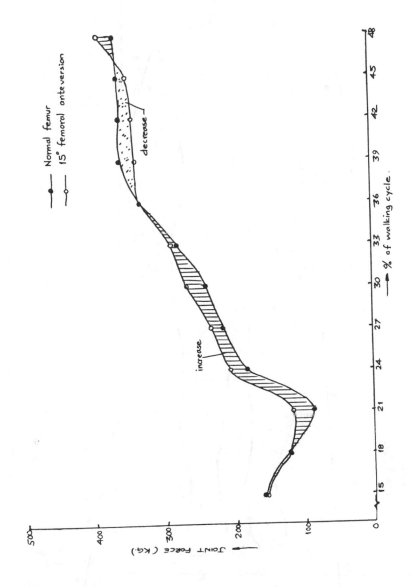

Figure D-13. Ankle joint reaction (R_z) with normal and anteverted femur.

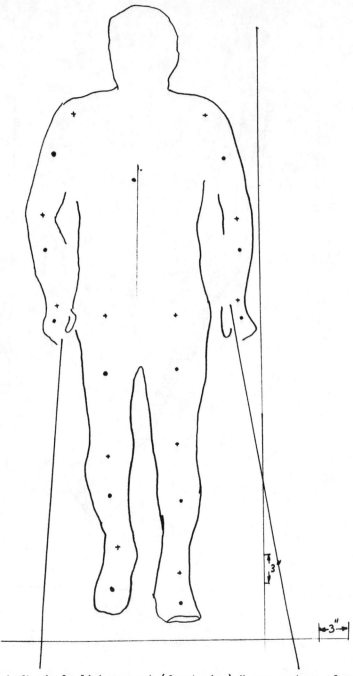

Figure D-14a. Left single limb support (front view) X are centers of masses of segments, + indicate joint locations. Thighs are parallel in sagittal view.

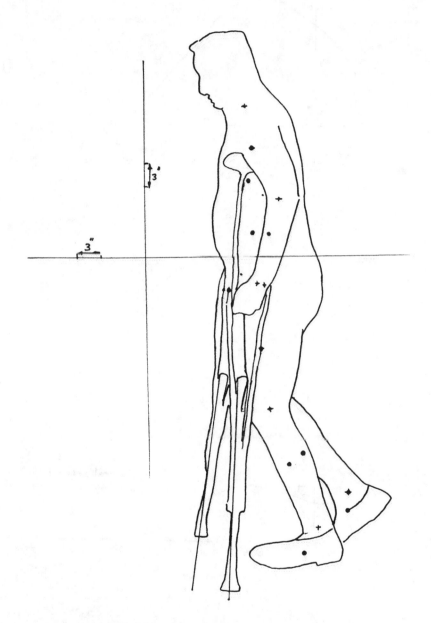

Figure D-14b. Single limb support -- sagittal view.

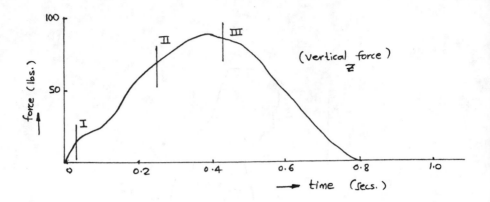

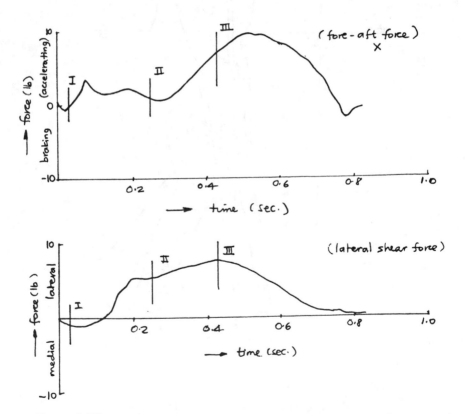

Figure D-15. Ground reaction forces for the painful side (left side)
I: left heel strike, II: begin right heel rise, III: left single limb
support.

Figure D-16. Center of pressure.

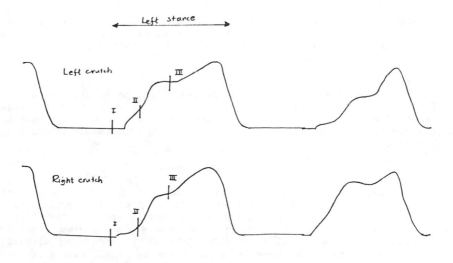

Figure D-17. Cane forces.

were inputted with no inertia forces. Note that in this case the resulting ground reaction forces are Rx = -0.665, Ry = .209, Rz = 37.86 whereas the actual measured values are Rx = 2.71, Ry = -2.9, Rz = 32.86 kg. In order to make the resulting ground reaction forces the same as the measured values one can assign inertia forces and/or torques on the segments. The spatial motion of the limbs during gait is capable of generating such forces and torques since the gait is a dynamic act. Another factor that will result in these dynamic forces is the friction between the case and the ground which effectively creates inertia forces and torques on the hand. Since the hand and the cane are treated as part of the torso, this means external forces on the torso. The single limb instant was thus analyzed in four different ways:

a. With the configuration as shown in Table VI and with the cane forces read off the measured cane force plots. The cane forces are applied at the hands, which are assumed integral with the torso or the pelvis. The cane force components are obtained by reading the cane force value from Fig. 17 and then resolving along the reference X,Y,Z axes.

b. Tilting the entire lower extremities about the right ankle joint with the new values of value (3) and (4) being 32° and 9.97° respectively; friction forces at the cane/ground transferred to the torso as an external force.

c. Keeping the configuration as in (a) above and assigning inertia forces and moments at the torso.

d. Keeping the configuration unchanged from (a) above and assigning inertia forces at the torso and at the left leg and the left foot (the swinging extremity) to result in the required ground support reactions on the right foot.

The results shown in the table were computed using a merit criterion $U = \Sigma F + 4M$ and are based on a body weight of 72.3 kg. To reflect the true weight of the patient (84.81 g) all the forces would have to be proportionately increased.

The altered gait pattern and the support from the canes results in lowered knee joint forces. This can be seen when the results for single limb support are compared to those from a normal subject at the same instant of walking cycle. This corresponds to around 39% of the walking cycle and is shown in Table VIII. The merit criterion used for obtaining these results was minimized

$$U = \Sigma F + 4M + kR_{z_{knee}} \qquad \text{where } R_{z_{knee}}$$

is the vertical force at the knee joint and indicates the necessity of minimizing the knee joint force. With a value of k = 0, the knee joint force is 445 kg which can be reduced to 327 kg with the weighting factor k greater than 3. Note that this represents the maximum possible reduction without modification of the gait or support from orthotic devices. With canes, the knee joint force can be significantly reduced from 327 kg to 113 kg.

VI. ANALYSIS OF THE UPPER EXTREMITIES DURING TYPING

A clinical investigation of the upper extremities during typing is

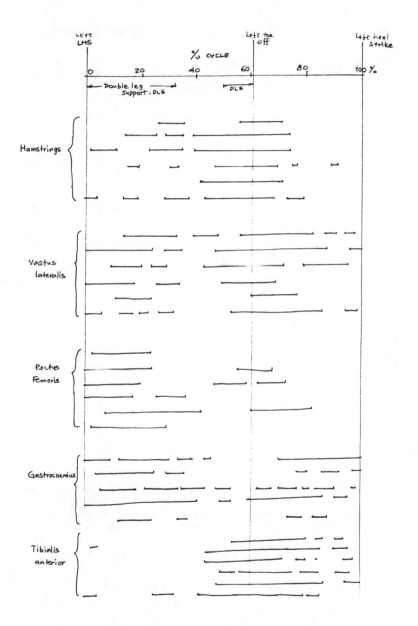

Figure D-18. EMG from major muscles for left walking cycle.

TABLE VI. Segmental configuration for the lower extremities

Value*	Left heel strike	Light heel rising	Single limb support
1	0	0	0
2	0	0	0
3	2	15	25
4	-2	-1	0
5	0	0	0
6	-19.2	-27	-25
7	25	28.5	9
8	0	0	0
9	0	0	0
10	-8	-7	-18
11	0	0	0
12	0	0	0
13	7	12	50
14	-7	-5	0
15	0	0	0
16	0	0	0

*'VALUE' array defines the rotations of the segments in the order listed in the lower extremities program (III).

TABLE VII. Results for muscle forces and joint reactions for right side.

	left heel strike	right heel rising	Single limb support a[1]	b[2]	c[3]	d[4]
Calf	2.9	27.0	14.1	46.4	25.6	0.9
Hamstrings	1.2	0	7.2	52.8	134.6	42.6
Quadriceps	0	25.5	29.1	57.4	62.7	18.2
Tibialis anterior	0	0	0	0	0	0
Hip joint:						
Rx	0.9	43.1	4.6	32.3	13.6	0.8
Ry	-1.8	-13.0	6.8	-14.3	21.2	2.2
Rz	9.7	144.3	87.7	166.1	239.9	119.6
Knee joint:						
Rx	-0.4	10.9	21.5	65.3	40.5	13.7
Ry	0.5	-1.6	-3.4	-39.9	-19.8	-6.8
Rz	9.7	144.3	87.7	166.1	239.9	112.6
Ankle joint:						
Rx	0.1	28.2	72.3	103.1	87.9	84.2
Ry	-0.1	-14.0	-10.4	-58.3	-37.5	-33.4
Rz	11.7	102.6	164.3	163.3	183.3	180.5
Ground reactions:						
Rx	0	0.27	-0.65	2.71	2.71	2.71
Ry	0.4	-2.05	2.09	-2.9	-2.9	-2.9
Rz	5.8	26.3	32.86	32.86	32.86	32.86
Cane force (right):						
Fx	0	11.0	-1.0	-0.76	-0.76	-1
Fy	0	1.7	3.3	3.84	3.84	3.3
Fz	0	7.5	19.6	19.6	19.6	19.6
Cane force (Left):						
Fx	0	-1.2	-2.2	-1.95	-1.95	-2.2
Fy	0	-0.2	-1.44	-0.94	-0.94	-1.44
Fz	0	4.3	18.1	18.1	18.1	18.1
Inertia forces at torso:						
Fx	0	0	3.85	0	0	4.15
Fy	0	0	-3.95	0	0	-4.75
Fz	0	0	1.74	1.74	0	1.74

(continued)

617

TABLE VII   (continued)

	left heel strike	right heel rising	Single limb support			
			a[1]	b[2]	c[3]	d[4]
Inertia moments at torso:						
MX	0	0	0	0	574.5	0
MY	0	0	0	0	399.2	0
MZ	0	0	0	0	0	0
U = F + 4M	18.8	408	241.9	392.8	870.9	1489
Center of support:						
X	2.5	10.8	15.2	15.2	1 .2	15.2
Y	-1.25	-2.5	-2.1	-2.1	-2.9	-2.9

[1]No inertia forces
[2]Body tilt at right ankle joint
[3]Inertia moments at torso
[4]Inertia force on left leg (x = -1.21, y = 3.45, z = 0) and left foot (x = -2.45, y = 2.34, z = 0).

TABLE VIII. Muscle forces (kg) and joint reactions (kg) at 39 percent of the cycle with modified objective function: minimize $U = \Sigma F_i + 4(\text{joint moments}) + kR_z$.

Muscle	k=0 Right leg	Left leg	k=1,2 Right leg	Left leg	k=3,4,5 and 10 Right leg	Left leg	with canes
Abductor group*	118.7	11.1	169.7	11.1	179.7	11.1	32.0
Adductor group	0	4.0	33.4	4.0	57.4	4.0	32.5
Calf group	172.5	3.7	172.5	3.7	129.8	3.7	31.7
Hamstrings	116.9	0	85.1	0	110.4	0	42.6
Quadriceps	0	2.6	1.8	2.6	23.8	2.6	18.2
Gracilis	0	0	0	0	0	0	25.4
Satorius	49.1	16.4	0	16.4	0	16.4	0
Iliopsoas	0	0	0	0	0	0	0
Gluteus maximus	0	0	0	0	0	0	0
Tibialis anterior	0	5.9	0	5.9	0	5.9	0
Tibialis posterior	0	0	0	0	0	0	34.3
Extensor digitorum longus	0	2.9	0	2.9	0	2.9	0
Extensor hallucis longus	90.5	1.5	90.5	1.5	38.9	1.5	0
Flexor digitorum longus	35.4	0	35.4	0	85.1	0	107.2
Flexor hallucis longus	0	0	0	0	0	0	0
Peroneus longus	0	0	0	0	0	0	0
Peroneus brevis	0	0	0	0	0	0	0
Peroneus tertius	0	0	0	0	63.4	0	0
Hip joint reaction: X	-22.2	12.4	-26.5	12.	-37.9	12.4	0.7
Y	-2.2	3.8	-30.1	3.8	-50.4	3.8	2.2
Z	304.6	18.4	341.7	18.	419.1	18.4	119.9
Knee joint reaction: X	-64.3	9.0	-67.9	9.0	-43.6	9.0	13.7
Y	20.7	0.8	12.8	0.8	12.5	0.8	-6.8
Z	445.3	23.8	366.4	23.8	326.8	23.8	112.6
Ankle joint reaction: X	-38.0	0.7	-38.0	0.7	-35.2	0.7	-84.1
Y	41.8	0.6	41.8	0.6	43.7	0.6	34.1
Z	359.7	12.8	359.7	12.8	377.9	12.8	180.4
Hip moment: X,Y,Z	0	0	0	0	0	0	0
Knee moment: X,Y,Z	0	0	0	0	0	0	0
Ankle moment (kg-cm), X,Y,Z	0	0	0	0	0	0	0

described here. The case in question involved a 20-year old female typist who complained of excruciating pain in the upper back in the shoulder region. This occurred when the patient inserted a new sheet of paper on the roll in an extensively used office model typewriter and pushed the knob to lock the paper on the role. The process was highly repetitious and several cycles of inserting a paper and pushing the knob were performed in a typical day's work. After several days on the work the typist complained of severe sudden pain in the upper back in the shoulder area. Following several days of recuperation when the act was resumed the pain seemed to appear again.

To investigate the case the upper extremity model as developed in Chapter IV was used. The computations are based on the computer program as listed in Appendix E. Since the reported occurrence of the pain was on the upper back region, it becomes necessary to include the muscles connecting the shoulder girdle to the spine. Accordingly, the upper extremities model was modified to include the trapezius, the serratus anterior, the rhomboids, and the levator scasulae muscles. The scapula was also included in the free body analysis along with the arm and the forearm. The location and models for these muscles is shown in Fig. 19a-e.

To represent the typing act the upper extremities configuration is described by a set of 15 angles ϕ_1, ϕ_2,...,ϕ_{15} as listed in Table IX. Once the elbow angle between the arm and the forearm is established by means of ϕ_{10}, the arm orientation with respect to the shoulder is defined by successive rotations ϕ_{13} thru ϕ_{15}. These represent respectively a rotation about the mediolateral y axis at the shoulder joint, followed by a rotation about the anterior posterior x axis, followed by a rotation about the vertical z axis. Since the exact orientation of the arm at the shoulder joint as defined by these three cannot be accurately established without adequate x-ray photographs, it was decided to select the angles ϕ_{14} and ϕ_{15} as variables. The muscle and joint forces therefore will be computed for three different arm configurations (a,b,c as listed in Table I) representing varying amounts of arm abduction at the shoulder joint. The angle $\phi_{14} = -20^0$ corresponds to minimum abduction while $\phi_{14} = -70^0$ corresponds to maximum abduction, with $\phi_{14} = -40^0$ being an intermediate value.

The external loading on the hand, while pushing the knob in, to lock the paper, is modelled by a force and a torque. The force is horizontal, directed along the y axis and the torque is about the Y axis as shown in Fig. 2. Using a force transducer the force required to push the knob in was measured to be around 4 lb. The torque was estimated and various combinations of force and torque with different magnitudes were explored using the model. The force is assumed to act on the hand at its center of mass. The merit criterion used was the same as established in Chapter V, that is, minimize the sum of all muscle forces, four times the joint moments and twice the tensile joint reactions.

The results with the different force and torque combinations are shown in Tables X through XII for the three arm abduction positions. As might be expected, the muscles in the vicinity of the shoulder joint show significant activity, namely, the pectoralis major, the deltoid, the supraspinatus, the infraspinatus, the teres major and minor, the trapezius and the serratus anterior. Since the reported pain was in the

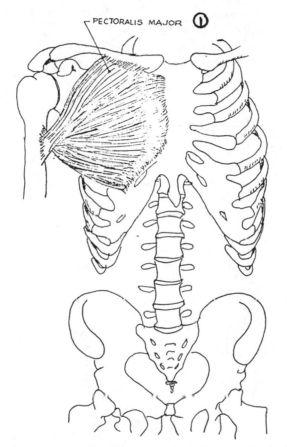

Figure D-19a.   The pectoralis major muscle.

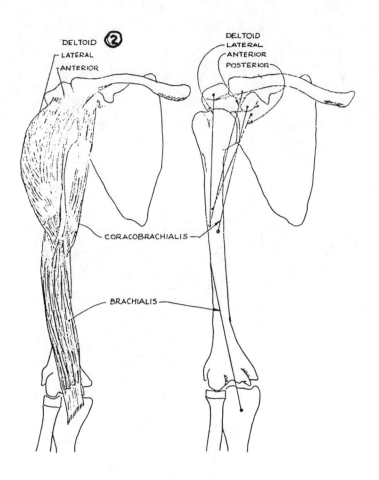

Figure D-19b. Muscle model for brachialis, coracobrachialis, and deltoid.

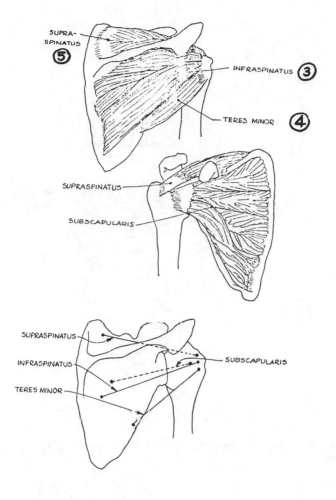

Figure D-19c. Muscle models for teres minor, subscapularis, infraspinatus and supraspinatus.

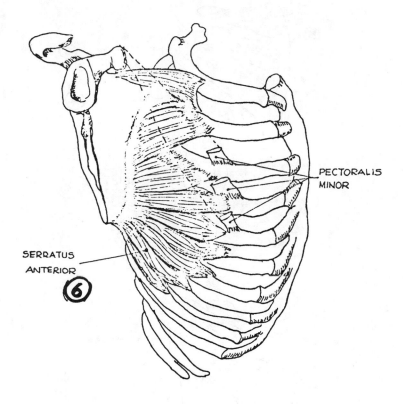

PECTORALIS
MINOR

SERRATUS
ANTERIOR
⑥

Figure D-19d. The serratus anterior and pectoralis
minor muscles.

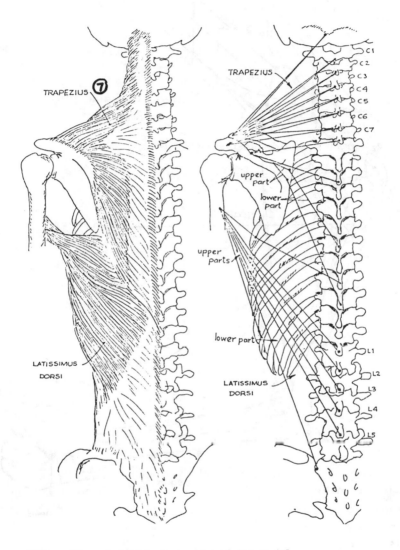

Figure 19e. Latissimus dorsi and its model;
trapezius and its model ('full').

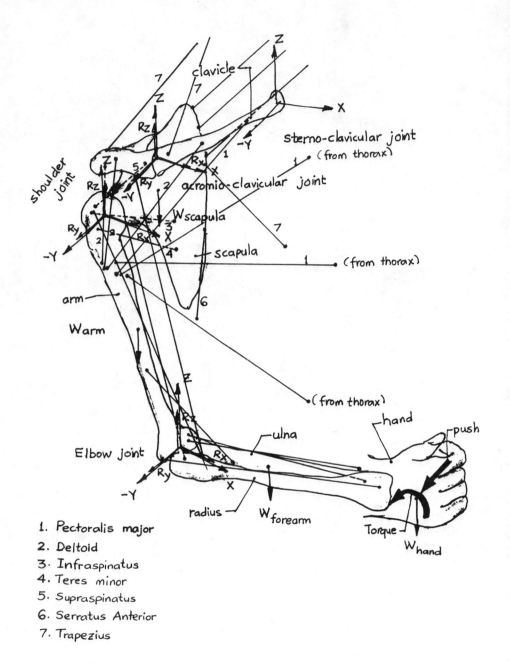

1. Pectoralis major
2. Deltoid
3. Infraspinatus
4. Teres minor
5. Supraspinatus
6. Serratus Anterior
7. Trapezius

Figure D-20. The upper extremities model for the analysis of typing.

TABLE IX. Segmental configuration for the typing act.

Angle ($°$)	Description	a (minimum abduction)	b	c (maximum abduction)
ϕ_1	Rotation of clavicle about Z axis	0	0	0
ϕ_2	Rotation of clavicle about Y axis	0	0	0
ϕ_3	Rotation of clavicle about X axis	-5	-5	-5
ϕ_4	Rotation of scapula about Z axis	0	0	0
ϕ_5	Rotation of scapula about Y axis	0	0	0
ϕ_6	Rotation of scapula about X axis	0	0	0
ϕ_7	Rotation of humerus about Y axis	-90	-90	-90
ϕ_8	Rotation of humerus about Z axis	0	0	0
ϕ_9	Rotation of ulna about Z axis	0	0	0
ϕ_{10}	Rotation of ulna about Y axis	-50	-50	-50
ϕ_{11}	Rotation of radius about Y axis	-50	-50	-50
ϕ_{12}	Rotation of hand about Y axis	-50	-50	-50
ϕ_{13}	Rotation of arm about Y axis	50	50	50
ϕ_{14}	Rotation of arm about X axis	-40	-40	-70
ϕ_{15}	Rotation of arm about Z axis	10	10	10

TABLE X. Results for muscle and joint forces with minimum abduction (ϕ_{14} = -20°).

Muscle & joint force		Force(kg) 0 Torque(kgcm) 0	0.9 -1.15	1.81 -1.15	2.27 -1.15	-0.9 -1.15	-1.81 +1.15	-2.27 1.15
Latissimus dorsi		0	0.3	11.5	16.2	0	0	0
Pectoralis major		4.1	33.6	61.9	76.0	0	0	0
Deltoid		41.9	21.7	29.1	37.9	34.4	39.2	41.2
Supraspinatus		0	0	0	0	0	0	0
Infraspinatus		9.4	20.2	38.7	48.9	14.8	25.8	31.2
Teres Major		0	0	0	0	0	0	0
Teres minor		0	0	0	0	0	0	0
Biceps		3.9	3.8	3.8	3.8	4.3	1.8	0
Triceps		0	0	0	0	0	0	0
Brachialis		5.9	8.0	9.5	10.5	1.4	0	0
Brachioradialis		0	0	0	0	0	0	0
Trapezius		0	0	0	0	2.9	8.5	8.0
Serratus Anterior		10.8	17.4	32.2	40.6	5.2	5.3	8.7
Rhomboid		0	0	0	0	0	0	1.7
Acromio-clavicular joint	x	-7.04	-6.61	-17.18	-20.35	3.35	0.83	1.68
	y	13.97	36.47	69.33	86.81	3.68	10.98	13.79
	z	-0.81	-11.04	-23.72	-34.5	-12.23	-17.65	-19.22
Shoulder joint	x	-35.98	-38.37	-65.58	-82.95	-42.05	-46.81	-49.10
	y	19.91	44 98	84.70	105.52	18.20	23.40	25.99
	z	27.75	19.07	16.99	16.75	27.95	21.17	17.91
Elbow joint	x	-9.19	-17.74	-31.86	-38.84	-14.03	-19.05	-21.73
	y	1.10	1.98	3.21	3.88	1.06	1.03	1.01
	z	4.55	5.44	5.39	5.46	0.69	-1.18	-1.64
Joint moments		0	0	0	0	0	0	0
U = ΣF + 4M		78.2	115.86	213.32	266.95	84.67	103.19	114.87

628

TABLE XI. Results for muscle and joint forces with intermediate abduction ($\phi_{14} = -40°$).

Muscle & joint force	Force(kg) Torque(kgcm)	0 -1.15	0.9 -1.15	1.81 -1.15	2.27 -1.15	-0.9 -1.15	-1.81 +1.15	-2.27 1.15
Latissimus dorsi		0	0	4.6	7.4	0	0	0
Pectoralis major ①		0	30.4	52.9	64.5	0	0	0
Deltoid ②		39.9	15.5	16.8	20.4	29.6	29.9	33.2
Supraspinatus ⑤		0	0	0	0	0	0	0
Infraspinatus ③		11.8	24.5	38.6	49.7	18.9	3.6	38.6
Teres Major		0	0	0	0	0	0	0
Teres Minor ④		0	0	0	0	0	0	0
Biceps		0.7	4.7	6.0	6.6	1.0	0	0
Triceps		0	0	0	0	0	0.9	4.0
Brachialis		3.3	8.8	12.1	13.9	0	0	0
Brachioradialis		6.7	0.0	0	0	0	0	0
Trapezius ⑦		0	0	0	0	2.8	8.8	0.1
Serratus Anterior ⑥		10.7	15.9	26.1	32.5	5.8	5.2	7.9
Rhomboid		0	0	0	0	0	0	1.4
Acromio-clavicular joint	x	-10.49	-5.76	-12.34	-14.3	3.65	1.01	1.46
	y	14.77	32.75	58.34	72.48	3.70	11.09	14.09
	z	0.48	-11.20	-20.64	-29.54	-12.53	-17.68	-19.29
Shoulder joint	x	-34.82	-37.77	-50.95	-63.83	-42.95	-46.12	-53.14
	y	22.21	45.58	75.16	94.00	23.99	28.34	32.89
	z	16.80	8.67	4.51	-0.55	15.90	4.40	4.20
Elbow joint	x	-10.58	-13.69	-28.57	-35.94	-12.22	-19.75	-25.08
	y	0.32	2.39	2.44	2.54	-1.00	-2.85	-2.04
	z	0.71	5.70	5.69	7.26	-1.43	-0.36	1.35
Joint moments		0	0	0	0	0	0	0
U = ΣF + 4M			73.91	105.41	178.02	222.26	82.20	100.64	117.30

629

TABLE XII. Results for muscle and joint forces with maximum abduction (ϕ_{14} = -70°).

Muscle & joint force	Force(kg) 0 / Torque(kgcm) 0	0.9 / -1.15	1.81 / -1.15	2.27 / -1.15	-0.9 / -1.15	-1.81 / +1.15	-2.27 / 1.15
Latissimus dorsi	0	0	0	0	0	0	0
Pectoralis major ①	2.1	30.3	49.4	59.5	0	0	0
Deltoid ②	32.5	20.3	27.3	31.5	17.2	36.9	46.4
Supraspinatus ⑤	0	0	0	0	0	0	0
Infraspinatus ③	0	0	0	0	34.6	0	0
Teres Major	0	0	4.4	3.2	0	0	0
Teres Minor ④	17.5	20.2	32.4	43.9	0	24.9	24.3
Biceps	0	4.1	7.9	9.3	0	0	0
Triceps	0	0	0	0	3.8	13.6	18.7
Brachialis	0	6.4	13.1	15.8	0	0	0
Brachioradialis	4.1	2.6	0	0	0	0	0
Trapezius ⑦	0	0	0	0	3.5	9.1	10.1
Serratus Anterior ⑥	10.3	15.9	23.1	28.3	7.5	5.3	6.71
Acromio-clavicular joint x	-8.63	-5.62	-8.70	-7.71	-0.61	1.12	1.18
y	16.72	33.17	54.30	66.19	9.28	11.2	14.24
z	-6.49	-13.57	-21.89	-31.67	-12.31	-17.68	-19.08
Shoulder joint x	-35.78	-37.43	-56.27	-65.51	-37.07	-50.12	-65.62
y	25.49	41.55	67.63	82.41	28.04	27.04	37.86
z	-2.44	1.31	-3.89	-13.57	-13.64	-5.51	1.18
Elbow joint x	-9.6	-12.97	-17.48	-23.89	-14.31	-23.4	-28.77
y	-1.46	3.36	8.26	9.45	0.74	7.56	10.41
z	-2.08	1.48	5.08	5.74	0.09	2.45	3.34
Joint moments	0	0	0	0	0	0	0
U = ΣF + 4M		72.2	105.8	166.2	204.67	78.8	104.82	123.74

upper back region, the muscles of particular concern are the supraspinatus, the infraspinatus, the teres major and minor and the serratus. For low angles of arm abduction (ϕ_{14} = -20⁰ and -40⁰) it appears that the infraspinatus is required to bear high forces. For the maximum abduction angle investigated (ϕ_{14} = -70⁰), on the other hand, the teres minor carries a significant force. The direction of the force applied on the hand, whether push or pull, does not seem to affect the muscle load sharing significantly.

It is possible that the extent of muscle activity in these shoulder girdle muscles required for overall equilibrium is greater than what the muscle can bear, resulting in pain. However, the configuration is important for any biomechanical analysis and as is evident from this study, the selected configuration must be accurately established to determine the exact muscle load sharing.

B. EXAMPLES UTILIZING THE INTERACTIVE PROGRAM

I. EFFECT OF GAIT PATTERN AND CANE SUPPORT ON HIP JOINT FORCE

The lower extremities model for walking was used in conjunction with the interactive computing program with graphics capabilities for the evaluation of joint forces for patients with locomotion disabilities. The relative orientation of the segments can be input from kinesiological data and the desired configuration can be manipulated interactively through the graphics capability of the program. The use of the program will be illustrated for the specific case of locomotion of a subject with hip pain where the voluntary adjustment to gait pattern and concomittant reduction in affected joint force pattern can be diagnosed conveniently.

The walking model used is the same as that used for the investigation of the normal gait. However, the analysis program is augmented by the interactive graphics program where the segmental orientations as obtained from experimental evaluation can be conveniently input and the desired posture can be readily manipulated. For the graphics display, the lower extremities are modelled by simple geometric solids articulated at the respective joints and positioned in space to attain the desired configuration.

The kinesiological data was obtained from a 61 year old male patient with degenerative hip joint disease on the left side, walking without a cane and walking with a cane in either of the hands respectively. The ground support reaction forces, the cane forces and the center of support location were also measured experimentally. The electromyographic activity from six major muscle groups on the affected extremity was also recorded. The monitored muscles were the hip abductors and adductors, the gluteus maximus, the hamstrings, the quadriceps and the calf.

The kinesiological data for segmental orientation obtained from front and side photographs (Fig. 21a) was used to generate a computer diagram of the body. A typical computer generated schematic view at the instant of single limb support is shown in Fig. 21b. Note that for convenience of analysis the right leg is treated as the affected side with all the information from the left side transferred to the right side. The right hip joint force therefore is one to be minimized in the model. The twelve angles $\phi_1, \phi_2, \ldots, \phi_{12}$ required to define the kinematic geometry as well as the cane forces and the ground support

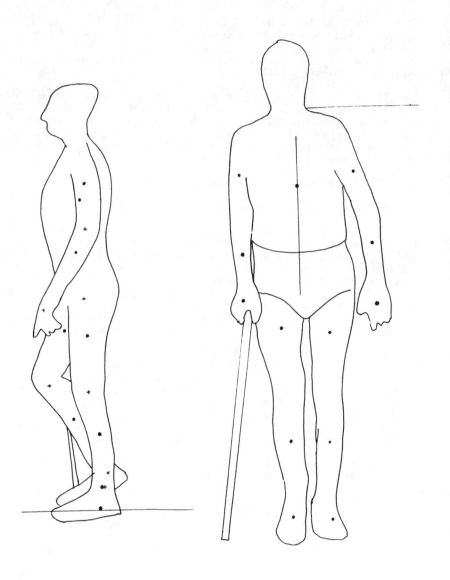

Figure D-21a. Side and front view at single limb support (right foot passing left foot). (o mass centers; + joint centers).

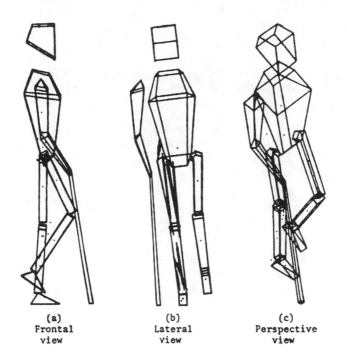

(a)	**(b)**	**(c)**
Frontal	Lateral	Perspective
view	view	view

Figure D-21b. Graphic display of the lower extremities at an instant in the walking cycle.

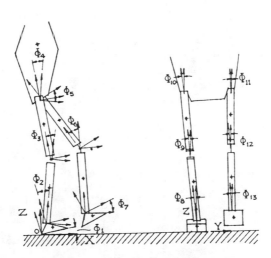

Figure D-22. Segmental orientations.

TABLE XIII. Segmental orientations, cane forces and ground reactions (left foot on the ground, right leg passing the left foot).

	No Cane	Cane in Right Hand	Cane in Left Hand
ϕ_1	0	0	0
ϕ_2	8	8	7
ϕ_3	-9	-7.5	-7
ϕ_4	15	15	15
ϕ_5	-39	-37	-40
ϕ_6	54	53	53.5
ϕ_7	3	5	3
ϕ_8	2	2	2.5
ϕ_9	0	0	0
ϕ_{10}	-2	0	-2
ϕ_{11}	0	0	0
ϕ_{12}	0	0	0
ϕ_{13}	0	0	0
Ground forces:			
F_x	1.28	1.28	1.49
F_y	-1.07	-1.71	-1.28
F_z	5.76	10.45	10.45
Cane force:			
F_x	0	1.28	0
F_y	0	-2.74	0
F_z	0	19.76	0

forces are illustrated in Fig. 22 and listed in Table XIII. The cane force directed along the cane is assumed to act on the upper torso or the hand carrying the cane.

THE COMPUTING PROCEDURE:

The process utilized for the calculation of muscle and joint forces is described in detail in the interactive program outline (see Appendix F). The process begins by defining the body segments and corresponding muscles with all their geometric information. The graphics capability can then be used to check the adequacy of the constructed model. Any required modifications can be easily undertaken in order to achieve the acceptable description of the configuration and muscle attachment points. The equations of equilibrium of the different body segments are automatically defined in terms of the unknown muscle forces and joint reactions and moments. The system of equations is then solved using linear programming to obtain the magnitudes of the muscle forces and joint reactions. The system removes the burden of writing the equations of equilibrium which can be a formidable task especially when several iterative geometrical changes are made. The computing procedure is outlined in Fig. 23.

RESULTS

The muscle forces in the six muscle groups and the hip joint vertical force on the affected side are tabulated for the three cases of walking without cane and walking with a cane in either hand (Table XIV). These are reported for a model body weight of 72.3 kg and should be proportionally changed to reflect the weight of the real subject. The predicted muscle force patterns agree well with the observed experimental data. The hip joint force is seen to be minimum when the cane is used in the hand on the opposite side from the affected hip, and is greater when no cane is used or when the cane is used in the ipsilateral side (same side as the abnormality). In the latter instant the patient gait appeared to be erratic and awkward, and the use of the cane was seen to be ineffective. In all of these calculations the merit criterion used was the same as that for normal walking, that is, minimize all muscle forces and four times the joint ligament forces. The results for the muscle forces and joint reaction at the same instant in a normal walking pattern are shown in Table XIV for comparison. Clearly, the normal hip joint force is considerably higher than that on the pathological hip and can be reduced only by altering the gait and through support from assistive devices.

II. USE OF INTERACTIVE MUSCULOSKELETAL MODEL IN SIMULATED ORTHOPEDIC SURGERY FOR TIBIAL OSTEOTOMY

The use of the interactive musculoskeletal computer model for diagnostic purposes was demonstrated above for the case of locomotion disability. The model can be equally valuable as a surgical tool also. The musculoskeletal model in conjunction with an interactive graphics display and analysis program can be used in simulating orthopedic surgical operations prior to their execution.

The tibial osteotomy of the right side for correcting valgus-varus deformities is selected for illustration. The system can serve as a practical tool for establishing beforehand the optimal biomechanical parameters for the surgery, such as the location of the wedge and the

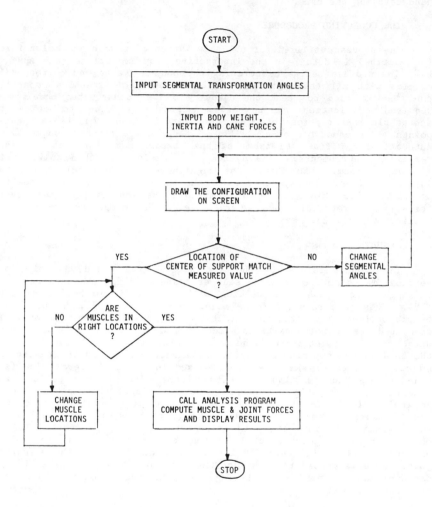

Figure D-23. The computing procedure.

TABLE XIV. Muscle and joint forces.

	Cane in Left Hand		Cane in Right Hand		No Cane		Normal	
	Left side	Right side	Left side	Right side	Left Side	Right side	Left side	Right side
Calf	48.6	0	58.5	0	54.2	0	92	0
Quadriceps	9.5	0	8.8	0	10.4	0	0	0
Hamstrings	8	5.1	6.9	3.5	9.4	4.6	42	36
Hip adductors	0	0	0	0	0	0	0	0
Hip abductors	33.1	4.8	11.9	2.3	17.5	2.0	97	13
Gluteus maximus	6.7	0	8.5	0	11.6	0	0	0
Hip joint: R_x	2.74	-2.41	-0.13	-0.55	-2.08	-2.88	5.7	-7.5
R_y	6.71	-0.39	1.25	-0.44	3.45	-0.33	-4.2	9.4
R_z	105.11	3.86	85.81	-0.33	106.91	5.01	241.6	35.5
Knee joint: R_x	13.4	-3.07	15.79	-0.88	14.89	-3.48	-32.7	-12.9
R_y	-4.06	-0.30	-4.58	-0.81	-4.23	-0.13	55.9	2.7
R_z	121.77	14.38	129.43	12.63	153.95	14.74	386.4	37.2
Ankle Joint: R_x	14.68	0.59	20.08	0.69	22.54	0.59	-28.2	-0.72
R_y	-8.82	0.06	-7.14	0.02	-8.53	0.07	57.2	0.02
R_z	155.81	0.15	143.18	0.17	153.89	0.11	469.9	0.6

637

angle of the cutting plane. The model can then be used for post—surgical quantitative assessment of the effects of alternate surgical procedures on the motion patterns and skeletal forces.

TIBIAL OSTEOTOMY:

Tibial osteotomy is a surgical procedure which is used to correct excessive valgus or varus deformities at the knee. Such deformities can be either congenital or acquired due to cartilage wear or osteoarthritis at the knee joint. The latter condition is prevalent among older people due to the mechanical wear occurring in the joint over a long period of time. Due to the valgus (knock—kneed) or varus (bowlegged) orientation, there is an imbalance in pressure distribution over the joint surface plateaus resulting in uneven wear. Typical valgus, varus and normal knee joints at one instant of the walking cycle are shown in Fig. 24. The lines of gravity for the same three knees are shown in Figs. 25 a—c. The surgical treatment essentially consists of removal of an appropriate wedge from the proximal end of the tibia just below the joint surface and then realigning and attaching the two tibial segments. In some cases the wedge can be removed from the distal end of the femur just above the joint surface in which case the procedure is known as a femoral osteotomy. A consequence of the operation is to favorably redistribute the condyular pressure. Another application of the surgery is to relocate the center of pressure and reduce excessive unilateral wear.

Geometric modelling for interactive analysis:

The lower extremities model used for the analysis of muscle forces and joint reactions during normal walking is utilized here. The model anthropometric dimensions however can be altered if need be, to reflect the actual dimensions of the surgical patient. The geometric modelling and graphical display for interactive analysis using the interactive program (see Appendix F) can be performed in the following stepwise manner:

Step 1.   Scaling and modification:

The pre—programmed lower extremity model is displayed interactively on an interative graphics terminal. This allows the physician to change the dimensions of the basic model to conform to the patient's anthropometric data.

Step 2.   Input gait pattern:

If any specific data in the pattern of walking is known it is inputted into the computer. Otherwise, a sorted typical normal walking pattern is used for description of the entire motion or for filling in any incomplete kinesiological data.

Step 3.   Display modified skeletal configuration:

The new three dimensional skeletal representation at different instants of the gait cycle are displayed. Typical displays in two and three dimensions might appear as shown in Fig. 26 a—c. This is from an instant in the normal walking cycle.

Step 4.   Modification of gait pattern:

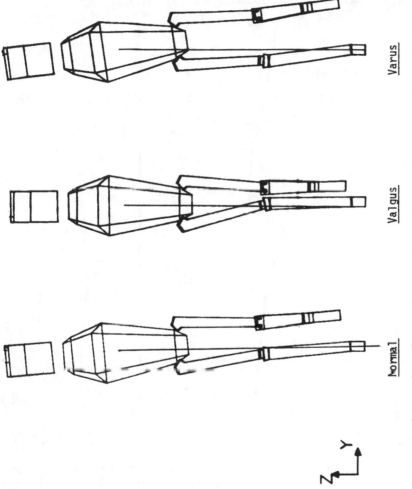

Figure D-24. Normal, valgus and varus knees showing the center of gravity lines at an instant in the walking cycle.

639

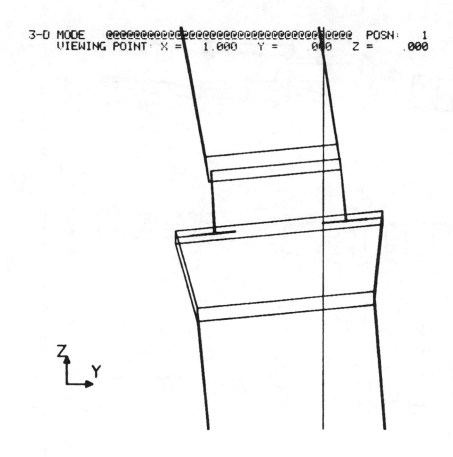

Figure D-25a.

Line of gravity in relation to the knee joint in normal position.

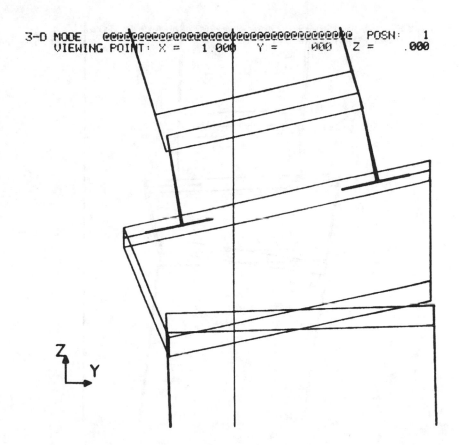

Figure D-25b.

Line of gravity in relation to the knee joint in valgus position.

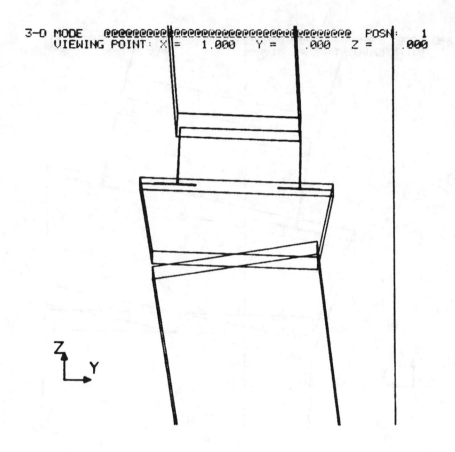

Figure D-25c. Line of gravity in relation to the knee joint in varus position.

The configuration or the walking pattern can be modified if necessary to closely simulate the patient's structure and walking pattern.

Step 5.   Modification of muscle attachment points:

The lines representing important muscle forces can be displayed for the selected posture. Muscle attachment points can be relocated to different sites, if necessary, to conform with the physician's judgment about the patient's musculoskeletal structure and anatomy. A typical display showing muscle lines appears in Fig. 27.

Step 6.   Display line of gravity:

The line of gravity can be established by computing the resultant center of mass and displayed on the simulated configuration. A hard copy of the configuration with the line of gravity can be saved for reference (See Figs. 24, 26).

Step 7.   Save pre-surgical data:

Once the musculoskeletal system and the pattern of motion has been defined satisfactorily, the patient's pre-surgical data are stored in the computer.

Step 8.   Display line of gravity in relation to the affected knee:

The knee in question can then be enlarged for viewing convenience and to quantitatively illustrate the location of the condylar contact zones in relation to the line of gravity. A hard copy of the enlarged knee is kept for reference (See Fig. 25).

Step 9.   Compute pre-surgical muscle and joint forces:

Once the configuration is established the program is executed to yield the detailed muscle force data and the joint reaction forces at the various joints. It should be noted here that the medial and the lateral condylar pressures at the affected knee are calculated separately.

Step 10.   Selection of trial wedge:

The enlarged knee model (from Step 8) is then recalled for display to allow the surgeon to select a trial tibial wedge. The detailed procedure is as outlined below:

i.   The location of the wedge along the longitudinal axis of the tibia is defined. This corresponds to the dimension OA in Fig. 28. A horizontal plane A'A' passing through the point A is displayed.

ii.   The part which represents the wedge is defined in terms of the mean thickness 't' together with the angle α which the wedge plane A'A" makes with the plane A'A' (Fig. 28).

iii. The computer would then move point A to point B (representing removal of the wedge from the tibia) by tilting the knee joint and the rest of the body above the wedge through an angle α.

iv.   The remainder of the body (pelvis, torso and left extremity) tilted

643

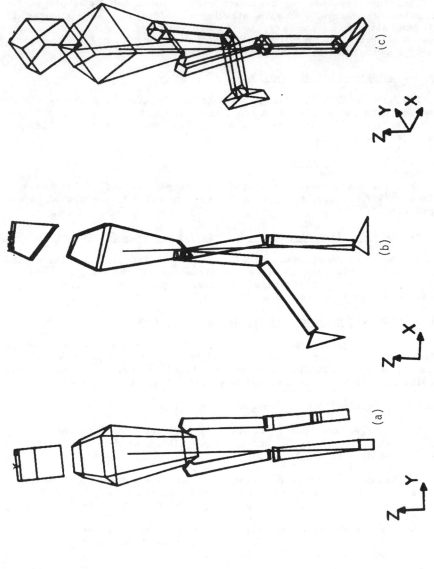

Figure D-26. Typical display of skeletal configuration. Normal walking cycle (a) front view, (b) side view, (c) perspective view.

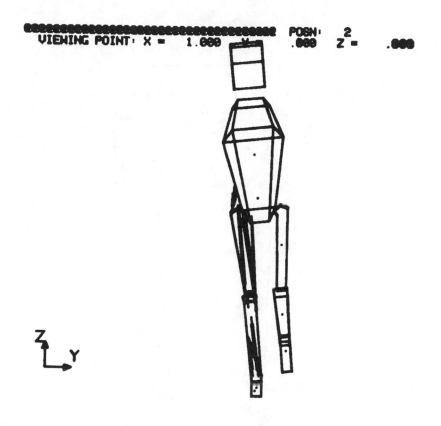

Figure D-27. Muscle force lines for the right extremity (normal knee, front view).

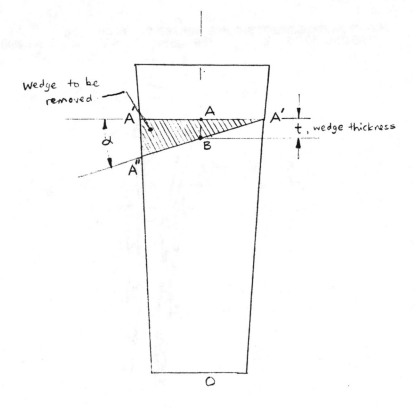

Figure D-28. Definition of wedge location, wedge thickness and wedge angle.

at the right hip joint in the coronal angle by the same angle as the wedge angle but in the opposite direction (Fig. 29 a & b). This will result in a decrease in the right hip abduction-adduction angle from the original position. Note that this automatic adjustment in configuration is done if the left extremity is off the ground, that is, the position represents single-legged support.

v.   If the line does not fall within the foot boundaries, that is, the resultant center of pressure location is outside the foot, the abduction-adduction angle at the right ankle joint can be altered. This is done by tilting the entire body at the right ankle joint so that the line of gravity is shifted to a desired or acceptable position relative to the foot. For double-legged support phase, any change at the knee joint due to removal of the wedge and consequent shortening of the tibia must be accommodated by flexion-extension in the contralateral knee joint in the sagittal plane.

vi. A 2- or 3-dimensional display of the posture can be automatically displayed on command to evaluate the posture after surgery at this instant of the gait cycle. If desired, an enlarged view of the knee can be drawn to observe the line of gravity in relation to the two condyles and to obtain a rough estimate of the pressure distribution (Fig. 30 a-c).

viii. The program for calculating the muscle and joint forces is then executed and a hard copy of the posture and the post surgical results is made for the record

ix. By inputting the wedge angle, the configuration at different intervals of the gait cycle can be automatically displayed for postural evaluation. It can be corrected if necessary to insure that the line of gravity falls in the expected region of support. The program is subsequently executed and hard copies of the post-surgical analytical results are kept for the record.

x.   If the surgeon is not satisfied with the results of the estimated wedge, he may try other wedge geometries (wedge thickness, location of the wedge, wedge angle, etc.) and repeat the procedure (i through ix) until an optimal selection is made.

Step 11.   Post operative gait evaluation of the patient:

In due course of time, a post operative gait evaluation of the patient can be conducted to compare the actual dynamic gait parameters to those predicted quasistatically by the computer analysis. Such comparisons can be helpful in future surgical decisions when optimizing the parameters for tibial osteotomy.

SUMMARY

The above examples serve to illustrate the importance of the musculoskeletal model as a diagnostic and surgical tool. The programs can be used both in the batch mode as well as the interactive mode. The batch mode programs are meant for the analysis of a certain section of the total human body (such as the lower extremities, the upper extremities, etc.) and often for analysis of specific acts or postures. The interactive program on the other hand is much more general in nature and can be applied to any section of the body once the appropriate

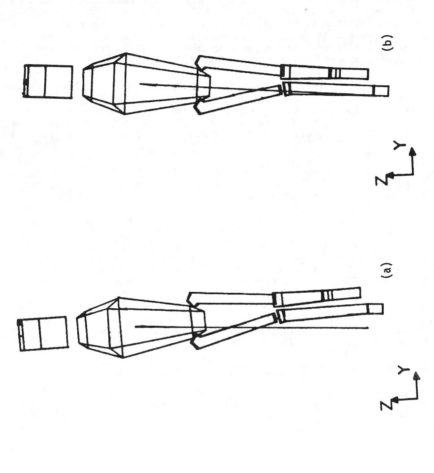

Figure D-29. Post-surgical adjustment in posture to contain the line of gravity within the foot support area: (a) removal of wedge results in unstable support; (b) reduction in hip and ankle adduction angle results in stable support.

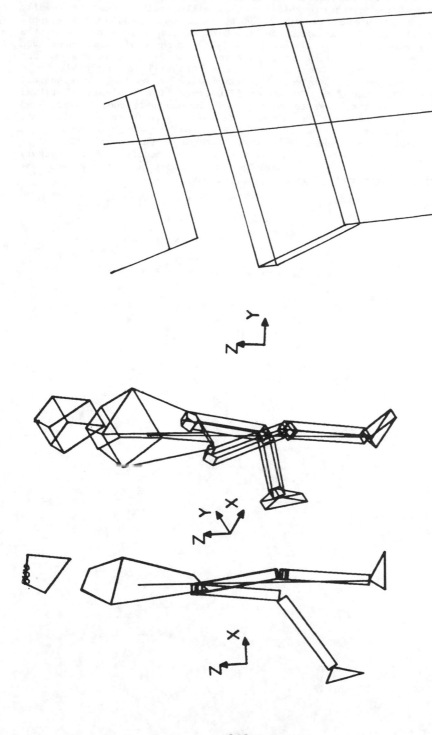

Figure D-30. Post-surgical walking posture: (a)-(b): side and perspective views; (c): line of gravity in relation to the knee.

649

segments, muscles and joints are defined. Its importance however has in its interactive computing ability which will permit a user to readily change the posture or configuration to be analyzed and in its graphic display ability for drawing the model on a graphics screen. In the graphics mode again one can interactively alter the configuration or even change the model topology to a certain extent as well as analyze the muscle load sharing. One instance where this will be invaluable is in pre- and post-surgical analysis for determining the optimal biomechanical parameters for surgery. This was illustrated for the case of tibial osteotomy. The total system would be an indispensable tool for pre- and post-surgical quantitative assessments of the effects of alternate surgical procedures on the motion patterns and the skeletal forces. It should result in considerable savings in terms of physicians' time since the surgical parameters as well as the projected benefits of the surgery would be immediately clear without having to resort to expensive and time consuming experimentation on a living patient.

Computer Program Listings
for the Musculoskeletal Models

PROGRAM FOR THE JAW

1. CAPABILITY

This program is for the analysis of the jaw during mastication. Six muscles on each side of the jaw are included. The jaw articulates with the skull through the temperomandibular (TMP) joints.

2. LIMITATION

The jaw motion consists of translations SX, SZ at the TMP joint (along x and z direction, x being anterior-posterios, z being up or down) and a rotation PHI about the y axis (medio-lateral). Thus, jaw movements are limited to be symmetric about the sagittal plane. The jaw is assumed massless.

The merit criterion is to minimize the sum of all muscle forces (F), two times the tensile components of the joint reaction (R) and four times the moments at the joint, $U = \Sigma F + 2R + 4M$.

3. INPUT

i. The translations of the jaw at the TMP joint ('SX', 'SZ', cm) and rotation about the Y-axis ('PHI', degrees).

ii. The bite force location on the mandible (coordinates, DX, DY, DZ) with respect to the TMP joint and the bite force magnitude ('BITEF', kg) and orientations ('PHIX', PHIY' degrees) measured with respect to the vertical.

4. OUTPUT

i. All the values inputted in (3) above.

ii. Output from call to the 'SIMPDX' routine with minimum objective value, etc.

iii. Values of all muscle forces and joint reactions and moments. These are printed in abbreviated form for each side of the jaw as follows:

No.	Abbreviation	Description
1	ANTMPR	Anterior Temporalis
2	POTMPR	Posterior Temporalis
3	MASST1	Masseter (superficial fibers)
4	MASST2	Masseter (deep fibers)
5	MDPTGD	Medial Pterygoid
6	LTPTGD	Lateral Pterygoid
7	TMPJRX	TMP joint reaction force along x
8	TMPJRY	TMP joint reaction force along y
9	TMPJRZ	TMP joint reaction force along z
10	TMPJMX	TMP joint moment about x
11	TMPJMY	TMP joint moment about y
12	TMPJMZ	TMP joint moment about z

5. ILLUSTRATION

A computer output for the solution to biting with open jaw (jaw opening = 17 mm) with a bite force of 6.9 kg is shown on page 7. The input values for this case are:

Translation of the jaw: SX = 0.8 cm (forward)

$$SZ = -0.7 \text{ cm (down}$$

Rotation of the jaw: PHI = 14.95 degrees

Coordinates of point of action of bite force: DX = 6.82 cm
DY = 0
DZ = -4.27 cm

Orientation of bite force: PHIY = 0 degrees
PHIY = 10. degrees

6. PROGRAM STRUCTURE

The 'MAIN' PROGRAM is used in conjunction with the subroutine 'PACT' which prints muscle forces and joint reactions.

7. LANGUAGE

FORTRAN V

8. MACHINE USED

UNIVAC 1110

9. ESTIMATED RUNNING TIME

4 secs

10. USING THE PROGRAM

In its present set up the program is meant to analyze biting with open jaw (specified by the translation 'S'X, "SZ' and rotation 'PHI' about y axis) with he bite force ('BITEF') applied at a desired location (DX, DY, DZ on the mandible) at a preselected orientation ('PHIX', 'PHIY'). The merit criterion is minimize $U = \Sigma F + 4 M$, ΣF being the sum of all muscle forces and ΣM being the sum of the moments.

i.   Changing angles, bit bone, etc.

Results for different jaw openings or different locations, magnitudes or orientations of the bite force can be readily obtained by inputting the new values. A typical input data, punched in free format might appear as shown below. This data was used to obtain the results for the illustration described above:

Card 2:   (DX,DY,DZ,PHIX,PHIY,BITEF,

PHIX,PHIY,
BITEF)   6.82,   0.0,   -4.27,   010,   10.09,   6.9

CARD 1:   (SX,SX, PHI)

0.8,   -0.7,   14.95

ii.   Changing the objective function

The statements that set up the objective function are numbered 208 through 213 in the 'MAIN' program listing. The thirty variables in the jaw program are numbered as follows:

No.	Description
1-6	right side muscle (see muscle list)
7-12	left side muscle (see muscle list)
13-15	positive X,Y,Z components of reaction at right TM joint
16-18	negative X,Y,Z components of reaction at right TM joint
19-21	positive X,Y,Z components of reaction at left TM joint
22-24	negative X,Y,Z components of reaction at left TM joint
25-27	positive X,Y,Z moment components
28-30	negative X,Y,Z moment components

The cost function is thus specified by the desired coefficient for each of the above 30 variables. Thus if only moments are to be minimized, with a weighting factor of 1,

COST (25), COST (20),..., COST (6) = 1. , etc.

iii. Changing muscle attachment points

The muscle attachment points for the six right side muscles are given by X1, Y1, Z1 (on the skull) and X2, Y2, Z2 (on the mandible). Thus X1(3), Y1(2), Z1(3) (origin) and X2(3),

653

Y2(3), Z1(3) insertion) represent the X,Y,Z coordinates of the masseter (superficial fibers) muscle on the skull and on the mandible respectively. If the coordinate attachment point is to be changed then the old values should be replaced by the new ones, immediately after line 77 in the 'MAIN' program. The old values as input in the program are:

$$X1(3) = 3.5$$
$$Y1(3) = -5.0$$
$$Z1(3) = -0.5$$
$$X2(3) = 1.0$$
$$Y2(3) = -4.5$$
$$Z2(3) = -4.7$$

For example, if the origin on the skull for the masseter muscle is to be changed to a new location whose coordinates are 5.0, -6.0, -1.0 (X,Y,Z), then, after line no. 77, the new data should be inserted as:

$$X1(3) = 5.0$$
$$Y1(3) = -6.0$$
$$Z1(3) = -1.0$$

Note however, that due to assumed symmetry about the sagittal plane, the coordinates of the masseter muscle on the left side of the jaw will also be changed to the new values.

iv. If desired the dimensions of the segment as well as weights can be scaled or proportioned to match the actual subject being analyzed. If the jaw and the skull dimensions are to be scaled by a factor F_1, then this can be done by adding the following lines of code after line no. 77:

$$DO\ XXX\ I + 1,6$$

$$X1(I) = X1(I)\ x\ F1$$
$$Y1(I) = Y1(I)\ x\ F1$$
$$Z1(I) = Z1(I)\ x\ F1$$
$$X2(I) = X2(I)\ x\ F1$$
$$Y2(I) = Y2(I)\ x\ F1$$
$$Z2(I) = Z2(I)\ x\ F1$$

XXX CONTINUE.

Alternately, the lines 67 through 72 where the initial data is first specified can be replaced by the new scaled data.

PROGRAM FOR THE UPPER EXTREMITIES

1. CAPABILITY

This program is capable of analyzing the upper extremities (the clavicle, the scapula, the arm, the forearm and the hand which is assumed as one rigid body) connected together by 34 muscles. Although the scapula, the clavicle and the thorax have been included in the model, they are treated as ground and are used only to establish the locations

```
**********                    **********
*****   ANALYSIS CF JAW DURING MASTICATION   *****
***************************************

POTATION OF JAW ABOUT Y AXIS =  14.95  DEGREES
TRANSLATION OF JAW ALONG X =    .80  CM.
TRANSLATION OF JAW ALONG Z =   -.70  CM.

OCATION OF POINT OF ACTION OF BITE FORCE...DX=  6.82     DY=    .00        DZ= -4.27
RIENTATION OF BITE FORCE WITH VERTICAL....PHIX=   .00PHIY= 10.00
ITE FORCE =   6.90 KG.


     F I N A L    O U T P U T    R E P O R T


     SOLUTION IS OPTIMAL
     MINIMUM OBJECTIVE          .3915159849346599+002
     SUM OF INFEASIBILITIES       .00
     SUM OF   /AZ=RHS/           .14-15
     SUM OF /DJ/ IN BASIS        .35-16
     MAXIMUM DJ IN BASIS         .69-17
     PIVOT   TOLERANCE           .10-06
     REDUCED COST TOLERANCE     -.10-05
     FEASIBILITY TOLERANCE       .10-05
     SOLUTION TIME  SECS         .10
     A- MATRIX  DENSITY          .6111
     TOTAL   ITERATIONS        11
     TOTAL   INVERSIONS         2
```

```
                        SOLUTION VECTOR
   INDEX     JX              X                          PI
     1       18      .5098577126693343+001     .6489330895610061+000
     2        3      .4261064404976416+001     .0000000000000000+000
     3        1      .5117580588369893+001     .2947688654145568+001
     4        9      .4261064404976416+001    -.2168404344971009-018
     5       24      .5098577126693343+001     .1184610822094979+001
     6        7      .5117580588369893+001     .4553649124439119-017
```

```
                    RIGHT SIDE      LEFT SIDE

      1 ANTMPR=      5.12            5.12
      2 POTMPR=       .00             .00
      3 MASST1=      4.26            4.26
      4 MASST2=       .00             .00
      5 MDPTGD=       .00             .00
      6 LTPTGD=       .00             .00
      7 TMPJRX=       .00             .00
      8 TMPJRY=       .00             .00
      9 TMPJRZ=     -5.10           -5.10
     10 TMPJMX=                      .00
     11 TMPJMY=                      .00
     12 TMPJMZ=                      .00
```

655

of the arm and forearm muscles that originate on these segments. The overall configuration is established by defining the motions of all the segments, while the equilibrium analysis is done only for the arm and the forearm (with the hand) which are connected at the elbow and the shoulder joints.

2. LIMITATION

Only the arm (humerus) and the forearm (ulna-radius) are analyzed and the remaining upper extremities segments (scapula, clavicle, thorax) are assumed ground. The arm and the forearm (with the hand) weigh 2.2 kg and 1.8 kg respectively and any external force is assumed to be applied at the wrist joint along a specified orientation.

3. INPUT

i.   The 13 angles (in degrees) used to define the overall configuration. The angles are input to the 'COORD' routine and used to compute the coordinates of all points of interest with respect to the ground frame (at the sternoclavicular joint). The data is input in free format and represents the motions of the segments with respect to ground frame as follows:

PHI	Description
1	Rotation of clavicle about Z axis
2	Rotation of clavical about Y axis
3	Rotation of clavicle about X axis
4	Rotation of scapula about Z axis
5	Rotation of scapula about Y axis
6	Rotation of scapula about X axis
7	Rotation of humerus about Y axis
8	Rotation of humerus about Y axis
9	Rotation of ulna about Z axis
10	Rotation of ulna about Y axis
11	Rotation of radius about Y axis
12	Rotation of hand about Y axis
13	Rotation of arm about X axis

ii.   The external pull force applied to the hand ('PULL', kg) and the orientation of the pull force about Y and Z axes ('PHIY', 'PHIZ' degrees)

4. OUTPUT

i.   The angles used above to set up the desired configuration.

ii.   Output from call to the 'SIMPDX' routine with minimum objective value, number of iterations, etc.

iii. Output from call to the 'SIMPDX' routine with minimum objective value, number of iterations, etc.

iv.   Values of all the muscle forces and joint reactions and moments. These are printed in the following abbreviated

656

forms:

No.	Abbreviation	Description
1	LTDOR1	Lastissimus dorsi (from lumbar spine)
2	LTDOR2	Lastissimus dorsi (from thorax)
3	LTDOR3	Lastissimus dorsi (from pelvis)
4	PCMAJ1	Pectoralis major (from thorax)
5	PCMAJ2	Pectoralis major (from thorax)
6	PCMAJ3	Pectoralis major (from clavicle)
7	DELTD1	Deltoid major (from clavicle)
8	DELTD2	Deltoid major (from scapula)
9	DELTD3	Deltoid major (from scapula)
10	SUPSPN	Supraspinatus
11	INFSPN	Infraspinatus
12	TERMAJ	Teres major
13	TERMIN	Teres minor
14	SUBSCP	Subscapularis
15	CORCBR	Coracobrachialis
16	BCPBRS	Biceps brachii (short head)
17	BCPBR1	Biceps brachii (long head)
18	BCPBR2	
19	BRACRD	Brachioradialis
20	TRCPS1	Triceps brachii (from scapula)
21	TRCPS2	Triceps brachii (from humerus)
22	TRCPS3	Triceps brachii (from humerus)
23	BRCHLS	Brachialis
24	ANCONS	Anconeus
25	SUPINR	Supinator
26	PRONTR	Pronator teres
27	EXCPRL	Extensor carpi radialis longus
28	EXCPRB	Extensor carpi radialis brevis
29	EXDGCM	Extensor digitor common
30	EXCPUL	Extensor carpi ulnaris
31	FXCPUL	Flexor carpi ulnaris
32	FXCPRD	Flexor carpi radialis
33	FXPOLL	Flexor pollicis longus
34	FXDGSB	Flexor digitorum superficialis
35	SHLDRX	Reaction force at shoulder joint along x
36	SHLDRY	Reaction force at shoulder joint along y
37	SHLDRZ	Reaction force at shoulder joint along z
38	ELBORX	Reaction force at elbow joint along x
39	ELBORY	Reaction force at elbow joint along y
40	ELBORZ	Reaction force at elbow joint along z
41	SHLDMX	Reaction moment at shoulder joint about x
42	SHLDMY	Reaction moment at shoulder joint about y
43	SHLDMZ	Reaction moment at shoulder joint about z
44	ELBOMX	Reaction moment at elbow joint about x
45	ELBOMY	Reaction moment at elbow joint about y
46	ELBOMZ	Reaction moment at elbow joint about z

5. STRUCTURE

The 'MAIN' program is augmented by the following subroutines:

'COORD':   for establishing the overall configuration and locations
            of points of interest with respect to ground frame

'DIRCOS':   for calculating the direction cosines of a muscle force

'MUSF':   for setting up a muscle force between two segments

'REAC':   for establishing joint reaction forces in terms of x,y,z
            components (positive and negative)

'MOME':   for establishing joint moments in terms of x,y,z
            components (positive and negative)

'PACT':   for printing the optimal muscle forces and joint reaction
            forces and moments

6. ILLUSTRATION

The optimal solution to the case when a pull force is applied to
the hand with the arm held horizontal is shown in Table 5-19.   The
configuration in this case is established by the rotations listed in
the table.

A pull force ('PULL') of 2.95 kg is applied at the wrist joint at
an angle 'PHIY' = 70° with respect to the vertical and 'PHIZ' = 2° with
respect to the Z axis.

7. LANGUAGE

FORTRAN V

8. MACHINE USED

UNIVAC 1110

9. ESTIMATED RUNNING TIME

6 secs

10. USING THE PROGRAM

a.   Changing segmental configuration

The overall configuration is established by reading a set of
13 angles to define the 'PHI' array.   These are read in free
format and for the case illustrated above might appear as:

```
-30.,  0.,  -5.,  0.,  0.,  0.,  -90.,  -8.,  7.,  0.,  0.,  0.,  30.
```

The first three angles represent successive Z,Y,X rotations
with respect to the ground frame for the clavicle.   In this

658

case it is assumed that PHI(1) = -30., PHI(2) = 2., PHI(3) = -5. define the position of the clavicle in the normal (standing erect) position. Similarly PHI(4) = PHI(5) = PHI(6) = 0 represent the successive Z,Y,X rotations of the scapula at the acromio-clavicular joint to bring it to the normal (standing erect) position. The humerus is then rotated about Y axis (PHI(y) = -90.) which brings it to a horizontal position from the normal vertical position and then adducted PHI(8) = -8°. The ulna, the radius and the hand are first rotated about Y axis through the same angle as the humerus (PHI(7)) and then abducted through PHI(9) to represent the misalignment between the longitudinal axes of the forearm and the arm (antecubital angle). They are then rotated about Y (angles PHI(10) through PHI(12)) to represent the amount of flexion at the elbow joint. The entire arm, forearm, hand complex is then rotated about the longitudinal axis of the arm (PHI(13)). Any desired segmental orientation can be readily attained by changing the input 'PHI' values.

b. Changing the external force

The external force, such as during pulling, is specified by its magnitude and orientation, and is assumed to be applied at the wrist. The data is input in free format in the form PHIY, PHIZ, PULL where

PHIY,PHIZ = direction of the pull force (in degrees) with respect to the Y and Z axes respectively

PULL = magnitude of the pull force (kg)

Thus, different values of the pull force or its orientation can be readily input.

c. Changing the merit function

The merit function used in the program is minimized

$U = \Sigma F + 2F + 4M$

where   F = all muscle forces
           R = tensile components of joint reactions
           M = joint moments

The objective function is specified by the 'COST' array by assigning the required cost coefficient for each variable. The 58 variables in the program represent muscle forces, joint reactions and moments in the following manner:

Variable No.	Description
1-34	muscle forces
35-40	reaction forces at shoulder joint
41-46	reaction forces at elbow joint
47-52	reaction moments at shoulder joint
53-58	reaction moments at elbow joint

To set up the merit function F + 2R + 4M, the 'COST' array should be defined by:

COST (1), COST (2), ..., COST (34) = 1.
COST (35), COST (36), ..., COST (46) = 2.
COST (47), COST (48), ..., COST (58) = 3.

PROGRAM FOR THE LOWER EXTREMITIES

1. CAPABILITY

This program is for the analysis of the lower extremities during static acts (leaping forward or backward, standing, squatting, etc.) as well as dynamic acts (walking, for example). The model for the lower extremities includes seven segments (the thigh, leg and foot for both sides, and the pelvis) with the feet treated as rigid segments. The pelvis represents the entire upper body, that is, the upper extremities and the upper torso.

2. LIMITATIONS

The weights of the segments and the locations of the center of masses are applicable to a total body weight of 72.3 kg. For analysis of subjects of dimensions different than those in the model appropriate changes should be made as explained later.

3. INPUT

i. The rotations of the joints of the lower extremities (thru the 'VALUE' array in the 'COORD' subroutine) are input as follows:

Value	Description
1	Rotation about Y for the right foot at the right ankle joint
2	Rotation about Z for the right foot at the right ankle joint
3	Rotation about Y for the right leg at the right ankle joint
4	Rotation about X for the right leg at the right ankle joint
5	Rotation about Z for the right leg at the right ankle joint
6	Rotation about Y for the right thigh at right knee joint
7	Rotation about Y for pelvis at right hip joint
8	Rotation about X for pelvis at right hip joint
9	Rotation about Z for pelvis at right hip joint

10	Rotation about Y for left thigh at left hip joint
11	Rotation about X for left thigh at left hip joint
12	Rotation about Z for left thigh at left hip joint
13	Rotation about Y for left leg at left knee joint
14	Rotation about Y for left foot at left ankle joint
15	Rotation about X for left foot at left ankle joint
16	Rotation about Z for left foot at left ankle joint

The angles represent relative orientation between segments.

ii. The inertia force components ('FX','PX','FZ') along X, Y and Z acting through the centroid and the inertia moment components ('FMX', "FMY', 'FMZ') about X, Y and Z for each of the seven segments. These are read in free format in the form:

FX, FY, FZ, FMX, FMY, FMZ

In the following order: right foot, right leg, right thigh, pelvis, left thigh, left leg and left foot, respectively.

4. OUTPUT

i. All the values of the joint angles input in Value 3 above to describe the configuration.

ii. The ground-to-foot support reactions as well as the location of the center of support with respect to the right ankle joint.

iii. Values of all muscle forces and joint reaction forces and moments. These are printed in abbreviated form as follows:

No.	Abbreviation	Description
1	GRACLS	Gracilis
2	ADLONG	Adductor longus
3	ADMAGA	Adductor magnus (adductor part)
4	ADMAGE	Adductor magnus (extensor part)
5	ADBREV	Adductor brevis
6	SEMITS	Semitendinosus
7	SEMIMB	Semimembranosus
8	BCPSLH	Biceps femoris (long head)
9	RECFEM	Rectus femoris
10	SARTOR	Sartorius
11	TFASLT	Tensor fasciaw latae
12	GLUMAX	Glutens maximum
13	ILIOPS	Illiopsoas
14	GLUMED	Gluteus medius
15	GLUMIN	Gluteus minimus
16	BCPSSH	Biceps femoris (short head)

17	VASTMD	Vaetus medialis
18	VASTIN	Vastus intermedius
19	VASTLT	Vastus lateralis
20	GASTRM	Gastrocnemius (medial head)
21	GASTRL	Gastrocnemius (lateral head)
22	GOLEUS	Soleus
23	TIBANT	Tibialis anterior
24	TBPOST	Tibialis posterior
25	EXDGLG	Extensor digitorum longus
26	EXHLLG	Extensor hallucis longus
27	FXDGLG	Flexor digitorum longus
28	FXHLLG	Flexor hallucis longus
29	PERONL	Peroneus longus
30	PERONB	Peroneus brevis
31	PERONT	Peroneus tertius
32	ANKLRX	Ankle joint reaction force along X
33	ANKLRY	Ankle joint reaction force along Y
34	ANKLRZ	Ankle joint reaction force along Z
35	ANKLMX	Ankle joint moment about X
36	ANKLMY	Ankle joint moment about Y
37	ANKLMZ	Ankle joint moment about Z
38	KNEERX	Knee joint reaction force along X
39	KNEERY	Knee joint reaction force along Y
40	KNEERZ	Knee joint reaction force along Z
41	KNEEMX	Knee joint moment about X
42	KNEEMY	Knee joint moment about Y
43	KNEEMZ	Knee joint moment about Z
44	HIPJRX	Hip joint reaction force along X
45	HIPJRY	Hip joint reaction force along Y
46	HIPJRZ	Hip joint reaction force along Z
47	HIPJMX	Hip joint moment about X
48	HIPJMY	Hip joint moment about Y
49	HIPJMZ	Hip joint moment about Z

5. ILLUSTRATION

To illustrate the application of the program, the standing erect posture is analyzed. This configuration is (see TABLE 5-1(b), page 166) defined by all the values of the joint angles ('VALUE' array) as being zero, that is,

$$VALUE(1), VALUE(2), ..., VALUE(16) = 0$$

Since this is a static case, the only external forces are the segment weights and inertia forces and moments are all zero. Therefore, for each segment FX = 0, FY = 0, FZ = W, FMX = 0, FMY = 0, FMZ = 0, where W is the segment weight (kg).

The foot, the leg, the thigh and the pelvis weights are taken as 1.1, 3.6, 7.45 and 48.0 kg respectively, for a total body weight of 72.3 kg, when both extremities are included. This data therefore will appear as follows:

	FX	FY	FZ	FMX	FMY	FMZ	
7	0.,	0.,	−1.1	0.,	0.,	0.	← Left foot
6	0.,	0.,	−3.6,	0.,	0.,	0.	← Left leg
5	0.,	0.,	−7.45,	0.,	0.,	0.	← Left thigh
4	0.,	0.,	−48.0,	0.,	0.,	0.	← Pelvis
3	0.,	0.,	−7.45,	0.,	0.,	0.	← Right thigh
2	0.,	0.,	−3.6,	0.,	0.,	0.	← Right leg
Card 1	0.,	0.,	−1.1,	0.,	0.,	0.	← Right foot

The optimal solution based on a merit criterion of minimizing all muscle forces and four times all joint moments given in C pter V.

For further illustration, the input configuration for the 10° static leaning is shown in Fig. 5-1(b), p. 166, and the corresponding optimal solution for muscle forces and joint reactions and moments is shown in TABLE 5-2, p. 167. The leaning posture is established by a pelvic tilt at the right hip joint, VALUE(7) = 10.0, and VALUE(10) = −10.0 to bring the left extremity to the erect position.

As with the standing erect posture, the leaning is a static activity and so all the inertia forces and moments are zero and the only external forces acting on the segments are the segment weights. Consequently, this data is identical to the one used for analyzing the standing erect position.

6. PROGRAM STRUCTURE

The 'MAIN' program is used in conjunction with the following subroutines:

'REAC': for establishing reaction forces at a joint in terms of the X,Y,Z components (both positive and negative)

'MOME': for establishing reaction moments at a joint in terms of the X,Y,Z components (both positive and negative)

'MUSF': for setting up a muscle force between two segments

'WRAP': for setting up a muscle force between two segments with wrapping on an intermediate segment

'DIRCOS': for calculating the direction cosines of a muscle force

'RACT': for printing muscle forces and joint reactions and moments after an optimal solution is obtained by call to the 'SIMPDZ' routine

'MATX': for defining a 4x4 matrix for either translation along one or the three axes or rotation about one of the three axes

663

'MATX2': for defining a 4x4 matrix for translations along the X,Y,Z directions as well as rotation about one of the three axes

'MATY2': for multiplying two 4x4 matrices

'MATY1': for multiplying a 4x4 matrix with a 4x1 matrix

'COORD': for establishing the configuration and locations of all points of interest with respect to ground frame.

7. LANGUAGE

FORTRAN V

8. MACHINE USED

UNIVAC 1110

9. ESTIMATED RUNNING TIME

8 seconds

10. RUNNING THE PROGRAM

a. Changing Configuration

The configuration is defined by 16 angles as listed under 'INPUT' above. These are punched in free format consecutively. Thus, the configuration for leaning forward 10° might appear as:

```
0.,0.,0.,0.,0.,0.,10,.0,.0.,-10.,0,0,0.,0.,0.,0.
```

Note that in both the standing erect and the leaning cases, the two feet are stationary and treated as ground. This means equilibrium equations for these two segments are not set up. This is defined by means of the 'NEQ' array, where NEQ(1) = 0 and NEQ(7) = 0 (the seven segments are numbered 1 through 7, beginning with the right foot). In situations where only one foot is on the ground (such as in walking), NEQ(1) = 0, NEQ(7) = 31, and the line No. 42 in the 'MAIN' PROGRAM used to define 'NEQ' array should be changed.

b. Changing External Forces Acting on the Segments

The external forces on each of the seven segments are defined by means of 'FX', 'FY', 'FZ', 'FMX', 'FMY', 'FMZ' which represent the force components acting at the centroid and the moment components along X,Y,Z, respectively.

For static situations, the inertia forces and moments are zero so that the only external forces acting on the segments are the segment weights. Therefore FX = FY = FMX = FMY = FMZ = 0, and FZ = W, where W is the weight of the segment. For a 72.3 kg body weight, the distribution among the extremities is:

664

```
          right and left foot:   1.1 kg (each)
          right and left leg:    3.6 kg (each)
          right and left leg:    7.45 kg (each)
                      pelvis:    48.0 kg
```

for a static case thus, the input data might appear as shown in
the illustration above. If the body weight is different than
the 72.3 kg assumed here then the segment weight distribution
can be readily changed.

c. Changing Merit Criterion

The merit criterion is established through the 'COST' array by
assigning the desired cost coefficient to a variable. The 134
variables representing muscle forces, joint reactions and
moments will have costs assigned through the COST(1),
COST(2),..., COST(134) values and denote the following:

Variable No.   Description

    1-31   Right side muscle forces
    32-62   Left side muscle forces
    63-68   Reaction forces at the right ankle joint
    69-74   Reaction forces at the right knee joint
    75-80   Reaction forces at the right hip joint
    81-86   Reaction forces at the right left hip joint
    87-92   Reaction forces at the left knee joint
    93-98   Reaction forces at the left ankle joint
    99-104   Reaction moments at the right ankle joint
    105-110   Reaction moments at the right knee joint
    111-116   Reaction moments at the right hip joint
    117-122   Reaction moments at the left hip joint
    123-128   Reaction moments at the left knee joint
    129-134   Reaction moments at the left ankle joint

Presently in the program the merit criterion is minimizes all
muscle forces (Variables 1 through 62) and four times all joint
moments (Variables 99 through 134). Accordingly,

        COST(1), COST(2),...,COST(62) = 1.0

     COST(99), COST(100),...,COST(134) = 4.0

The merit function therefore can be readily changed by
assigning the desired cost coefficient to a variable through the
'COST' array.

PROGRAM FOR THE HAND

1. CAPABILITY

This program can be used for the analysis of the upper extremities
(arm and forearm) along with the segmental model of the hand. The hand
is defined by a total of 20 segments including the metacarpals and the

phalanges for the five fingers as well as the carpals which are lumped
into a single segment connected to the five metacarpals at the distal end
and the ulna-radius at the proximal end. The segments are numbered as
follows:

Number	Segment Name and Abbreviation
1	Carpals (CARPALS)
2	Metacarpal I (MTCRP1)
3	Proximal phalax I (PROXP1)
4	Distal phalanx I (DISTP1)
5	Metacarpal II (MTCRP2)
6	Proximal phalanx II (PROXP2)
7	Middle phalanx II (MIDLP2)
8	Distal phalanx II (DISTP2)
9	Metacarpal III (MTCRP3)
10	Proximal phalanx III (PROXP3)
11	Middle phalanx III (MIDLP3)
12	Distal phalanx III (DISTP3)
13	Metacarpal IV (MTCRP4)
14	Proximal phalanx IV (PROXP4)
15	Middle phalanx IV (MIDLP4)
16	Distal phalanx IV (DISTP4)
17	Metacarpal V (MTCRP5)
18	Proximal phalanx V (PROXP5)
19	Middle phalanx V (MIDLP5)
20	Distal phalanx V (DISTP5)
21	Ulna-radius (ULNRAD)
22	Humerus (Humrus)
23	Scapula (SCAPLA)
24	Clavicle (CLVCLE)
25	Thorax (THORAX)

Although the scapula, the clavicle and the thorax have been included
in the upper extremity model they are used only to establish the
attachment points of muscles connecting the arm and the forearm. Thus,
for force analysis they are excluded from the hand model. There are a
total of 279 rows (constraints) and 493 columns (variables) in the
program. The overall segmental configuration with respect to ground
frame is established by specifying the successive X,Y and Z rotations of
eah segment with respect to ground frame.

The model can be used to analyze such finger actions as pinching,
pulling, etc., which are simulated by force acting on the distal or
middle phalanges.

The main program creates the linear programming problem in terms of
the constraints, the columns or variables, the objective function and the
right hand side and writes the data structure on a file ('15') for
subsequent execution by the FMPS-LP package.

2. LIMITATIONS

666

The coordinates of muscle attachment points or joint locations, the lengths of the segment, etc. are as described in the 'DATA' element of the program. Any required modifications to the model dimensions or segment weights would entail changes in the program as will be explained later. The arm weight is taken to be 2.2 kg and the forearm weight, 1.2 kg when modelled as a single rigid segment is now treated as massless since the exact distribution among the different segments cannot be readily ascertained.

The merit criterion is the minimized sum of all muscle forces, four times the joint moments and two times the tensile joint reaction forces (at the arm and the forearm). This again can be readily modified if need be.

3. INPUT

i. The X, Y, Z rotations of the twenty five segments with respect to ground frame defined by 'PHIX', 'PHIY', PHIZ' rays, read in free format. The carpals and the thorax are treated as ground so that PHIX(1), PHIY(1), PHIZ(1) for the carpals and PHIX(25), PHIY(25), PHIZ(25) for the thorax may be set at any arbitrary values and will not be used in the program.

ii. The coordinates of the muscle attachments, joint locations, muscle and joint names, etc. as listed in the 'DATA' element of the program (lines 6 through 201). This is appended for execution by the program after the segmental configuration has been defined.

iii. The external forces acting on the distal or middle phalanges in terms of the X, Y, Z components along the ground frame as well as the location of the point of action of the forces on each segment. The input is by means of the statement (line 414 of the 'MAIN' program):

READ, FX, FY, FZ, J

where FX,FY,FZ = X,Y,Z components respectively of the force acting on the segment

J = location in the coordinate data values ('XIR', 'YIR', 'ZIR' arrays) corresponding to the point of action of the force.

(J = 33, point of action on DISTP2,
34, point of action on DISTP3,
53, point of action on DISTP4,
182, point of action on DISTP5)

4. OUTPUT

i. All the segmental rotations as defined in (3-i) above.

ii. The input force components on the phalanges.

iii. The optimal solution obtained by call to the FMPS-LP package. The optimal solution consists of the optimal objective function value, the constraint row types and the right hand side and the values of the columns or variables (which correspond to the muscle forces, joint reactions and moments) at optimum. There are a total of 279 rows in the hand problem (including the objective row) and a total of 493 columns. The columns (and the rows) are identified by a maximum of 89 alphanumeric characters which have been selected to abbreviate the muscle force names or joint names (on the following pages).

5. ILLUSTRATION

Table II shows the computer output solution to the analysis of finger pull where the forearm is horizontal palm facing up and a force is applied on the distal phalanx of the second finger (DISTP2). The input configuration representing the relative orientations of the upper extremity and the hand segments is as defined in Table I. The forces acting on the distal phalanges are also shown in Table I--in this case, the third, the fourth and the fifth distal phalanges have no forces acting on them (FX=FY=FZ=0) while the second distal phalanx has a force given by FX = 4.5095, FY = -0.5197, FZ = 0 kg. (See lines 202 through 205 in the 'DATA' element) read in free format. Actually this represents a total force of 10 lb (= 4.535 kg) which when resolved gives the X,Y,Z components as above.

The merit function in this case was to minimize all muscle forces, two times the tensile components of the reactions (at the elbow and the shoulder joint) and four times all the joint moments.

As can be seen from Table II, the active muscles for the second finger are the flexor digitorum profundus (FXDP21 = FXDP22 = FXDP23 = FXDP24 = FXDP25 = 9.02 kg), the flexor digitorum superficialis (FXDS25 = FXDS26 = FXSS27 = FXDS 28 = FXDS29 = 1.51 kg) and the muscles at the wrist, the extensor carpi radialis longus (EXCRL1 = EXCRL2 = 5.20 kg) and the flexor carpi radialis (FXCPR1 = FXCPR2 = RXCPR3 = 8.56 kg).

6. PROGRAM STRUCTURE

The 'MAIN' program is required in conjunction with other subroutines, namely:

'MSUF':   for establishing a muscle force between two segments

'REAC':   for establishing reaction components at a joint between two segments

'MOMENT':   for establishing moment components at a joint between two segments

'MUSFG':   for setting up a muscle between a body and ground

'GRONDR':   for setting up reaction forces at a joint between a body and ground

'GRONDM':   for setting up reaction moments at a joint between a body and ground

'REACON':   for setting up a reaction between two segments, constrained along a desired orientation

668

A. MUSCLE FORCES:

No.	Column or Variable Name	Description	Segments and points connected			
			Point 1	Segment 1	Point	Segment
1	ABPOLL	Abductor pollicis longus	11	ULNRAD	17	MTCRP1
2	ADPOL1	Adductor pollicis	20	CARPLS	21	PROXP1
3	ADPOL2	Adductor pollicis	22	MTCRP2	21	PROXP1
4	ADPOL3	Adductor pollicis	23	MTCRP3	21	PROXP1
5	ADPOL4	Adductor pollicis	24	MTCRP3	21	PROXP1
6	OPNPOL	Opponens pollicis	25	CARPLS	26	MTCRP2
7	DORI11	Dorsal Interosseous	27	MTCRP1	28	PROXP2
8	DORI15	Dorsal Interosseous	58	MTCRP2	28	PROXP2
9	DORI21	Dorsal Interosseous	59	MTCRP2	60	PROXP3
10	VOLI11	Volar Interosseous	65	MTCRP2	66	PROXP2
11	DURI30	Dorsal Interosseous	87	MTCRP3	60	PROXP3
12	DORI34	Dorsal Interosseous	89	MTCRP3	90	PROXP3
13	DORI40	Dorsal Interosseous	90	PROXP3	114	MTCRP4
14	VOLI40	Volar Interosseous	115	MTCRP4	116	PROXP4
15	DORI44	Dorsal Interosseous	119	MTCRP4	120	PROXP4
16	EXCRPU	Extensor carpi ulnaris	133	ULNRAD	134	MTCRP5
17	FXCRPU	Flexor carpi ulnaris	135	ULNRAD	136	CARPLS
18	DORI50	Dorsal Interosseous	146	MTCRP5	120	PROXP4
19	VOLI50	Volar Interosseous	150	MTCRP5	151	PROXP5
20	ABDGMN	Abductor digiti minimi	154	CARPLS	155	PROXP5
21	FXDGMN	Flexor digiti minimi	156	CARPLS	155	PROXP5
22	ODGMN1	Oppenens digiti minimi	157	CARPLS	158	MTCRP5
23	ODGMN2	Oppenens digiti minimi	157	CARPLS	159	MTCRP5
24	ODGMN3	Oppenens digiti minimi	157	CARPLS	160	MTCRP5
25	BRRDLS	Brachio radialis	228	HUMRUS	221	ULNRAD
26	BRCHLS	Brachialis	229	HUMRUS	222	ULNRAD
27	BCPSSH	Biceps brachii (short head)	230	SCAPLA	223	ULNRAD
28	TRCPB1	Triceps brachii	234	SCAPLA	224	ULNRAD
29	TRCPB2	Triceps brachii	235	HUMRUS	224	ULNRAD
30	TRCPB3	Triceps brachii	236	HUMRUS	224	ULNRAD
31	ANCONS	Anconeus	237	HUMRUS	225	ULNRAD
32	SUPINT	Supinator	238	HUMRUS	226	ULNRAD
33	PRONTR	Pronator teres	239	HUMRUS	227	ULNRAD
34	LTDOR1	Latissimus dorsi	248	THORAX	253	ULNRAD
35	LTDOR2	Latissimus dorsi	249	THORAX	254	ULNRAD
36	LTDOR3	Latissimus dorsi	250	THORAX	254	ULNRAD
37	PCMAJ1	Pectoralis major	251	THORAX	255	ULNRAD
38	PCMAJ2	Pectoralis major	252	THORAX	255	ULNRAD

(continued)

A. MUSCLE FORCES: (continued)

No.	Column of Variable Name	Description	Segments and points connected			
			Point 1	Segment 1	Point 2	Segment 2
39	PCMAJ3	Pectoralis major	256	CLVCLE	255	ULNRAD
40	DELTD1	Deltoid	257	CLVCLE	258	ULNRAD
41	DELTD2	Deltoid	259	SCAPLA	260	ULNRAD
42	DELTD3	Deltoid	261	SCAPLA	262	ULNRAD
43	SUPRSP	Supraspinatus	263	SCAPLA	264	ULNRAD
44	INFRSP	Infraspinatus	265	SCAPLA	266	ULNRAD
45	TRMAJR	Teres major	267	SCAPLA	268	ULNRAD
46	TRMINR	Teres minor	269	SCAPLA	270	ULNRAD
47	SUBSCP	Subscapularis	271	SCAPLA	272	ULNRAD
48	CORACO	Coracobrachialis	273	SCAPLA	274	ULNRAD
49	EXPLL1	Extensor pollicis longus	1	ULNRAD	2	MTCRP1
50	EXPLL2	Extensor pollicis longus	2	MTCRPL	3	PROXP1
51	EXPLL3	Extensor pollicis longus	3	PROXP1	4	DISTP1
52	FXDLL1	Flexor pollicis longus (from ulna-radius)	5	ULNRAD	6	CARPLS
53	FXDLL2	Flexor pollicis longus (from ulna-radius)	6	CARPLS	7	MTCRP1
54	FXDLL3	Flexor pollicis longus (from ulna-radius)	7	MTCRP1	8	PROXP1
55	FXDLL4	Flexor pollicis longus (from ulna-radius)	9	PROXP1	10	DISTP1
56	FXPLB1	Extensor pollicis brevis	11	ULNRAD	12	MTCRP1
57	FXPLB2	Extensor pollicis brevis	12	MTCRP1	13	PROXP1
58	FXPLB1	Flexor pollicis brevis	14	CARPLS	15	MTCRP1
59	FXPLB2	Flexor pollicis brevis	15	MTCRP1	16	PROXP1
60	ABPLB1	Abductor pollicis brevis	18	CARPLS	15	MTCRP1
61	ABPLB2	Abductor pollicis brevis	15	MTCRP1	19	PROXP1
62	DORI12	Dorsal Interosseous	27	MTCRP1	28	PROXP2
63	DORI13	Dorsal Interosseous	29	PROXP2	30	MIDLP2
64	DORI14	Dorsal Interosseous	31	MIDLP2	32	DISTP2
65	FXDS21	Flexor digitorum superficialis (from ulna-radius)	40	ULNRAD	41	CARPLS
66	FXDS22	Flexor digitorum superficialis (from ulna-radius)	41	CARPLS	42	MTCRP2
67	FXDS23	Flexor digitorum superficialis (from ulna-radius)	42	MTCRP2	43	PROXP2
68	FXDS24	Flexor digitorum superficialis (from ulna-radius)	43	PROXP2	44	MIDLP2
69	FXDP21	Flexor digitorum profundus	40	ULNRAD	45	CARPLS

(continued)

A. MUSCLE FORCES: (continued)

No.	Column or Variable Name	Description	Segments and points connected			
			Point 1	Segment 1	Point 2	Segment 2
70	FXDP22	Flexor digitorum profundus	45	CARPLS	42	MTCRP2
71	FXDP23	Flexor digitorum profundus	42	MTCRP2	43	PROXP2
72	FXDP24	Flexor digitorum profundus	43	PROXP2	46	MIDLP2
73	FXDP25	Flexor digitorum profundus	46	MIDLP2	47	DISTP2
74	EXIPM1	Extensor indicis proprius (central band)	48	ULNRAD	49	MTCRP2
75	EXIPM2	Extensor indicis proprius (central band)	49	MTCRP2	50	PROXP2
76	EXIPM3	Extensor indicis proprius (central band)	50	PROXP2	51	MIDLP2
77	EXIPR1	Extensor indicis proprius (radial band)	48	MTCRP2	52	PROXP2
78	EXIPR2	Extensor indicis proprius (radial band)	49	MTCRP2	52	PROXP2
79	EXIPR3	Extensor indicis proprius (radial band)	52	PROXP2	30	MIDLP2
80	EXIPR4	Extensor indicis proprius (radial band)	31	MIDLP2	32	DISTP2
81	EXIPU1	Extensor indicis proprius (ulnar band)	48	ULNRAD	49	MTCRP2
82	EXIPU2	Extensor indicis proprius (ulnar band)	49	MTCRP2	56	PROXP2
83	EXIPU3	Extensor indicis proprius (ulnar band)	56	PROXP2	57	MIDLP2
84	EXIPU4	Extensor indicis proprius (ulnar band)	31	MIDLP2	32	DISTP2
85	DORI16	Dorsal interosseous	58	MTCRP2	28	PROXP2
86	DORI17	Dorsal interosseous	29	PROXP2	30	MIDLP2
87	DORI18	Dorsal interosseous	31	MIDLP2	32	DISTP2
88	DORI22	Dorsal interosseous	59	MTCRP2	60	PROXP3
89	DORI23	Dorsal interosseous	61	PROXP3	62	MIDLP3
90	DORI24	Dorsal interosseous	63	MIDLP3	64	DISTP3
91	VOLI12	Volar interosseous	65	MTCRP2	67	PROXP2
92	VOLI13	Volar Interosseous	68	PROXP2	57	MIDLP2
93	VOLI14	Volar Interosseous	31	MIDLP2	32	DISTP2
94	FXDS31	Flexor digitorum superficialis (from ulna-radius)	40	ULNRAD	72	CARPLS
95	FXDS32	Flexor digitorum superficialis (from ulna-radius)	72	CAPPLS	73	MTCRP3

(continued)

A. MUSCLE FORCES: (Continued)

No.	Column or Variable Name	Description	Point 1	Segment 1	Point 2	Segment 2
			Segments and points connected			
96	FXDS33	Flexor digitorum superficialis (from ulna-radius)	73	MTCRP3	74	PROXP3
97	FXDS34	Flexor digitorum superficialis (from ulna-radius)	74	PROXP3	75	MIDLP3
98	FXDP31	Flexor digitorum profundus	76	ULNRAD	77	CARPLS
99	FXDP32	Flexor digitorum profundus	77	CARPLS	73	MTCRP3
100	FXDP33	Flexor digirorum profundus	73	MTCRP3	74	PROXP3
101	FXDP34	Flexor digitorum profundus	74	PROXP3	78	MIPLP3
102	FXDP35	Flexor digitorum profundus	78	MIDLP3	79	DISTP3
103	DORI31	Dorsal Interosseous	87	MTCRP3	88	PROXP3
104	DORI32	Dorsal Interosseous	61	PROXP3	62	MIDLP3
105	DORI33	Dorsal Interosseous	63	MIDLP3	64	DISTP3
106	DORI35	Dorsal Interosseous	89	MTCRP3	91	PROXP3
107	DORI36	Dorsal Interosseous	92	PROXP3	93	MIDLP3
108	DORI37	Dorsal Interosseous	63	MIDLP3	64	DISTP3
109	FXDS41	Flexor digitorum superficialis (from ulna-radius)	95	ULNRAD	96	CARPLS
110	FXDS42	Flexor digitorum superficialis (from ulna-radius)	96	CARPLS	97	MTCRP4
111	FXDS43	Flexor digitorum superficialis (from ulna-radius)	97	MTCRP4	98	PROXP4
112	FXDS44	Flexor digitorum superficialis (from ulna-radius)	98	PROXP4	99	MIDLP4
113	FXDP41	Flexor digitorum profundus	101	ULNRAD	101	CARPLS
114	FXDP42	Flexor digitorum profundus	101	CARPLS	97	MTCRP4
115	FXDP43	Flexor digitorum profundus	97	MTCRP4	98	PROXP4
116	FXDP44	Flexor digitorum profundus	98	PROXP4	102	MIDLP4
117	FXDP45	Flexor digitorum profundus	102	MIDLP4	103	DISTP4
118	DORI41	Dorsal Interosseous	114	MTCRP4	91	PAOXP3
119	DORI42	Dorsal Interosseous	92	PROXP3	93	MIDLP3
120	DORI43	Dorsal Interosseous	63	MIDLP3	64	DISTP3
121	VOLI41	Volar Interosseous	115	MTCRP4	117	PROXP4
122	VOLI42	Volar Interosseous	118	PROXP4	109	MIDLP4
123	VOLI43	Volar Interosseous	110	MIDLP4	111	DISTP4
124	DORI45	Dorsal Interosseous	119	MTCRP4	120	PROXP4
125	DORI46	Dorsal Interosseous	121	PROXP4	122	MIDLP4
126	DORI47	Dorsal Interosseous	110	MIDLP4	111	DISTP4

(continued)

A. MUSCLE FORCES: (Continued)

No.	Column or Variable Name	Description	Segments and points connected			
			Point 1	Segment 1	Point 2	Segment 2
127	FXDS51	Flexor digitorum superficialis (from ulna-radius)	95	ULNRAD	175	CARPLS
128	FXDS52	Flexor digitorum superficialis (from ulna-radius)	125	CARPLS	176	MTCRP5
129	FXDS53	Flexor digitorum superficialis (from ulna-radius)	126	MTCRP5	127	PROXP5
130	FXDS54	Flexor digitorum superficialis (from ulna-radius)	127	PROXP5	128	MIDLP5
131	FXDP51	Flexor digitorum profundus	129	ULNRAD	130	CARPLS
132	FXDP52	Flexor digitorum profundus	130	CARPLS	126	MTCRP5
133	FXDP53	Flexor digitorum profundus	126	MTCRP5	127	PROXP5
134	FXDP54	Flexor digitorum profundus	127	PROXP5	131	MIDLP5
135	FXDP55	Flexor digitorum profundus	131	MIDLP5	132	DISTP5
136	DORI51	Dorsal Interosseous	146	MTCRP5	120	PROXP4
137	DORI52	Dorsal Interosseous	121	PROXP4	122	MIDLP4
138	DORI53	Dorsal Interosseous	110	MIDLP4	111	DISTP4
139	VOLI51	Volar Interosseous	150	MTCRP5	152	PROXP5
140	VOLI52	Volar Interosseous	153	PROXP5	141	MIDLP5
141	VOLI53	Volar Interosseous	142	MIDLP5	143	DISTP5
142	EDMNM1	Extensor digiti minimi	161	ULNRAD	137	MTCRP5
143	EDMNM2	Extensor digiti minimi	137	MTCRP5	138	PROXP5
144	EDMNM3	Extensor digiti minimi	138	PROXP5	139	MIDLP5
145	EDMNR1	Extensor digiti minimi	161	ULNRAD	137	MTCRP5
146	EDMNR2	Extensor digit minimi	137	MTCRP5	140	PROXP5
147	EDMNR3	Extensor digiti minimi	140	PROXP5	141	MIDLP5
148	EDMNR4	Extensor digiti minimi	142	MIDLP5	143	DISTP5
149	EDMNU1	Extensor digiti minimi	161	ULNRAD	137	MTCRP5
150	EDMNU2	Extensor digiti minimi	137	MTCRP5	144	PROXP5
151	EDMNU3	Extensor digiti minimi	144	PROXP5	145	MIDLP5
152	EDMNU4	Extensor digiti minimi	142	MIDLP5	143	DISTP5
153	FXDS25	Flexor digitorum superficialis (from humerus)	240	HUMRUS	40	ULNRAD
154	FXDS26	Flexor digitorum superficialis (from humerus)	40	ULNRAD	41	CARPLS
155	FXDS27	Flexor digitorum superficialis (from humerus)	41	CARPLS	42	MTCRP2
156	FXDS28	Flexor digitorum superficialis (from humerus)	42	MTCRP2	43	PROXP2

(continued)

A. MUSCLE FORCES: (continued)

No.	Column or Variable Name	Description	Segments and points connected			
			Point 1	Segment 1	Point 2	Segment 2
157	FXDS29	Flexor digitorum superficialis (from humerus)	43	PROXP2	44	MIDLP2
158	FXDS35	Flexor digitorum superficialis (from humerus)	240	HUMRUS	40	ULNRAD
159	FXDS36	Flexor digitorum superficialis (from humerus)	40	ULNRAD	72	CARPLS
160	FXDS37	Flexor digitorum superficialis (from humerus)	72	CARPLS	73	MTCRP3
161	FXDS38	Flexor digitorum superficialis (from humerus)	73	MTCRP3	74	PROXP3
162	FXDS39	Flexor digitorum superficialis (from humerus)	74	PROXP3	75	MIDLP3
163	FXDS45	Flexor digitorum superficialis (from humerus)	240	HUMRUS	95	ULNRAD
164	FXDS46	Flexor digitorum superficialis (from humerus)	95	ULNRAD	96	CARPLS
165	FXDS47	Flexor digitorum superficialis (from humerus)	96	CARPLS	97	MTCRP4
166	FXDS48	Flexor digitorum superficialis (from humerus)	97	MTCRP4	98	PROXP4
167	FXDS49	Flexor digitorum superficialis (from humerus)	98	PROXP4	99	MIDLP4
168	FXDS55	Flexor digitorum superficialis (from humerus)	240	HUMRUS	95	ULNRAD
169	FXDS56	Flexor digitorum superficialis (from humerus)	95	ULNRAD	125	CARPLS
170	FXDS57	Flexor digitorum superficialis (from humerus)	125	CARPLS	126	MTCRP5
171	FXDS58	Flexor digitorum superficialis (from humerus)	126	MTCRP5	127	PROXP5
172	FXDS59	Flexor digitorum superficialis (from humerus)	127	PROXP5	128	MIDL5
173	EXCRU1	Extensor carpi ulnaris	243	HUMRUS	133	ULNRAD
174	EXCRU2	Extensor Carpi ulnaris	133	ULNRAD	134	MTCRP5
175	FXCRU1	Flexor carpi ulnaris	244	HUMRUS	135	ULNRAD
176	FXCRU2	Flexor carpi ulnaris	135	ULNRAD	136	CARPLS
177	FXPLL5	Flexor pollicis longus	245	HUMRUS	5	ULNRAD

(continued)

674

A. MUSCLE FORCES: (continued)

No.	Column or Variable Name	Description	Point 1	Segment 1	Point 2	Segment 2
			Segments and points connected			
178	FXPLL6	Flexor pollicis longus (from humerus)	5	ULNRAD	6	CARPLS
179	FXPLL7	Flexor pollicis longus (from humerus)	6	CARPLS	7	MTCAP1
180	FXPLL8	Flexor pollicis longus (from humerus)	7	MTCRP1	8	PROXP1
181	FXPLL9	Flexor pollicis longus (from humerus)	9	PROXP1	10	DISTP1
182	EXCRL1	Extensor carpi radialis longus	241	HUMRUS	35	ULNRAD
183	EXCRL2	Extensor carpi radialis longus	35	ULNRAD	36	MTCRP2
184	EXCRB1	Extensor carpi radialis brevis	242	HUMRUS	70	ULNRAD
185	EXCRB2	Extensor carpi radialis brevis	70	ULNRAD	71	MTCRP3
186	EXCRB3	Extensor carpi radialis brevis	242	HUMRUS	70	ULNRAD
187	EXCRB4	Extensor carpi radialis brevis	70	ULNRAD	69	MTCRP2
188	FXCPR1	Flexor carpi radialis	246	HUMRUS	37	ULNRAD
189	FXCPR2	Flexor carpi radialis	37	ULNRAD	38	CARPLS
190	FXCPR3	Flexor carpi radialis	38	CARPLS	39	MTCRP2
191	EDGM20	Extensor digitorum (central band)	247	HUMRUS	48	ULNRAD
192	EDGM21	Extensor digitorum (central band)	48	ULNRAD	49	MTLRP2
193	EDGM22	Extensor digitorum (central band)	49	MTCRP2	50	PROXP2
194	EDGM23	Extensor digitorum (central band)	50	PROXP2	51	MIDLP2
195	EDGR20	Extensor digitorum (radial band)	247	HUMRUS	48	ULNRAD
196	EDGR21	Extensor digitorum (radial band)	48	ULNRAD	49	MTCRP2
197	EDGR22	Extensor digitorum (radial band)	49	MTCRP2	52	PROXP2
198	EDGR23	Extensor digitorum (radial band)	52	PROXP2	30	MIDLP2
199	EDGR34	Extensor digitorum (radial band)	31	MIDLP2	32	DISTP2
200	EDGU20	Extensor digitorum (ulnar band)	247	HUMRUS	48	ULNRAD

(continued)

A. MUSCLE FORCES: (continued)

No.	Column or Variable Name	Description	Segments and points connected			
			Point 1	Segment 1	Point 2	Segment 2
201	EDGU21	Extensor digitorum (ulnar band)	48	ULNRAD	49	MTCRP2
202	EDGU22	Extensor digitorum (ulnar band)	49	MTCRP2	56	PROXP2
203	EDGU23	Extensor digitorum (ulnar band)	56	PROXP2	57	MIDLP2
204	EDGU24	Extensor digitorum (ulnar band)	31	MIDLP2	32	DISTP2
205	EDGM30	Extensor digitorum (central band)	247	HUMRUS	80	ULNRAD
206	EDGM31	Extensor digitorum (central band)	80	ULNRAD	81	MTCRP3
207	EDGM32	Extensor digitorum (central band)	81	MTCRP3	82	PROXP3
208	EDGM33	Extensor digitorum (central band)	82	PROXP3	83	MIDLP3
209	EDGR30	Extensor digitorum (radial band)	247	HUMRUS	80	ULNRAD
210	EDGR31	Extensor digitorum (radial band)	80	ULNRAD	81	MTCRP3
211	EDGR32	Extensor digitorum (radial band)	81	MTCRP3	84	PROXP3
212	EDGR33	Extensor digitorum (radial band)	84	PROXP3	62	MIDLP3
213	EDGR34	Extensor digitorum (radial band)	63	MIDLP3	64	DISTP3
214	EDGU30	Extensor digitorum (ulnar band)	247	HUMRUS	80	ULNRAD
215	EDGU31	Extensor digitorum (ulnar band)	80	ULNRAD	81	MTCRP3
216	EDGU32	Extensor digitorum (ulnar band)	81	MTCRP3	85	PROXP3
217	EDGU33	Extensor digitorum (ulnar band)	85	PROXP3	86	MIDLP3
218	EDGU34	Extensor digitorum (ulnar band)	63	MIDLP3	64	DISTP3
219	EDSM40	Extensor digitorum (central band)	274	HUMRUS	104	ULNRAD

(continued)

676

A. MUSCLE FORCES: (continued)

No.	Column or Variable Name	Description	Segments and points connected			
			Point 1	Segment 1	Point 2	Segment 2
220	EDGM41	Extensor digitorum (central band)	104	ULNRAD	105	MTCRP4
221	EDSM42	Extensor digitorum (central band)	105	MTCRP4	106	PROXP4
222	EDGM43	Extensor digitorum (central band)	106	PROXP4	107	MIDLP4
223	EDGR40	Extensor digitorum (radial band)	247	HUMRUS	104	ULNRAD
224	EDGR41	Extensor digitorum (radial band)	104	ULNRAD	105	MTCRP4
225	EDGR42	Extensor digitorum (radial band)	105	MTCRP4	108	PROXP4
226	EDGR43	Extensor digitorum (radial band)	108	PROXP4	109	MIDLP4
227	EDGR44	Extensor digitorum (radial band)	110	MIDLP4	111	DISTP4
228	EDGU40	Extensor digitorum (ulnar band)	247	HUMRUS	104	ULNRAD
229	EDGU41	Extensor digitorum (ulnar band)	104	ULNRAD	105	MTCRP4
230	EDGU42	Extensor digitorum (ulnar band)	105	MTCRP4	112	PROXP4
231	EDGU43	Extensor digitorum (ulnar band)	112	PROXP4	113	MIDLP4
232	EDGU44	Extensor digitorum (ulnar band)	110	MIDLP4	111	DISTP4
233	EDGM50	Extensor digitorum (central band)	247	HUMRUS	104	ULNRAD
234	EDGM51	Extensor digitorum (central band)	104	ULNRAD	137	MTCRP5
235	EDGM52	Extensor digitorum (central band)	137	MTCRP5	138	PROXP5
236	EDGM53	Extensor digitorum (central band)	138	PROXP5	139	MIDLP5
237	EDGR50	Extensor digitorum (radial band)	247	HUMRUS	104	ULNRAD
238	EDGR51	Extensor digitorum (radial band)	104	ULNRAD	137	MTCRP5
239	EDGR52	Extensor digitorum (Radial band)	137	MTCRP5	140	PROXP5

677

A. MUSCLE FORCES: (continued)

No.	Column or Variable Name	Description	Segments and points connected			
			Point 1	Segment 1	Point 2	Segment 2
240	EDGR53	Extensor digitorum (radial band)	140	PROXP5	141	MIDLP5
241	EDGR54	Extensor digitorum (radial band)	142	MIDLP5	143	DISTP5
242	EDGU50	Extensor digitorum (ulnar band)	247	HUMRUS	104	ULNRAD
243	EDGU51	Extensor digitorum (ulnar band)	104	ULNRAD	137	MTCRP5
244	EDGU52	Extensor digitorum (ulnar band)	137	MTCRP5	144	PROXP5
245	EDGU53	Extensor digitorum (ulnar band)	144	PROXP5	145	MIDLP5
246	EDGU54	Extensor digitorum (ulnar band)	142	MIDLP5	143	DISTP5
247	BCPSL1	Biceps brachii (long head)	231	SCAPLA	232	HUMRUS
248	BCPSL2	Biceps brachii (long head)	233	HUMRUS	223	ULNRAD

B. JOINT MOMENTS: The joint moment name is obtained by suffixing the joint name with '1P', '2P', '3P', or '1N', '2N', '3N' to denote the positive X,Y,Z components or the negative X,Y,Z components respectively.

No.	Column or Variable Name	Joint Description
1	PP1MC1	Proximal phalanx I/ metacarpal I
2	DP1PP1	Distal phalanx I/ prox. phalanx I
3	PP2MC2	Prox. phalanx II/metacarpal II
4	MP2PP2	Middle phalanx II/prox. phalanx II
5	DP2MP2	Distal phalanx II/middle phalanx II
6	PP3MC3	Prox. phalanx III/metacarpal III
7	MP3PP3	Middle phalanx III/prox. phalanx III
8	DP3MP3	Distal phalanx III/middle phalanx III
9	DD4MC4	Prox. phalanx IV/metacarpal IV
10	MP4PP4	Middle phalanx IV/prox. phalanx IV
11	DP4MP4	Distal phalanx IV/middle phalanx IV
12	PP5MC5	Prox. phalanx V/metacarpal V
13	MP5PP5	Middle phalanx V/prox. phalanx V
14	DP5MP5	Distal phalanx V/middle phalanx V
15	MC1CRP	Metacarpal I/carpals
16	MC2CRP	Metacarpals II/carpals
17	MC3CRP	Metacarpals III/carpals
18	MC4CRP	Metacarpals IV/carpals
19	MC5CRP	Metacarpals V/carpals
20	CRPURD	Carpals/ulna-radius
21	ELBOWR	Ulna-radius/humerus
22	SHOLDR	Humerus/scapula (ground)

C. JOINT REACTIONS

a. Constrained joint reactions: defined by the orientations of
the reaction about Z axis (ϕ_1)
and Y axis (ϕ_2).

No.	Joint Name	Description	Point	Constraint angles	
				ϕ_1	ϕ_2
1	MC4MC5	metacarpals IV/V	185	100.	-30.
2	MC3MC4	metacarpals III/IV	184	70.	5.
3	MC3MC2	metacarpals III/II	183	-120.	-30.

b. Constrained joint reactions: defined by the coordinates of
the two points along which the
reaction is directed.

No.	Joint Name	Description	Point 1	Point 2 (VX,VY,VZ)
1.	DP1PP1	distal/prox. phalanx I	55	3
2	PP1DP1	distal/prox. phalanx I	191	3
3	PP1MC1	prox. phalanx/metacarpal I	192	2
4	MC1PP1	prox. phalanx/metacarpal I	193	2
5	DP2MP2	distal/middle phalanx II	194	7
6	MP2DP2	distal/middle phalanx II	195	7
7	MP2PP2	middle/prox. phalanx II	196	6
8	PP2MP2	middle/prox. phalanx II	197	6
9	PP2MC2	prox. phalanx/metacarpal II	198	5
10	MC2PP2	prox. phalanx/metacarpal II	199	5
11	DP3MP3	distal/middle phalanx III	173	11
12	MP3DP3	distal/middle phalanx III	188	11
13	MP3PP3	middle/prox. phalanx III	172	10
14	PP3MP3	middle/prox. phelanx III	187	10
15	PP3MC3	prox. phalanx/metacarpal III	171	9
16	MC3PP3	prox. phalanx/metacarpal III	186	9
17	DP4MP4	distal/middle phalanx IV	200	15
18	MP4DP4	distal/middle phalanx IV	201	15
19	MP4PP4	middle/proximal phalanx IV	202	14
20	PP4MP4	middle/proximal phalanx IV	203	14
21	PP4MC4	prox. phalanx/metacarpal IV	204	13
22	MC4PP4	prox. phalanx/metacarpal IV	205	13
23	DP5MP5	distal/middle phalanx V	206	19
24	MP5DP5	distal/middle phalanx V	207	19
25	MP5PP5	middle/prox. phalanx V	208	18
26	PP5MP5	middle/prox. phalanx V	209	18
27	PP5MC5	prox. phalanx/metacarpal V	210	17
28	MC5PP5	prox. phalanx/metacarpal V	211	17

c. Partially constrained reactions: defined by the X,Y, or Z component along the required direction (positive or negative)

No.	Joint Name	Description	Reaction Components	Point
1	DP1PP1	distal/prox. phalanx I	±Z	165
2	PP1MC1	prox. phalanx/metacarpal I	±Z	164
3	DP2MP2	distal/middle phalanx II	±Y	169
4	MP2PP2	middle/prox. phalanx II	±Y	168
5	PP2MC2	prox. phalanx/metacarpal II	±Y	167
6	DP3MP3	distal/middle phalanx III	±Y	147
7	MP3PP3	middle/prox. phalanx III	±Y	148
8	PP3MC3	prox. phalanx/metacarpal III	±Y	149
9	DD4MP4	distal/middle phalanx IV	±Y	177
10	MP4PP4	middle/prox. phalanx IV	±Y	176
11	PP4MC4	prox. phalanx/metacarpal IV	±Y	175
12	DP5MP5	distal/middle phalanx V	±Y	181
13	MP5PP5	middle/prox. phalanx V	±Y	180
14	PP5MC5	prox. phalanx/metacarpal V	±Y	179
15	MC1CPR	metacarpal I/carpals	X;Y	212
16	CRPMC1	metacarpal I/carpals	X,-Y	216
17	MC1CRP	metacarpal I/carpals	±Z	163
18	MC2CRP	metacarpal II/carplas	X,-Z	213
19	CRPMC2	metacarpal II/carpals	X,Z	217
20	MC2CRP	metacarpal II/carpals	±Y	166
21	MC3CRP	metacarpal III/carpals	X,-Z	170
22	CRPMC3	metacarpal III/carpals	X,Z	189
23	MC3CRP	metacarpal III/carpals	±Y	162
24	MC4CRP	metacarpals IV/carpals	X,-Z	214
25	CRPMC4	metacarpal VI/carpals	X,Z	218
26	MC4CRP	metacarpal IV/carpals	±Y	174
27	MC5CRP	metacarpal V/carpals	X,-Z	215
28	CRPMC5	metacarpal V/carpals	X,Z	219
29	MC5CRP	metacarpal V/carpals	±Y	178
30	CPRD1U	carpals/ulna-radius	-X,Y,Z	124
31	CPRD1R	carpals/ulna-radius	-X,Y,Z	190
32	CPRD2U	carpals/ulna-radius	-X,Y,-Z	94
33	CPRD2R	carpals/ulna-radius	-X,-Y,-Z	123

d. Unconstrained reactions: defined by all three X,Y,Z components in both positive and negative directions

No.	Joint name	Description	Point
1	ELBOWR	ulna-radius/humerus	220
2	SHOLDR	humerus/scapula (ground)	VX(20),VY(20),VZ(20)

681

'REAC2':   for setting up partially constrained reactions (components only along desired X,Y,Z directions)

'REAC3':   for establishing a reaction between two segments, constrained along a desired direction defined by two points

'EQUATE': for establishing the equilibrium equations for a segment

'EQUAL':   for equating two muscle force components

'DIRCOS': for calculating the direction of cosines of a force

'RHS2':   for establishing the right hand side for a segment with known external forces

"Data':   data used to define the coordinates of muscle attachment point, muscle and joint names, etc.

The 'MAIN' program in conjunction with the above routines creates the data structure for the linear programming problem. This data structure, in terms of the names and types of rows, names of columns and their entries in the respective rows, as well as the objective row the right hand side, etc. is written in the required format on to a file assigned as '15' which is then read for execution by the 'FMPS-LP' optimization package.

7. LANGUAGE

FORTRAN V

8. MACHINE USED

UNIVAC 1110

9. ESTIMATED RUNNING TIME

The compliation of the 'MAIN' program and the additional routines and execution to obtain the required LP data structure requires approximately 7 seconds. Subsequent execution to obtain an optimum solution using the FMPS-LS package requires approximately 34 seconds.

10. USING THE PROGRAM

The input to the program consists of the configuration data, the muscle and joint reaction data, and the external force data.

The configuration can be readily changed by inputting the new values through the 'PHIX', 'PHIY', 'PHIZ' arrays. These values are read in free format, (see lines 73-75 in the 'MAIN' program) the 'PHIX' values for the 25 segments read first, followed by the 'PHIY' and the 'PHIZ' values. As an example, the lines 1-5 in the 'DATA; element defines the configuration described in the illustration earlier.

The merit function can be conveniently changed by specifying the cost coefficients in the objective function for the three types of variables, namely the muscle forces, the joint reaction forces and the joint moments. In the 'MAIN' program the cost coefficients for these three types of variables are defined by means of 'COSTF', 'COSTR', 'COSTM', respectively. Since no provision has been made to permit changing their values outside the program (such as through a 'READ' statement) and necessary changes must be made in the 'MAIN' element.-

To change the cost coefficient of <u>all</u> muscle forces from an initial of 1 to a new value, lines 234 and 266 where the 'COSTF' is assigned the value must be replaced by the new values. Similarly to change the cost coefficient of the joint moments, line 285 should be replaced by 'COST = new value'.

The muscles and the muscle attachment points are as defined in Section 4 above. To change the attachment point of a muscle axis system must be changed. For example, the 'ABPOLL' muscle connects point '11' to pont '17' on the   . The coordinates of point '11' are given by X1(11), Y1(11), Z1(11) with respect to body-fixed axes and the point lies on segment number JBODY(11). Similarly point '176' lies on a segment whose number is JBODY(17) and whose coordinates with respect to body-fixed axes are X1(17), Y1(17), Z1(17). When the overall configuration is established through matrix transformation, the same two points are defined by X1R(11), Y1R(11), Z1R(11) and X1R(17), Y1R(17), Z1R(17), respectively. To change the corrdinates of the point with respect to the body-fixed axes new values must be defined prior to the matrix transformation. This can be done immediately following the statement in the 'MAIN' program where the coordinate data is read (line 94).

PROGRAM FOR THE SPINE, THE UPPER AND THE LOWER EXTREMITIES

1. CAPABILITY

This program is for the analysis of the spine during static and dynamic activities such as stooping and lifting weights, impact and inertia effects, etc. The comprehensive model includes the uppwer and the lower extremities along with the spine and the head. The spine is modelled by fourteen segments including the head--seven cervical vertebrae, five lumbar vertebrae and the twelve thoracic vertebrae along with the rib cage modelled as one segment. The hands and the foot are also included as rigid segments. There are a total of 24 segments numbered as follows:

Segment No.	Segment
1	Right leg
2	Right thigh
3	Pelvis
4	Lumbar vertebra, L5
5	L4
6	L3
7	L2
8	L1
9	Thorax
10	Cervical vertebra, C7
11	C6
12	C5
13	C4

14	C3
15	C2
16	C1
17	Skull
18	Right clavicale
19	Right scapula
20	Right humerus
21	Right ulna
22	Right radius
23	Right hand
24	Right feet

Although the ulna and the radius have been treated as two different segments, this is done only for the purpose of establishing the points of attachments on the two segments separately. For force analysis, however both segments are lumped into a single segment. The foot (both feet) is treated as ground and the reference ground frame is selected midway between the two ankle joints.

There are a total of 105 constraint rows and 431 columns or variables (muscle forces, joint reactions and moments) in the overall optimization formulation.

2. LIMITATIONS

The muscle forces in the model are defined for only one side of the sagittal plane and therefore the model can be used only for the analysis of actions that retain symmetry about this plane.

3. INPUT

i. The X,Y,Z rotations of the twenty-four segments in the model (with respect to ground frame). These are defined through the 'PHIX', 'PHIY', 'PHIZ' arrays as follows:

'PHIX', 'PHIY', 'PHIZ'	Segment
1	Right leg
2	Right thigh
3	Pelvis
4	Fifth lumbar vertebra, L5
5	L4
6	L3
7	L2
8	L1
9	Thorax
10	Seventh cervical vertebra, C7
11	C6
12	C5
13	C4
14	C3
15	C2
16	C1
17	Skull
18	Right clavicle
19	Right scapula
20	Right arm
21	Right ulna

```
22                    Right radius
23                    Right hand
24                    Right foot
```

ii.   The coordinates of muscle attachment points and joint
      locations, muscle names, etc. as listed in 'DATA' element.

iii. The angles along which the intervertebral reactions at the
      articular facets are constrained.

iv.   The external forces along the X,Y,Z axes (such as inertia
      forces during impact or vibration) acting on the segments at
      their centers of masses.

4.   OUTPUT

i.   The X,Y,Z rotations of the twenty four segments input above.

ii.   The values of muscle forces and joint reactions. These are
      printed in abbreviated form on the following pages.

5.   ILLUSTRATION

For the case of illustration, the static standing erect posture (no
dynamic forces) will be considered. The configuration for this case is
as defined in Ch. IV.   These values are read in free format the 'PHIZ'
angles for the 24 segments defined first, followed by the 'PHIY' and the
'PHIZ' angles.

The intervertebral reactions at the discs are constrained to act
along pre-defined orientations. These angles are assumed to be the
average of the tilts (about Y axis) in the sagittal plane for two
adjacent vertebrae measured with respect to vertical:

No.	Joint Name	Segment 1 Name	Segment 1 PHIY'	Segment 2 Name	Segment 2 PHIY'	Reaction constraint angle
1	C2C3JT	C2	10	C3	10	10
2	C3C4JT	C3	10	C4	15	12.5
3	C4C5JT	C4	15	C5	15	15
4	C5C6JT	C5	15	C6	15	15
5	C6C7JT	C6	15	C7	20	17.5
6	C7TXJT	C7	20	Thorax	25	22.5
7	TXLIJT	Thorax	-15	L1	-15	-15
8	L1L2JT	L1	-15	L2	-14	-14.5
9	L2L3JT	L2	-14	L3	-10	-12
10	L3L4JT	L3	-10	L4	5	-2.5
11	L4L5JT	L4	5	L5	18	11.5
12	L5PVJT	L5	18	Pelvis	35	26.5

Similarly, the constraint angles at the articular facet joints for the
vertebral column during standing erect posture are given by:

No.	Joint name	Segment 1	Segment 2	Reaction constraint angle (degrees)
1	C3C2FACT	C3	C2	150
2	C3C4FACT	C3	C4	330
3	C4C5FACT	C4	C5	335
4	C5C6FACT	C5	C6	335
5	C6C7FACT	C6	C7	335
6	C7TXFACT	C7	Thorax	340
7	L1TXFACT	L1	Thorax	75
8	L1L2FACT	L1	L2	255
9	L2L3FACT	L2	L3	256
10	L3L4FACT	L3	L4	260
11	L4L5FACT	L4	L5	275
12	L5PVFACT	L5	Pelvis	288

The angles are measured with respect to vertical and represent the orientation of the reaction force on segment 1.

Since the case being analyzed is a static activity, there are no inertia forces, and the only forces acting on a segment are the segment weights. Therefore FX = FY = 0, and FZ = W, where W = weight of the segment in kg. The forces FX, FY, FZ and the locations of the center of mass for each segment are input in the following order:

No.	FX	FY	FZ	Location of Center of Mass J	Description
1	0	0	-7.2	501	Right thigh
2	0	0	-3.8	502	Right leg
3	0	0	-4.0	503	Scapula
4	0	0	-2.2	504	Right arm
5	0	0	-1.2	505	Right forearm
6	0	0	-0.6	506	Right hand
7	0	0	-5.4	490	Skull
8	0	0	-8.9	491	Thorax
9	0	0	-18.	492	Pelvis

The optimal results in this case using a merit function of minimizing all muscle forces, two times the tensile components of joint reactions at the upper extremity joints, and four times all joint moments are given in Chapter VI.

6. PROGRAM STRUCTURE

The 'MAIN' program essentially creates a data structure in the standard format for subsequent execution by the FMPS-LP (a large scale linearing programming algorithm). The standard-data structure consists of the names and types of the constraint rows followed by the column (variables) names and their entries inthe specified constraint rows and then the 'RHS' side entries. The following subroutines are used in augmentation with the 'MAIN' program:

A. MUSCLE FORCES

No.	Column or Variable Name	Description	Segments and points connected			
			Point 1	Segment 1	Point 2	Segment 2
1	SEMCAPC4	Semispinalis capitis	1	SKUL	251	C4
2	SEMCAPC5	Semispinalis capitis	2	SKUL	252	C5
3	SEMCAPC6	Semispinalis capitis	3	SKUL	253	C6
4	SEMCAPC7	Semispinalis capitis	4	SKUL	254	C7
5	SEMCAPTX	Semispinalis capitis	5	SKUL	255	THRX
6	SPLCAPC7	Splenius capitis	6	SKUL	256	C7
7	SPLCPATX	Splenius capitis	7	SKUL	257	THRX
8	LNSSCPC3	Longissimus capitis	8	SKUL	258	C3
9	LNSSCPC4	Longissimus capitis	9	SKUL	259	C4
10	LNSSCPC5	Longissimus capitis	10	SKUL	260	C5
11	LNSSCPC6	Longissimus capitis	11	SKUL	261	C6
12	LNSSCPC7	Longissimus capitis	12	SKUL	262	C7
13	LNSSCPTX	Longissimus capitis	13	SKUL	263	THRX
14	RCPOSTMJ	Rectus capitis posterior major	14	SKUL	264	C2
15	RCPOSTMN	Rectus capitis posterior minor	15	SKUL	265	C1
16	OQCAPSUP	Oblique capitis superior	16	SKUL	266	C1
17	RECAPANT	Rectus capitis anterior	17	SKUL	267	C1
18	RECAPLAT	Rectus capitis lateralis	18	SKUL	268	C1
19	LNGCAPC3	Longus capitis	19	SKUL	269	C3
20	LNGCAPC4	Longus capitis	20	SKUL	270	C4
21	LNGCAPC5	Longus capitis	21	SKUL	271	C5
22	LNGCAPC6	Longus capitis	22	SKUL	272	C6
23	STCLMDTX	Sterno cleido mastoid	23	SKUL	273	THRX
24	STCLMCC1	Sterno cleido mastoid	24	SKUL	274	RCLV
25	TRPEZCLV	Trapezius	109	SKUL	359	RCLV
26	LCRVC2TX	Longissimus cervicis	25	C2	275	THRX
27	LCRVC3TX	Longissimus cervicis	26	C3	276	THRX
28	LCRVC4TX	Longissimus cervicis	27	C4	277	THRX
29	LCRVC5TX	Longissimus cervicis	28	C5	278	THRX
30	LCRVC6TX	Longissimus cervicis	29	C6	279	THRX
31	LCRVC2C7	Longissimus cervicis	30	C7	280	THRX
32	LCRVC3C7	Longissimus cervicis	31	C3	281	C7
33	LCRVC4C7	Longissimus cervicis	32	C4	282	C7
34	LCRVC5C7	Longissimus cervicis	33	C5	283	C7
35	LCRVC6C7	Longissimus cervices	34	C6	284	C7
36	ILCVC4TX	Iliocostalis cervicis	35	C4	285	THRX
37	ILCVC5TX	Iliocostalis cervicis	36	C5	286	THRX
38	ILCVC6TX	Iliocostalis cervicis	37	C6	287	THRX

(continued)

A. MUSCLE FORCES (continued)

No.	Column or Variable Name	Description	Segments and points connected			
			Point 1	Segment 1	Point 2	Segment 2
39	SPCVC1TX	Splenius cervicis	38	C1	288	THRX
40	SPCVC2TX	Splenius cervicis	39	C2	289	THRX
41	SPCVC3TX	Splenius cervicis	40	C3	290	THRX
42	SMCVC2TX	Semispinalis cervicis	41	C2	291	THRX
43	SMCVC3TX	Semispinalis cervicis	42	C3	292	THRX
44	SMCVC4TX	Semispinalis cervicis	43	C4	293	THRX
45	SMCVC5TX	Semispinalis cervicis	44	C5	294	THRX
46	SMCVC6TX	Semispinalis cercicis	45	C6	295	THRX
47	SMCVC7TX	Semispinalis cervicis	46	C7	296	THRX
48	MLCVC2C5	Multifidus cervicis	47	C2	297	C5
49	MLCVC2C6	Multifidus cervicis	48	C2	298	C6
50	MLCVC3C6	Multifidus cervicis	49	C3	299	C6
51	MLCVC3C7	Multifidus cervicis	50	C3	300	C7
52	MLCVC4C7	Multifidus cervicis	51	C4	301	C7
53	MLCVC4CX	Multifidus cervicis	52	C4	302	THRX
54	MULTIFC1	Multifidus cervicis	53	C5	303	THRX
55	MULTIFC2	Multifidus cervicis	54	C5	304	THRX
56	MULTIFC3	Multifidus cervicis	55	C6	305	THRX
57	MULTIFC4	Multifidus cervicis	56	C6	306	THRX
58	MULTIFC5	Multifidus cervicis	57	C7	307	THRX
59	MULTIFC6	Multifidus cervicis	58	C7	308	THRX
60	RTCVC2C4	Rotatores cervicis	59	C2	309	C4
61	RTCVC3C4	Rotatores cervicis	60	C3	310	C4
62	RTCVC3C5	Rotatores cervicis	61	C3	311	C5
63	RTCVC4C5	Rotatores cervicis	62	C4	312	C5
64	RTCVC4C6	Rotatores cervicis	63	C4	313	C6
65	RTCVC5C6	Rotatores cervicis	64	C5	314	C6
66	RTCVC5C7	Rotatores cervicis	65	C5	315	C7
67	RTCVC6C7	Rotatores cervicis	66	C6	316	C7
68	RTCVC6TX	Rotatores cervicis	67	C6	317	THRX
69	RTCVC7TX	Rotatores cervicis	68	C7	318	THRX
70	ROTCERVS	Rotatores cervicis	69	C7	319	THRX
71	LNCLC1C5	Longus colli	70	C1	320	C5
72	LNCLC2C5	Longus colli	71	C2	321	C5
73	LNCLC3C5	Longus colli	72	C3	322	C5
74	LNCLC4C5	Longus colli	73	C4	323	C5
75	LNCLC1C6	Longus colli	74	C1	324	C6
76	LNCLC2C6	Longus colli	75	C2	325	C6
77	LNCLC3C6	Longus colli	76	C3	326	C6

(continued)

A. MUSCLE FORCES (continued)

No.	Column or Variable Name	Description	Segments and points connected			
			Point 1	Segment 1	Point 2	Segment 2
78	LNCLC4C6	Longus colli	77	C4	327	C6
79	LNCLC1C7	Longus colli	78	C1	328	C7
80	LNCLC2C7	Longus colli	79	C2	329	C7
81	LNCLC3C7	Longus colli	80	C3	330	C7
82	LNCLC4C7	Longus colli	81	C4	331	C7
83	LNCLC1TX	Longus colli	82	C1	332	THRX
84	LNCLC2TX	Longus colli	83	C2	333	THRX
85	LNCLC3TX	Longus colli	84	C3	334	THRX
86	LNCLC4TX	Longus colli	85	C4	335	THRX
87	LNCLC5TX	Longus colli	86	C5	336	THRX
88	LNCLC6TX	Longus colli	87	C6	337	THRX
89	LNCLSUP1	Longus colli	88	C1	338	C3
90	LNCLSUP2	Longus colli	89	C2	339	C4
91	LNCLSUP3	Longus colli	90	C1	340	C5
92	LNCLSUP4	Longus colli	91	C2	341	C3
93	LNCLSUP5	Longus colli	92	C1	342	C4
94	LNCLSUP6	Longus colli	93	C2	343	C5
95	SCALANT1	Scalenus anterior	94	C3	344	THRX
96	SCALANT2	Scalenus anterior	95	C4	345	THRX
97	SCALANT3	Scalenus anterior	96	C5	346	THRX
98	SCALANT4	Scalenus anterior	97	C6	347	THRX
99	SCALMED1	Scalenus medius	98	C1	348	THRX
100	SCALMED2	Scalenus medius	99	C2	349	THRX
101	SCALMED3	Scalenus medius	100	C3	350	THRX
102	SCALMED4	Scalenus medius	101	C4	351	THRX
103	SCALMED5	Scalenus medius	102	C5	352	THRX
104	SCALMED6	Scalenus medius	103	C6	353	THRX
105	SCALMED7	Scalenus medius	104	C7	354	THRX
106	SCALPST1	Scalenus posterior	105	C5	355	THRX
107	SCALPST2	Scalenus posterior	106	C6	356	THRX
108	SCALPST3	Scalenus posterior	107	C7	357	THRX
109	OQCAPINF	Oblique capitis inferior	108	C1	358	C2
110	TRPZC1CL	Trapezius	110	C1	359	RCLV
111	TRPZC2CL	Trapezius	111	C2	359	RCLV
112	TRPZC3CL	Trapezius	112	C3	359	RCLV
113	TRPZC4CL	Trapezius	113	C4	359	RCLV
114	TRPZC5SC	Trapezius	121	C5	371	RSCP

(continued)

A. MUSCLE FORCES (continued)

No.	Column or Variable Name	Description	Segments and points connected			
			Point 1	Segment 1	Point 2	Segment 2
115	TRPZC6SC	Trapezius	122	C6	372	RSCP
116	TRPZC7SC	Trapezius	123	C7	373	RSCP
117	TRPZTXSC	Trapezius	124	THRX	374	RSCP
118	TRPZTX01	Trapezius	188	THRX	438	RSCP
119	TRPZTX02	Trapezius	228	THRX	478	RSCP
120	RHOMC6SC	Rhomboid	125	C6	375	RSCP
121	RHOMC7SC	Rhomboid	126	C7	376	RSCP
122	RHOMBOD1	Rhomboid	127	THRX	377	RSCP
123	RHOMBOD2	Rhomboid	128	THRX	378	RSCP
124	RHOMBOD3	Rhomboid	129	THRX	379	RSCP
125	LEVSCPC1	Levator scapulae	130	C1	380	RSCP
126	LEVSCPC2	Levator scapulae	131	C2	381	RSCP
127	LEVSCPC3	Levator scapulae	132	C3	382	RSCP
128	LEVSCPC4	Levator scapulae	133	C4	383	RSCP
129	RECTABDO	Rectus abdominis	134	THRX	384	PLVS
130	EXTABDO1	External abdominal oblique	135	THRX	385	PLVS
131	EXTABDO2	External abdominal oblique	136	THRX	386	PLVS
132	INTABDOM	Internal abdominal oblique	137	THRX	387	PLVS
133	QUADLMTX	Quadratus lumborum	138	THRX	388	PLVS
134	QUADLML1	Quadratus lumborum	139	THRX	388	PLVS
135	QUADLML2	Quadratus lumborum	140	L2	388	PLVS
136	QUADLML3	Quadratus lumborum	141	L3	388	PLVS
137	QUADLML4	Quadratus lumborum	142	L4	388	PLVS
138	MULTLUM1	Multifidus lumborum	143	THRX	393	L1
139	MULTLUM2	Multifidus lumborum	144	THRX	394	L1
140	MULTLUM3	Multifidus lumborum	145	THRX	395	L2
141	MULTLUM4	Multifidus lumborum	146	THRX	396	L2
142	MULTLUM5	Multifidus lumborum	147	THRX	397	L3
143	MULTLUM6	Multifidus lumborum	148	THRX	398	L3
144	MLLMTXL4	Multifidus lumborum	149	THRX	399	L4
145	MLLML1L4	Multifidus lumborum	150	L1	400	L4
146	MLLML1L5	Multifidus lumborum	151	L1	401	L5
147	MLLML2L5	Multifidus lumborum	152	L2	402	L5
148	MLLML5PV	Multifidus lumborum	153	L5	403	PLVS
149	RTLMTXL1	Rotatores lumborum	154	THRX	404	L1
150	RTLMTXL2	Rotatores lumborum	155	THRX	405	L2
151	RTLML1L3	Rotatores lumborum	156	L1	406	L3
152	RTLML2L4	Rotatores lumborum	157	L2	407	L4

(continued)

A. MUSCLE FORCES (continued)

No.	Column or Variable Name	Description	Point 1	Segment 1	Point 2	Segment 2
153	RTLML3L5	Rotatoris lumborum	158	L3	408	L5
154	IL1ACUSS	Iliacus	409	PLVS	410	RTH1
155	ERECTOR1	Erectoris spinae	164	THRX	414	PLVS
156	ERECTOR2	Erectoris spinae	164	THRX	415	PLVS
157	ERECTRL1	Erectores spinae	166	L1	416	THRX
158	ERECTRL2	Erectores spinae	167	L2	417	THRX
159	ERECTRL3	Erectores spinae	168	L3	418	THRX
160	ERECTRL4	Erectores spinae	169	L4	419	THRX
161	ERECTRL5	Erectores spinae	170	L5	420	THRX
162	ERECTOR3	Erectores spinae	171	PLVS	421	THRX
163	ERECTOR4	Erectores spinae	172	PLVS	422	THRX
164	PECTMAJ2	Pectoralis major	185	THRX	434	RARM
165	PECTMAJ3	Pectoralis major	186	THRX	434	RARM
166	PECTMINR	Pectoralis minor	187	THRX	437	RARM
167	GRACILIS	Gracilis	114	PLVS	364	RLEG
168	ADLONGUS	Adductor longus	115	11	365	RTH1
169	ADMAGNUS	Adductor magnus	116	PLVS	366	RTH1
170	ADMAGNEX	Adductor magnus	117	PLVS	367	RTH1
171	ADBREVIS	Adductor brevis	118	PLVS	368	RTH1
172	SEMITEND	Semitendinosus	119	PLVS	369	RLEG
173	SEMIMEMB	Semimembranosus	120	PLVS	370	RLEG
174	BICEPFLH	Biceps femoris (long head)	206	PLVS	456	RLEG
175	SARTORUS	Sartorius	230	PLVS	480	RLEG
176	GLUTMAX1	Gluteus maximus	231	PLVS	481	RTH1
177	GLUTMEDS	Gluteus medius	233	PLVS	483	RTH1
178	GLUTMINS	Gluteus minimus	234	PLVS	484	RTH
179	STRHYDCL	Sternohyoid	508	SKUL	509	RCLV
180	STRHYDTX	Sternohyoid	508	SKUL	510	THRX
181	STRTHYRD	Sternothyroid	360	SKUL	361	THRX
182	LATDORTX	Latissimus dorsi	178	THRX	428	RARM
183	LATDORPV	Latissimus dorsi	179	PLVS	429	RARM
184	SUBCLAVS	Subclavius	180	THRX	430	RCLV
185	SERRANT1	Serratus anterior	181	THRX	431	RSCP
186	SERRANT2	Serratus anterior	432	THRX	238	RSCP
187	SERRANT3	Serratus anterior	183	THRX	433	RSCP
188	PECTMAJ1	Pectoralis major	184	RCLV	434	RARM
189	DELTOID1	Deltoid	189	RCLV	439	RARM
190	DELTOID2	Deltoid	190	RSCP	440	RARM

(continued)

A. MUSCLE FORCES   (continued)

No.	Column or Variable Name	Description	Segments and points connected			
			Point 1	Segment 1	Point 2	Segment 2
191	DELTOID3	Deltoid	191	RSCP	441	RARM
192	SUPRSPIN	Supraspinatur	192	RSCP	442	RARM
193	INFRSPIN	Infraspinatus	193	RSCP	443	RARM
194	TERESMAJ	Teres major	194	RSCP	444	RARM
195	TERESMIN	Teres minor	195	RSCP	445	RARM
196	SUBSCAPL	Subscapularis	196	RSCP	446	RARM
197	CORACOBR	Coracobrachialis	197	RSCP	447	RARM
198	BICEPSSH	Biceps brachii (short hd.)	198	RSCP	448	RRAD
199	TRICEPS1	Triceps brachii	200	RSCP	450	RULN
200	TRICEP52	Triceps brachii	201	RARM	451	RULN
201	TRICEP53	Triceps brachii	202	RARM	452	RULN
202	BRACHILS	Brachialis	203	RARM	453	RULN
203	ANCONEUS	Anconeus	204	RARM	454	RULN
204	SUPINATR	Supinator	205	RARM	455	RHND
205	PRONATOR	Pronator teres	207	RARM	457	RRAD
206	EXPOLLNG	Extensor pollicis longus	210	RULN	460	RHND
207	EXPOLLBV	Extensor pollicis brevis	211	RRAD	461	RHND
208	EXCRPRDL	Extensor carpi radialis longus	212	RARM	462	RHND
209	EXCRPRDB	Extensor carpi radialis brevis	213	RARM	463	RHND
210	EXDIGITR	Extensor digitorum	214	RARM	464	RHND
211	EXCRPUL1	Extensor carpi ulnaris	215	RARM	465	RHND
212	EXCRPUL2	Extensor carpi ulnaris	216	RULN	466	RHND
213	FXCRPUL1	Flexor carpi ulnaris	217	RARM	467	RHND
214	FXCRPUL2	Flexor carpi ulnaris	218	RULN	468	RHND
215	FXCRPRAD	Flexor carpi radialis	219	RARM	469	RHND
216	FXPOLLL1	Flexor pollicis longus	220	RARM	470	RHND
217	FXPOLLL2	Flexor pollicis longus	221	RRAD	471	RHND
218	FXDIGTS1	Flexor digitorum superf.	222	RARM	472	RHND
219	FXDIGTS2	Flexor digitorum superf.	223	RULN	473	RHND
220	FXDIGTS3	Flexor digitorum superf.	224	RRAD	474	RHND
221	FXDIGTPF	Flexor digitorum profundus	225	RULN	475	RHND
222	ABPOLLL1	Abductor pollicis longus	226	RULN	476	RHND
223	ABPOLLL2	Abductor pollicis longus	227	RRAD	477	RHND
224	BRACHRAD	Brachioradialis	237	RARM	487	RRAD
225	BICEPFSH	Biceps femoris (short head)	235	RTHI	485	RLEG
226	GASTRCMH	Gastrocnemius	236	RTHI	486	RFUT
227	GASTRCLH	Gastrocnemius	240	RTHI	486	RFUT
228	SOLEUSTB	Soleus	241	RLEG	486	RFUT

(continued)

A. MUSCLE FORCES   (continued)

No.	Column or Variable Name	Description	Segments and points connected			
			Point 1	Segment 1	Point 2	Segment 2
229	SOLEUSFB	Soleus	242	RLEG	486	RFUT
230	TIBANTER	Tibialis anterior	243	RLEG	493	RFUT
231	TIBPOSTR	Tibialis posterior	244	RLEG	494	RFUT
232	EXDIGTRL	Extensor digitorum longus	245	RLEG	495	RFUT
233	EXHALLSL	Extensor hallucis longus	246	RLEG	496	RFUT
234	FXDIGTRL	Flexor digitorum longus	247	RLEG	497	RFUT
235	FXHALLSL	Flexor hallucis longus	248	RLEG	498	RFUT
236	PERONESL	Peroneus longus	249	RLEG	499	RFUT
237	PERONESB	Peroneus brevis	250	RLEG	500	RFUT
238	PERONEST	Peroneus tertius	389	RLEG	390	RFUT
239	ILIOPSLI	Iliopsoas	159	L1	409	PLVS
240	ILIOPSO1	Iliopsoas	409	PLVS	410	RTHI
241	ILIOPSL2	Iliopsoas	160	L2	409	PLVS
242	ILIOPSO2	Iliopsoas	409	PLVS	410	RTHI
243	ILIOPSL3	Iliopsoas	161	L3	409	PLVS
244	ILIOPSO3	Iliopsoas	409	PLVS	410	RTHI
245	ILIOPSL4	Iliopsoas	162	L4	409	PLVS
246	ILIOPSO4	Iliopsoas	409	PLVS	410	RTHI
247	ILIOPSL5	Iliopsoas	163	L5	409	PLVS
248	ILIOPO5	Iliopsoas	409	PLVS	410	RTHI
249	LATDORL1	Latissimus dorsi	173	L1	423	THRX
250	LATDORO1	Latissimus dorsi	423	THRX	232	RARM
251	LATDORL2	Latissimus dorsi	174	L2	424	THRX
252	LATDORL2	Latissimus dorsi	424	THRX	232	RARM
253	LATDORL3	Latissimus dorsi	175	L3	425	THRX
254	LATDORO3	Latissimus dorsi	425	THRX	232	RARM
255	LATDORL4	Latissimus dorsi	176	L4	426	THRX
256	LATDORO4	Latissimus dorsi	426	THRX	232	RARM
257	LATDORL5	Latissimus dorsi	177	L5	427	THRX
258	LATDORO5	Latissimus dorsi	427	THRX	232	RARM
259	RECTFEM1	Rectus Femoris	411	PLVS	435	RTHI
260	RECTFEM2	Rectus Femoris	435	RTHI	436	RLEG
261	TFASLAT1	Tensor fasciae latae	391	PLVS	392	RTHI
262	TFASLAT2	Tensor fasciae latae	392	RTHI	412	RLEG
263	GLUTMAX2	Gluteus maximus	413	PLVS	392	RTHI
264	GLUTMAX3	Gluteus maximus	392	RTHI	412	RLEG
265	BICEPSL1	Biceps brachii (long head)	199	RSCP	449	RARM
266	BICEPSL2	Biceps brachii (long head)	239	RARM	489	RRAD
267	VASMED	Vastus medialis	458	RTHI	435	RTHI
			435	RTHI	436	RLEG
268	VASINT	Vastus intermedius	208	RTHI	435	RTHI
			435	RTHI	436	RLEG
269	VASLAT	Vastus lateralis	165	RTHI	435	RTHI
			435	RTHI	436	RLEG

B. JOINT REACTIONS:

    a. Unconstrained joints: Where all three X,Y,Z (both positive and negative) components are set up. The reaction names are obtained by suffixing the joint name with '4P', '5P', '6P' to denote positive X,Y,Z components, and '4N', '5N', '6N' to denote negative components.

No.	Joint Name	Description	
1	SKLLC1	Skull/C1	(reactions about X,Z only)
2	CIC2JT	C1/C2	(reactions about X,Z only)
3	HIPPJT	Pelvis/thigh	(reactions about X,Z only)
4	STCLJT	Thorax/clavicle	(reactions about X,Z only)
5	ANKLJT	Leg/foot	(reactions about X,Z only)
6	KNEEJT	Leg/thigh	(reactions about X,Z only)
7	ACRMCL	Scapula/clavicle	(reactions about X,Z only)
8	SHOLDER	Humerus/scapula	(reactions about X,Z only)
9	ELBOWJ	Ulna/humerus	(reactions about X,Z only)
10	WRISTJ	Radius/hand	(reactions about X,Z only)

    b. Constrained joints: where the reaction force is constrained to act along a predefined orientation.

No.	Joint Name	Description
1	C2C3JT	C2/C3
2	C3C4JT	C3/C4
3	C4C5JT	C4/C5
4	C5C6Jt	C5/C6
5	C6C7JT	C6/C7
6	C7TXJT	C7/Thorax
7	TXL1JT	Thorax/L1
8	L1L2JT	L1/L2
9	L2L3JT	L2/L3
10	L3L4JT	L3/L4
11	L4L5JT	L4/L5
12	L5PVJT	L5/PV
13	C3C2FACT	Articular facet between C3/C2
14	C3C4FACT	Articular facet between C3/C4
15	C4C5FACT	Articular facet between C4/C5
16	C5C6FACT	Articular facet between C5/C6
17	C6C7FACT	Articular facet between C6/C7

(continued)

b. Constrained joints: (continued)

No.	Joint Name	Description
18	C7TXFACT	Articular facet between C7/Thorax
19	L1TXFACT	Articular facet between L1/Thorax
20	L1L2FACT	Articular facet between L1/L2
21	L2L3FACT	Articular facet between L2/L3
22	L3L4FACT	Articular facet between L3/L4
23	L4L5FACT	Articular facet between L4/L5
24	L5PVFACT	Articular facet between L5/Pelvis

c. Joint moments: The moment names are obtained by suffixing the joint name with '1P', '2P', '3P' to denote positive X,Y,Z components and with '1N', '2N', '3N' to develop negative X,Y,Z components respectively.

No.	Joint Name	Description	
1	SKLLC1	Skull/C1	(moments about Y only)
2	C1C2JT	C1/C2	(moments about Y only)
3	HIPPJT	Pelvis/thigh	
4	STCLJT	thorax/clavicle	
5	ANKLJT	leg/foot	
6	KNEEJT	leg/thigh	
7	ACRMCL	scapula/clavilce	
8	SHOLDR	humerus/scapula	
9	ELBOWJ	ulna/humerus	
10	WRISTJ	radius/hand	
11	C2C3JT	C2/C3	(moments about Y only)
12	C3C4JT	C3/C4	(moments about Y only)
13	C4C5JT	C4/C5	(moments about Y only)
14	C5C6JT	C5/C6	(moments about Y only)
15	C6C7JT	C6/C7	(moments about Y only)
16	C7TXJT	C7/Thorax	(moments about Y only)
17	TXL1JT	Thorax/L1	(moments about Y only)
18	L1L2JT	L1/L2	(moments about Y only)
19	L2L3JT	L2/L3	(moments about Y only)
20	L3L4JT	L3/L4	(moments about Y only)
21	L4L5JT	L4/L5	(moments about Y only)
22	L5PVJT	L5/Pelvis	(moments about Y only)

'GRONDM': for setting up moments about desired axes between a segment and ground.

'GRONDR': for setting up reactions along desired axes between a segment and ground.

'RHS2': for setting up the right hand side for a segment with known external forces.

'WRAP': for setting up a muscle with several components that wrap at a common point on a segment and insert along a common orientation (e.g., the quadriceps group).

'MOMENT': for establishing moment components about X,Y,Z (both positive and negative) between two segments.

'REAC': for establishing reaction components along X,Y,Z (both positive and negative) between two segments.

'EQUAL': for equating forces in two components of a muscle that wraps at a point.

'EQUATE': for establishing the equilibrium equation constraints for a segment.

'MUSF': for setting up a muscle force between two components.

'RHSSYM': for setting up the right hand side for a segment which is symmetric about a plane.

'RCNSYM': for setting up a constrained reaction between two segments which are symmetric about a plane.

'MOMSYM': for establishing moments between two segments which are symmetric about a plane.

'RECSYM': for setting up reactions between two segments which are symmetric about a plane.

'MUSSYM': for setting up a muscle force between two segments symmetric about a plane.

'DIRCOS': for calculating the direction consines of a muscle force.

7. ESTIMATED RUNNING TIME

The compilation of the elements and execution to obtain the linear programming problem in standard form requires approximately 8 seconds. Subsequent execution using the FMPS-LP routine to obtain an optimal solution requires approximately 45 secs with a basis specified.

8. LANGUAGE

FORTRAN V

9. MACHINE USED

696

10. USING THE PROGRAM

Although the application of the program was illustrated for the case of static standing erect posture, the program can be readiloy adapted to other situations such as stooping and lifting weights, impact and vibration analysis, etc. In all cases first, the desired configuration must be established by specifying the rotations for the 24 segments in the model. Next, the angles at which the intervertebral reaction both at the discs as well as at the articular facets are constrained must be defined. And finally, the external forces acting on the segments in terms of X,Y and Z components with respect to the ground frame should be defined. Note that the model can only be applied to actions that retain symmetry about the sagittal plane. This is because muscles on only side of the sagittal plane are defined for segments which are symmetric with respect to this plane (the skull, all the vertebrae, the thorax, and the pelvis) and only three equilibrium equations ($\Sigma FX = 0$, $\Sigma FZ = 0$, $\Sigma MY = 0$) have been established for these segments.

A. APPLICATION TO STOOPING AND LIFTING WEIGHTS: The stooping posture is defined by the following configuration:

Segment No.	PHIX	PHIY	PHIZ
1	0	-6	0
2	0	-6	0
3	0	0	0
4	0	40	0
5	0	40	0
6	0	45	0
7	0	48	0
8	0	50	0
9	0	52	0
10	0	65	0
11	0	55	0
12	0	55	0
13	0	47	0
14	0	47	0
15	0	45	0
16	0	45	0
17	0	45	0
18	-25	0	10
19	10	52	-10
20	0	0	0
21	-12	0	0
22	0	0	105
23	-20	0	105
24	0	0	0

The intervertebral reactions at the discs are constrained at the orientations:

No.	Joint Name	Reaction constraint angle (degrees)
1	C2C3JT	46
2	C3C4JT	47
3	C4C5JT	51
4	C5C6JT	55
5	C6C7JT	60
6	C7TXJT	67.5
7	TXL1JT	51
8	L1L2JT	49
9	L2L3JT	46.5
10	L3L4JT	42.5
11	L4L5JT	40
12	L5PVJT	37.5

At the articular facets the constraint angles for the reactions are given by:

No.	Joint name	Constraint angles
1	C3C2FACT	187
2	C3C4FACT	367
3	C4C5FACT	367
4	C5C6FACT	375
5	C6C7FACT	375
6	C7TXFACT	385
7	L1TXFACT	140
8	L1L2FACT	320
9	L2L3FACT	318
10	L3L4FACT	315
11	L4L5FACT	310
12	L5PVFACT	310

The external forces acting on the various segments are:

No.	FX	FY	FZ	J	Segment
1	0	0	-7.2	501	thigh
2	0	0	-3.8	502	leg
3	0	0	-4.0	503	scapula
4	0	0	-2.2	504	arm
5	0	0	-1.2	505	foreamm
6	0	0	W*	506	hand
7	0	0	-5.4	490	skull
8	0	0	-8.9	491	thorax
9	0	0	-18.0	492	pelvis

W* represents the total weight on the hand. If no weights are lifted, W = −0.6 which is the weight of the hand alone. If a total weight of 100 lb (45.35 kg) is lifted then for each hand, the lifted weight is 45.35/2 = 22.68 kg, so that W = −(22.68+0.6) = −23.28, assumed to act through the center of mass of the hand.

From physiological observations it has been found that during strenuous activity, the abdomen develops an internal pressures which effectively relieves the total load on the vertebral column. The internal abdominal pressure can be included in the analysis if the area on which this pressure acts (the diaphragm) and the location of the resultant force between the pelvis and the thorax is known.

B.   APPLICATION TO IMPACT AND VIBRATION ANALYSIS

In these cases the effect of impact (resulting in horizontal inertia forces on the segments) or base oscillations (resulting in vertical inertia forces on the segments) on muscle load sharing can be studied. The posture selected in this case will be the standing erect posture, the configuration for which already has been defined. The intervertebral reaction constraint angles do not change also, so that the only change from the static erect standing posture analysis will be in the external forces acting on the segments. For example, for sudden impact while travelling forward or backward FX = ±W/g A, where W is the segment weight and A represents the acceleration or deceleration. The body usually will not be in the standing erect posture in impact cases and hence the lower extremities including the pelvis may be assumed grounded. Similarly base excitation (assumed to be sinusoidal at a certain frequency) will result in vertical inertia forces and in this case FZ = −W/g (1±k) where k represents the factor of body weight by which the total vertical force is changed.

PROGRAM FOR THE FOOT

1.   CAPABILITY

This program is for the comprehensive analysis of the foot, with the various foot segments treated as individual rigid bodies. The program is for the analysis of the lower extremities, with the pelvis, the right and left thighs, the right and left legs and the right and left feet included in the model. Each foot consists of thirteen segments, the seven tarsals (talus, calcaneus, navicular, cuboid, the three cuneiforms), the five metatarsals and the five toes which are in effect lumped as one segment. With the inferior extensor retinacula for both feet included in the model, there are thus a total of 33 segments in the model, numbered as follows:

Segment No.	Segment
1	Right leg (RLEG)
2	Right thigh (RTHI)
3	Pelvis (PLVS)
4	Right Retinaculum (RRET)
5	Right Talus (RTAL)
6	Right Calcaneus (RCAL)
7	Right Navicular (RVAV)
8	Right Cuboid (RCUB)
9	Right cuneiform I (RCF1)

699

10	Right cuneiform II (RCF2)
11	Right cuneiform III (RCF3)
12	Right metatarsal I (RMT1)
13	Right metatarsal II (RMT2)
14	Right metatarsal III (RMT3)
15	Right metatarsal IV (RMT4)
16	Right metatarsal V (RMT5)
17	Right toes (RLFT)
18	Left thigh (LTH1)
19	Left leg (LLEG)
20	Left retinaculum (LRET)
21	Left talus (LTAL)
22	Left calcaneus (LCAL)
23	Left snavicular (LNAV)
24	Left cuboid (LCUB)
25	Left cuneiform I (LCF1)
26	Left cuneiform II (LCF2)
27	Left cuneiform III (LCF3)
28	Left metatarsal I (LMT1)
29	Left metatarsal II (LMT2)
30	Left metatarsal III (LMT3)
31	Left metatarsal IV (LMT4)
32	Left metatarsal V (LMT5)
33	Left toes (LLFT)

The extrinsic and intrinsic muscles for each extremity are represented by 168 muscle forces, in addition to the working ligaments which are modelled by 25 ligament forces. The joint reactions are modelled by both constrained and unconstrained reaction forces and moments. There are a total of 29 joints on each extremity where the X, Y, and Z moment components are set up along with unconstrained reaction forces at some of the joints (the hip, the knee, the ankle, retinaculum/leg, retinaculum/calcaneus). At the remaining joints the reactions are constrained to act along certain selected orientations depending on the shape or geometry of the two articulating surfaces. In addition, at some of the tarsal joints the X, Y, or Z reaction components, that are established are constrained in sense only.

The ground-to-foot support reactions are included by means of X, Y, Z components at eight locations under each foot and under the five metatarsal heads, under the toes and under the medial and lateral parts of the heel.

The program actually generates the overall constraint matrix consisting of the entires of the variables. This matrix continaing all required information such as names of constraint rows, types of constraints (equality or inequality), names of the variables (columns) along with their respective non-zero entries inthe different constraint rows (as well as in the objective function) and the right hand side vector entries is written on to a file (numbered '15'). This file is then read by the FMPS-LP optimization package for subsequent execution.

2. LIMITATION

Since the foot is now modelled as a collection of individual segments the distribution of the mass of the foot among these segments cannot be reacily determined. Since the foot mass is a small proportion of the overall body weight (foot wt. = 1.0 kg, while total body wt = 72.3

kg) the foot is treated as massless with the total body weight equal to 70.3 kg. The weight of the pelvis, the thigh and the leg are assumed to be 48.3, 7.2 and 3.8 kg, respectively.

The relative motion between the foot segments is insignificant compare to the motions at the other joints (such as at the hip or at the knee and therefore the seven tarsals are treated as one unit for the purpose of kinematics. The unit articulates with the leg at the talo-tibial joint and with the five metatarsals at the tarso-metatarsal joints. The five metatarsals however are allowed their freedom of motion at the tarso-metatarsal joints by specifying their orientations with respect to these joints.

3. INPUT

a.　The coordinates of muscle attachment points as listed under "DATA FOR FOOT PROGRAM."

b.　The rotations of the various segments with respect to the ground frame. The ground frame is selected to be at the right ankle joint with X representing anterior posterior, Y being medio-lateral and Z being superior-inferior. The thirty three segments are each given a successive Z,Y,X rotations with respect to ground frame described by 'PHIZ', 'PHIY' and 'PHIZ' arrays in the program as follows:

I.　'PHIZ', 'PHIY', 'PHIZ' for the segment

1. Right leg and retinaculum
2. Right thigh
3. Pelvis
4. Right tarsals (talus, calcaneus, navicular, cuboid, three cuneiforms)
5. Right metatarsal I
6. Right metatarsal II
7. Right metatarsal III
8. Right metatarsal IV
9. Right metatarsal V
10. Right toes
11. Left thigh
12. Left leg
13. Left tarsals
14. Left metatarsal I
15. Left metatarsal II
16. Left metatarsal III
17. Left metatarsal IV
18. Left metatarsal V
19. Left toes

These angles are defined in the program by reading in free format the 'PHIX' values, the 'PHIY' values and the 'PHIZ' values respectively for all the segments.

c.　The muscle attachment points, muscle names as lissted in "DATA FOR FOOT PROGRAM".

d.　The joint names and coordinates of the joints as listed in "DATA FOR FOOT PROGRAM".

4. OUTPUT

i. One of the program output is the values of the segmental rotations used to establish the configuration. The 'PHIX', 'PHIY', 'PHIZ' angles for all the 33 segments are listed.

ii. The location of the overall center of support, in terms of its X and Y coordinate with respect to the right ankle joint.

iii. The output from the FMPS-LP optimization routine listing the names of the constraint rows followed by the optimal solution with merit cost assigned to each variable such as muscle force, joint reacitons and moments, etc. These are printed in abbreviated form as follows:

A. MUSCLE FORCES

No.	Column or Variable name	Description	Point 1	Segment 1	Point 2	Segment 2
1	GRACLS	Gracilis	1	PLVS	2	RLEG
2	ADLONG	Adductor longus	3	PLVS	4	RTH1
3	ADMAGA	Adductor magnus (adductor)	5	PLVS	6	RTH1
4	ADMAGE	Adductor magnus (extensor)	7	PLVS	8	RTH1
5	ADBREV	Adductor brevis	9	PLVS	10	RTH1
6	SEMITS	Semitendinosus	11	PLVS	12	RLEG
7	SEMIMB	Semimembranosus	13	PLVS	14	RLEG
8	BICFLH	Biceps femoris (long head)	15	PLVS	16	RLEG
9	SARTOR	Sartorius	20	PLVS	21	RLEG
10	GLMAX1	Gluteus maximus	25	PLVS	26	RTH1
11	ILPSOS	Iliopsoas	28	PLVS	29	RTH1
12	GLUMED	Gluteus medius	30	PLVS	31	RTH1
13	GLUMIN	Gluteus minimus	32	PLVS	33	RTH1
14	BICFSH	Biceps femoris (short head)	34	RTH1	16	RLEG
15	GASTMH	Gastrocnemius (medial)	38	RTH1	39	RCAL
16	GASTLH	Gastrocnemius (lateral)	40	RTH1	39	RCAL
17	SOLEST	Soleus	41	RLEG	39	RCAL
18	SOLESF	Soleus	42	RLEG	39	RCAL
19	TPOCF1	Tibialis posterior	47	RLEG	48	RCF1
20	TPONAV	Tibialis posterior	47	RLEG	49	RNAV
21	EDGB41	Extensor digitorum brevis	90	RCAL	60	RLFT
22	FXD1G5	Flexor digiti minimi	102	RMT5	103	RLFT
23	DORI11	Dorsal interosseous	112	RMT1	113	RLFT
24	DORI21	Dorsal interosseous	114	RMT2	113	RLFT
25	DORI22	Dorsal interosseous	115	RMT2	116	RLFT

(continued)

A. MUSCLE FORCES (continued)

No.	Column or Variable Name	Description	Point 1	Segment 1	Point 2	Segment 2
26	DORI31	Dorsal interosseous	117	RMT3	116	RLFT
27	DORI32	Dorsal interosseous	118	RMT3	119	RLFT
28	DORI41	Dorsal interosseous	120	RMT4	119	RLFT
29	DORI42	Dorsal interosseous	121	RMT4	122	RLFT
30	DORI51	Dorsal interosseous	123	RMT5	122	RLFT
31	VOLI31	Volar interosseous	124	RMT3	125	RLFT
32	VOLI41	Volar interosseous	126	RMT4	127	RLFT
33	VOLI51	Volar interosseous	128	RMT5	129	RLFT
34	EDGB31	Extensor digitorum brevis	90	RCAL	59	RLFT
35	TFASL1	Tensor fascial latae	22	PLVS	23	RTH1
36	TFASL2	Tensor fascial latae	23	RTH1	24	RLEG
37	GMAXT1	Glutcus maximus	27	PLVS	23	RTH1
38	GMAXT2	Gluteus maximus	23	RTH1	24	RLEG
39	TBANT1	Tibialis anterior	43	RLEG	44	RRET
40	TANCFL	Tibialis anterior	44	RRET	45	RCF1
41	TBANT2	Tibialis anterior	43	RLEG	44	RRET
42	TANMT1	Tibialis anterior	44	RRET	46	RMT1
43	TPOST1	Tibialis posterior	47	RLEG	50	RTAL
44	TPOCF2	Tibialis posterior	50	RTAL	51	RCF2
45	TPOST2	Tibialis posterior	47	RLEG	50	RTAL
46	TPOCF3	Tibialis posterior	50	RTAL	52	RCF3
47	TPOST3	Tibialis posterior	47	RLEG	50	RTAL
48	TPOMT2	Tibialis posterior	50	RTAL	53	RMT2
49	TPOST4	Tibialis posterior	47	RLEG	50	RTAL
50	TPOMT3	Tibialis posterior	50	RTAL	54	RMT3
51	TPOST5	Tibialis posterior	47	RLEG	50	RTAL
52	TPOMT4	Tibialis posterior	50	RTAL	55	RMT4
53	EXDL21	Extensor digitorum longus	56	RLEG	57	RRET
54	EXDL22	Extensor digitorum longus	57	RRET	58	RLFT
55	EXDL31	Extensor digitorum longus	56	RLEG	57	RRET
56	EXDL32	Extensor digitorum longus	57	RRET	59	RLFT
57	EXDL41	Extensor digitorum longus	56	RLEG	57	RRET
58	EXDL42	Extensor digitorum longus	57	RRET	60	RLFT
59	EXDL51	Extnesor digitorum longus	56	RLEG	57	RRET
60	EXDL52	Extensor digitorum longus	57	RRET	61	RLFT
61	EXHL11	Extensor hallucis longus	62	RLEG	63	RRET
62	EXHL12	Extensor hallucis longus	63	RRET	64	RLFT
63	FXDL21	Flexor digitorum longus	65	RLEG	66	RTAL
64	FXDL22	Flexor digitorum longus	66	RTAL	67	RMT2

(continued)

703

A. MUSCLE FORCES (continued)

No.	Column or Variable Name	Description	Point 1	Segment 1	Point 2	Segment 2
65	FXDL23	Flexor digitorum longus	67	RMT2	68	RLFT
66	FXDL31	Flexor digitorum longus	65	RLEG	66	RTAL
67	FXDL32	Flexor digitorum longus	66	RTAL	69	RMT3
68	FXDL33	Flexor hallucis longus	69	RMT3	70	RLFT
69	FXDL41	Flexor hallucis longus	65	RLEG	66	RTAL
70	FXDL42	Flexor hallucis longus	66	RTAL	71	RMT4
71	FXDL43	Flexor hallucis longus	71	RMT4	72	RLFT
72	FXDL51	Flexor hallucis longus	65	RLEG	66	RTAL
73	FXDL52	Flexor hallucis longus	66	RTAL	73	RMT5
74	FXDL53	Flexor hallucis longus	73	RMT5	74	RLFT
75	FXHL11	Flexor hallucis longus	75	RLEG	76	RTAL
76	FXHL12	Flexor hallucis longus	76	RTAL	77	RCAL
77	FXHL13	Flexor hallucis longus	77	RCAL	78	RMT1
78	FXHL14	Flexor hallucis longus	78	RMT1	79	RLFT
79	PRLNG1	Peroneus longus	80	RLEG	81	RCAL
80	PRLNG2	Peroneus longus	81	RCAL	82	RCUB
81	PRLCF1	Peroneus longus	82	RCUB	83	RCF1
82	PRLNG3	Peroneus longus	80	RLEG	81	RCAL
83	PRLNG4	Peroneus longus	81	RCAL	82	RCUB
84	PRLMT1	Peroneus longus	82	RCUB	84	RMT1
85	PRBRV1	Peroneus brevis	80	RLEG	85	RCAL
86	PRBRV2	Peroneus brevis	85	RCAL	86	RMT5
87	PRTRT1	Peroneus tertius	87	RLEG	57	RRET
88	PPTRT2	Peroneus tertius	57	RRET	88	RMT5
89	EDGB21	Extensor digitorum brevis	90	RCAL	92	RCF3
90	EDGB22	Extensor digitorum brevis	92	RCF3	58	RLFT
91	EHLB11	Extensor hallucis brevis	90	RCAL	91	RMT2
92	EHLB12	Extensor hallucis brevis	91	RMT2	64	RLFT
93	FXDB21	Flexor digitorum brevis	93	RCAL	67	RMT2
94	FXDB22	Flexor digitorum brevis	67	RMT2	68	RLFT
95	FXDB31	Flexor digitorum brevis	93	RCAL	69	RMT3
96	FXDB32	Flexor digitorum brevis	69	RMT3	70	RLFT
97	FXDB41	Flexor digitorum brevis	93	RCAL	71	RMT4
98	FXDB42	Flexor digitorum brevis	71	RMT4	72	RLFT
99	FXDB51	Flexor digitorum brevis	93	RCAL	73	RMT5
100	FXDB52	Flexor digitorum brevis	73	RMT5	74	RLFT

(continued)

A. MUSCLE FORCES (continued)

No.	Column or Variable Name	Description	Point 1	Segment 1	Point 2	Segment 2
101	ABDHL1	Abductor hallucis longus	94	RCAL	95	RMT1
102	ABDHL2	Abductor hallucis longus	95	RMT1	96	RLFT
103	ADHAL1	Adductor hallucis	97	RMT2	98	RMT1
104	ADHAL2	Adductor hallucis	98	RMT1	99	RLFT
105	ADHAL3	Adductor hallucis	100	RMT3	98	RMT1
106	ADHAL4	Adductor hallucis	98	RMT1	99	RLFT
107	ADHAL5	Adductor hallucis	101	RMT4	98	RMT1
108	ADHAL6	Adductor hallucis	98	RMT1	99	RLFT
109	ABDG51	Abductor digiti minimi	104	RCAL	105	RMT5
110	ABDG52	Abductor digiti minimi	105	RMT5	106	RLFT
111	FXHLB1	Flexor hallucis brevis	107	RCUB	108	RMT1
112	FXHLB2	Flexor hallucis brevis	108	RMT1	96	RLFT
113	FXHLB3	Flexor hallucis brevis	109	RCF3	108	RMT1
114	FXHLB4	Flexor hallucis brevis	108	RMT1	96	RLFT
115	FXHLB5	Flexor hallucis brevis	107	RCUB	98	RMT1
116	FXHLB6	Flexor hallucis brevis	98	RMT1	99	RLFT
117	FXHLB7	Flexor hallucis brevis	109	RCF3	98	RMT1
118	FXHLB8	Flexor hallucis brevis	98	RMT1	99	RLFT
119	QDPL21	Qudratus plantae	110	RCAL	67	RMT2
120	QDPL22	Qudratus plantae	67	RMT2	68	RLFT
121	QDPL31	Qudratus plantae	110	RCAL	69	RMT3
122	QDPL32	Qudratus plantae	69	RMT3	70	RLFT
123	QDPL41	Qudratus plantae	110	RCAL	71	RMT4
124	QDPL42	Qudratus planaaa	71	RMT4	72	RLFT
125	QDPL51	Qudratus plantae	110	RCAL	73	RMT5
126	QDPL52	Qudratus palntae	73	RMT5	74	RLFT
127	QDPL23	Qudratus plantae	111	RCAL	67	RMT2
128	QDPL24	Qudratus plantae	67	RMT2	68	RLFT
129	QDPL33	Qudratus plantae	111	RCAL	69	RMT3
130	QDPL34	Qudratus plantae	69	RMT3	70	RLFT
131	QDPL43	Qudratus plantae	111	RCAL	71	RMT4
132	QDPL44	Qudratus plantae	71	RMT4	72	RLFT
133	QDPL53	Qudratus plantae	111	RCAL	73	RMT5
134	QDPL54	Qudratus plantae	73	RMT5	74	RLFT
135	PLNT10	Plantar aponeurosis	130	RCAL	108	RMT1
136	PLNT11	Plantar aponeurosis	108	RMT1	96	RLFT
137	PLNT12	Plantar aponeurosis	130	RCAL	67	RMT2
138	PLNT13	Plantar aponeurosis	67	RMT2	68	RLFT
139	PLNT14	Plantar aponeurosis	130	RCAL	69	RMT3

(continued)

A. MUSCLE FORCES (continuted)

No.	Column or Variable Name	Description	Point 1	Segment 1	Point 2	Segment 2
140	PLNT15	Plantar aponeurosis	69	RMT3	70	RLFT
141	PLNT16	Plantar aponeurosis	130	RCAL	71	RMT4
142	RLNT17	Plantar aponeurosis	71	RMT4	72	RLFT
143	PLNT18	Plantar aponeurosis	130	RCAL	73	RMT5
144	PLNT19	Plantar aponeurosis	73	RMT5	74	RLFT
145	PLNT20	Plantar aponeurosis	131	RCAL	108	RMT1
146	PLNT21	Plantar aponeurosis	108	RMT1	96	RLFT
147	PLNT22	Plantar aponeurosis	131	RCAL	67	RMT2
148	PLNT23	Plantar aponeurosis	67	RMT2	68	RLFT
149	PLNT24	Plantar aponeurosis	131	RCAL	69	RMT3
150	PLNT25	Plantar aponeurosis	69	RMT3	70	RLFT
151	P1NT26	Plantar aponeurosis	131	RCAL	71	RMT4
152	PLNT27	Plantar aponeurosis	71	RMT4	72	RLFT
153	PLNT28	Plantar aponeurosis	131	RCAL	73	RMT5
154	PLNT29	Plantar aponeurosis	73	RMT5	74	RLFT
155	PLNT30	Plantar aponeurosis	132	RCAL	108	RMT1
156	PLNT31	Plantar aponeurosis	108	RMT1	96	RLFT
157	PLNT32	Plantar aponeurosis	132	RCAL	67	RMT2
158	PLNT33	Plantar aponeurosis	67	RMT2	68	RLFT
159	PLNT34	Plantar aponeurosis	132	RCAL	69	RMT3
160	PLNT35	Plantar aponeurosis	69	RMT3	70	RLFT
161	PLNT36	Plantar aponeurosis	132	RCAL	71	RMT4
162	PLNT37	Plantar aponeurosis	71	RMT4	72	RLFT
163	PLNT38	Plantar aponeurosis	132	RCAL	73	RMT5
164	PLNT39	Plantar aponeurosis	73	RMT5	74	RLFT
165	RECFEM	Quadriceps group	17	PLVS	18	RTH1
165	RECFEM	Quadriceps group	18	RTHI	19	RLEG
166	VASMED	Quadriceps group	35	RTHI	18	RTH1
166	VASMED	Quadriceps group	18	RTHI	19	RLEG
167	VASINT	Quadriceps group	36	RTHI	18	RTH1
167	VASINT	Quadriceps group	18	RTHL	19	RLEG
168	VASLAT	Quadriceps group	37	RTHI	18	RTH1
168	VASLAT	Quadriceps group	18	RTHI	19	RLEG

B. JOINT MOMENTS: The joint components along X,Y,Z are defined by
suffixing '1P', '2P', '3P' to denote X,Y,Z components (positive) and '1N', '2N', '3N' to denote X,Y,Z components (negative).

No.	Joint name	Description
1	TIBTAL	Right tibia/talus
2	KNEEJT	Right knee joint (leg/thigh)
3	HIPPJT	Right hip joint (thigh/pelvis)
4	RETLEG	Right retinaculum/leg
5	RETCAL	Right retinaculum/calcaneus
6	TALCAL	Right talus/calcaneus
7	TALNAV	Right talus/navicular
8	CBDCAL	Right cuboid/calcaneus
9	NAVCF1	Right navicular/cuneiform I
10	NAVCF2	Right navicular/cuneiform II
11	NAVCF3	Right navicular/cuneiform III
12	CBDVF3	Right cuboid/cuneiform III
13	CBDMT4	Right cuboid/metatarsal IV
14	CBDMT5	Right cuboid/metatarsal V
15	CF1CF2	Right cuneiform I/cuneiform II
16	CF2CF3	Right cuneiform II/cuneiform III
17	MT1CF1	Right metatarsal I/cuneiform I
18	MT2CF1	Right metatarsal II/cuneiform I
19	MT2CF2	Right metatarsal II/cuneiform II
20	MT2CF3	Right metatarsal II/cuneiform III
21	MT2MT3	Right metatarsal II/metatarsal III
22	MT3CF3	Right metatarsal III/cuneiform III
23	MT3MT4	Right metatarsal III/metatarsal IV
24	MT4MT5	Right metatarsal IV/metatarsal V

(continued)

707

B. JOINT MOMENTS: (continued)

No.	Joint name	Description
25	MTLLFT	Right metatarsal I/toes
26	MT2LFT	Right metatarsal II/toes
27	MT3LFT	Right metatarsal III/toes
28	MT4LFT	Right metatarsal IV/toes
29	MT5LFT	Right metatarsal V/toes
30	LTBTAL	Left tibia/talus
31	LKNEJT	Left knee joint (thigh/leg)
32	LHIPJT	Left hip joint (thigh/pelvis)
33	LRTLEG	Left retinaculum/leg
34	LRTCAL	Left retinaculum/calcaneus
35	LTLCAL	Left talus/calcaneus
36	LTLNAV	Left talus/navicular
37	LCBCAL	Left cuboid/calcaneus
38	LNVCF1	Left navicular/cuneiform I
39	LNVCF2	Left navicular/cuneiform II
40	LNVCF3	Left navicular/cuneiform III
41	LCBCF3	Left cuboid/cuneiform III
42	LCBMT4	Left cuboid/metatarsal IV
43	LCBMT5	Left cuboid/metatarsal V
44	LF1CF2	Left cuneiform I/cuneiform II
45	LF2CF3	Left cuneiform II/cuneiform III
46	LM1CF1	Left metatarsal I/cuneiform I
47	LM2CF1	Left metatarsal II/cuneiform I
48	LM2CF2	Left metatarsal II/cuneiform II
49	LM2CF3	Left metatarsal II/cuneiform III
50	LM2MT3	Left metatarsal II/metatarsal III
51	LM3CF3	Left metatarsal III/cuneiform III
52	LM3MT4	Left metatarsal III/metatarsal IV
53	LM4MT5	Left metatarsal IV/metatarsal V
54	LM1LFT	Left metatarsal I/toes
55	LM2LFT	Left metatarsal II/toes
56	LM3LFT	Left metararsal III/toes
57	LM4LFT	Left metatarsal IV/toes
58	LM5LFT	Left metatarsal V/toes

C. JOINT REACTIONS

a. <u>Unconstrained reactions</u>: defined by suffixing '4P', '5P', '6P' to the joint name   denote positive X,Y,Z components, and '4N', '5N', '6N' to the joint name to denote negative X,Y,Z components. The coordinates of the joint are given by X1R(POINT), Y1R(POINT), Z1R(POINT).

No.	Joint name	Description	Point
1	TIBTAL	Right ankle joint (tibia/talus)	135
2	KNEEJT	Right knee joint (leg/thigh)	136
3	HIPPJT	Right hip joint (thigh/pelvis)	137
4	RETLEG	Right retinaculum/leg	138
5	RETCAL	Right retinaculum/calcaneus	139
6	LTBTAL	Left ankle joint (tibia/talus)	309
7	LKNEJT	Left knee joint (leg/thigh)	310
8	LHIPJT	Left hip joint (thigh/pelvis)	311
9	LRTLEG	Left retinaculum/leg	312
10	LRTCAL	Left retinaculum/calcaneus	313

b. <u>Constrained reactions</u>: constrained to act along predefined orientations.

No.	Joint name	Description	Constraint angles (°)		Point
			PHI1	PHI2	
1	CALTALR	Right calcaneus/talus	-40	-50	140
2	TALNAVR	Right talus/navicular	15	22	141
3	CALCBDR	Right calcaneus/cuboid	0	0	142
4	NAVCF12	Right navicular/cuneiform I	10	30	143
5	NAVCF2R	Right navicular/cuneiform II	-20	0	144
6	NAVCF3R	Right navicular/cuneiform III	-50	20	145
7	CBDCF3R	Right cuboid/cuneiform III	45	-30	146
8	CBDMT4R	Right cuboid/metatarsal IV	-21	34	147
9	CBDMT5R	Right cuboid/metatarsal V	0	0	148
10	CF2CF1R	Right cuneiform II/cuneiform I	60	5	149
11	CF3CF2R	Right cuneiform III/cuneiform II	75	-25	150
12	CF1MT1R	Right cuneiform I/metatarsal I	10	20	151
13	MT2CF1R	Right metatarsal II/cuneiform I	0	-90	152

(continued)

No.	Joint name	Description	Constraint angles (°)		Point
			PHI1	PHI2	
14	CF2MT2R	Right cuneiform II/metatarsal II	-22.5	23	153
15	CF3MT2R	Right cuneiform III/metatarsal II	60	0	154
16	MT3MT2R	Right metatarsal III/metatarsal II	90	-20	155
17	CF3MT3R	Right cuneiform III/metatarsal III	-33	36	156
18	MT4MT3R	Right metatarsal IV/metatarsal III	90	-30	157
19	MT5MT4R	Right metatarsal V/metatarsal IV	90	-40	158
20	MT1LFTR	Right metatarsal I/toes	10	0	159
21	MT2LFTR	Right metatarsal II/toes	0	0	160
22	MT3LFTR	Right metatarsal III/toes	0	0	161
23	MT4LFTR	Right metatarsal IV/toes	0	0	162
24	MT5LFTR	Right metatarsal V/toes	0	0	163
25	CALT11R	Right calcaneus/talus	-40	-90	347
26	CALTL2R	Right calcaneus/talus	0	-70	348
27	CALTL3R	Right calcaneus/talus	35	-45	349
28	TALTB1R	Right talus/tibia	0	-90	350
29	TALTB2R	Right talus/tibia	0	-90	351
30	TALNV1R	Right talus/navicular	0	0	352
31	TALTBYR	Right talus/tibia	90	0	413
32	CALTALL	Left calcaneus/talus	40	-50	314
33	TALNAVL	Left talus/navicular	-15	22	315
34	CALCBDL	Left calcaneus/cuboid	0	0	316
35	VANCF1L	Left navicular/cuneiform I	-10	30	317
36	NAVCF2L	Left navicular/cuneiform II	20	0	318
37	NAVCF3L	Left navicular/cuneiform III	50	20	319
38	CBDCF3L	Left cuboid/cuneiform III	-45	-30	320
39	CBDMT4L	Left cuboid/metatarsal IV	21	34	321
40	CBDMT5L	Left cuboid/metatarsal V	0	0	322
41	CF2CF1L	Left cuneiform II/cuneiform I	-60	5	323
42	CF3CF2L	Left cuneiform III/cuneiform II	-75	-25	324
43	CF1MT1L	Left cuneiform I/metatarsal I	-10	20	325
44	MT2CF1L	Left metatarsal II/cuneiform I	0	-90	326
45	CF2MT2L	Left cuneiform II/metatarsal II	22.5	23	327
46	CF3MT2L	Left cuneiform III/metatarsal II	-60	0	328
47	MT3MT2L	Left metatarsal III/metatarsal II	-90	-20	329
48	CF3MT3L	Left cuneiform III/metatarsal III	33	36	330
49	MT4MT3L	Left metatarsal IV/metatarsal III	-90	-30	331
50	MT5MT4L	Left metatarsal V/metatarsal IV	-90	-40	332
51	MT1LFTL	Left metatarsal I/toes	-10	0	333
52	MT2LFTL	Left metatarsal II/toes	0	0	334
53	MT3LFTL	Left metatarsal III/toes	0	0	335
54	MT4LFTL	Left metatarsal IV/toes	0	0	336

(continued)

b. <u>Constrained reactions</u>: (continued)

No.	Joint name	Description	Constraint angles (°) PHI1	PHI2	Point
55	MT5LFTL	Left metatarsal V/toes	0	0	337
56	CALTL1L	Left calcaneus/talus	40	-90	369
57	CALTL2L	Left calcaneus/talus	0	-70	370
58	CALTL3L	Left calcaneus/talus	-35	-45	371
59	TALTB1L	Left talus/tibia	0	-90	372
60	TALTB2L	Left talus/tibia	0	-90	373
61	TALNV1L	Left talus/navicular	0	0	374
62	TALTBYL	Left talus/tibia	-90	0	414

c. Partially constrained reaction: defined in terms of X, Y or Z components along require directions. Their names are obtained by suffixing '4P', '5P', '6P' to the joint name to denote positive X,Y,Z components and '4N', '5N', '6N' to denote negative X,Y,Z components respectively.

No.	Joint name	Description	Reaction Components	Point
1	RTLNV1	Right talus/navicular	-X,-Y,Z	391
2	RTLNV2	Right talus/navicular	-X,Z	392
3	RTLNV3	Right talus/navicular	-X,Y,Z	393
4	RCLCBD	Right calcaneus/cuboid	-X,-Y	394
5	RCLTL1	Right calcaneus/talus	-X,-Y,-Z	395
6	RCLTL2	Right calcaneus/talus	-X,-Y,-Z	396
7	RCLTL3	Right calcaneus/talus	-X,Y,-Z	397
8	RCLTL4	Right calcaneus/talus	X,-Y,-Z	398
9	RCLTL5	Right calcaneus/talus	-X,Y,-Z	399
10	RCLTL6	Right calcaneus/talus	X,-Y,-Z	400
11	LTLNV1	Left talus/navicular	-X,Y,Z	401
12	LTLNV2	Left talus/navicular	-X,Z	402
13	LTLNV3	Left talus/navicular	-X,-Y,Z	403
14	LCLCBD	Left calcaneus/cuboid	-X,Y	404
15	LCLTL1	Left calcaneus/talus	-X,Y,-Z	405
16	LCLTL2	Left calcaneus/talus	-X,Y,-Z	406
17	LCLTL3	Left calcaneus/talus	-X,-Y,-Z	407
18	LCLTL4	Left calcaneus/talus	X,Y,-Z	408
19	LCLTL5	Left calcaneus/talus	-X,-Y,-Z	409
20	LCLTL5	Left calcaneus/talus	X,Y,-Z	410

D. LIGAMENT FORCES: These are treated like muscles except that
the cost coefficients for the ligaments in
the merit criterion is 0.5 whereas muscles
are assigned a value 1.

No.	Column or Variable Name	Description	Segments and points connected			
			Point 1	Segment 1	Point 2	Segment 2
1	PLNT40R	Right plantar aponeurosis	130	RCAL	78	RMT1
2	PLNT41R	Right plantar aponeurosis	130	RCAL	67	RMT2
3	PLNT42R	Right plantar aponeurosis	130	RCAL	69	RMT3
4	PLNT43R	Right plantar aponeurosis	130	RCAL	71	RMT4
5	PLNT44R	Right plantar aponeurosis	130	RCAL	73	RMT5
6	PLNT45R	Right plantar aponeurosis	131	RCAL	78	RMT1
7	PLNT46R	Right plantar aponeurosis	131	RCAL	67	RMT2
8	PLNT47R	Right plantar aponeurosis	131	RCAL	69	RMT3
9	PLNT48R	Right plantar aponeurosis	131	RCAL	71	RMT4
10	PLNT49R	Right plantar aponeurosis	131	RCAL	73	RMT5
11	PLNT50R	Right plantar aponeurosis	132	RCAL	78	RMT1
12	PLNT51R	Right plantar aponeurosis	132	RCAL	67	RMT2
13	PLNT52R	Right plantar aponeurosis	132	RCAL	69	RMT3
14	PLNT53R	Right plantar aponeurosis	132	RCAL	71	RMT4
15	PLNT54R	Right plantar aponeurosis	132	RCAL	73	RMT5
16	LIGLP1R	Right long plantar ligament	353	RCAL	354	RMT3
17	LIGLP2R	Right long plantar ligament	353	RCAL	355	RMT4
18	LIGLP3R	Right long plantar ligament	353	RCAL	356	RMT5
19	LIGLP4R	Right long plantar ligament	353	RCAL	357	RCUB
20	LIGSP1R	Right short plantar ligament	358	RCAL	359	RCUB
21	LIGSP2R	Right short plantar ligament	360	RCAL	361	RCUB
22	LIGCN1R	Right calcaneo-navicular lig.	362	RCAL	363	RNAV
23	LIGCN4R	Right calcaneo-navicular lig.	367	RCAL	368	RNAV
24	PLNT40L	Left plantar aponeurosis	304	LCAL	252	LMT1
25	PLNT41L	Left plantar aponeurosis	304	LCAL	241	LMT2
26	PLNT42L	Left plantar aponeurosis	304	LCAL	243	LMT3
27	PLNT43L	Left plantar aponeurosis	304	LCAL	245	LMT4
28	PLNT44L	Left plantar aponeurosis	304	LCAL	247	LMT5
29	PLNT45L	Left plantar aponeurosis	305	LCAL	252	LMT1
30	PLNT46L	Left plantar aponeurosis	305	LCAL	241	LMT2
31	PLNT47L	Left plantar aponeurosis	305	LCAL	243	LMT3
32	PLNT48L	Left plantar aponeurosis	305	LCAL	245	LMT4
33	PLNT49L	Left plantar aponeurosis	305	LCAL	247	LMT5
34	PLNT50L	Left plantar aponeurosis	306	LCAL	252	LMT1
35	PLNT51L	Left plantar aponeurosis	306	LCAL	241	LMT2

(continued)

712

D. LIGAMENT FORCES: (continued)

No.	Column or Variable Name	Description	Point 1	Segment 1	Point 2	Segment 2
			Segments and points connected			
36	PLNT52L	Left plantar aponeurosis	306	LCAL	243	LMT3
37	PLNT53L	Left plantar aponeurosis	306	LCAL	245	LMT4
38	PLNT54L	Left plantar aponeurosis	306	LCAL	247	LMT5
39	LIGLP1L	Left long plantar ligament	375	LCAL	376	LMT3
40	LIGLP2L	Left long plantar ligament	375	LCAL	377	LMT4
41	LIGLP3L	Left long plantar ligament	375	LCAL	378	LMT5
42	LIGLP3L	Left long plantar ligament	375	LCAL	379	LCUB
43	LIGSP1L	Left short plantar ligament	380	LCAL	381	LCUB
44	LIGSP2L	Left short plantar ligament	382	LCAL	383	LCUB
45	LIGCN1L	Left calcaneo-navicular lig.	384	LCAL	385	LNAV
46	LIGCN4L	Left calcaneo-navicular lig.	389	LCAL	390	LNAV
47	LIGCN2R	Right calcaneo-navicular lig.	364	RCAL	365	RTAL
48	LIGCN3R	Right calcaneo-navicular lig.	365	RTAL	366	RNAV
49	LIGCN2L	Left calcaneo-navicular lig.	386	LCAL	387	LTAL
50	LIGCN3L	Left calcaneo-navicular lig.	387	LTAL	388	LNAV

E. GROUND-TO-FOOT REACTIONS: The ground-to-foot reactions are defined in terms of X and Y components (positive and negative) and Z component (positive) acting at eight locations under each foot. The names of the reaction components follows the same convection as defined for unconstrained reactions.

No.	Column or Variable Name	Description	Reaction Components	Point
1	MT1GRD	under right metatarsal I	±X,±Y,Z	164
2	MT2GRD	under right metatarsal II	±X,±Y,Z	165
3	MT3GRD	under right metatarsal III	±X,±Y,Z	166
4	MT4GRD	under right metatarsal IV	±X,±Y,Z	167
5	MT5GRD	under right metatarsal V	±X,±Y,Z	168
6	CALGRD	under right calcaneus	±X,±Y,Z	169
7	LFTGRD	under right toes	±X,±Y,Z	170
8	CALGDR	under right calcaneus	±X,±Y,Z	411
9	LMT1GD	under left metatarsal I	±X,±Y,Z	338
10	LMT2GD	under left metatarsal II	±X,±Y,Z	339
11	LMT3GD	under left metatarsal III	±X,±Y,Z	340
12	LMT4GD	under left metatarsal IV	±X,±Y,Z	341
13	LMT5GD	under left metatarsal V	±X,±Y,Z	342
14	LCALGD	under left calcaneus	±X,±Y,Z	343
15	LLFTGD	under left toes	±X,±Y,Z	344
16	CALGDL	under left calcaneus	±X,±Y,Z	412

The muscles names listed above are suffixed with the letter 'R' to denote right side muscles. The left side muscles names are formed by suffixing the letter 'L' and connect points obtained by adding 174 to the points listed above. Thus, 'GRACLSR' is the right side gracilis muscle connecting point no. 1 and point no. 2 while 'GRACLSL' is the left side gracilis muscle connecting point no. 175 and point no. 176.

5. ILLUSTRATION

A computer output for the solution to the leaning forward configuration is shown on page (26). The configuration is defined by the 'PHIX', 'PHIY', 'PHIZ', rotations for the 33 segments as shown on page 24. The center of support in this posture lies at a distance X = 4.1 cm forward of the right ankle joint and Y = 4.8 cm medial. The posture is symmetric about the sagittal plane.

The optimal solution is obtained using the merit criterion, minimize all muscle forces, four times all joint moments and 0.5 times all ligament forces. The output list only the variables in the optimal basis; variables omitted are equal to zero.

6. PROGRAM STRUCTURE

The 'MAIN' program is used along with 12 other subroutines as follows:

'MUSF':   establishes a muscle force between two segments.

'REAC':   establishes X,Y,Z components (positive and negative) of reaction force at a joint.

'MOMENT'': establishes X,Y,Z components (positive and negative) of reaction moment at a joint.

'GRONDR':   establishes X,Y,Z components along required directions of reaction forces between two segments, one of which is ground.

'GRONDM':   establishes X,Y,Z components along required directions of reaction moments between two segments, one of which is ground.

'RHS2':   establishes the right hand side for a segment if the X,Y,Z components of the known force are specified at a desired point on the segment.

'REAC2':   establishes X,Y,Z components of reaction force at a joint along required directions.

'REACON':   establishes a constrained reaction at a joint if the desired orientation is specified.

'WRAP':   sets up a group of muscles that wraps around a segment at a common point.

'EQUAL':   for setting two muscle forces equal in cases where the muscle wraps around a segment and is represented by two or more components.

'EQUATE':   sets up the equilibrium constraint rows for a segment.

'DATA':   data file for the foot program.

7. LANGUAGE

FORTRAN V

8. MACHINE

UNIVAC 1110

9. ESTIMATED RUNNING TIME

The compilation of the FORTRAN elements requires about 3.6 secs., while execution of the program to obtain the constraint matrix in the FMPS-LP format requires approximately 6.5 secs. The actual optimization for the leaning forward case required about two minutes with a full basis provided as a starting point for the optimization process.

10. USE OF THE PROGRAM

In its present set up the program can be applied for the analysis of static activities with both feet on the ground. The merit function is minimize $U = \Sigma F_m + 4 \Sigma M + 0.5 \Sigma F_L$

where ΣF_m = muscle forces

ΣM = joint moments

ΣF_L = ligament forces.

The program can be readily modified to analyze other activities or postures or to change the objective function and so on.

a. Changing the segmental configuration:

The segmental configuration is defined by the 'PHIX', PHIY', PHIZ' arrays containing the X,Y,Z rotations of the segments with respect to ground frame. The configuration for symmetrical standing and leaning forward for example was as defined on page 505. In this case both feet are on the ground.

If the configuration to be analyzed is such that only one foot is on the ground then the ground reaction forces on the foot which is off the ground should be eliminated. This can be done by deleting the statements in the 'MAIN' program which are used to establish the ground-to-foot support reactions. Any data that is read for establishing these reactions (such as names of the reactions and their location) should also be eliminated from the 'DATA' code.

b. Changing the merit criterion:

The merit criterion consists of cost coefficients assigned to the different variables and can be generally expressed as:

minimize $U = C_m \Sigma F_m + C_L \Sigma F_L + C_R \Sigma R + C_M \Sigma M$

where ΣF_M = sum of all muscle forces

ΣF_L = sum of all working ligaments

ΣR = sum of all joint reactions

ΣM = sum of all joint moments,

and C_M, C_L, C_R, C_M are the respective cost coefficients assigned to the above.

In the leaning forward analysis, $C_M = 1$, $C_L = 0.5$, $C_R = 0$ and $C_M = 4$. Any merit criterion can be readily formulated by assigning the desired values to these cost coefficients. In the 'MAIN' program where the muscles and ligament forces, joint reactions and moments are established, they are assigned the cost coefficients by means of 'COSTF', 'COSTR', 'COSTM' variables. In addition, reaction forces or moments that are constrained in direction (by establishing only the required X,Y,Z component in the proper directions) can be assigned the cost coefficients through the 'COSTRC' array as follows:

COSTRC(1) = cost coefficient for X component
COSTRC(2) = cost coefficient for Y component
COSTRC(3) = cost coefficient for Z component

c. Changing muscle attachment points, locations of joint, etc.

A mucle is established by specifying the coordinates of the two pints and the segments or bodies on which the points lie. To change a muscle attachment point, whose coordinates are stored in the 'X1', 'Y1', 'Z1' arrays (and 'X1R', 'Y1R', 'Z1R' arrays after transformation) and the segment number in the 'JBODY' array the change should be made in the input 'DATA' where these values are read. Alternately, changes can also be made in the 'MAIN' program immediately following statement number 15 when the coordinate information is input.

d. Changing joint reaction constraint angles:

The reactions at the joints of the foot are constrained to act along predefined orientations (defined by 'PHI1', and 'PHI1'') about Z and Y axes, respectively. These constraints affect the overall muscle load sharing, and in particular, the distribution of the ground reaction forces among the various sites under the foot. The geometry of the articulating surfaces of the tarsals may actually be used to differentiate between normal and pathological feet.

The joint reaction orientations can be changed in the 'DATA' element where these orientations are read by the program. Note that only the right foot joint constraint angles are specified and the left foot joint reaction constraint angles are assumed to be identical due to symmetry.

Interactive Computer Modeling and Graphics Program for the Musculoskeletal System

1. CAPABILITY

The complexity of the human musculoskeletal structure, and in general, all biomechanical systems, can make the modelling and analysis a cumbersome procedure. What is required therefore, is a <u>general interative computing and graphical display system</u> that should permit modelling of any multi-segmented, inter-connected rigid body structure to investigate different conditions and complex dynamic activities. The interactive capability should allow the user to modify the model and assist in the iterative nature of the modelling process. The system will thus remove the burden of writing the equations of equilibrium both for static and dynamic analysis including inertia effects. Because of the geometric complexities of the musculoskeletal structure, the interactive graphics capability will make it convenient to visually check the model and make any desired geometrical changes.

A comprehensive program which incorporates the above capabilities has been developed by R. J. Williams[1], and its implementation and application to certain segments of the human musculoskeletal structure will be illustrated in the following. Additional details about the program can be found in Refs. [1-3].

2. THE PROGRAM STRUCTURE

The computer system is divided into three major segments (see Fig. 1). The functions performed within each segment are related and thus provide convenient divisions for discussing the organization of the system.

A. MODEL PREPARATION SEGMENT

This segment is primarily concerned with initially defining the

---

[1]Control Data Corporation, Minneapolis, MN.

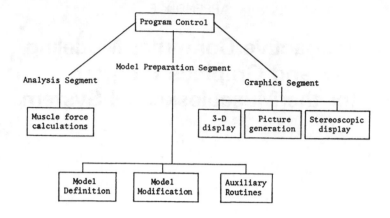

Figure F-1. Organization of the interactive computer graphics system.

model of a mechanical structure to the system or modifying a previously defined model. The information to be provided during model definition is the kinematic topology of the structure, that is:

i.   the rigid bodies included in a model.

ii.   description of joints connecting the bodies indicating the types of intersegmental motions such as rotation, translation, etc.

iii.   the actuators (muscle forces) connecting the rigid bodies along with their points of attachment.

iv.   forces acting on the structure.

The modifications that can be made to a model can be of three types:

i.   additions
ii.   changes
iii.   deletions

and allow for conveniently changing a previously defined model without redescribing the entire model. This is especially important when operating in an interactive computing environment.

The model definition and modification is done by means of a problem-oriente command language for communicating the input data to the computer system. The general format for the command language statement is:

Command name/data pertinent to the command/

The command name alerts the system to the type of data which appears between slashes (/). A set of typical commands used to define as well as change the model is given below. The action produced by these commands will depend upon the mode of operation the system is currently in (INPUT, ADD, CHANGE, or DELETE). The description given for each command generally applies to the INPUT mode:

Command	Description
BODY	Define a body segment in the model
BALL HINGE ROTATE FREE TRANSFORM	Commands used to define the joints between the bodies
POINT	Define a reference point on a body segment
MUSCLE	Define a muscle
TITLE	Establish a descriptive title for the model (< 36 letters)
GRAVITY	Define the gravity vector in terms of its components in a reference coordinate system
FORCE TORQUE	Define forces and torques which represent known loads applied to the structure
REACTION MOMENT	Refer to joint reaction or moments
EQUALITY INEQUALITY	Define equality or inequality constraints relating unknown forces to one another
UNFO UNTO	Define the unknown forces (UNFO) or torques (UNTO) other than muscles in the structure

The auxiliary routines included in this segment consists of a set of three routines which facilitate the operation of the Model Preparation Segment:

a. A routine to obtain the listing of all the components which have been defined for the model (such as bodies, joints, actuator forces, etc.) along with a description of each (command; ECHO).

b. A routine to provide a print out of the description of one particular component of a model. This is useful in an interative mode to obtain the current description of a model component when a modification is being contemplated.

c. A routine to save the entire data structure describing a model on a mass storage device. This facilitates easy and inexpensive recreation of the model from stored data rather than redefining the model component by component (command: WRITE).

B.   THE ANALYSIS SEGMENT

This segment is used to determine the unknown actuator or muscle forces, joint reaction forces and moments in the structure being analyzed. The first step involved in obtaining the solution is to formulate the equations of equilibrium for the model. These are formulated and evaluated numerically by a general algorithm programmed into the system. The actual form of the equations is determined from the data used to define a model during the model definition phase. Therefore, these equations do not have to be written by the user. During the course of the calculations, the system must be supplied with the kinematic data which defines the configuration, the relative velocities and accelerations for dynamic situations and magnitudes of the applied forces acting on the parts of the structure.

The system is made to enter the analysis segment by means of the command EXECUTE (or EXEC for short).

C.   THE GRAPHICS SEGMENT

Because of the geometric complexity of many mechanical structures, the interactive graphics capability of the system is an invaluable addition to the modelling procedure. This segment transforms the numerical model of the structure into a visible entity in a graphics display screen, thus allowing the user to check visually the adequacy of the structure created. In addition, many types of interactive graphic modifications can be made to the model, the results of which are immediately evident on the screen.

The graphics segment consists of three parts as shown in Fig. 1:

   i.   The Picture Generation part of the graphics segment
        provides the capability of producing pictures of the rigid
        bodies defined in a model. These pictures consist of a
        collection of straight lines which represent an outline of
        the surface of the segments ("wire frame" pictures).

   ii. The 3-D Display part of the graphics segment is used to
        obtain a three dimensional (3-D) orthogonal projection of
        the model. The projected view can be in any direction as
        specified by the user. The information displayed consists
        of any pictures of rigid bodies which have been generated,
        along with the actuators, joint reaction forces and
        applied forces which have been defined for a model.

   iii. The Stereoscopic Display part of the graphics segment is
        used to generate two stereoscopic images of a model on the
        screen with a specified picture width, eye separation and
        picture separation. When viewed through a stereoscope,
        these projections produce a true 3-D image in the brain
        allowing the uwer to perceive depth. The command GRAPHICS
        (or GRAP, for short) is used to get the system in the
        graphics mode.

MODEL DEFINITION AND MODIFICATION SEGMENT

A mechanical system in general consists of a series of connected rigid bodies (B1,B2,..., in Fig. 2) undergoing a predetermined dynamic

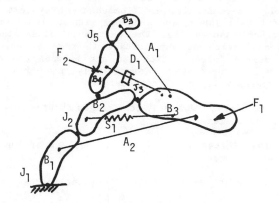

Figure F-2. A general mechanical system of connected
rigid bodies.

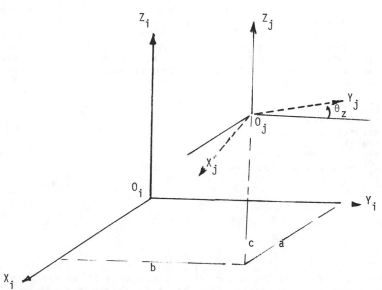

Figure F-3. Rotation about Z axis following a translation (a,b,c).

motion described as relative motion between the bodies at the joints. The joints connecting the rigid bodies are denoted as J1,J2,..., etc. The system can be acted upon with known externally applied forces (F1,F2) as well as gravitational and inertial forces of the rigid bodies. Springs (S1) (where the force is proportional to the net relative displacement of the two ends) and dampers (D1) (where the force is proportional to the net relative velocity of the two ends) can also be included in the system. Conceptually, in biomechanical systems, the muscle has properties of both a spring and a damper with a high level of complexity. It is further assumed that the system is controlled or maintained in dynamic equilibrium at any instant in time by actuator force (A1,A2) (such as muscle forces) which can act between the rigid bodies or within the joints of the system. Consequently, there are three parts of the mechanical system which need to be defined:

a. the kinematic topology, that is, the rigid bodies and the joints between the bodies.

b. the known external forces and moments, as well as springs and dampers which act between the rigid bodies.

c. the unknown force actuators and joint constraint forces.

The kinematic topology is defined by establishing a local cartesian coordinate system for each rigid body. The coordinate system is assumed to remain fixed to the body and moves with it. The mass characteristics (mass, center of gravity, moments and products of inertia) are defined with respect to the local coordinate system. The joints between the bodies must then describe how the coordinate systems of adjacent rigid bodies move with respect to each other. The total motion can be described by a series of relative displacements and rotations. The (4x4) transformation matrix is used to describe these relative motions. Four elementary matrices are used to describe simple as well as complex joints in a step by step fashion:

1. Rotation about the X-axis
2. Rotation about the Y-axis
3. Rotation about the Z-axis
4. Translation only.

For example, the rotation about Z-axis can be described by:

$$
T_z = \begin{bmatrix} 1 & 0 & 0 & 0 \\ a & \cos\theta_z & -\sin\theta_z & 0 \\ b & \sin\theta_z & \cos\theta_z & 0 \\ c & 0 & 0 & 1 \end{bmatrix}
$$

which represents the translation of the origin (a,b,c) followed by a rotation θ_z about an axis parallel to the Z axis, (see Fig. 3). This matrix can be used to define a revolute joint between two rigid bodies. The translation-only matrix describes a relative displacement between two parallel coordinate systems.

The angle in the rotational matrices and any one of the displacement directions in the translation-only matrix may be defined to be a variable quantity. Consequently, combinations of these elementary matrices can be grouped together to define any joint. A spherical joint can be described by a combination of three variable rotational matrices.

The input to the computer system to define kinematic topology, muscle forces, etc., is in the form of command language statements as outlined in the following.

GENERAL RULES FOR COMMAND LANGUAGE STATEMENTS

The general format of the command statements is:

command name/data for this command/

The command name is the first entry in any line and can begin anywhere in the line. The purpose of the name is to either direct the control of the system or alert the system as to how to interpret the data. The data is set off in the line by two slashes (/).

The system searches the statement for the <u>first four letters</u> in the line or until it encounters a slash. It then constructs a word from these letters and compares it with the pre-defined vocabulary. If a match is found, the system takes the appropriate action. If no match is found, the system will print a message to that effect and request another statement. Since only four letters are used to interpret the command name, those longer than four letters can be abbreviated.

The data is usually a list of alphanumeric entries. How the data is interpreted is dependent on the command name. The following rules apply:

i. Each entry in the list of data is separated by a comma. If two commas appear in succession, the entry which would have appeared between them is treated as a numeric zero or an alphabetic blank.

e.g., /NAM1,NAM2,,2.4,5.3/ is equivalent to
       /NAM1,NAM2,0.0,2.4,5.3/ assuming the third entry to be
       numeric

ii. If a list of data is terminated with a slash before it is completed, any entries not specified are treated as blanks or zeroes.

e.g., /NAM1,NAM2,1.1,2.4/ is equivalent to
       /NAM1,NAM2,1.1,2.4,0.0/ provided three numeric entries
       are needed. This is a convenient way of entering
       trailing zeroes.

iii. Alphabetic names in the data can be four letters long or less. If they are longer, only the first four are considered.

iv. If blanks appear within a numeric entry, it is assumed that a decimal point is located at the position of the first blank. Leading and trailing blank are ignored, and the sign + or − is considered part of a number. Blanks are removed from alphabetic data.

e.g., /N A M 1,NAM2,+1.1,2 4,=5.3/ is equivalent to
       /NAM1,NAM2,+1.1,2.4,−5.3/

723

v.   Numeric data can be in integer or decimal format. The system converts each entry to the proper form before interpreting and storing it.

CONTROL STATEMENTS FOR THE MODEL PREPARATION SEGMENT

A model is initially defined and/or modified in the Model Preparation Segment.   The modifications can be one of three types: additions, deletions, and changes. This segment is set up to operate by first setting what is called the mode of operation. This is done by entering one of the four commands: INPUT(I), ADD(A), DELETE(D), or CHANGE(C). The commands may be abbreviated by their first letters.   The INPUT command initializes the ring data structure and prepares the system to accept statements which define a new model. The other three commands instruct the system to interpret the following commands as modifications to the model currently defined in the structure.   Hence the command statements which refer to particular model components have different meanings and different formats depending on which mode of operation the system is in.   Thus, the general flow of information to the system might appear as:

    INPUT   (Initialize data structure and prepare to accept
      .   model definition statements)
      .
      .
Model Definition Statement
      .
      .
      .
    CHANGE, ADD or DELETE (Prepare to modify the model)
      .
      .
      .
      .
Model Modification Statements
      .
      .
      .
      .
[Enter the Graphics Segment]
      .
      .
      .
      .

    1.   The 'BODY' Statement:

        The body statement is used to define a rigid body.

        BODY/name 1, name 2, $\overline{X},\overline{Y},\overline{Z}$ mass, $I_{xx},I_{yy},I_{zz},I_{xy},I_{xz},I_{yz}$/

        name 1: name of the rigid body ($\leq$4 alphanumeric characters)

        name 2: name of a previously defined rigid body. Body
                 "name 1" is defined with respect to body "name 2".
                 For the very first BODY statement, "name 2" refers to
                 fixed or ground frame.

724

$\overline{x}, \overline{y}, \overline{z}$: coordinates of the center of gravity with respect to local coordinates (body-fixed coordinates)

mass: mass of the rigid body

$I_{xx}, \ldots, I_{yz}$: moments and products of inertia

When a series of BODY statements are entered in the input mode, the computer system builds the structure in the order in which the bodies are defined. Thus, the very first BODY statement indicates the ground body or frame and the model is then constructed sequentially. This eliminates the possibility of modifying the topology once established. If any modification is required, for example, adding an extra body or allowing an additional freedom at a joint, the entire model should be redefined component by component.

Further, no closed kinematic loops can be defined since a body segment can be defined only once. In biomechanical systems however in many instances closed loops do occur, such as in the stance phase of walking where both feet are on ground and the lower extremities form a closed loop; or in the pinch mechanism where the tips of two fingers are brought in pinch contact. In such cases, the closed loop may be defined except for one joint which closes the loop. In place of this joint, the reactions which would normally occur at this joint are modelled by unknown forces.

The use of the BODY statement is illustrated below by means of the human lower extremities model (shown in Fig. 4).

The pelvis is considered ground or fixed, and the model is built sequentially by specifying the thigh, the leg, the foot, and the patella.

BODY/THI,PLVS,0.,-.015,-.181,7.7,.3416,.3399,.00224,,,.0209/
    BALL/HIP,X,Y,Z,0.,-.085,0./

BODY/LEG,THI,,,-.177,3.85,.1619,.1619,.0001/HINGE/KNEE,Y,Z,+1,
    -.02,0.,.005,-.418/

BODY/FOOT,LEG,.0566,,-.0146,1.44,,.07265,,,-.03165/BALL/ANKLE,
    X,Z,Y,0,0,-.434/

BODY/PTLA,THI/TRAN/T1,RY,V,-.007,.005,-.39/TRAN/T2,T,CR,.064/

GRAVITY/0,0.-9.8067/

The first BODY statement defines the 'THI' (Thigh) with respect to 'PLVS' (pelvis). Since this is the very first BODY statement, 'PLVS' represents ground frame. Additional information in this statement defines:

$$\left.\begin{array}{l} \overline{X} = 0 \\ \overline{Y} = -.015 \text{ m} \\ \overline{Z} = -.181 \text{ m} \end{array}\right\} \text{ centroid of thigh w.r.t. } X_t, Y_t, Z_t \text{ axes}$$

$$m = 7.7 \text{ KG}$$

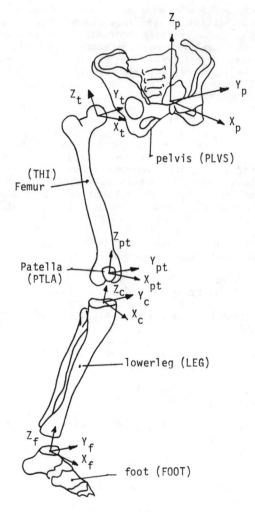

Figure F-4. A model of the human lower extremity.

$$\left.\begin{array}{l} I_{xx} = 0.3416 \text{ kgm}^2 \\ I_{yy} = .3399 \text{ kgm}^2 \\ I_{zz} = .00224 \text{ kgm}^2 \end{array}\right\} \text{ Moments of inertia}$$

$$\left.\begin{array}{l} I_{xy} = 0 \\ I_{xz} = 0 \\ I_{yz} = .0209 \text{ kgm}^2 \end{array}\right\} \text{ Products of inertia}$$

Note that the information in the statement is separated by ',' and a blank between two ','s implies 0. Similarly, if a statement is terminated by means of a '/' before all the parameters are specified, then the parameters omitted are assumed 0. For example, the fourth BODY statement implies that the patella ('PTLA') is connected to the thigh and is assumed massless.

VARIATIONS TO THE 'BODY' STATEMENT

The format shown above for the BODY statement is generally used in the INPUT mode when the model definition has to be initiated. The format for changing certain previously defined parameters for a body (such as location of center of mass, the mass or the moments or products of inertia) is:

BODY/name 1, $\overline{X}, \overline{Y}, \overline{Z}$, mass, $I_{xx}, I_{yy}, I_{zz}, I_{xy}, I_{xz}, I_{yz}/$ (CHANGE mode)

where the parameters defined are the new values. Similarly when information about a previously defined body is to be output, the format is:

BODY/name of the body/   for a single body

or

BODY/ALL/   if all the defined segments are to be retrieved.

Note that this format is used only in the FIND mode.

2. The GRAVITY Statement:

Defines the gravity vector.

GRAVITY/$G_{xx}, G_{yy}, G_{zz}/$

where G_{xx}, G_{yy}, G_{zz} are the gravity vector components in the reference coordinate system

Note that if not specified the value $(0., 0., -9.80566 \text{ m/sec}^2)$ is used by default and all other units in the model (length, inertia, etc.) must be in metric. For non-metric units, this vector must be set to the right value. In the FIND mode the format for obtaining the gravity vector

727

is: GRAV.

The gravity vector determines the orientation of the model in the gravitational field and establishes the directions of the weights of the rigid bodies.

3.   The BALL Statement

Defines a ball joint between two previously defined bodies

    BALL/name, axis 1, [axis 2], [axis 3],X,Y,Z/

        name: name of the ball joint

        axis 1⎫
        axis 2⎬   The axes of rotation signified by an X,Y,
        axis 3⎭   or Z. Axis 2 and 3 are optional.
        X,Y,Z: the constant translation of the joint center.

The BALL statement thus defines a ball and socket joint with three degrees of rotational freedom. The order in which the transformations are carried out are specified by the order in which the axes are described in the BALL statement. For example in the lower extremities model described earlier:

    BODY/THI,PLVS,0.,-.012,-.181,7.7,.3416,.3399,.00224,,,.0209/

    BALL/HIP,X,Y,Z,0.,-.085,0./

a ball joint has been established between the pelvis and the thigh. The translations to the thigh ball center (where the thigh coordinate axes are defined) are X = 0., Y = -.085 m, Z = 0 and the motion at the joint consists of a series of rotations in the order X,Y,Z. If the last one (or two axes) were to be omitted from the statement then the ball joint would have only two (or one) degrees of freedom at the joint.

In general therefore, the 'BALL' statement represents a series of transformations in the order specified and the system will expect a value for each one when the data for the configuration is requested in the EXEC mode.

Besides defining the order of rotations of the hip joint, the BALL statement automatically establishes a set of six generalized coordinates and joint reaction forces and moments. These are located at the body-fixed coordinate axes system and are parallel to them. The positive components of the hip joint reaction for example are as shown in Fig. 5d. Figures 5a, b and c represent the intermediate positions of the thigh-fixed coordinate axes system after successive X,Y,Z rotations. The naming convention for the joint reactions and moments is as follows:

1.   The reaction force component names are formed by taking the first three letters of the joint name and prefixing 4, 5, or 6 to denote the X,Y,Z component respectively. For example the 'HIP' joint forces will be '4HIP', '5HIP', '6HIP', along X,Y,Z respectively acting parallel to the thigh-fixed coordinates axes, on the thigh (see Fig. 3d). The reactions on the pelvis will be equal and opposite.

728

2.   The reaction moment component names are formed by taking the first three letters of the joint name and prefixing 1, 2 or 3 to denote the X,Y,Z component, respectively.   For example, the 'HIP' joint moment will be '1HIP', '2HIP', '3HIP' about X,Y,Z respectively as shown in Fig. 3(d).   The reactions on the pelvis will be equal and opposite.

The same names are used for the negative components as well as and the system automatically keeps track of the sign during the analysis.

VARIATIONS TO THE BALL STATEMENT

In the CHANGE mode, the translation coordinates of the joint location can be changed:

BALL/name, X,Y,Z/   (CHANGE mode)

If information regarding a ball joint is to be retrieved:

BALL/name   (FIND mode)

This will print out information such as the segments connected by the joint, the degrees of freedom at the joint, location of the joint, etc.

3.   The HINGE Statement:

Defines a hinge joint between two segments.

HINGE/name, rotational axis, translational axis, ± 1,r,X,Y,Z/

name: name of the joint

rotational axis: name of the rotational axis about which joint motion occurs (X,Y or Z)

translational axis: name of the translation axis along which the radius of the joint is measured (X,Y, or Z)

± 1:   directional indicator

r: radius of the joint

X,Y,Z: the initial constant translation location the center of rotation.

Using the knee joint as an example, a hinge has been defined between the leg and the thigh as follows:

BODY/LEG,THI,,,-.177,3.85,.1619,.1619,.0001/

HINGE/KNEE,Y,Z,+1,-.02,0.,.005,-.418/'

The center of rotation of the knee joint is defined by X = 0., Y = .005, Z = -.418 m   with respect to the thigh-fixed coordinate axes system.   The rotation at the joint occurs about the Y axis of the intermediate coordinate system, followed by a translation along the z-axis of the rotated system (see Fig. 6).   Note that the HINGE statement

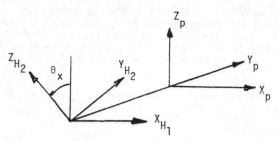

a) Rotation about the x-axis.

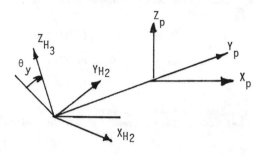

b) Rotation about the y-axis.

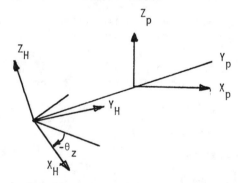

c) Rotation about the z-axis.

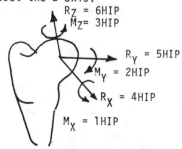

d) Hip joint reaction names.

Figure F-5. Successive rotation at hip joint defined by the BALL statement and the joint reactions names.

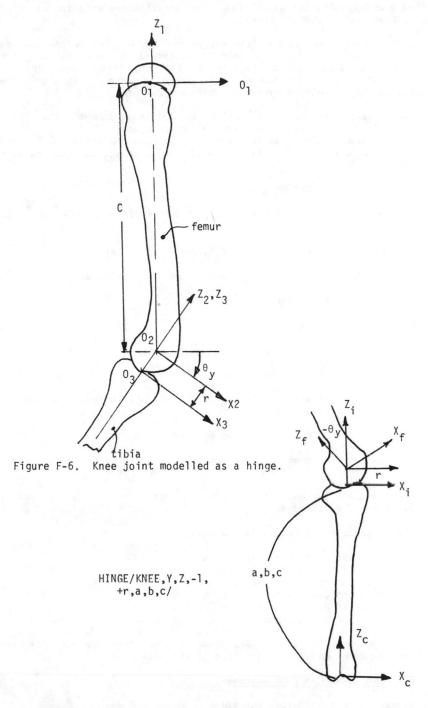

Figure F-6. Knee joint modelled as a hinge.

HINGE/KNEE,Y,Z,-1,
    +r,a,b,c/

Figure F-7. Traversing a hinge joint in the "negative" direction.

is thus equivalent to two successive transformations--a rotation about an axis followed by a constant translation.

A set of generalized coordinates, joint reaction forces and moments are also established when the 'HINGE' statement is used. The joint reaction force components act parallel to the body fixed axes. In case of the 'KNEE' joint for example, the reactions are located and are parallel to the coordinate system attached to the leg.

The directional indicator indicates the position of the joint contact relative to the intermediate axes system. If the lower extremities model had been defined starting up from the ankle joint, then the KNEE joint statement (see Fig. 7) will have a directional indicator = -1 as shown.

VARIATIONS OF THE 'HINGE' STATEMENT

To change the translational axis or the location of the center of rotation or the joint radius,

> HINGE/name, trans. axis, rX,Y,Z/   (CHANGE mode)

> HINGE/name   (FIND mode)

5.   The FREE Statement

Defines a free joint.

> FREE/name,±1/

name: name of the joint

±1:   define equations of equilibrium with reactions and moments at the joint

-1:   no equations to be defined at the joint

In the FIND mode,   FREE/name/   retrieves the joint description

The statement is used to describe the most general type of joint, with six degrees of freedom.

The order of the joint variables are:

1.   translation in the X-direction
2.   translation in the Y-direction
3.   translation in the Z-direction
4.   rotation about X-axis
5.   rotation about Y-axis
6.   rotation about Z-axis

All six are considered variables of the structure and must be specified when the model configuration is established.

6.   The TRANSFORMATION Statement

Defines a transformation matrix:

TRANSFORMATION/name,type,diff.,X,Y,Z,θ/

(or simply, TRAN/name,type,diff.,X,Y,Z,θ/

name: name of the transformation matrix

type: the type of transformation specified by one of the
          following:

    RX - rotation about the X-axis
    RY - rotation about the Y-axis
    RZ - rotation about the Z-axis
    TX - translation along the X-axis
    TY - translation along the Y-axis
    TZ - translation along the Z-axis
     T - translation only matrix (can only be a constant matrix)
  diff - differentiation of the matrix by:

         C - constant matrix
        CE - constant matrix and six equations of equilibrium
        CR - constant matrix, three force equations of equilibrium
             and reaction forces
CRM or CMR - constant matrix, six equations of equilibrium,
             reactions and moments
         V - variable matrix
        VE - ...
        VR - ...
VRM or VMR - ...
     X,Y,Z - constant translation to the origin of the defined
             coordinate system
         θ - value of the angle for a constant rotation (radians)

Variations - To change the constant translation to the origin of a
defined coordinate system or the value of the angle for a constant
rotational transformation matrix,

          TRAN/name,X,Y,Z/   (CHANGE mode)

          TRAN/name/ ⎫
                     ⎬   (FIND mode)
          TRAN/ALL/ ⎭

    The TRAN statement provides a great deal of flexibility in defining
the complex motions between segments.   The statement provides many
options for defining the type of freedom at a joint—rotational or
translational as well as whether reaction forces and moments are to be
established at a joint.

    This is accomplished by defining the **type** and **diff** entries in the
general statement.   The **type** entry relates to the type of motion to be
permitted at the joint—rotation or translation along the various axes.
The **diff** entry has two functions.   First, it is used to determine whether
or not the transformation matrix is a constant matrix (C) or if it
contains a variable (V) of the structure.   The second function of the
diff entry refers to establishing generalized coordinates, and thus
equations of equilibrium and joint reaction forces and moments.   There
are four options to this part:

C or V	alone implies that no generalized coordinates are defined
CE or VE	causes six generalized coordinates to be defined along with six equations of equilibrium but no joint reaction forces or moments are defined. This option is useful when it is convenient to define the coordinate system for a rigid body at a location which does not correspond to a joint.
CR or VR	establishes three translational generalized coordinates along with three corresponding reaction forces. This option can be used when the rigid body defined can be considered to be a point mass and rotational motion is of no consequence. The net effect of this description is a set of three equations in the model, which state simply that the sum of the forces in all three directions at the point mass are equal to zero.
(VRM or VMR) or (CRM or CMR)	establishes a normal joint with six generalized coordinates along with six equations of equilibrium, joint reaction forces and moments.

Note that the generalized coordinates when established, are parallel to the coordinate system defined with the TRAN statement.

As an example consider the joint between the patella and the thigh in the lower extremities model:

```
BODY/PTLA,THI/
TRAN/T1,RY,V,-.007,.005,-.39/
TRAN/T2,T,CR,.064/
```

The patella ('PTLA') is assumed to rotate at a constant radius about a line through the thigh as shown in Fig. 8. The muscles which attach to the patella are assumed to pass through a single point 'A'. The patella thus can be treated as a point mass requiring three equations of equilibrium for force balance only. The matrix 'T1' defines a variable rotation about the Y-axis (RY,V) at a coordinate system located at the center of rotation of the patella, followed by a constant translation T2 along the X-axis (T,CR).

Similarly the hip (ball) and the knee (hinge) joints can be defined by sets of TRAN statements instead of using the BALL or HINGE statements. The hip joint could have been defined by a set of three TRAN statements having **type** and **diff** entries of: (RX,V), (RY,V), and (RX,VRM). Similarly the knee joint could have been defined by two TRAN statement having **type** and **diff** entries of: (RY,V), (T,CRM).

7. The COST Statement

Sets the cost coefficients of the joint reaction forces and moments.

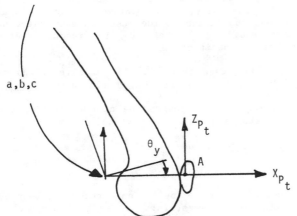

Figure F-8. Coordinate system used to establish the position of the patella.

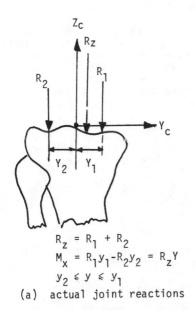

$$R_z = R_1 + R_2$$
$$M_x = R_1y_1 - R_2y_2 = R_zY$$
$$y_2 \leqslant y \leqslant y_1$$

(a) actual joint reactions

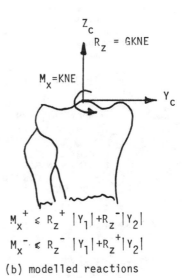

$$M_x^+ \leqslant R_z^+ |Y_1| + R_z^- |Y_2|$$
$$M_x^- \leqslant R_z^- |Y_1| + R_z^+ |Y_2|$$

(b) modelled reactions

Figure F-9. Joint reactions at the knee joint.

```
      COST/REAC,cost coefficient value/
   or
      COST/MOME,cost coefficient value/
```

The first statement sets the cost coefficient for the joint reaction
forces; the second statement can be used to set the cost coefficients for
the joint moments. Note that the use of these statements results in **all**
the joint moments set to cost coefficient values used in the statement.
For example, COST/REAC,2.0/ would result in all joint reaction forces
assigned a cost value of 2 in the merit criterion.

If these statements are not used then all the joint reactions are
assigned a value = 400 by default (consistent with metric units).

Variations - If it is required to change the cost coefficients of
all joint reaction forces or all joint reaction moments,

```
      COST/REAC,cost coefficient value/ ⎫
                                         ⎬  (CHANGE mode)
      COST/MOME,cost coefficient value/ ⎭
```

Similarly to find the values of the cost coefficients that have been set
for the reactions and moments the statement is:

```
              COST                     (FIND mode)
```

It is possible that all reaction forces are not to be set to the
same cost value; similarly, all reaction moments are not to be set to the
same cost value. This can be accomplished in two ways:

a. By using the COST statement before the joint reactions and
   moments of a joint are established, (that is, before a 'BALL',
   'HINGE', 'TRAN', etc., type statements which establish the
   reactions and moments at a joint) and again after the reactions
   and moments are established to set the COST to the original
   value. For example, suppose in the lower extremities example,
   the hip joint reaction costs are to be set at 4. This could be
   done as follows:

```
         BODY/THI,PLVS,.../
         BODY/HIP,.../
   ①  →           COST/REAC,4.0/
         BODY/LEG,THI,.../
         HINGE/KNEE,.../
   ②  →           COST/REAC,2.0/
         BODY/FOOT,LEG,.../
            etc.
```

① causes the joint(s) following this statement to be assigned a
   reaction cost of 4.0 (i.e., at the knee joint)

② resets the cost value to the original default value of 2.

   Note that the hip joint reactions have a COST = 2 since this is
   the preset default value.

b. Alternately, joint reaction costs can be modified in the CHANGE
   mode by using the REACTION and MOMENT statements for each joint

separately. These statements have the capability of assigning a desired cost coefficient not only to a specific joint but also to individual components of the reaction (along X,Y or Z, in either positive or negative direction) at that joint. These statements are discussed in the following.

8. The REACTION and MOMENT Statements:

Used to change or add reaction forces and moments.

(used in CHANGE or ADD modes only)

REACTION
MOMENT /joint name, axis, cost, DI/

or

REAC
MOME /joint name, axis, cost, DI/

joint name: previously defined joint, using BALL, HINGE, FREE or TRAN type statements, or a point where a reaction has been set up (such as through a REAP statement)

axes: axis along which the reaction or moment is defined (x,y, or z)

cost: the value of the cost coefficient of the reaction or moment

DI: Direction Indicator used to determine which reaction force or moment is being referenced

It can take the values 1,-1 or ;0:

1 - positive reaction component
-1 - negative reaction component
0 - both components.

Variations: To delete a previously assigned reaction or moment at a joint:

REAC
MOME /joint name, axis, DI/   (in DELETE mode)

To find an assigned reaction or moment,

REAC
MOME /ALL/   (in FIND mode)

or

REAC
MOME /ALL/   (in FIND mode)

In the ADD mode, these statements are used to add back to the model reaction forces or moments that have previously been DELETED.

Example: To illustrate how the REAC or MOME statements are used let

737

us consider the lower extremities model.

To change the cost coefficient of the hip joint reaction (X component, both directions) one can use (in the CHANGE mode)

$$\left.\begin{array}{l} \text{REAC/HIP,X,4.0.0/} \\ \text{or} \quad \text{REAC/HIP,X,4./} \end{array}\right\} \text{Sets cost to a value} = 4$$

Similarly, to change the cost coefficient of the positive Y component of the knee joint reaction:

REAC/KNEE,Y,4.0,1/   Sets cost to a value = 4

To change the cost coefficient of the moments at the ankle joint about Z axis, both components,

$$\left.\begin{array}{l} \text{MOME/ANKL,Z,100.,0/} \\ \text{or} \quad \text{MOME/ANKL,Z,100./} \end{array}\right\} \text{sets cost to a value} = 100$$

To delete the Z component (both directions, since the patella is free to slide along these directions) and the negative X component at the femur-patella joint one can use (in the DELETE mode):

REAC/T2,Z/
REAC/T2,X-1/,

T2 being the name of the final transformation matrix which defines the coordinate system attached to the patella and sets up the reactions.

9.   The JTTOL Statement

Defines the joint tolerance to a moment.

JTTOL/name,joint,name,moment,reaction,+X,-X/

name: name of the joint tolerance

joint name: name of the joint or transformation where the reaction and moment are defined

moment: the moment which is allowed to vary, (MX,MY,MZ)

reaction: the reaction component which is allowed to move to produce the moment (RX,RY,RZ)

+X: the limit the "reaction" can move in the positive direction so as to produce the "moment".

-X: the negative limit

Variations

MODE	FORMAT
CHANGE	JTTOL/name, +X,-X/

738

```
        FIND                          JTTOL/name/
                                       JTTOL/ALL/
        DELETE                         JTTOL/name/
        ADD                            JTTOL/name,joint name, moment,
                                          reaction, +X,-X/
```

The JTTOL statement is used to define the ability of a joint to resist a moment. It is used when the joint is physically wide and thus the resultant joint reaction force is not necessarily restricted to act at a point. For example, the hip joint, which is a ball and socket joint cannot resist a moment in any direction and the reactions act at the center of the ball. The knee joint on the other hand, can physically resist a moment about the anterior-post posterior (x axis) direction (see Fig. 9a & b). The total joint reaction force is the sum of the forces on the two condyles, (R_1+R_2). In the normal modelled condition there is no relationship between the moment M_x and the reaction R_z; if the moment M_x is zero, this means that the reaction R_z is forced to act at the center of the joint. To model this situation more realistically, the JTTOL statement can be used to form two inequality constraints of the form shown in Fig. 9b. The superscripts + and − refer to the positive and negative components of the moment and reaction force. The inequality constraints essentially state that the joint reaction (which will either be the positive or negative component) is free to act anywhere along the y-axis from $-y_2$ to $+y_1$.

    For example, the knee joint in the lower extremities model can be assigned a joint tolerance by

            JTTOL/KTOL,KNEE,MX,RZ,.015,-.015/.

meaning that the vertical reaction can lie anywhere between −.015 to .015 m. In general, the following combinations of moments and reactions are allowed (and are the only ones which have physical meanings):

            MX-RY; MX-RZ; MY-RX; MY-RZ; MZ-RX; MX-RY.

    In the CHANGE mode, the JTTOL statement can be used to alter the limits, in both positive and negative directions, on how far the reaction force can move to produce the moment. In the DELETE mode a joint can be made free of any such constraint.

10. The CONSTRAINT Statement

    Lists any and all equations of equilibrium (constraint equations) defined at a joint.

            CONSTRAINT/name/   or   CONSTRAINT/ALL/

        name: name of a joint or transformation where the equations
                 are located.

    This statement can only be used in the FIND mode.

11. The POINT Statement

    Defines a point on a rigid body or in the reference coordinate system.

```
                    POINT/name, body, X,Y,Z/
```

name: name of the point ($\leq$ 4 alphanumeric characters)

body: name of the rigid body (or the ground) on which the point is located

X,Y,Z: coordinates of the point in the coordinate system attached to the rigid body

Variations

MODE	FORMAT
CHANGE	POINT/name,X,Y,Z
DELETE	POINT/name
ADD	POINT/name,body,X,Y,Z/
FIND	POINT/name/
	POINT/ALL/

The POINT statement is used to locate and name points on the rigid bodies (including ground or fixed reference body), so they can be easily referred to in other command statements.

12. The MUSCLE Statement

Defines a muscle component between two previously defined points

```
            MUSCLE/name,point name, point name/
```
or
```
            MUSC/name,point name, point name/
```

name: name of the muscle ($\leq$4 alphanumeric characters)

point name: name of the muscle attachment point

Variations

MODE	FORMAT
CHANGE	MUSC/name, cost value/
DELETE	MUSC/name/
ADD	MUSC/name,ptname,pt.name/
FIND	MUSC/name/
	MUSC/ALL/

A muscle is specified by the muscle name and the two points it connects. The muscle exerts a tensile force on the two segments along a line connecting the two points. The muscle is assigned a value of 1.0 in the merit criterion by default and if desired this cost coefficient can be changed in the CHANGE mode using the Statement shown above.

The example for defining a muscle is illustrated in Fig. 10. The soleus muscle (SOLE) connects point P1 on the leg to point P2 on the foot:

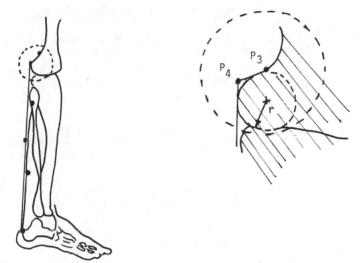

Figure F-10. Models of the soleus muscle (SOLE) and the lateral head of the Gastrocnemius muscle (GASL).

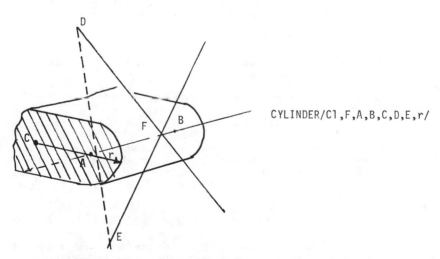

CYLINDER/Cl,F,A,B,C,D,E,r/

Figure F-11. The cylindrical material constraint.

POINT/P1,LEG,x_1,y_1,z_1/
POINT/P2,FOOT,x_2,y_2,z_2/
MUSC/SOLE,P1,P2/

13. The CYLINDER or LINE Statement

Defines a geometric constraint around which a muscle wraps

CYLINDER
  LINE /name, point name, point 1, point 2, material
         ref. point, ref. point, ref. point, radius/

or

CYL1/name, point name, point 1, point 2, material
       ref. point, ref. point, radius/radius/

name:	name of the constraint.
point name:	name of a point whose location is determined.
point 1:	name of a point on the LINE or axis of the cylinder.
point 2:	name of a second point on the LINE or axis of the cylinder.

material reference point: name of a point which is considered to be
                          on the material side of the constraint.

ref. point (2):   the names of the reference points which are used
                  to determine the point of wrapping.

radius: radius of the cylinder (not required for the LINE
          Statement)

In general, the EQUA and INEQ statements allow for the definition of added constraints in the linear programming problem which relate various unknowns to one another. An inequality may appear in the form

$$F1 \leqslant rF2,$$

F1 and F2 being the two unknown variables being related and r being the value of the "ratio" in the command statements.

Thus if $F \leqslant 2 F2$, then this can be defined by:

$$INEQ/F1,LE,F2,2./$$

Similarly, an unknown variable (such as a muscle force or other unknown forces defined by using UNFO statements) can also be related to a constant value by specifying the name 'CONSTANT' for name 2, and the value of the constant for the ratio. For example, if $F1 \leqslant 40.$, then this can be defined by:

$$INEQ/F1,LE,CONSTANT,40./$$

The formats for the EQUA and INEQ statements in other modes of operation have been shown under "variations". The "number" used in these statements refers to a particular constraint. Since the constraints are not given names, they are referred to in the order in which they are defined. The first one is number 1, second is 2 and so on. If one of them is deleted at any time, the number used to refer to those after the deleted one is decreased by one.

For example, suppose there are three inequality constraints in the list of input command statements as follows:

INEQ/F1,type,F2,ratio/
INEQ/F3,type,F4,ratio/
INEQ/F5,type,F6,ratio/

Note that these statements need not appear consecutively, but they may be embedded in the list of command statements and the three statements above merely reflect the order in which the three appear. Thus, the three constraints will be numbered consecutively 1, 2, 3 and if it is required to delete say the second one, then in the DELETE mode,

INEQ/2/   (INEQ/F3,type ...)

will delete the second one (INEG/F3,type,.../) and the third one will now be numbered 2, and all the following constraints will have their numbers decreased by 1.

15. The REAP Statement

Defines a common reaction point where three or more muscles meet

REAP/name,body,x,y,z/

    name: name of the common reaction point.
    body: name of the body used as a reference to define the
             point.
    x,y,z: coordinates of the point on the referenced rigid body.

Variations:

REAP/name,X,Y,Z/   (CHANGE mode)
REAP/name/   (FIND mode)
REAP/name/   (DELETE mode)

The REAP statement is used to model muscles that have more than one component which wrap at a common point. For example, the tensor fasciae latae (TFSL) muscle and the superficial fibers of the gluteus maximus (GUMI) meet at the greater trochanter on the femur (point R3) and insert by the common iliotibial tract (1LTB) into the let (see Fig. 13). Consequently there will be reaction forces on the greater trochanter due to the wrapping. The point of wrap can be defined as a reaction point using the REAP statement:

$$REAP/R3,THI,X_{R_3},Y_{R_3},Z_{R_3}/$$

With the statement the program automatically defines

    i.   a point with the name specified in the statement (R3)

    ii.   reaction forces at the defined point in terms of X,Y,Z
          components parallel to the coordinate axes system of the body
          on which the point is located (X_t,Y_t,Z_t for the THI)

    iii. a constant translation-only matrix which defines the coordinate

743

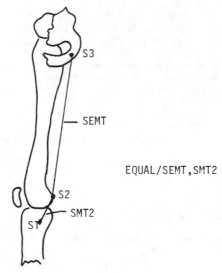

SEMT

S3

S2

SMT2

S1

EQUAL/SEMT,SMT2

Figure F-12. Semitendinosus muscle modeled as two components.

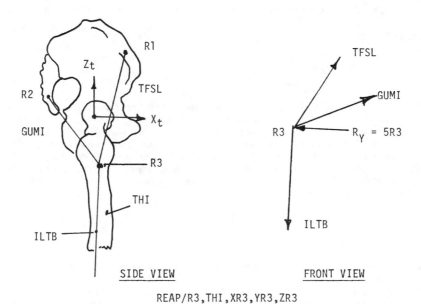

REAP/R3,THI,XR3,YR3,ZR3

Figure F-13. Three force components meeting at a reaction point.

744

system at the point--that is, the translation vector in the matrix contains the X,Y,Z coordinates of the point (X_R, Y_R, Z_R).

MODE	FORMAT
CHANGE	CYI/name,radius/
DELETE	CYLI LINE /name/
ADD	Same as the CYLINDER LINE statement in INPUT mode
FIND	CYLI LINE /name/
	CYLI LINE /ALL/

The CYLINDER or LINE statement is useful when it is required to model a muscle which wraps around an intervening segment between its points of origin and insertion, for example, the gastrocnemius muscle (see Fig. 10). This muscle arises from the condyles and inserts into the foot. In certain postures such as during standing erect or when the leg is not flexed with respect to the thigh at the knee joint, the muscle cannot be modelled by a direct line (connecting point P3 and P2) due to obstruction by the condyles. Instead the true or equivalent origin may be assumed at point P4 which is the intersection of two tangents to the curved condyle. The CYLI/LINE statements essentially determines if there is an obstruction between the point of origin and insertion. In the presence of an obstruction the equivalent attachment point P4 is computed.

The use of the CYLI statement is illustrated in Fig. 11 where a muscle connects points D and E with wrapping at point F on the interposed cylinder of radius r whose axis is defined by points A and B. The point F is determined by the intersection of the two tangents to the cylinder from reference points E and D, such that, the distance DFE is a minimum. Point C defines where the material is for the constraint, so that it can be determined on which side of the cylinder the line wraps.

The CYLI statement in the gastrocnemius muscle model would appear as:

CYLI/GASL,P4,P5,P6,P7,P3,P2,/MUSC/GASL,P4,P2/

Note that P4 must be defined with a POINT statement as being located on the THI (its coordinates are arbitrary) since the CYLI statement does not define this point. The GASL muscle is then defined as acting betweem P4 and P2.

If the line DE in Fig. 11 does not intersect the cylinder, point F is chosen as the midpoint of this line. Relating this to the gastrocnemius muscle this would mean that P4 would be located approximately midway down the leg if the muscle unwraps. This is still accurate, since the force in the muscle is a vector quantity acting between points of attachment and therefore the actual point along the vector where the force is applied is not important. Point P4 would still

745

be considered attached to the thigh for directional and other calculations.

14. The EQUALITY/INEQUALITY Statement

Used to establish an equality or inequality constraint relationship between two unknown forces

INEQUALITY/name 1, type, name 2, ratio/

(or   INEQ/name 1, name 2, ratio)
(or EQUA/name 1, name 2, ratio/)

name 1: name of a muscle or other unknown force component

type: sense of the equality; can take one of the three
         values:

         LE   <
         EQ for   =   type constraints, respectively.
                     >

name 2: name of a second muscle or unknown force component or
         'CONSTANT'

ratio: value of the ratio of the magnitudes of the forces
         referred to by name 1 and name 2 (name 1/name2) or the
         value of the constant if name 2 is 'CONSTANT'.

Note that the EQUALITY statement achieves the same purpose as INEQUALITY statement with EQ type specified. Also, in the EQUA statement if the ratio is 1. This may be omitted and the statement might then appear as EQUA/name 1, name 2/ ratio = 1 by default.

Variations

MODE	FORMAT
ADD	same as above
CHANGE	EQUA/number, ratio/
DELETE	EQUA/number/
FIND	EQUA/number/
	EQUA/ALL/

An illustration of the EQUA statement is shown for the semitendinsus muscle model in Fig. 12. This muscle originates on the pelvis (point S3) and inserts on the leg (point S1) with wrapping on the thigh bone point S2). With S1, S2 and S3 defined using POINT statements, two muscle components for the semitendiniosus can be defined, one connecting S2 to S3 (SEMT) and the other connecting S1 to S2 (SMT2). Since both components actually represent the same muscle, both can be equated by:

EQUA/SEMT,SMT2/

In this manner the effects of the reaction between the muscle and intermediate bone are included without defining the reactions explicitly. Point S2 can be either specified as a fixed point or could

746

be determined by using a cylinder or line constraint. The second method automatically takes into account the effects of unwrapping.

iv. a point mass (a rigid body) with the same name as the point with three generalized coordinates and three equations of equilibrilum. The reaction forces created at the point use the sae convention as that for defining joint reactions (i.e., 4R3, 5R3, GR3 name the X,Y,Z components, respectively). Since all the three reaction force components are established for each reaction point defined, those that are not wanted must be deleted by using the REAP statement in the DELETE mode. For the case illustrated (Fig. 13) all, except the negative y component must be deleted. Once the reaction point has been defined, its name can be used the same as any other point in the statements defining the muscles.

16. The UNFO Statement

Defines an unknown force component

UNFO/name,point 1,point 2,(point 3),cost,DI/

name: name of the force component

point 1, point 2, point 3: names of points. (When point 3 is not specified (blank), the force acts between point 1, and point 2. When point 3 is specified, it is the point of application for an externally applied force and a vector directed from point 1 to point 2 indicates its positive direction).

cost: the value of the cost coefficient for this force in the merit ceriterion.

DI: directional indicator used to determine whether the force is unidirectional (like a muscle) or bidirectional. It can take the values 1, −1 or 0:

1 positive direction only (i.e., only tensile force, like a muscle)

−1 negative direction only

0 force can act in both directions.

Variations:

MODE	FORMAT
ADD	Same as above
CHANGE	UNFO/name,cost,DI/
DELETE	UNFO/name/
FIND	UNFO/name/
	UNFO/ALL/

The UNFO statement is used to define forces other than muscles which are applied to the structure or act between two rigid bodies of the

747

structure.

If the unknown force acts between two rigid bodies of the structure (such as a muscle force) then the third point (point 3) is specified as a blank. The DI indicates the directional type of the force whether tensile alone (trying to pull the two bodies together) or compressive alone (trying to push the two bodies apart), or in both directions. Thus, if Points 1 and 2 are on two different segments, a muscle force could also be defined by using the UNFO statement:

$$UNFO/name,point\ 1,point\ 2,,cost,1/$$

As can be seen this statement offers the option of assigning the desired cost coefficient also.

If the force is externally applied to the structure, three points are required (see Fig. 14). The side force applied to the foot during bicycling is defined as unknown force by means of:

$$POINT/P1,FOOT,x_1,y_1 z_1/$$

$$POINT/P2,FOOT,x_2,y_2,z_2/$$

$$UNFO/FY,P1,P2,P2,0.,0/$$

which defines an unknown force capable of acting in both directions acting at point P2 and in the direction P1 to P2. Note that P1, P2 need not be defined as lying on the same body as the point of application. Suppose the direction of the force on the foot is to be given by two points on the ground reference frame (in this case, the pelvis) then the two points can be defined on the pelvis. The direction of the force acting at P3 on the foot then will be based on these two points.

17. The UNTO Statement

Defines an unknown externally applied torque

$$UNTO/name,point\ 1,point\ 2,body,cost,DI/$$

name: name of the torque

points 1,2:   names of two points which establish the positive direction of the torque vector (right hand screw rule).

body: name of th rigid body upon which the torque acts.

cost: the value of the cost coefficient for this torque in the merit criterion.

DI: Directional indicator used to determine the direction in which the torque acts. It can take the values +1, -1, or 0:

1   positive direction only
-1   negative direction only
0   both directions

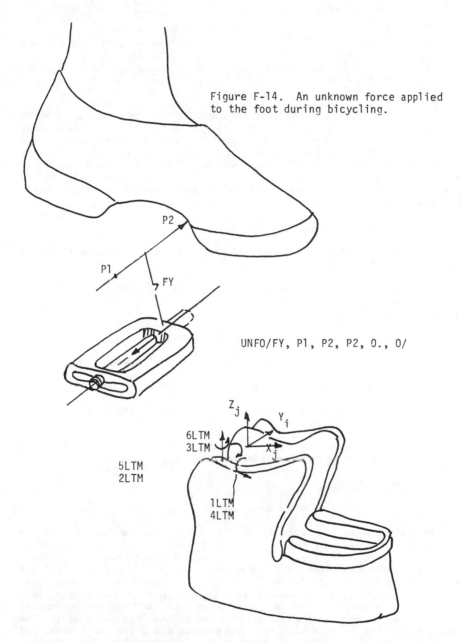

Figure F-14. An unknown force applied to the foot during bicycling.

UNFO/FY, P1, P2, P2, 0., 0/

Figure F-15. The jaw model showing the use of the ground reaction statement.

Variations:

MODE	FORMAT
ADD	Same as above
CHANGE	UNTO/name,cost,DI/
DELETE	UNTO/name/
FIND	UNTO/name/
	UNTO/ALL/

The UNTO statement is similar to the UNTO statement and is used to establish an unknown torque on the structure or between two rigid bodies. Since the torque is defined to be acting on a rigid body rather than at a particular point on the body, "point 3" in the UNFO statement is replaced by the name of the body on which the torque is to be defined. Points 1 and 2 are used to define the positive sense of the torque using the right hand screw rule, with the torque vector being directed from point 1 to point 2.

18. The GROUND REACTION Statement

Establishes unknown ground reaction bases and torques which act a point on a rigid body.

GROUND. REACTION/name,body,x,y,z,cost 1, cost 2/

(or GROU/.../)

name: name of the gound reaction point.

body: rigid body upon which the reactions act.

x,y,z: coordinates of the point on the rigid body.

cost 1:   value of the cost coefficient of the unknown forces which are defined.

cost 2:   value of the cost coefficient of the unknown torques which are defined.

Variations:

In ADD. mode: same as above

The ground reaction force is used to define unknown reaction forces and torques which act between a body and ground where there is no joint defined. In this case, these forces and torques would have to be defined using three UNFO and three UNTO statements. However, the same effect can be accomplished by using the GROU statement. The unknown reaction foces ae defined to act parallel to the axes in the ground (fixed) coordinate system. The moments act about these same axes in the same manner as joint reaction moments act about various body-fixed coordinate axes. The naming convention is also similar. The numbers 1,2,3 for the moments and 4,5,6 for the forces (corresponding to X,Y,Z components, respectively) are prefixed to the first three letters of the name used in the GROU statement. To refer to the forces and torques defined by the GROU statement and to make modifications to them, the UNFO and UNTO statements should be used. The POINT statement can be used to modify the point defined with this statement.

As an illustration, the jaw bone (JAW) is defined with respect to the skull (SKUL) which is considered to be ground in Fig. 15. The relationship between the coordinate system of these bodies as described in the TRAN statement, is a variable rotation about the y-axis. The entry VE in the TRAN statement establishes six generalized coordinates in the direction of the X_j, Y_j, Z_j axes but no joint reactions are defined. The joint reactions at the temporomanidbular (TM) joints are defined with the GROU statements and are shown for the left and the right TM joints as follows:

```
BODY/JAW,SKUL/
TRAN/ROTY,RY,VE/
GROU/LTMJ,JAW,X,-Y,Z,2.0,400./
GROU/RTMJ,JAW,X,Y,Z,2.0,400./
```

19. The FORCE Statement

Defines a known force applied to the structure.

FORCE/name,point 1,point 2,(point 3)/

or FORC/.../)

name: name of the defined force

points 1,2: names of two points

point 3: name of a point which is the point of application for an externally applied force. If not specified (left blank), the force acts between point 1 and point 2. If present, a vector directed from point 1 to point 2 indicates the positive direction of an externally applied force at point 3.

Variations:

MODE	FORMAT
CHANGE	Same as above
ADD	Same as above
DELETE	FORCE/name/
FIND	FORCE/name/ FORCE/ALL/

The force statement can be used to define an externally applied force on the structure or a force acting between the rigid bodies. For an externally applied force, the first two points in the FORCE statement determine the direction of the force vector while the third point is the point of application of the force. The magnitude of the force is provided in the execution mode as well as described later. A positive value for the force indicates that it is acting in the direction point 1 to point 2 while a negative value indicates it is acting in the opposite direction. If the third point in the FORCE statement (point 3) is absent the force is assumed to be acting between the two points specified. A positive value for the force represents a tensile force (trying to pull the two points together) while a negative value for the force represents a compressive force.

20. The TORQUE Statement

Defines an externally applied torque

TORQUE/name,point 1,point 2, body/
(or TORQ/.../)

name: name of the define torque

points 1,2:   names of two points used to determine the positive direction of the torque vector.

body: name of the rigid body to which the torque is applied.

Variations:

MODE	FORMAT
CHANGE	Same as above
ADD	Same as above
DELETE	TORQ/name/
FIND	TORQUE/name/ TORQUE/ALL/

The TORQ statements is used to define an externally applied torque acting on one of the rigid bodies of the structure. The positive direction of the torque is established by the right hand rule using a vector directed from point 1 to point 2. The value of the torque (positive or negative) is supplied when the equations of equilibrium are evaluated in the execution mode.

21. The SPRING Statement

Defines a linear spring in the STRUCTURE

SPRING/name,point 1,point 2,k_s,FL/

name: name of the defined spring.

points 1,2:   names of the points between which the spring acts.

k_s: value of the spring constant for the spring.

FL: free length of the spring.

Variations:

MODE	FORMAT
ADD	Same as above
CHANGE	SPRI/name,k_s,FL/

DELETE	SPRJ/name/
FIND	SPRI/name/
	SPRJ/ALL/

A spring is an actuator which generates a force dependent on the net change in its length from free condition. The spring is defined to act between two points (points 1,2) and the force in the spring is determined by the spring constant, the free length and the actual length of the spring in the structure for a particular configuration.

22. The DAMPER Statement

Defines a viscous damper in the structure

DAMPER/name,point 1,point 2,C/
(or DAMP/.../)

name: name of the defined damper.

points 1,2: names of the points between which the damper acts.

C: value of the damping coefficient for the damper (note: the force in the damper α relative velocity of the two ends, C being the constant of proportionality).

Variations:

MODE	FORMAT
ADD	Same as above
CHANGE	DAMP/name,C/
DELETE	DAMP/name/
FIND	DAMP/name/
	DAMP/ALL/

The viscous damper produces a force which is proportional to the relative velocity of its two ends, and will exhibit a force during dynamic activity only. The damper is defined by the two points (point 1,point 2) and in the analysis the force is computed by obtaining the velocity of the two ends.

THE 'AUXILIARY' ROUTINES

The command statements above are used in the model preparation segment of the program. In addition, certain auxiliary routines and additional command statements are also available in this segment, namely, FIND, ECHO, READ, and WRITE, which can be used at the same level of control as INPUT, ADD, DELETE and CHANGE.

1. The FIND Statement:

The FIND statement puts the system into a mode of operation which allows the user to obtain a description of any component defined in a model. The format to the command statements to retrieve the various components have already been described above. Most of the components can be found by using the name given to them when they were defined. In many of these statements, the name of the component can be replaced with the word ALL. When this is done, a description of all the components of the type specified by the command name will be obtained. For example, if the following sequence of statements is entered:

FIND (if not already in the FIND mode)
POINT/ALL/

a listing of all the points will be obtained.

The information displayed for each component type using the FIND mode appears as shown in Table 1. THe information shown corresponds to that used in the command statements which define the components.

2. The ECHO Statement

The ECHO statement is used to obtain a listing of all the components in a model. The output produced with this statement can be in one of two formats. One is suitable for interactive terminal use as it consists of short lines and the information is more or less abbreviated. This form is obtained by entering the statement:

ECHO/SHORT/   or   ECHO/S/

The output is of a form similar to that shown in Table I.

The second format in which a comprehensive listing of the model data is obtained is:

ECHO/LONG/   or   ECHO/L/

The long ECHO output for the lower extemities model applied to bicycling, for example, is shown in Table II.

3. The READ and WRITE Statements

These command statements offer the option of saving a created model on a mass storage device and recreating from previously saved data structure.

The data structure is written on to a mass storage device, (e.g., a disk) with an unformatted FORTRAN write statement. The procedure for setting up the area on the device will be dependent on the methods used at a particular computer installation. However, the operation of the routine which does the writing remains the same as long as an unformatted write statement can be used. The data structure is saved by:

WRITE/i/

where is is the number $(1 < i < 99)$ used to reference the data file where the data is to be written. This operation leaves the current data structure unchanged.

In order to recall the saved data structure the following statement is used:

READ/i/

TABLE I. Format of the Output Lines for the FIND and ECHO/SHORT/Modes

BODY** name 1   name 2 $\overline{X}$ $\overline{Y}$ $\overline{Z}$ m
    I_{XX} I_{YY} I_{ZZ} I_{XY} I_{XZ} I_{YZ}

GRAVITY** G_{XX} G_{YY} G_{ZZ}

TRAN**   jt. name type. diff. variable no. x y z θ

POINT**   name   body   x y z

DAMPER** name pt. 1   pt. 2 C_D

SPRING** name pt. 1   pt. 2 k_s fl_s

TORQUE** name pt. 1   pt. 2   body

FORCE**   name pt. 1   pt. 2   pt. 3

UNFO[1]   name pt. 1   pt. 2   pt. 3   cost coef. ($\pm$ 1,0)

UNTO**   name pt. 1   pt. 2   body   cost coef. ($\pm$ 1,0)

REACTION FORCE** name eq. no.[2] cost of pos. component
                      cost of neg. comp. ($\pm$ 1,0)

(IN) EQUALITY CONSTRAINT** name 1 $\begin{matrix} LE \\ EQ \\ GE \end{matrix}$ name 2 (CONS) ratio

COST COEF** cost coef. of reac. forces cost coef. of
                 moment reactions

CONSTRAINT EQUATION ** no. joint name   type[3]

jt. tolerance** name moment reaction   +limit   -limit

CYLINDER** name pt. name pt 1   pt 2 mat'l. ref. pt. ref. pt
             ref. pt. radius

LINE** name pt. name   pt. 1 pt. 2   mat'l ref. pt
       ref. pt.   ref. pt.

1. Muscles are included in this category.
2. Corresponds to the number of the CONSTRAINT EQUATION in which it appears. A negative number indicates that it has been DELETED.
3. Same as the numbering system used to name joint reactions.

TABLE II. LISTING OF LEG MODEL DATA

GRAVITATIONAL VECTOR

GXX = 0.0000
GYY = 0.0000
GZZ = -9.80671.00

LINK INFORMATION (BASE = GROU)

I	J	XBAR	YBAR	ZBAR	MASS	IXX	IYY	IZZ	IXY	IXZ	IYZ
PLVS GROU		0.000	0.000	0.000	0.000	0.000	0.000	0.000	0.000	0.000	0.000
THI PLVS		0.000	-1.500-02	-1.810-01	7.700+00	3.416-01	3.399-01	2.240-03	0.000	0.000	2.090-02
CALF THI		0.000	0.000	-1.770-01	3.850+00	1.619-01	1.619-01	1.000-04	0.000	0.000	0.000
FOOT CALF		5.660-02	0.000	-1.460-02	1.440+00	7.265-02	7.265-02		0.000	-3.165-02	0.000
1 PTLA THI		0.000	0.000	0.000	0.000	0.000	0.000	0.000	0.000	0.000	0.000
1 P62 THI		0.000	0.000	0.000	0.000	0.000	0.000	0.000	0.000	0.000	0.000
1 REDL FOOT		0.000	0.000	0.000	0.000	0.000	0.000	0.000	0.000	0.000	0.000
1 REHL FOOT		0.000	0.000	0.000	0.000	0.000	0.000	0.000	0.000	0.300	0.000
1 RTBA FOOT		0.000	0.000	0.000	0.000	0.000	0.000	0.000	0.000	0.300	0.000
1 RPRT FOOT		0.000	0.000	0.000	0.000	0.000	0.000	0.000	0.000	0.000	0.000

TRANSFORMATION MATRICES

	TYPE[2]	DIFF[3]	DOF	SPEC[4]	DERVD	X	Y	Z	ANG
PLVR	1	1	1	1	+++ LINK PLVS + WRT + LINK GROU +++	0.000	0.000	6.000-02	0.000
PLVT	4	0				0.000	0.000		0.000
HIP	1	1	2	2	+++ LINK THI + WRT + LINK PLVS +++	0.000	-8.500-02	0.000	0.000
HIP	2	2	3	3		0.000	0.000	0.000	0.000
HIP	3	3	4	4	+++ LINK THI + WRT + LINK THI +++	0.000	0.000	0.000	0.000
KNEE	2	2	5	5	+++ LINK CALF + WRT + LINK THI +++	0.000	5.000-03	-4.180-01	0.000
KNEE	4	0				0.000	0.000	-2.000-02	0.000
ANKL	1	1	6	6	+++ LINK FOOT + WRT + LINK CALF +++	0.000	0.000	-4.340-01	0.000
ANKL	3	3	7	7		0.000	0.000	0.000	0.000
ANKL	2	2	8	8		0.000	0.000	0.000	0.000
T21	2	2	9	9	+++ LINK PTLA + WRT + LINK THI +++	-7.000-03	5.000-03	-3.900-01	0.000
T23	4	0				6.400-02	0.000	0.000	0.000
1 P62	4	0			+++ LINK P62 + WRT + LINK THI +++	0.000	-7.000-02	-4.000-02	0.000
1 REDL	4	0			+++ LINK REDL + WRT + LINK FOOT +++	2.000-02	-8.000-03	-7.000-03	0.000
1 REHL	4	0			+++ LINK REHL + WRT + LINK FOOT +++	2.500-02	4.000-03	0.000	0.000
1 RTBA	4	0			+++ LINK RTBA + WRT + LINK FOOT +++	2.500-02	1.000-02	0.000	0.000
1 RPRT	4	0			+++ LINK RPRT + WRT + LINK FOOT +++	2.000-02	-1.000-02	-1.000-02	0.000

POINTS OF INTEREST

NAME	LINK	X	Y	Z
P1	PLVS	3.500-02	-3.500-02	-3.500-02
P2	CALF	1.000-02	3.000-02	-4.100-02
P3	PLVS	5.000-02	-3.300-02	-1.500-02
P4	THI	1.500-02	-5.000-03	-2.350-01
P5	PLVS	-2.000-02	-7.000-02	-5.500-02
P6	THI	0.000	4.500-02	-3.950-01
P7	PLVS	0.000	-6.000-02	-5.500-02
P8	THI	1.000-02	-5.000-03	-1.950-01
P9	PLVS	4.000-02	-4.000-02	-2.500-02
P10	THI	5.000-03	-1.500-02	-1.350-01
P11	PLVS	-4.500-02	-7.000-02	-4.000-02
P12	CALF	1.000-02	3.000-02	-4.100-02
P80	THI	-2.000-02	3.500-02	-4.150-01
P81	THI	-2.000-02	0.000	-4.150-01
P82	THI	0.000	3.500-02	-4.150-01
P83	THI	-2.000-02	3.500-02	-4.150-01
P13	PLVS	-3.800-02	-8.000-02	-3.000-02
P14	CALF	-2.000-02	3.000-02	-2.600-02
P15	PLVS	-4.500-02	-7.000-02	-4.000-02
P16	CALF	-2.500-02	-3.500-02	-4.100-02
P26	THI	5.000-03	-5.000-03	-2.850-01
P27	CALF	-2.500-02	-3.500-02	-4.100-02
P20	PLVS	4.000-02	-1.100-01	1.000-01
P84	PLVS	-7.500-02	-4.500-02	7.000-02
P120	PLVS	-8.000-02	-1.000-02	0.000
P21	CALF	3.000-02	-3.500-02	-2.100-02
P62	P62	0.000	0.000	0.000
P22	PLVS	-4.500-02	-7.000-02	0.000
P23	THI	0.000	0.000	0.000
P67	THI	-1.500-02	-3.500-02	-6.500-02
P210	THI	0.000	0.000	0.000
P211	THI	0.000	-4.000-02	-1.800-02
P212	THI	1.000-02	0.000	0.000
P55	PLVS	-3.000-02	-1.100-01	1.100-01
P56	THI	-1.500-02	-5.500-02	-1.500-02
P57	PLVS	2.000-02	-1.000-01	8.500-02
P58	THI	0.000	-6.000-02	-1.500-02
P24	PLVS	-1.000-02	-8.000-02	1.000-01
P25	THI	-1.000-02	-1.000-02	-7.000-02
P63	PLVS	2.500-02	-7.500-02	0.000
P85	PLVS	2.500-02	-7.500-02	0.000
P86	PLVS	2.500-02	-5.500-02	0.000
P87	PLVS	0.000	-7.500-02	0.000
P18	PLVS	4.500-02	-1.100-01	8.500-02
P76	THI	4.500-02	-5.000-03	-5.000-02
P19	CALF	0.000	4.000-02	-2.000-02
P77	THI	0.000	4.500-02	-3.000-01
P17	PLVS	2.500-02	-1.000-01	2.500-02
P28	THI	2.500-02	-5.000-03	-2.350-01
P29	THI	3.000-02	-1.500-02	-2.150-01
P30	THI	2.500-02	-2.500-02	-2.350-01
P59	PTLA	0.000	0.000	0.000
P60	CALF	4.500-02	0.000	-4.300-02

P31	THI	-2.000-02	2.000-02	-4.150-01
P32	FOOT	-4.500-02	-1.500-02	-4.700-02
P68	THI	1.410-02	2.000-02	-4.076-01
P69	THI	1.410-02	4.000-02	-4.076-01
P70	THI	3.410-02	2.000-02	-4.076-01
P71	THI	5.000-03	3.000-02	-3.739-01
P33	THI	-2.000-02	-2.000-02	-4.200-01
P34	FOOT	-4.500-02	-1.500-02	-4.700-02
P72	IHI	2.000-03	-2.000-02	-4.086-01
P73	THI	2.000-03	0.000	-4.086-01
P74	THI	2.200-02	-2.000-02	-4.086-01
P75	THI	5.000-03	-2.000-02	-3.839-01
P35	CALF	-3.000-02	-3.000-02	-8.200-02
P36	FOOT	-4.500-02	-1.500-02	-4.700-02
P101	CALF	-5.000-03	-1.000-02	-1.050-01
P102	FOOT	-4.500-02	-1.500-02	-4.700-02
P39	CALF	-2.500-02	2.000-02	-4.340-01
P40	FOOT	3.600-02	1.300-02	-2.400-02
[1] REDL	REDL	0.000	0.000	0.000
P41	CALF	0.000	-2.500-02	-2.640-01
P42	FOOT	1.140-01	-2.600-02	-4.600-02
[1] REHL	REHL	0.000	0.000	0.000
P43	CALF	0.000	-2.000-02	-2.520-01
P44	FOOT	1.180-01	-1.200-02	-3.900-02
[1] RTBA	RTBA	0.000	0.000	0.000
P37	CALF	2.000-02	0.000	-1.500-01
P38	FOOT	5.400-02	1.300-02	-1.700-02
P45	CALF	-2.500-02	1.500-02	-4.340-01
P46	FOOT	0.000	9.000-03	-2.300-02
P47	CALF	-1.800-02	-5.000-03	-4.280-01
P48	FOOT	-1.800-02	0.000	-1.200-02
P49	CALF	-1.500-02	-2.800-02	-4.540-01
P50	FOOT	0.000	-2.800-02	-4.300-02
P51	CALF	-1.500-02	-2.800-02	-4.540-01
P52	FOOT	8.000-03	-2.800-02	-3.600-02
[1] PPRT	RPRT	0.000	0.000	0.000
P53	CALF	-1.000-02	-2.000-02	-3.200-01
P54	FOOT	6.000-02	-3.900-02	-4.600-02
P91	FOOT	2.348-01	0.000	-3.030-02
P90	FOOT	1.757-01	0.000	-1.588-01
P64	FOOT	1.410-01	0.000	-6.500-02
P200	FOOT	1.500-02	-2.500-02	-2.500-02
P201	FOOT	1.800-02	-2.400-02	-2.400-02
P202	FOOT	1.300-02	-2.400-02	-2.300-02
P203	CALF	0.000	2.650-02	-4.280-01
P204	CALF	0.000	2.500-02	-4.240-01
P205	CALF	0.000	2.800-02	-4.320-01
P206	FOOT	2.800-02	3.000-03	1.000-03
P207	FOOT	3.300-02	0.000	0.000
P300	THI	0.000	5.000-03	-4.280-01
P301	CALF	-2.500-02	0.000	-5.000-03
P213	THI	0.000	0.000	0.000
P214	THI	2.000-02	-5.000-02	-3.500-02
P215	THI	-1.500-02	-5.000-02	0.000
P125	PLVS	-3.000-02	-8.000-02	-4.500-02
P126	THI	-5.000-03	5.000-03	-3.100-01
P127	PLVS	-6.000-02	-5.000-02	8.000-02

NAME	NO.	DIRECTIONAL PTS PT1 TO PT2		POINT OF APPLICATION
FV	1	P90	P64	P64
FH	2	P64	P91	P64

UNKNOWN FORCE ELEMENTS

NAME	PT1	PT2	PT3	COST	
GRCL	P1	P2		1.00+00	1[6]
ADLN	P3	P4		1.00+00	1
ADME	P5	P6		1.00+00	1
ADMA	P7	P8		1.00+00	1
ADBR	P9	P10		1.00+00	1
SEMT	P11	P83		1.00+00	1
SMT2	P83	P12		0.00	1
SEMM	P13	P14		1.00+00	1
BIFL	P15	P16		1.00+00	1
BIFS	P26	P27		1.00+00	1
TFSL	P20	P62		1.00+00	1
GUMI	P84	P62		1.00+00	1
GIT2	P120	P62		1.00+00	1
ILTB	P62	P21		0.00	1
GUMF	P22	P23		1.00+00	1
GLMD	P55	P56		1.00+00	1
GLMN	P57	P58		1.00+00	1
ILAC	P25	P63		1.00+00	1
SART	P18	P76		1.00+00	1
SRT2	P19	P77		0.00	1
RECF	P17	P59		1.00+00	1
VSMD	P28	P59		1.00+00	1
VSIN	P29	P59		1.00+00	1
VSLT	P30	P59		1.00+00	1
PETT	P59	P60		0.00	1
GASM	P31	P32		1.00+00	1
GASL	P33	P34		1.00+00	1
SOLE	P35	P36		1.00+00	1
SOLT	P101	P102		1.00+00	1
TIBP	P39	P40		1.00+00	1
EDL1	P41	REDL		1.00+00	1
EDL2	REDL	P42		0.00	1
EDL3	REDL	P204		0.00	1
EDL4	REDL	P201		0.00	1
EDL5	REDL	P202		0.00	1
EDL6	REDL	P205		0.00	1
EDL7	REDL	P206		0.00	1
EDL8	REDL	P207		0.00	1
EHL1	P43	REHL		1.00+00	1
EHL2	P44	REHL		0.00	1
EHL3	REHL	P204		0.00	1
EHL4	REHL	P200		0.00	1
EHL5	REHL	P201		0.00	1
EHL6	REHL	P202		0.00	1

EHL7	REHL	P203	0.00	1
EHL8	REHL	P205	0.00	1
IBA1	P37	RTBA	1.00+00	1
TBA2	P38	RTBA	0.00	1
TBA3	RTBA	P204	0.00	1
TBA4	RTBA	P200	0.00	1
TBA5	RTBA	P201	0.00	1
TBA6	RTBA	P202	0.00	1
TBA7	RTBA	P203	0.00	1
TBA8	RTBA	P205	0.00	1
FLDL	P45	P46	1.00+00	1
FLHL	P47	P48	1.00+00	1
PRNL	P49	P50	1.00+00	1
PRNB	P51	P52	1.00+00	1
PRT1	P53	RPRT	1.00+00	1
PRT2	P54	RPRT	0.00	1
PRT3	RPRT	P204	0.00	1
PRT4	RPRT	P201	0.00	1
PRT5	RPRT	P202	0.00	1
PRT6	RPRT	P205	0.00	1
PRT7	RPRT	P206	0.00	1
PRT8	RPRT	P207	0.00	1
CURL	P300	P301	1.00+00	1
ADE2	P125	P126	1.00+00	1
GLD2	P127	P56	1.00+00	1

REACTION FORCES

NAME	NEQ[7]	COST		
4HIP	4	0.00	0.00	0
5HIP	5	0.00	0.00	0
6HIP	6	0.00	0.00	0
4KNE	10	2.00+00	2.00+00	0
5KNE	11	2.00+00	2.00+00	0
6KNE	12	2.00+00	0.00	0
4ANK	16	2.00+00	2.00+00	0
5ANK	17	2.00+00	2.00+00	0
6ANK	18	2.00+00	0.00	0
4T23	19	0.00	0.00	0
5T23	20	1.00+00	1.00+00	0
6T23	-21[8]	0.00	0.00	0
4P62	-22	0.00	0.00	0
5P62	23	0.00	0.00	0
6P62	-24	0.00	0.00	0
4RED	-25	0.00	0.00	0
5RED	-26	0.00	0.00	0
6RED	-27	0.00	0.00	0
4REH	-28	0.00	0.00	0
5REH	-29	0.00	0.00	0
6REH	-30	0.00	0.00	0
4RTB	-31	0.00	0.00	0
5RTB	-32	0.00	0.00	0
6RTB	-33	0.00	0.00	0
4RPR	-34	0.00	0.00	0
5RPR	-35	0.00	0.00	0
6RPR	-36	0.00	0.00	0

MOMENT REACTIONS

NAME	NEQ[7]	COST		
1HIP	1	4.00+02	4.00+02	0
2HIP	2	4.00+02	4.00+02	0
3HIP	3	4.00+02	4.00+02	0
1KNE	7	4.00+02	4.00+02	0
2KNE	8	4.00+02	4.00+02	0
3KNE	9	4.00+02	4.00+02	0
1ANK	13	4.00+02	4.00+02	0
2ANK	14	4.00+02	4.00+02	0
3ANK	15	4.00+02	4.00+02	0

CONSTRAINT EQUATIONS

NO.	JOINT	TYPE[9]
1	HIP	1
2	HIP	2
3	HIP	3
4	HIP	4
5	HIP	5
6	HIP	6
7	KNEE	1
8	KNEE	2
9	KNEE	3
10	KNEE	4
11	KNEE	5
12	KNEE	6
13	ANKL	1
14	ANKL	2
15	ANKL	3
16	ANKL	4
17	ANKL	5
18	ANKL	6
19	T23	4
20	T23	5
21	T23	6
22	P62	4
23	P62	5
24	P62	6
25	REDL	4
26	REDL	5
27	REDL	6
28	REHL	4
29	REHL	5
30	REHL	6
31	RTBA	4
32	RTBA	5
33	RTBA	6
34	RPRT	4
35	RPRT	5
36	RPRT	6

EQUALITY AND INEQUALITY CONSTRAINTS

NO.	FORCE1		FORCE2	RATIO(F1/F2)/CONSTANT
1	SEMT	.EQ.	SMT2	1.000+00
2	SART	.EQ.	SRT2	1.000+00
3	EDL1	.EQ.	EDL2	1.000+00
4	EHL1	.EQ.	EHL2	1.000+00
5	TBA1	.EQ.	TBA2	1.000+00
6	PRT1	.EQ.	PRT2	1.000+00

CYLINDER AND LINE (MATERIAL) CONSTRAINTS

NAME	TYPE	POINT	DIRECTIONAL POINTS		MATERIAL SIDE P.T.	REFERENCE POINTS		RADIUS
GUMF	CYLI	P23	P210	P211	P212	P22	P67	2.500-02
GASM	CYLI	P31	P66	P69	P70	P71	P32	3.480-02
GASL	CYLI	P33	P72	P73	P74	P75	P34	2.480-02

NOTES

1. Created with a REAP command statement.
2. A numerical code for the _type_ designation for a transformation matrix: RX-1; RY-2; RZ-3; and T,TX,TY,TZ-4.
3. A numerical code for the _diff_ designation for a transformation matrix. A constant matrix is noted with a 0.
   Variable matrices are: RX-1; RY-2; RZ-3; TX-4; TY-5;
   TZ-6.
4. Indicates the numerical order of the variables.
5. Muscles and other unknown forces are included in one list.
6. The direction indicator for an unknown force.
7. The number of the constraint equation in which the reaction
   force or moment appears.
8. A minus sign indicates a deleted reaction force or moment.
9. A numerical code indicating the type of constraint equation.
   The same coding numbers used for naming the joint reactions
   are also used here.

This statement would appear in place of the INPUT statement when the model is to be recreated in the computer system. Following the READ statement the model can be analyzed or modified.

4.   Additional Statements

   a.   The TITLE Statement:

The TITLE statement is used to assign a descriptive title to the model. The statement has the word TITLE in the first five spaces of the line followed by a 36 character or less description of the model. This statement can appear in the INPUT, ADD, and CHANGE modes of operation. The title assigned with this statement appears in the output listing obtained with the ECHO/L/ statement and with the output from the Analysis and Graphics segments.

   b.   The PRINT Statement

This statement instructs the program to print (ON) or not to print (OFF) the context of the input command statements as they are processed. The format for this statement is:

$$\text{PRINT/ON} \quad \text{or} \quad \text{PRINT/OFF/}$$

With this capability the input statements can be printed when the system is operating in a noninteractive envionment. The default condition is OFF. This statement may be entered at any time.

   c.   The STOP Statement:

This statement terminates the operation of the system and is accomplished by merely entering STOP.

THE MODEL ANALYSIS SEGMENT

The analysis segment of the computer system is concerned with the solution to the LP (linear programming) problem to obtain the numerical values for the defined unknown variables (muscle, forces, joint reactions and moments, etc.) in the model.

The general form of the LP problem states:

$$\text{minimize or maximize} \quad U = \sum_{i=1}^{n} c_i x_i \quad \text{subject to}$$

$$\sum_{i=1}^{n} zj_i x_i = b_j \quad j = 1, 2, \ldots, m \text{ (linear constraint equations)}$$

where $x_i \geq 0$.

U is known as the merit or objective function and represents a linear weighted sum of all the unknowns.

In a musculoskeletal system x_i represent the muscle forces, joint reaction forces and moments and any other unknown forces and torques that have been defined. The values of c_i in the criterion are the cost coefficients which have been set for the variables when they were defined or subsequently modified. The constraint equations consist mainly of the equations of equilibrium for the musculoskeletal structure as well as other defined constraints (such as by means of EQUAL, INEQUAL, JTTOL Statements). Needless to say, since the equations of equilibrium are a subset of the total constraints and since they are equality constraints, the final solution will always satisfy equilibrium. In the computer system the inequality constraints defined in the model are automatically converted to equality constraints by inclusion of slack variables. The slack variables are included in the merit criterion with cost coefficients equal to zero. The final optimal solution is obtained by means of the simplex algorithm.

OPERATION OF THE ANALYSIS SEGMENT:

The analysis is initiated by entering the command EXECUTE (or EXEC, or x for short) after the model has been defined and/or modified. Once in the EXEC mode of operation, the system begins to prompt for the information required to solve the problem:

i.   The position number or range of position numbers and increment:

The first prompt appears as:

ENTER POSITION NO. OF RANGE OF POSN NOS AND INCREMENT

The system can be used to analyze one configuration of the model at a time or a serie of configurations. The solution for a particular configuration is identified by a POSN NO. which accompanies the final result. If only one config. is to be analyzed, the proper response to the prompt is the number which the user desires to correspond to that config. If a consecutive series of configs. is to be analyzed, the no. corresponding to the first one, followed by the number corresponding to the last one separated by a comma should be entered. For example, if configurations 1, 2 and 3 are to be analyzed, the proper response is: 1,3: If the nos. corresponding to the configs. are not consecutive, but are separated by a constant (e.g., 1,3,5,7), the no. of the first, the last and the increment between any two configs. should be entered thus: 1,7,2 resulting in configurations 1,3,5,7 to be analyzed.

The only other alternate response at this point is to enter a blank line. This will terminate the EXEC mode of operation.

ii. The configuration of the model:

The second prompt appears as:

ENTER VALUE(S) OF VARIABLES (___VALUE(S))POSN. NO. = ____

The values of the variables of the model are entered in the order they were in the joint definition statements. For example, in the lower extremities model definition part earlier, the order would be: the angular rotations of the hip joint about the X, Y and axes; the angle at the knee; followed by the angles at the ankle in proper order; and finally he angle which locates the patella. Each value is separated by a comma and the same rules for decoding numerical data in the command statements apply. The value of the angles should be in <u>radians</u>. Length dimension of a variable should be consistent with those used for the rest of the model measurements.

iii. The velocities:

The next prompt from the system is simply:

VELOCITIES

Following this, the velocity of the variables are entered in the same order as discussed above. If all the zero (such as in a static case), then a blank line should be entered.

iv. The accelerations:

The response to the prompt ACCELERATIONS is the values of the accelerations of the variables of the model in the same order as discussed above. If all are zero (such as in a static case) then a blank line should be entered.

The three prompts (ii)-(iv), with their corresponding responses establish the total config. of the structure for the analysis.

v. The known forces:

The values of the known forces acting on the structure is requested by the following prompt for each force:

ENTER VALUE OF FOCRCE Name POSN NO. _____

where Name is the name used in the FORCE command statement. The position no. is shown as a reference. The value of the force should appear in the first 20 spaces of the response line, and a decimal point should be present to avoid misinterpretation of the data. If no forces have been defined in a model, this prompt will not appear. The force units should be consistent with those used for the rest of the model; the default units are NEWTONS.

vi. The known torques:

The final information requested is the value of the torques acting on this structure. For each one, the prompt appears as:

ENTER VALUE OF TORQUE Name POSN. NO._____

The format of the response is the same as that for inputting the force values. The default units for the torques are NEWTON-METER.

One all the information above is supplied, the system will formulate the LP problem and solve it. The results for each config. (POSN NO.) are displayed in tabular form as follows:

```
PROBLEM TITLE:  (36 character descriptive title)
RESULTS FOR POSN. NO.:  _____
VALUE OF THE MERIT CRITERION:  _____

        FORCE       VALUE
        name        magnitude
```

The TITLE is the same as that specified by the TITLE Statement. The values displayed are for non-zero vaiables only; those not shown ae zero.

If a series of configs. are analysed, the system will begin prompting for the data specifying the config., forces, and torques for the next position number. If only one configuration is to be analyzed, the EXEC mode is terminated after the solution is printed.

AUXILIARY INPUT AND OUTPUT ROUTINES

The normal operation of the Analysis Segment can be modified by including special input and output routines in the system. For large models or where several configurations are to be analyzed the configuration position, velocity and acceleration data can be read in by using special input routines. These are FORTRAN subroutines and help automate the data transfer between the system and data storage devices.

a.   SUBROUTINE SPECOS 9IPOSN,Q,QD,QDD,N) Devices.

This subroutine is used for describing the model configuration. The routine must have a

DIMENSION Q(1), QD(1), QDD(1)

statement in the beginning. The arguments in the routine stand for:

IPSOSN –   the position number for the configuration being requested.

Q,QD,QDD –   arrayus containing the values of the variables (Q) of the model, their velocities (QD) and accelerations (QDD). These are defined in this routine.

N –   number or variables

The remainder of the routine can consist of any legal FORTRAN statements.

b.   FUNCTION FNON (ITH,NAME,Q,QD,QDD,IPOSN)
     FUNCTION TNON (ITH,NAME,Q,QD,QDD,IPOSN)

The values of the known forces and torques are entered in FUNCTION subroutines called FNON and TNON. The arguments Q, QD, QDD, IPOSN have the same meaning as above, and the first two entries are used to identify the force or torque whose value is to be defined. These routines are called once for each force and torque defined in the model. To refer to the forces and torques in the numerical order in which they were defined, the argument ITH is used. ITH equals 1 for the first force, 2 for the second, and so on. The forces and torques are numbered separately. They can be referred to by name with the argument NAME, which contains the

766

four letter (or less) name of the force or torque. Since these are FUNCTION subroutines, the value of the force or torque is returned in a variable corresponding to the function name. Therefore, somewhere in the routines, statements such as:

$$FNON =$$
$$TNON =$$

must appear.

    c.   SUBROUTINE OUTPUT

Additional output information can be retrieved from the system in a routine called OUTPUT, which is called immediately after the results for the analysis of a configuration are listed. The routine can be used to obtain the positions, velocities and accelerations of the points defined in the model, the list of unknown variables in a model and their values corresponding to the configuration just analyzed.

The data for positions, velocities, and acelerations for points can be obtained by:

```
CALL PTPOS1 (NAME,X,Y,Z)
CALL PTPOS2 (NAME,X,Y,Z,VX,VY,VZ)
CALL PTPOS3 (NAME,X,Y,Z,VX,VY,VZ,AX,AY,AZ)
```

The arguments in these statements refer to:

    NAME –   name of the point.

    X,Y,Z –   coordinates of the point in the reference coordinate system.

    VX,VY,VZ –   velocity components in the reference coordinate system.

    AX,AY,AZ –   acceleration components.

The direction cosines (XCOS,YCOS,ZCOS) of a vector direted from one point (X1,Y1,Z1) to another (X2,Y2,Z2) and the distance between the two points (XL) can be determined with the statement.

    CALL DIRCOS (X1,Y1,Z1,X2,Y2,Z2,XCOS,YCOS,ZCOS,XL)

The relative velocity between two points is therefore:

$$VREL = (VX1-VX2)*XCOS + (VY1-VY2)*ZCOS$$

This expression yields a positive velocity whenb the points are approaching one another.

The list of unknowns in a model can also be obtained in the OUTPUT routine. To do this the following labeled common areas must be included in the routine:

```
COMMON/RINGD/DATA(1)
COMMON/RINGL/LIST(1)
COMMON/LINSTF/ICONST(7),IANS,IVARNM,NVAR
```

767

The names of the unknowns are stored in the LIST array beginning at location IVARNM and continuing for the next NVAR locations. Consequently, the following statements will write the list on data file 15.

```
ILAST = IVARNM + NVAR -1
WRITE(15)(LIST(J),I = IVARNM,ILAST)
```

The value corresponding to these unknowns, and in the same order, are contained in the DATA array. They begin at location IANS, therefore, the following statements will write the values of the NVAR unknowns:

```
ILAST = IANS + NVAR - 1
WRITE(15) (DATA(I),I=IANS, ILAST)
```

If an unknown can act in two directions, such as a joint reaction force, its name occurs in the list twice. The two names are consecutive and the first refers to the positive component, and the second to the negative one. This information can be used in reference to the values of the unknown to determine the sign of the values, since they are all stored as positive numbers.

THE GRAPHICS SEGMENT

The graphics segment serves as a visual aid in the modelling process and permits creation and display of pictures of the model as well as modification to the model interactively. The picture display wire-frame drawings of the rigid bodies (bones) along with the defined muscles, joint reactions and known applied forces.

General operation of the graphics segment:

The operation of the graphics segment depends partially on the type of graphics terminal being used. The three general aspects needed for a storage tube terminal with this system are:

1. a keyboard which functions like a standard teletype terminal,
2. a display screen, and
3. a graphic input procedure which allows a point on the screen to be identified by positioning a curser at its location and transfers its coodinates to the computer.

The computer system has been implemented on a UNIVAC 1110 computer with a Tektonix 4010 graphics terminal. In order to use other terminals or computer-terminal combinations some software modifications of the system may be required.

With a Tektronix 4010 terminal whenever graphic input data is required, two hairlines, one vertical and one horizontal, become visible on the screen. The positions of the hairlines are controlled by two thumb wheels on the keyboard. By manipulating the thumb wheels, any point on the screen determined by the intersection of the hairlines can be referenced. To transmit the location of the point to the computer, any alphanumeric key (e.g., A,B,C,1,2,...) on the keyboard is pressed. If requests for graphics inputs are to be terminated, the key E (on the Tektronix terminal) is pressed. When this key is used, the system will ignore the graphic input and, depending on the situation, will usually terminate the current series of requests for graphic input data.

In order to enter the Graphics Segment of the System, the command GRAPHICS (GRAP) is entered. This command can be entered at the same level of control as the commands INPUT, ADD, EXEC,..., etc. Once in the graphics mode, the system is controlled by a series of command statements as will be described in the following. These command statements have been limited to three letters in length, and must appear in the first three spaces of a line. A list of graphics commands and their meanings is shown in Table III.

TABLE III List of Graphics Commands

Command

SKE	X	X	Establish new sketch plane
3-D	X	X	3-D display
SEC		X	Sectioned view
STE		X	Stereo view
CON	X	X	Graphical model construction
PLO		X	Plot
KPO	X	X	Key in point coordinates
POI	X		Define points graphically
FPO	X	X	Find point coordinates
DPO	X	X	Delete points
TPO		X	Define temporary point
MOV	X	X	Move a point graphically
LIN	X	X	Define a series of lines
SLI		X	Define single line segments
DLI	X	X	Delete a line
MSU		X	Define a muscle graphically
SMU		X	Show a list of muscles
SMU		X	Find a muscle name
DMU		X	Delete a muscle
WIN	X	X	Window the screen
FUL	X	X	Obtain full size display
RED	X	X	Redraw the display
REO		X	Reorient the 3-D display
CHA		X	Define section lines
REF	X		Change reference sketch plane
DUP	X		Duplicate current sketch plane
DUC	X		Duplicate to connect sketch plane
OPT	X	X	Optimize drawing process
SAV	X	X	Save current picture data
REC	X	X	Recall saved picture
DEL	X	X	Delete recalled picture data
ERA		X	Erase a line or muscle
VIS		X	Make a line or muscle visible
END	X	X	End graphics segment

Establishing a Sketch Plane

The method used to generate pictures of the segments in a model consists of drawing them one plane at a time. The planes are chosen parallel to one of the three planes determined by the axes of the coordinate system fixed to the rigid body representing the segment. For example, two planes used to produce a picture of the human mandible are shown in Fig. 16. These planes are parallel to the x_j-y_j plane of the jaw coordinate system at two different locations along the $-z_j$ axis.

To sketch the drawing of a segment on the graphics screen, the screen must be defined to represent one of the reference planes. To initiate the process, the command SKE is entered, following which the system issues a series of prompts to define the plane represented by the screen (whether xy, yz or zx) as well as the name of the segment. The first prompt requests the name of the body segment, and the response would be the name of any rigid body defined in the model preparation segment. Note that the graphics segment can be used only after the kinematic topology of a model has been defined. The second prompt requests the plane for the sketch and indicates the choice of the allowable responses.

The third prompt represents a scale-factor, which is the number of model length units represented by one inch on the screen. The choice will depend on the units used in the model (default is meters) and the accuracy required for the sketch. A factor of .05 for example, would mean each screen inch represents .05 meters.

The coordinates of the sketch origin are requested next. The numerical data supplied in response to the next prompt specifies the location of one point on the screen (i.e., the sketch origin). The stretch origin data also determines the location of the plane represented by the screen along the axis perpendicular to the specified plane. For example, to draw the jaw as shown in Fig. 16, the series of prompts and responses might appear as:

```
→    SKE
     ENTER NAME OF BODY SEGMENT
→    JAW
     DEFINE PLANE FOR SKETCH (XY,YZ,ZX)
→    XY
     ENTER SCALE FACTOR
→    .03
     ENTER (X,Y,Z) COORDINATES OF THE SKETCH ORIGIN
→    .05,0.,-.08
     DIGITIZE TWO POINTS ON THE X AXIS (SMALLER ONE FIRST, AND ONE
     POINT ON THE POSITIVE Y AXIS)
```

The responses are prefixed with →, while the others ae requests or prompts from the system. In this case the sketch origin is at .05, 0., -.08, meaning that the XY plane is -.08 meters from the body fixed axes system. The point on the screen corresponding to the sketch origin and the orientation of the coordinate system in the screen plane is specified with the graphic input of three points in response to the final prompt. The three points are shown in Fig. 17 numbered in the order in which they were entered. The first two points (1,2) determine the orientation of the X axis and the third point specifies one point on the positive Y-axis. After this the screen is erased and the reference coordinate axes defined earlier are shown. The point defined by the intersection of the two lines (1) is the sketch origin which has been assigned the value X = .05, Y = 0. for the jaw. The drawing process can begin at this time by

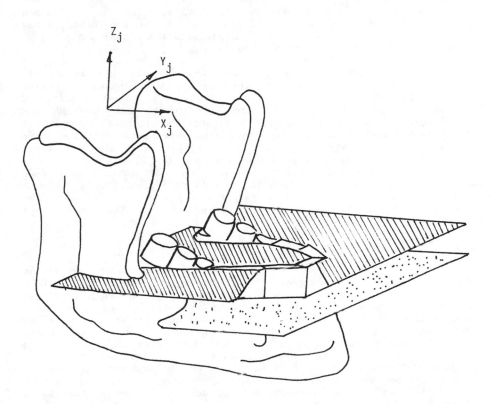

Figure F-16. Two reference planes used to construct the picture of the human mandible.

defining points and lines in the specified plane by using command statements such as KPO, POJ, LIN, etc. as will be explained later. Any points or lines previously defined in the plane described and lying within the screen area will also show up at the time the display is drawn.

Once the contents of one plane are defined, other planes can be drawn. This can be done in two ways. One consists of entering the command SKE and repeating the above process by defining a new sketch plane. The other option consists of using the command REF which permits the location of the plane to be changed without having the screen erased. Following this command, a lilne is printed identifying the current plane and its location along the axis. The location of the new plane can then be entered and drawing can begin in the new sketch plane. The procedure is especially convenient for generating planes which are similar and where the graphic information of one plane can be used as a reference for a second plane.

Sometimes entire planes or portions of planes of a picture are the same. For example, the pictorial lines in the planes shown in the jaw figure (Fig. 16) are the same but occur in two $x_i y_j$ planes at different constant z_j locations. In such cases the commands DUP and DUC are convenient to use. The command DUP duplicates the pictorial lines in the plane represented by the display screen in another parallel plane. The command DUC, not only duplicates the pictorial lines of the current screen plane but also adds lines between the two planes at their end points, thus connecting the two planes. When these commands are used, the system requests the location of the duplicate plane (along the axis perpendicular to the defined screen plane). The definition of the current screen plane, as well as, the contents of that plane remain unchanged.

The 3-D display

The 3-D display part of the Graphics Segment is used to produce an orthogonal projection of a model on the screen. The view shown consists of the muscles and their points of attachments, joint reaction forces, known force components and the lines which have been defined to represent the pictures of the segments. The 3-D display can also be used to define additional points, lines, muscles, etc. to complete the drawing process or to modify existing information. To distinguish the muscles from pictorial lines, the former are represesnted as thick lines (actually, a series of three lines drawn very close to one another) Externally applied forces and joint reaction forces are represented by small arrows which indicate the positive direction of action, the arrow heads being located at the points of application.

The model can be viewed from any direction as specified by the user by means of a "reference viewing point". This point is assumed located in the fixed reference coordinate system of a model and the viewing direction (or the direction in which the model is to be projected) is determined by this reference point and the origin of the coordinate system. In addition to the projecting direction, the orientation of the screen about the viewing direction must be specified. This is done by specifying the axis of the reference coordinate system which appears vertical to a person standing at the refeence viewing point and looking towads the origin. Thus there are six allowable orientations, with either ±x, ±y or ±z axes appearing vertical on the screen.

772

The 3-D display is initiated by the command 3-D. The response to the first prompt, requesting the vertical axis and reference viewing point determines the viewing direction and screen orientation. The second prompt requests the position number, which has the same meaning as that in the EXEC mode. The next series of prompts are the same as in the EXEC mode and request data to establish the configuration. Following this will be a request for the names of the muscles to be shown. A list of up to 20 names can be entered, or the words ALL or NONE. The latter two responses will display all of the muscles defined in the model or name of them, respectively.

When all the information requested above is furnished, the screen will be erased and a display of the model will appear. The display also contains the title of the model, the reference position number, and the viewing point. The vertical axis for the display can be determined from the coordinate axes drawn in the lower left hand corner, parallel to the fixed reference coordinate system.

The scale of the display picture is determined automatically. By calculating the coordinates of all the points in the model in the projection plane of the screen, the screen is scaled such that all points will lile within its borders. This may cause the picture to be scaled down so that all the detail is not visible. In such a case, the windowing capability in the Graphics Segment allows any portion of the screen to be enlarged to the same size as the screen.

The viewing direction can be changed by means of the REO command. A new viewing point and vertical axis can be entered following this command. The screen is erased and a new picture is drawn. The configuration of the model or the muscles displayed remain unchanged.

Sectioned news of the model are also possible by means of the SEC command. This command can be entered before or after a 3-D display picture has been produced. If entered prior to the 3-D command, the system will request for information to eotablish the 3-D display first. If the screen already contains a 3-D display, only the data required to define the sectioned view will be requested. The prompt for this appears as:

ENTER NAME OF CUTTING PLANE AND ITS LOCATION AND SECTION
LINE ANGLE AND SEPARATION.

The cutting plane can be parallel to any one of the planes defined by any pair of axes of the reference coord. system. It can be located anywhere along the third axis. In order to specify which side of the plane is to be blocked out (not shown) the name of the plane is pefixed with a + or -. The plus sign (+) indicates that all pictorial information in the half space which lies on the side of plane in the positive direction of the axis perpendicular to the plane will not be shown. The negative sign (-) implies the opposite. Therefore, six different cutting planes can be defined ±xy, ±zx, ±yz. The section lines are defined by specifying their sepaation and orientation on the viewing screen. The separation or perpendicular distance between the lines is given in units of screen inches. The angular orientation which is specified in units of degrees, is measured from the horizontal on the screen, positive upward and negative downward. Once a sectioned view has been produced on the screen, it can be treated the same as a 3-D display view. To show the entire picture again, the command 3-D should be entered.

The Stereoscopic Display

This part of the Graphic Segment allows stereo views of a segment and thus provides a way to perceive depth from a two-dimensional display screen. This is accomplished by drawing two separated perspective views of a model generated from different viewing points. These separate views can be combined in a stereoscope so that they appear to the eyes to be originating from the same place.

To produce a seteroview a 3-D display must be produced first. The command STE is then entered following which the system will prompt for the "picture width", "picture separation", and "eye separation" data. The picture width corresponds to the size of the picture desired in inches on the screen. This can be at most, one half the screen width. The picture separation distance is the amount the two perspective views will be separated on the screen. The eye separation is the distance between the two viewing points; a value which seems to produce good results is 2.5 in. By varying this parameter the effective depth perception can be altered. Increasing this distance the depth perception is enhanced while a decrease in this distance is equivalent to moving further away from an object reducing the depth perception.

Display Picture Components

The manipulation of picture components, that is, creation, deletion and modification of points, pictorial lines, and lines representing muscles, is accomplished through a combination of three-letter inputs, graphical inputs and numerical data. The action required of the user fo the various commands depends on which mode (sketch or 3-D display) the System is in. Note that the stereo display is for viewing only, and no picture components can be referenced in this mode.

1] Points (POI,KPO,FPO,DPO,MOV,TPO)

The point is the basic element in the Graphics Segment. Lines and muscles are both defined by their end points.

Points can be added to a model in either the sketch or 3-D Display mode. In the sketch mode point can be defined graphically by using the command POI. Following the entry of this command, the graphic input device of the terminal is enabled. The graphic cursor is positioned at desired locations on the screen and its location is transmitted to the computer which adds a point to the data structure. A point is also drawn at the specified location simultaneously. The system will continue to request graphic input until the user terminates these requests (by pressing the 'E' key on the Tektronix terminal).

Points can also be defined by keying in their coordinates by using the KPO command. Following the input of this command the system will wait for a line of input containing the data defining the point. In the Sketch mode, the input data are the two coordinates of the point in the plane oof the screen as indicated by the heating of the display (XY, YZ or ZX). Thus, if the screen is defined to represent a ZX plane of the rigid body, the first number keyed in will be the Z coordinate, which the second number would be the X coordinate. Following the input of the numerical data, a point will be drawn on the screen at the given coordinates (if it appears in the area of the plane shown on the screen)

774

and the system will wait for more data. Data will continued to be accepted until a blank line is sent which terminates the requests. In the 3-D Display mode, the numerical data input is more extensive. It contains the name of the rigid body upon which the point is defined, followed by the X,Y,Z coordinates of the point in the local coordinate system of the named body. Note that whenever a point is defined in the Graphics mode it is assigned a four-letter name internally.

The command FPO is used to graphically determine the coordinates of a point drawn on the screen. The graphic cursor is positioned on the point whose coordinates are desired and the information is transmitted to the computer. The computer responds by printing the name of the point, the rigid body on which it is located and its local coordinates. This command can be used in either the sketch or 3-D mode. If the system is in 3-D mode, the coordinates of the point in the reference coordinate system are also printed. If the cursoir is not psitioned close enough to the point, a message "NO POINT FOUND" is printed. If in Sketch mode, coordinates of the point where the cursor is, are also printed with the message, even though no point exists there.

Points can be deleted by using the DPO command, and positioning the cursor over the point to be deleted. When a point is deleted, a small cross is drawn at the point on the screen. The request for graphic input can be terminated in the usual way ("E" on the Tektronix terminal).

A temporary point can be defined in the graphics mode at a desired location just to "see how it looks". This is done by using the command TPO, which can be used only in the 3-D mode. Following this command the point is defined by inputting the name of the rigid body and local X,Y,Z coordinates of the point. A small cross is drawn as the location of the point. The description of the temporary point is changed each time the TPO command is entered.

The command MOV is used to graphically move a point from one location to another. The cursor is set at the point already defined whose coordinates are to be changed and this information is sent to the computer. The cursor is then positioned at the location where the point is to be moved. If in Sketch mode, the second graphic nput position can be anywhere on the screen. If it is the location of a previously defined point, the coordinates of the first point will be set equal to the second one. If no point is found at the second position, the coordinates of this position are derived and assigned to the first point. In the 3-D mode the second graphic input position following the MOV command may or may not correspond to a previously defined point. If no point exists in the second position, the coordinates of this position are derived assuming that the point is being moved in a plane parallel to the screen plane. If the second position corresponds to a previously defined point or a temporary point, the two points must be located on the same rigid body.

2] Pictorial lines (LIN,SLI,DLI)

Line drawings or wire-frame drawings of rigid bodies are produced by means of pictorial lines. The lines are defined gaphically by specifying their end points.

The command LIN can be used either in the Sketch or 3-D mode. The first line is connected to the first pair of end points. All subsequent lines are defined by specifying one point with the last defined end point of the previous line acting as the other end of the line. Thus, a series of connected lines is generated. To teminate one series of lines and start another, the request for more graphic input is terminated ('E' key on the Tektronix) and the command LIN is reentered again. In the Sketch mode the terminal points of the lines need not be previously defined. When a graphic input position is sent to the computer it checks to see if a point has been defined at that position. If one exists, the system uses that as an end point of the line. If not, a new point is defined and used as an end point of the line. Therefore, in the sketch mode the drawing of the lines can proceed in a "free hand" fashion. In the 3-D mode the points must be previously defined or be located on a previously defined line since there is no other way of deriving the unique coordinates of an arbitrary point on the screen. If the second point lies on a line, a new point is defined and the referenced line is broken into two lines. Therefore, three line segments will meet at the new point.

Independent line segments connecting pairs of points can be drawn using the command SLI. Following the entry of this command, the graphic device is activated and the system is prepared to receive pairs of points which define the lines. This command is useful in the 3-D mode since lines generally connect points in different planes and thus, are not a series of connected lines.

Lines can be deleted by using the command DLI. The line to be deleted is identified to the system by positioning the cursor over it. A small cross is drawn on the lilne indicating the line deleted.

3]   Muscles (MUS,DMU,SMU,FMU)

Muscles can be added or deleted as well as viewed in the pictures drawn in the Graphics segment.

Muscles can be added using the command MUS. The system waits for the entry of the name of the muscle following which the two end points are specified graphically. The system then waits for the name of another muscle. The process can be terminated by entering a blank line instead of a name. Muscles are added to the model just as if it were defined in the model preparation segment with the command MUSCLE.

Muscles can be deleted by using DMU and positioning the cursor over the muscle to be deleted. Upon successful completion of the deletion a small cross is drawn on the muscle line. The system will continue to request graphic input until terminated in the usual way ('E' key).

Muscles can be displayed by using the command SMU. As mentioned earlier, the muscles shown in the 3-D display view are specified by a list of names entered before the picture is drawn. In order to display other muscles SMU can be used. The system waits for a list of names (up to twenty) and then displays the muscle when the list is entered. The names entered here are added to the previous list of muscle names until that total equals twenty. These muscles are shown each time the current view is redrawn.

The command FMU is used to retrieve the name of a muscle displayed. To do so, the cursors are positioned over the muscle and the system responds by printing the name of the muscle on the screen.

Requests for additional graphic inputs can be terminated in the usual way.

4] __Hidden Lines Removal__ (ERA,VIS)

In a 3-D display normally all lines in the sketch of a rigid body would be visible. However, to facilitate viewing of solid objects a hidden line removal capability has been added to the Graphics segment. Note that removal of a line is not the same as deletion. When a line is removed from a picture it will not be shown when that view is redrawn but will reappear if the orientation were to be changed. Deletion on the other hand implies removal of a line permanently.

To erase or remove lines from a picture the command ERA is used which activates the graphic cursor. Two or three points on a line must be identified to remove it or a portion of it from the picture. The first point identifies the line in the data structure and must be on that portion of the line which will be removed. The next two points identified, define the boundaries of the invisible portion of the line. If one of these is an end point of the line, one end of the line will be removed otherwise the middle portion of the line is removed. To remove an entire line from the picture, instead of identifying its two end points as the second and third input points, the second point should be the same as the first. In this case, a third point is not needed and the entire line is removed. The points identified are marked with a cross (+) once the system has located the lines and performed the necessary calculations for determining the invisible portions of the lines. An illustration of using the ERA command to achieve picture 18(b) from 18(a) is shown. In this case line 1 is removed completely (input points: 1^1, 2^1) while lines 2,3,4 are removed partially. To remove end part of line 2, the three points are $1^2,2^2,3^2$ while to remove a middle part of line 4, the three input points are $1^4,2^4,3^4$.

It should be mentioned that __only one__ portion of a line can be removed using the ERA command. Obviously in some instances a single line may have to be broken into many alternating visible and invisible segments. This can be done by first removing the entire line from the picture and then defining new line segments over it in the visible portions which can be deleted later if necessary.

ERA can also be used to remove portions of muscle lines and the arrows representing joint reaction forces. When the first point is identified the system searches to find first a line, then a muscle and finally a reaction force. If nothing is found a message to that effect is printed. The procedure for removing a portion of a muscle is the same as that for a line. When a reaction force arrow is located, the entire arrow is removed from the picture and no further input is required.

If a portion of a line or a muscle or a reaction force arrow is inadvertently removed from a picture, it can be added back using the command __VIS__ and identifying graphically a point on the component which is to be shown in its entirety.

5] __Additional Graphics Capability__ (RED,WIN,FUL)

The commands which control the operation of the graphics segment can be grouped into two classes. One class deals with establishing the view on the display screen (SKE,3-D, REO,SEC,STE, etc.) and the other is

777

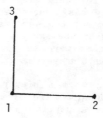

Figure F-17. Sketching the reference axes for drawing a picture. XY plane, line 12 denotes X axis, 13 denotes Z axis.

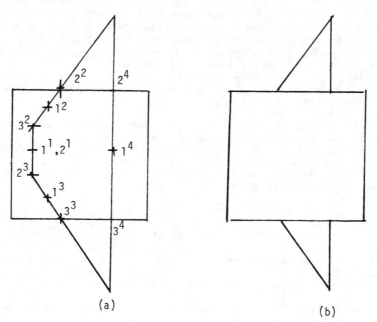

(a) (b)

Figure F-18. Use of EPRA command to remove hidden lines.

concerned with manipulating various pictorial components (POI,KPO,LIN,etc.). Additional graphics capabilities deal with manipulation of the entire picture and provide helpful aids in operation of this segment.

i. <u>Redrawing a picture</u> (RED)

The RED command can be used in either Sketch or 3D-mode and is used to redraw the current view shown on the display screen. Both the scale and orientation of the 3-D display, and the scale and plane definition in the sketch mode remain unchanged when the picture is redrawn. The RED command is mainly used to remove clutter from the screen such as previously entered commands and data. It is also useful for viewing the current screen contents after changes have been made without re-establishing the view with a SKE or 3-D comment and their input requirements. It is mandatory to use the RED command during the manual removal of hidden lines because the projecting direction of the display remains the same.

ii. <u>Windowing or enlarging a portion of the sreen</u> (WIN,FUL)

Windowing refers to the ability to magnify a portion of the current display such that this smaller part fills the entire screen. The portion of the display to be windowed is specified by locating two oppoite corners of a rectangle on the screen.

The command WIN causes the graphic input device of the terminal to be activated. Using the cursors, the lower left corner and the upper right corner of the portion of the screen which is to be blown up are entered graphically. The system rescales this smaller portion such that it fills the screen with no distortion. Either the height or the width will be the governing dimension used to obtain the new scale factor depending upon which dimension meets the screen boundary first.

The contents of the screen when the windowed view is drawn will reflect changes in the same manner as during the redrawing (RED) process. Once the view has been drawn the picture components can be manipulated the same as they were in the full size view. This includes the ability to window any part of the view again.

The full size view can be obtained back by using the command <u>FUL</u>. Any changes made to the windowed portion will be reflected in the full size drawing.

iii. <u>Saving pictorial data</u> (SAV,REC,DEL)

The space required in the data structure to store pictorial data (points and lines) can be quite extensive. Since this data is not needed during the Analysis Segment of the system, the capability of storing pictorial data on a data file (SAV) has been included in the system. In addition the capability to recall this data and integrating this into the data structure (REC) as well as deleting the data (DEL) has also been provided.

To save the picture data the command SAC is used. The system will prompt for a data file number where the data is to be stored. The numbered data file should have been assigned previously. The system will write on this file the location of the end points of the lines, the names of the rigid bodies on which the points are located and the names of the points.

To recall picture data that has been previously saved, REC is used. Following the entry of this command the system will prompt for a file number where the data is stored.

To delete the picture data the command DEL is used. This removes from the ring data structure everything that has been added since the current entry into the graphics segment. It is especially advantageous for removing pictorial data which has been REC added.

iv. Obtaining a permanent copy of the screen contents (PLO)

A permanent copy of the screen contents can be obtained either by using a hardcopy unit attached to the graphics terminal or by a plot on a digital plotter.

To obtain a plot the command PLO is used. The system will then issue two prompts. The first one is to determine whether or not a stereo view is desired (response: 3-D or STE). The second prompt is for the size of the plot. The plot will be placed in a square area equal to the plot size specified. Because of the variability of plotting routines at different computer installations, the plotter information is written on a data file numbered 12, in unformatted Fortran. A post processing routine must then be used to read this file and produce the plot using installation provided plot routines.

v. Ending a graphics session (END)

To exit from the graphics segment and return to another segment of the system, the command END is used. Note that if picture data is to be removed (DELeted) from the data structure, it should be removed before terminating the current graphics session. The data which is present in the data structure upon leaving the Graphics Segment cannot be otherwise removed. Once the END command is entered the graphics session is terminated. The screen is eased and the system will prompt with the term MODE?

vi. Optimizing the drawing process (OPT)

The pictorial lines in a 3-D display are drawn on the screen in the order as they were defined. For complex diagrams where the lines have been defined by the SLI command or LIN commands the drawing process can take substantial time. Some saving in time can be obtained if this process can be optimized by rearranging the order in which the lines are drawn to eliminate maqny of the invisible moves. An algorithm to eliminate unnecessary moves has been incorporated in the system and is used to determine the optimum drawing path.

The drawing process is optimized by using the command OPT. The number of paths generated is printed on the screen when the algorithm is finished.

An abbreviated list of the Graphics commands discussed above is given in Table 3.

Additional Capabilities of the Computer System

1] Model Proportioning (PROP)

This capability allows the dimension of a segment to be changed proportionally in the three directions of the axes of the coordinate system attached to it. Five components must be considered when a model dimension have to be changed. Three of these, the inertia characteristics of the rigid bodies, the location of the points on a rigid body, and the translational vectors of matrices, have a well-defined relationship when making proportional changes. The radii of cylindical material constraints may be measured in an arbitrary direction, and are, therefore, not included in the proportioning algorithm. Similarly the limits associated with JTTOL statements are also not included in this algorithm.

A model is proportioned by specifying three proportionality constants (X_p, Y_p, Z_p) along X,Y,Z axes respectively for a rigid body. The proportioning capability has been included in the system as part of the model preparation segment. To access this capability, the command PROP is entered. This may be done at the same level of control as the commands CHANGE, ADD, DELETE, etc. Once the proportioning mode, the system will prompt for the name of the rigid body and the X,Y,Z proportionality constants. An example line of input might be:

SEG1,.8,.5,1.15

where SEG1 is the name of the rigid body and $X_p = 0.8$, $Y_p = 0.5$, $Z_p = 1.15$ meaning that it will be proportioned to be larger in the Z direction $(Z_p = 1.15)$ and smaller in the X,Y directions. Following this entry the system will proportion or scale the body by the factors input and print a reminder that the radii of cylindrical material constraints are not changed nor are the limits for the JTTOL statements. These changes will have to be made in the CHANGE mode. The system will requet for a line of data for another body. If no other bodies are to be proportioned, the word STOP is entered and the system will return to the control section and issue the prompt MODE? At this time another control statement (e.g., CHANGE,EXEC,GRAP, etc. may be entered).

2] Interactive Graphical Construction of Model (CON)

This capability allows one to define models or parts of models interactively on a graphic display screen. This is useful for defining rigid bodies which are similar to or are mirror images of previously defined bodies, thus reducing the amount of imput data. The parts of a model of a rigid body which are constructed using this capability are: the rigid body itself (except for its inertia characteristics), the joint connecting the rigid body to another one in the model; points on the body, and pictorial lines. The capability of defining musles graphically already exists in the Graphics segment. The term **new body** refers to the rigid body which is being added to the model. The term **old body** refers

781

to a rigid body previously defined in a model which is similar to the one being added. The new body will have the same points and pictoial lines defined for its as the old body has. The new one may be proportioned relative to the old body and/or it may be a mirror image of the old one. The term **reference body** refers to the body to which the new body is connected. The joint defined between the reference body and the new body will be similar to the joint between the old body and whichever body it is connected to. A different description of the joint may be included if desired.

The general scheme for graphical construction is illustrated in Fig. 19. A body SEG1 has been preeviously defined in the conventional way and two additional bodies SEG2 and SEG3 (both proportionally similar to SEG1) are to be added. By specifying a plane of projection XZ and the location of the origin O_1 on the screen, an orthogonal projection of the picture of SEG1 can be produced. Using SEG1 as a reeference body, new body SEG2 can be defined as being attached to it. SEG1 is also the old body in this case since SEG2 is similar to it. Next the joint center O_2 on SEG1 can be specified graphically in the X and Z directions while the third coordinate is determined from the location of the screen plane in the y direction. The joint which is defined at O_2 will be the same as at O_1. The joint is duplicated by considering each transformation matrix separately. The joint reactions, as well as, the equations of equilibrium are added to the model at the new joint also. If the joint is to be reproduced exactly, only the name of the transformation matrices need by supplied. If the new joint is different than the one at O_1, provisions have been made for defining an arbitrary joint at the new joint center location.

Once the joint has been defined and the variables in its specified, the points and the lines for SEG2 can be added. They are proportionately

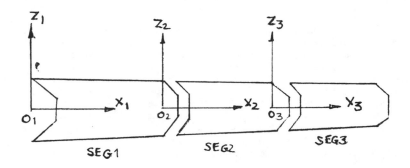

Figure F-19. Graphical construction of a model.

duplicated duplicated from those of SEG1 and picture of SEG2 is then drawn. Using the proportionality constants, the coordinates of points on the body as well as the translational vectors in the transformation matricse of the new joint are also obtained. After SEG2 has been defined, SEG3 is constructed. This time SEG3 is the new body, SEG1 is the old body and SEG2 is the reference body. In general, the old body may be any rigid body that has been defined in a model while the reference body must be one that is shown on the screen. The joint center

O_3 can now be defined graphically. The system determines the location of O_3 with respect to SEG2 coordinate system and points and lines are duplicated for SEG3. Note that by defining the screen to be located at different values, new joint centers can actually be defined in different planes or in three dimensions. Also, the joint may allow spatial motion.

Operation of Interactive Graphical Model Construction (CON)

This capability has been included as part of the graphics segment. It is equivalent to the picture generation part or the 3-D display part and is accessed by entering the command CON anytime the system is in the graphics mode.

Upon entering the command the first prompt issued is

ENTER: NEW BODY [,SIM/MIR,OLD BODY]

The terms new body and old body were explained earlier. The brackets [ ] indicate optional entries in the input data. If the new body is neither similar to nor a mirror image on any other rigid body, only its name need be entered. If the new body is similar to another defined body, the second entry is SIM (or MIR if it is a mirror image). Referring to Fig. 19, the input required to construct SEG2 will be:

SEG2,SIM,SEG1

If a mirror image is requested in the first line, the second prompt is:

ENTER REFLECTION PLANE

There are six possible entries to this prompt -- XY,YX,XZ,ZX,YZ,ZY. The points on the new body will be created such that the sign of one coordinate is reversed. For example, if XY or YX is the reflection plane, the sign of Z coordinate of all the points is reversed. The next prompt is:

ENTER REFERENCE BODY NAME

This is the name of a previously defined body in the model. In the example discussed earlier, SEG1 is the first reference body and SEG2 is the second one.

The next prompt is used to define the screen plane. This is established parallel to one of the planes defined by the axes of the coordinate system of the first reference body. The prompt appear as:

ENTER SCREEN PLANE (LEVEL AXIS LETTER FIRST)

One of the six responses listed above for defining a reference plane is entered here. To define the X,Z plane (see Fig. 19) the response is XZ, X being the level axis. Following this the next prompt is

LOCATE ORIGIN OF REF BODY

The origin is located by the graphic cursor. Any point on the screen may be established as the origin. A scale factor (no. of model units per screen inch) is requested next by

ENTER SCALE FACTOR (MODEL UNITS SCREEN IN.)

Once this data is entered, the screen is fully defined and the picture of an orthogonal projection of the reference body is drawn. The next prompt is:

REESCALE? (Y/N)

This provides an option for rescaling the screen in case the scale factor was not appropriately selected before resulting in the body appearing too large or too small. The response may be either Y (yes) or N (no). In case a rescaling is desired, the response Y is followed by the new scale factor. The screen is erased and a new picture of the reference body is drawn.

With the reference body shown on the screen the location of the joint center for the joint between it and the new body can be established. This is requested with the prompt:

LOCATE ORIGIN OF NEW BODY -- (X,Y, or Z)1 = XX·XXXX

and the point is established using the graphic cursor. The location of the screen plane is shown in the prompt. If the joint center is not in this plane, it can be changed. This is done by pressing 'E' key which will cause the system to ignore the graphic data and issue a prompt:

ENTER NEW REF PLANE OR NEW ORIGIN

There are two ways to respond to this prompt. If one number is entered, it is interpreted as the new location of the screen plane. If three numbers are entered, they are interpreted as the location of the new origin in the coordinate system of the reference body. In the first case, the prompt requesting the graphic location of the origin is reissued and when the origin is defined a small cross is drawn at its location on the screen.

The next prompt is for the proportionality constants relating the new body to the old body:

ENTER PROPORTIONALITY CONSTANTS   new/old

The proportionality constants x_p, x_p, z_p are entered.

The joint between the reference body and the new body is defined next. The system will print a description of the transformation matrices describing the joint between the old body and the rigid body it is connected to. An example of the printed output might be:

TRNI RY VRM 0.05 0.00 0.01 0.01
ENTER NAME OR NEW MATRIX DESCRIPTION

The first line is a description of the transformation matrix with the terms as defined in the TRAN command statement. The same two lines are printed for each matrix in the joint. If the matrices in the new joint are the same as the printed description, only a new name (four letters or less) need be supplied. If a different matrix is to be defined, it is described in the same manner as with a TRAN command The

784

command name and the slashes should not appear, however, and the input in general might appear as:

Name, type, diff., X,Y,Z,θ

The system will continue issuing prompts for new matrices until the word STOP is entered. This ends the joint description.

The system will define all of the points and lines for the new body. Before drawing the picture it will request the value of the variables in the joint just described. It does this by issuing a prompt which has the name of the matrix and the type of the variable requested. For example, the prompt

TRNI RY

is requesting the value for a variable angle (rotation about Y axis) in the matrix TRNI. The angles are specified in units of degrees. If a blank line is entered in response to this prompt the variable quantity is not changed. When all of the variables in a joint have been defined, the picture of the new body is drawn on the screen and a new prompt is issued:

ENTER OPTION

The five options that can now be entered are:

a. STOP: to terminate the model construction. The system remains in graphics mode and so a 3-D display may be requested at this time.

b. CON: to construct another rigid body. However the screen is still defined as a plane of the first reference body and the only rigid bodies that can be used as reference bodies are those shown on the screen. If a different reference body is required, the STOP option followed by the CON command should be used to initiate the drawing procedure. Five new bodies can be defined during any one model construction session.

c. REP: to reposition the rigid bodies shown on the screen with respect to each other. Prompts requesting the values of variables in all of the joints are issued as described above. A blank line in response indicates that the variable will not be changed. When all variables have been supplied the picture is redrawn without erasing the previous screen contents allowing a series of positions to be viewed on the screen.

d. RED: to erase the screen and the piccture at its last specified configuration to be redrawn.

e. MOV: to move the location of a joint center graphically. The joint center is first identified graphically with the cursor which is marked with a cross. The system will then issue prompts to define the new position.

Note that in the graphical model construction although rigid bodies are added their inertia characteristics are not added. This must be added using the CHANGE mode. Similarly muscles must be added in the Graphics mode, since the names of the points are not known. They can however be obtained by getting a listing of them in either the FIND mode or with an ECHO statement.

APPLICATION OF THE INTERACTIVE PROGRAM TO BICYCLING

To demonstrate the application of the interactive program, the lower extremities model will be developed and investigated for muscle load sharing during bicycling. The development of the model entails establishing the kinematic topology, locating the muscle attachments and determining the external forces (pedal forces) acting on the foot.

The lower extremities can be modelled by four rigid bodies--the pelvis (PLVSO, the femur (THI), the lower leg consisting of the tibia and the fibula (CALF), and the foot (FOOT). The coordinatee systems for these segments have been shown in Fig. 4. The reference or ground coordinate system is selected at the center of rotation of the pelvis with respect to the bicycle seat. It was seen from experimental studies that the pelvis remains vertical in the side view but undergoes a rocking motion with respect to the seat in the front view. Hence the reference coordinate system is placed at the center of rotation, directly below a line connecting the centers of the hip joint (Fig. 20).

The rocking motion of the pelvis with respect to the ground is described by a variable rotation about the x-axis:

BODY/PLVS,GROUND/
TRAN/T1,TX,V/

Since there are no muscles crossing this joint, no equations of equilibrium are defined, and inertia characteristics of the pelvis can be accordingly omitted.

The hip joint between the pelvis and the femur is treated as a ball joint with X,Y,Z order of rotation. The femur is assumed to move in a plane parallel to the XZ ground plane and the first rotation is therefore about X axis, followed by rotation about Y-axis and then about Z-axis to represent rotation of the femur about its long axis (resulting in the foot being moved in or out). The knee joint is defined by a HINGE statement and the ankle joint with a BALL statement. The patella (PTLA) is located with two TRAN statements.

Muscles are modelled by defining their points of attachments. Muscles which wrap around boney segments are modelled using the material constraint statements. This method is used to model the glutens maximus (fibers inserting on the femur), the semitendinosus, the gastrocnemius (both heads) and the ilio-psoas. The sartorius muscle originating on the pelvis and inserting on the tibia cannot be modelled by a direct straight line. Instead it is modelled as two parts (SART,SRT2), one in the vicinity of its origin and the other in the vicinity of its insertion which are then equated.

The extensor retinaculum at the ankle which prevents the extensor muscles from bowstringing is also included in the model. A representative model of the tibialis anterior is shown in Fig. 21. A

786

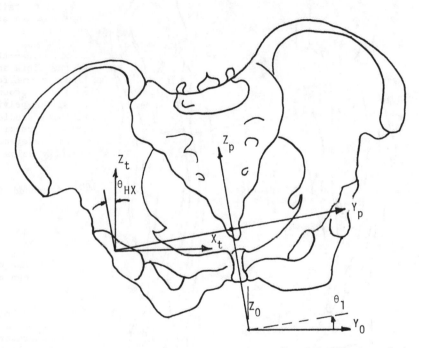

Figure F-20. Reference coorinate system (Y_0,Z_0) and the pelvic motion (θ_1).

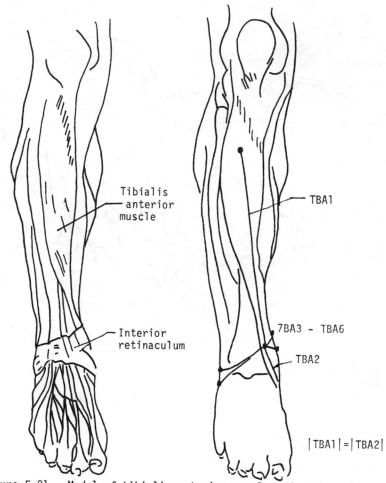

Figure F-21. Model of tibialis anterior muscle and retinaculum.

Tibialis
anterior
muscle

Interior
retinaculum

TBA1

7BA3 - TBA6

TBA2

|TBA1|=|TBA2|

point is chosen in the coordinate system of the foot where the muscle changes direction and goes under the retinanculum. This point is defined as a reaction point (REAP) and the tibialis anterior muscle is divided into two components each having equal magnitude. Since the forces defined with the reacion point do not exist physically they are deleted and replaced with four unknown force components (TBA3–TBA6)—two of which are attached to the foot and the other two to the tibia. The unknown force components (UNFO) are defined to act in only one direction (tension) and they have zero cost coefficient in the merit criterion. The result is to add three equations and five unknown force components for each muscle.

The bicycling configuration is shown in Fig. 22. It is defined by the three angles $(\theta_{H_x}, \theta_{H_y}, \theta_{H_z})$ at the hip joint, the angle θ_k at the knee joint and three angles $(\theta_{A_x}, \theta_{A_y}, \theta_{A_y})$ at the ankle joint. The pedal angle θ_p, the crank angle (θ_c) and the pelvis rocking angle θ_1 as well as the pedal forces on the foot were measured experimentally, and were used to determine the angles at the knee, hip and ankle (as well as velocities and accelerations). Data was collected at different loads and speeds of bicycling at every 15° of crank rotation. This forms the input data for the analysis segment and were ead in by using special SPECDS and FNON routines at various positions to be analyzed. The positions are numbered starting at zero crank angle (position no. 1) and increasing as the angle increases (position no. 2, $\theta_c = 15°$, etc.).

An example of the results of two positions is shown in Table 4. The names of the forces are the four letter abbreviations of muscles as well as the system generated names of the joint reactions. The magnitude of the forces are in Newtons and moments in Newton meters.

APPLICATION OF THE INTERACIVE PROGRAM TO THE JAW MODEL

The use of the interactive program for the analysis of the jaw is illustrated here. The model is a single one with only one segment involved, the jaw (JAW) which is connected to the skull (SKUL) through the two temporomandibular joints. The reference axes on the skull are selected at a point midway between the two joints with X being anterior-posterior, Y being medio-lateral and Z being up-down. The motion of the jaw with respect to the skull consists of translations along X and Z (SX and SZ matrices and a rotation about the Y axis (ROTY). The reactions at the two joints are defined by means of the GROUND REACTION command statements with a cost of 2 to the reaction forces and a cost of 400 to the reaction moments. There are twelve muscles, since on each side which are defined by their points of origin and insertion on the two segments. The external forces which simulate biting forces during mastication are modelled by forces FX and FZ parallel to the X and Z reference axes respectively acting at a defined point (P28) on the jaw. A listing of the Input command statements is shown in Table 5.

In the analysis mode, the requirement inputs will be the translations SX, SZ and the rotation ROTY along with the values of the external forces FX and FZ. For static cases the accelerations and velocities are zero.

An output produced by the ECHO/4 statement is shown in Table 6. A sample analysis follows this listing, obtained by the EXEC command

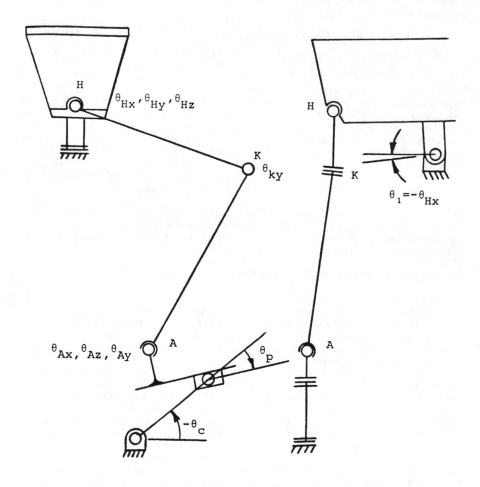

Figure F-22. Bicycling Configuration (schematic of the leg model as a closed loop mechanism.

TABLE 4.   Examples of standard output format showing the results
           for two positions.

PROBLEM TITLE: LEG OFFSET MERIT CRITERION CHANGED RESULTS FOR
POSITION NO. 1
VALUE OF THE OBJECTIVE FUNCTION I.89064438+003

FORCE	VALUE
SEMM	4.40472I+01
GUMF	1.124470+03
VSMD	7.451709+02
VSIN	7.451709+02
VSLT	7.451709+02
PETT	2.425825+03
GASM	9.900441+02
SOLE	2.801841+02
TIBP	3.473118+00
CURL	1.114882+03
ADE2	1.505592+03
GLD2	1.027815+03
FY	-7.645661+01
4HIP	-7.920420+02
5HIP	-1.102408+03
6HIP	-3.396045+03
5KNE	-5.786597+01
6KNE	-4.291698+03
5ANK	-3.626060+02
6ANK	-1.636962+03
4T23	2.569910+03
5T23	2.372994+02
1KNE	-2.751787+01

POSITION NO.   4   OBJECTIVE FUNCTION I.20371495+003

GUM1	2.924045+01
GIT2	3.885113+00
1LT8	2.583804+01
GASM	3.615496+02
SOLE	1.023581+02
SOLT	4.627452+0?
FLHL	7.432238+01
CURL	9.043702+01
ADE2	1.156992.+03
GLD2	1.040862+03
FY	-6.367898+01
4HIP	-2.380550.+02
5HIP	-7.221422+02
6HIP	-2.085750+03
5KNE	-3.871555+01
6KNE	-6.758482+02
4ANK	1.119838+02
5ANK	-2.461324+02
6ANK	-1.293995+03
5P62	-2.278936+01
1KNE	-1.187612+01

791

TABLE 5 Input Command Statements for the Jaw Model.

```
INPUT
TITLE ANALYSIS OF JAW DURING MASTICATION
BODY/JAW,SKUL/
TRAN/SX,TX,V/
TRAN/SZ,TZ,VE/
TRAN/ROTY,RY,V/
GROU/LTMJ,JAW,,-.045,,2.,400./
GROU/RTMJ,JAW,,.045,,2.,400.0/
POINT/P 1,SKUL,.015,-.045,.053/
POINT/P 2,JAW,.033,-.036,-.005/
MUSC/TPAR,P1,P2/
POINT/P 3,SKUL,-.045,-.057,.048/
POINT/P 4,JAW,.029,-.036,.003/
MUSC/TPPR,P3,P4/
POINT/P 5,SKUL,.035,-.05,-.005/
POINT/P 6,JAW,.01,-.045,-.047/
MUSC/MSSR,P5,P6/
POINT/P 7,SKUL,.02,-.055,-.001/
POINT/P 8,JAW,.015,-.045,-.027/
MUSC/MSDR,P7,P8/
POINT/P 9,SKUL,.023,-.02,-.006/
POINT/P10,JAW,.008,-.038,-.047/
MUSC/MPTR,P9,P10/
POINT/P11,SKUL,.03,-.03,-.002/
POINT/P12,JAW,.005,-.039/
MUSC/LPTR,P11,P12/
POINT/P13,SKUL,.015,.045,.053/
POINT/P14,JAW,.033,.036,-.005/
MUSC/TPAL,P13,P14/
POINT/P15,SKUL,-.045,.057,.048/
POINT/P16,JAW,.029,.036,.003/
MUSC/TPPL,P15,P16/
POINT/P17,SKUL,.035,.05,-.005/
POINT/P18,JAW,.01,.045,-.047/
MUSC/MSSL,P17,P18/
POINT/P19,SKUL,.02,.055,-.001/
POINT/P20,JAW,.015,.045,-.027/
MUSC/MSDL,P19,P20/
POINT/P21,SKUL,.023,.02,-.006/
POINT/P22,JAW,.008,.038,-.047/
MUSC/MPTL,P21,P22/
POINT/P23,SKUL,.03,.03,-.002/
POINT/P24,JAW,.005,.039/
MUSC/LPTL,P23,P24/
POINT/P25,SKUL/
POINT/P26,SKUL,.01/
POINT/P27,SKUL,,,.01/
POINT/P28,JAW,.0682,,-.0427/
FORCE/FX,P26,P25,P28/
FORCE/FZ,P27,P25,P28/
END
```

792

```
'*********************************************************
      TABLE 6        ANALYSIS OF JAW DURING MASTICATION
'*********************************************************

   GRAVITATIONAL VECTOR
                            GXX =    0.0000
                            GYY =    0.0000
                            GZZ =   -9.8067+00


   LINK INFORMATION   (BASE = SKUL)

     . I     J     XBAR   YBAR          ZBAR         MASS     IXX
     JAW    SKUL   0.000  0.000         0.000        0.000    0.000


   TRANSFORMATION MATRICES

          TYPE    DIFF    DOF    SPEC    DERVD     X        Y
   SX      4       4       1      1                0.000    0.000
   SZ      4       6       2      2                0.000    0.000
   ROTY    2       2       3      3                0.000    0.000
                +++ LINK JAW  + WRT + LINK SKUL +++


   POINTS OF INTEREST

          NAME    LINK        X              Y              Z
          LTMJ    JAW      0.000          -4.500-02       0.000
          RTMJ    JAW      0.000           4.500-02       0.000
          P1      SKUL     1.500-02       -4.500-02       5.300-02
          P2      JAW      3.300-02       -3.600-02      -5.000-03
          P3      SKUL    -4.500-02       -5.700-02       4.800-02
          P4      JAW      2.900-02       -3.600-02       3.000-03
          P5      SKUL     3.500-02       -5.000-02      -5.000-03
          P6      JAW      1.000-02       -4.500-02      -4.700-02
          P7      SKUL     2.000-02       -5.500-02      -1.000-03
          P8      JAW      1.500-02       -4.500-02      -2.700-02
          P9      SKUL     2.300-02       -2.000-02      -6.000-03
          P10     JAW      8.000-03       -3.800-02      -4.700-02
          P11     SKUL     3.000-02       -3.000-02      -2.000-03
          P12     JAW      5.000-03       -3.900-02       0.000
          P13     SKUL     1.500-02        4.500-02       5.300-02
          P14     JAW      3.300-02        3.600-02      -5.000-03
          P15     SKUL    -4.500-02        5.700-02       4.800-02
          P16     JAW      2.900-02        3.600-02       3.000-03
          P17     SKUL     3.500-02        5.000-02      -5.000-03
          P18     JAW      1.000-02        4.500-02      -4.700-02
          P19     SKUL     2.000-02        5.500-02      -1.000-03
          P20     JAW      1.500-02        4.500-02      -2.700-02
          P21     SKUL     2.300-02        2.000-02      -6.000-03
          P22     JAW      8.000-03        3.800-02      -4.700-02
          P23     SKUL     3.000-02        3.000-02      -2.000-03
          P24     JAW      5.000-03        3.900-02       0.000
          P25     SKUL     0.000           0.000          0.000
          P26     SKUL     1.000-02        0.000          0.000
```

IYY	IZZ	IXY	IXZ	IYZ
0.000	0.000	0.000	0.000	0.000

Z	ANG
0.000	0.000
0.000	0.000
0.000	0.000

P27	SKUL	0.000	0.000	1.000-02
P28	JAW	6.820-02	0.000	-4.270-02

EXTERNALLY APPLIED FORCES

		DIRECTIONAL PTS		POINT OF
NAME	NO.	PT1 TO PT2		APPLICATION
FX	1	P26	P25	P28
FZ	2	P27	P25	P28

UNKNOWN FORCE ELEMENT

NAME	PT1	PT2	PT3	COST	
4LTM			LTMJ	2.00+00	0
5LTM			LTMJ	2.00+00	0
6LTM			LTMJ	2.00+00	0
4RTM			RTMJ	2.00+00	0
5RTM			RTMJ	2.00+00	0
6RTM			RTMJ	2.00+00	0
TPAR	P1	P2		1.00+00	1
TPPR	P3	P4		1.00+00	1
MSSR	P5	P6		1.00+00	1
MSDR	P7	P8		1.00+00	1
MPTR	P9	P10		1.00+00	1
LPTR	P11	P12		1.00+00	1
TPAL	P13	P14		1.00+00	1
TPPL	P15	P16		1.00+00	1
MSSL	P17	P18		1.00+00	1
MSDL	P19	P20		1.00+00	1
MPTL	P21	P22		1.00+00	1
LPTL	P23	P24		1.00+00	1

UNKNOWN TORQUE ELEMENT

NAME	PT1	PT2	LINK	COST	
1LTM			JAW	4.00+02	0
2LTM			JAW	4.00+02	0
3LTM			JAW	4.00+02	0
1RTM			JAW	4.00+02	0
2RTM			JAW	4.00+02	0
3RTM			JAW	4.00+02	0

CONSTRAINT EQUATIONS

NO.	JOINT	TYPE
1	SZ	1
2	SZ	2
3	SZ	3
4	SZ	4
5	SZ	5
6	SZ	6

MODE?
EXEC

LINEAR PROGRAM SIZE== 36 VARIABLES   6 EQUATIONS
ENTER POSITION NO. OR RANGE OF POSN NOS AND INCREMENT
ENTER VALUE(S) OF THE VARIABLES (3 VALUE(S))   POSN,
VELOCITIES   NO.
ACCELERATIONS
ENTER VALUE OF FORCE FX   POSN. NO. =   1
ENTER VALUE OF FORCE FZ   POSN. NO. =   1
PROBLEM TITLE :   ANALYSIS OF JAW DURING MASTICATION
RESULTS FOR POSITION NO.   1
VALUE OF THE OBJECTIVE FUNCTION 6.7914919$+001

FORCE	VALUE
6LTM	=9.430182+00
6RTM	=9.430182+00
TPAR	4.475378+00
MSSR	1.062172+01
TPAL	4.475378+00
MSSL	1.062172+01

MODE?
STOP
BYE

795

statement. The Inputs ① - ⑤ are as follows:

①	.0005,-.005,.0925
②	blank
③	blank
④	9.2733
⑤	7.2321

① The first input represents the values of the variables in the transformation matrices. Here the three variables are SX, SZ, ROTY representing translation along X, translation along Z and rotation about Y, respectively. Note that the values input are in the order in which they were defined in the INPUT mode. Here SX = .0005 m, XZ = -.0005 m, ROTY = .0925 radians.

② The second input is for velocities. Since the case analyzed is static all velocities are zero. This can be input by either a blank line or card or by specifying them to be zero by 0.,0.,0.

③ The accelerations are zero and so this can be input by a blank line.

④ The fourth input is the force FX. This is the force acting at point P28 on the jaw, directed along the X axis of the fixed reference axis system. Here FX = 9.2733 Neutons.

⑤ The fifth input is the force FZ. This is the force acting at point 28 on the jaw, directed along the Z axis of the fixed reference system. Here FZ = 7.2321 Newtons. The resulting solution indicates that the active muscles on the left side are TPAL(=4.475 N) and MSSL(=10.621 N) and on the right side, TPAR (=4.475)N) and MSSRC = 10.621 N). The joint reactions consist of vertical component only at each of the two temporomandibular joints and equals 9.43 N acting down. The criterion used here was: $\Sigma F + 2R + 400 M$.

BIBLIOGRAPHY

1. Williams, R. J.:   Interactive modelling and analysis of dynamic systems with application to the musculoskeletal structure.   PhD Thesis, University of Wisconsin-Madison, 1976.

2. Williams, R. J. and Seireg, A.:   Interactive computer modelling of the musculoskeletal system.   IEEE Transactions on Biomed. Engg., BME-24, p. 213, 1977.

3. Williams, R. J. and Seireg, A: Interactive modelling and analysis of open or closed loop dynamic systems with redundant actuators.   J. Mech. Design, 101:407, 1979.

The Use of the Model for Control of Movement by Functional Electrical Stimulation and Muscle Response Feedback

INTRODUCTION

The history of application of electrical stimulation for inducing muscle contraction is more than one and one half centuries old. However, it was not until recent years that electrical stimulation has been utilized for "active" therapeutic devices and orthotics. It has been investigated as a means of rehabilitation for regaining, at least partially, voluntary movement over the lost muscle function of the disabled (Functional Electrical Stimulation, FES).

In 1960, Liberson developed the electrical personal stimulator as a correctional device for "drop foot." It was synchronized with the cadence by a foot-switch. The basic principle has been used in the field of orthotic devices.

Very useful and comprehensive bibliographical reviews can be found in the papers by McNeal and Reswick (1976) and Vodonik, et al., (1981). However, only recently can we find some research on finely-controlled electrical stimulators to address the needs of more advanced and intricate therapies.

Based on the type of electrodes used, functional electrical stimulation can be classified into three groups; over the skin stimulation, percutaneous stimulation, and totally implanted stimulation. he percutaneous stimulation should be used by a well-trained therapist because it utilizes needle electrodes. The totally implanted stimulation requires a surgical operation and should be used only for the disabled who definitely need it and can greatly benefit from it. After the operation the stimulator is much easier for the patient to handle, and it can be expected to produce the most consistent output due to avoiding the skin-electrode problems.

In the feasibility study discussed here, we utilize over the skin stimulation. The sensation of pain, which is one of the main problems of the over the skin stimulation, can be eliminated by applying the results of the studies by Gracinin, (1975); and Moreno-Aranda and Seireg, (1981).

The latter method employs a specially modulated high frequency signal which is electro-physiologically reasonable. The recommended stimulation parameters are as illustrated in Fig. 1.

Carrier frequency	ω_1 = 8,000 to 10,000 Hz
On-off frequency	$2\pi/\tau_2$ = 40 to 120 Hz
Duty cycle	τ_3/τ_2 = .2
Stimulation period	τ_6 = 1.5 sec.
Rest period	τ_7 = 4.5 sec.

When this stimulation regime was applied it was found to induce strong muscle contraction with minimum discomfort which are controllable by the stimulation current.

Since the actuation of the extremities is a redundant system, the superficial muscles needed to produce the action as well as the force patterns needed for thse muscles are calculated using the musculoskeletal model of the particular body segment. The stimulation required to produce these force patterns can be established accordingly and applied to the different muscles at the proper phase. The actual muscle forces can therefore be estimated by calibrating the processed electrical response of the stimulated muscle. These can in turn be used as the feedback control signals.

Based on the hypothesis that the electrical response has a linear relationship with the muscle force, at least in a quasistatic condition, the feedback from the muscle, is used to insure accurate generation of a predetermined movement. Postion and velocity feedback can also be used in addition to the muscle feedback as a safety measure.

STIMULATION TECHNIQUE

The stimulation technique, used here, is direct neuromuscular stimulation, where the stimulating electrodes are placed on the muscular belly. Almost all the stimulators, in clinical use for functional electrical stimulation (FES) are represented by the continuous lines in Figure 2. However, in this type of approach there is little control on the muscular activities of the stimulated muscle, and muscle force output may not be consistent.

The study reported here utilizes the electrical responses of the simulated muscle, and some of the properties processed electrical resonse (PER) to monitor muscular activities, as shown by one of the broken lines in Fig. 2.

STIMULATOR DESIGN

Referring to Fig. 1, a 8.8 KHz carrier frequency with an on-off frequency equal to 45 HZ and a duration (τ_1, Fig. 1) = 2.0 msec. is chosen in order to make the relationship between the processed electrical response and the induced torque on the joint approximately linear.

The multichannel electrical stimulator designed for this test is shown in Fig. 3. It is used in conjunction with a microcomputer for programming the voltage (or current) level of the stimulating wave

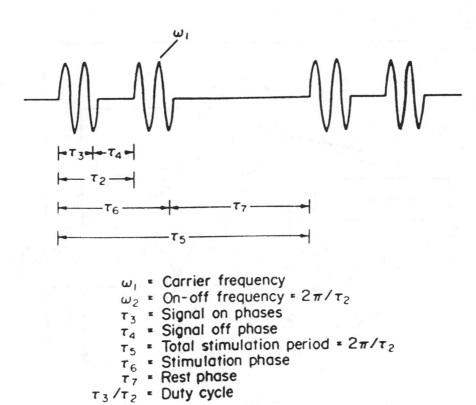

$$\omega_1 = \text{Carrier frequency}$$
$$\omega_2 = \text{On-off frequency} = 2\pi/\tau_2$$
$$\tau_3 = \text{Signal on phases}$$
$$\tau_4 = \text{Signal off phase}$$
$$\tau_5 = \text{Total stimulation period} = 2\pi/\tau_2$$
$$\tau_6 = \text{Stimulation phase}$$
$$\tau_7 = \text{Rest phase}$$
$$\tau_3/\tau_2 = \text{Duty cycle}$$

Figure G-1. Illustration of the signal pattern used for over the skin muscle stimulation.

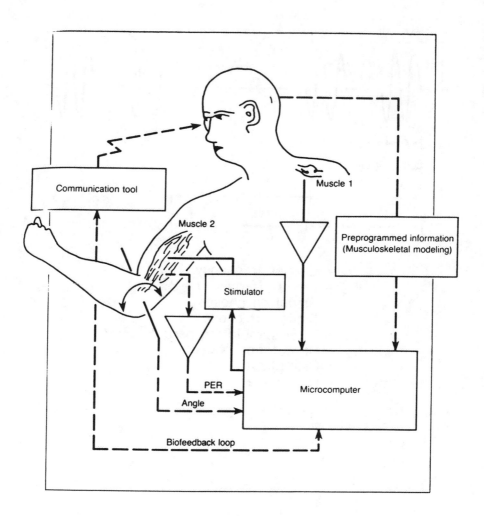

Figure G-2. Microcomputer-based electrical stimulator.

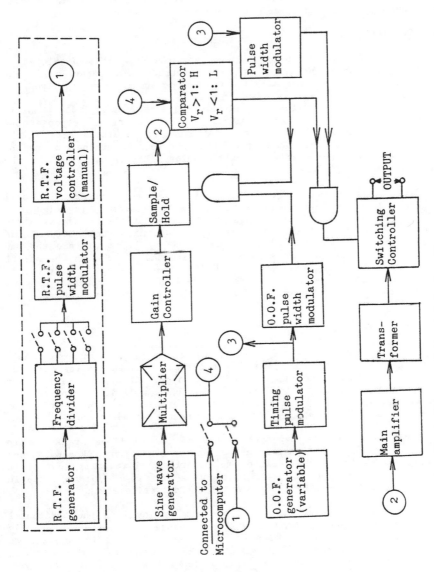

Figure G-3. Block diagram of the total circuit.

required for each muscle to produce a particular force at a particular length of muscle. An induction coil is used with a feedback loop, and a reed relay which reduces the noise associated with sensing the electrical response on the belly of the stimulated muscle.

The output stage amplifier, Fig. 4, has a switching gate controller, utilizing fast switching reed relays, to protect the recording amplifier from large stimulating voltage and skin polarization, and noise. It also has a feedback loop transformer to protect against change of load.

ELECTRODE TECHNIQUE

A type of floating electrode as shown in Fig. 5 is developed to minimize the motion artifact of the electrode-skin interface. The electrodes are made of pure silver supported on sponge soaked jelly.

By utilizing the switching control using a reed relay it is possible to monitor almost the entire waveform of the electrical response just beneath the stimulating electrodes. Although the starting part of the wave is noisy because of bouncing off the reed relay contacts, the method is very useful for comparatively small muscles. The distance between the electrodes was about 2.54 cm.

Gate control by the reed relay is used to protect the small electrical response (0.1 to 10 mV) from the large stimulating signal (0 to 60 V peak to peak), and consequently it is possible to use the two-electrode method. This gate control was inserted before the preamplifier, enabling use of a standard EMG amplifier. Since the raw electrical response wave is strongly periodic and depends on the on-off frequency, a specially designed peak-hold circuit with an integrator is placed just after the rectifier stage. Using this processor it was possible to obtain a faster rise time, which is very important for the puprose of this study.

THE EXPERIMENTAL STUDY

The first objective of this study is to obtain more reliable information for estimating the static and dynamic states of the electrically stimulated muscle, and to apply this information for developing methodology to control global muscular contraction. Fortunately, it was possible to develop the technique for monitoring the processed electrical response (PER) during muscle stimulation.

Here, based on the hypothesis that PER has a linear relationship with muscular force, at least in a quasistatic condition, PER can be used as a feedback signal, in addition to position and velocity control in generating a predetermined basic movement. This is a critical step in the control of movement by functional electrical stimulation.

In the performed tests, the lower arm is moved in the horizontal plane so that moment of inertia will not change with motion. The only muscle used here is the M. Biceps. This means that it is possible to control a slow movement of the reloaded elbow joint with a single muscle in this special case. This is not the general way of controlling the movement of human joints.

However, it was adequate for illustrating the feasibility of the proposed control system. The computer model of the upper extremity is

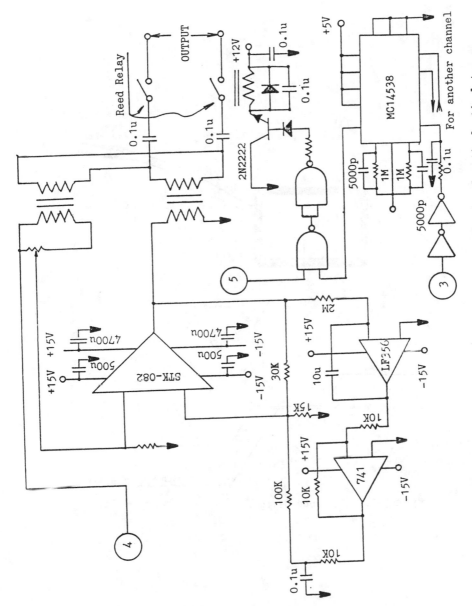

Figure G-4. Circuit diagram of the output stage amplifier of the electrical stimulator.

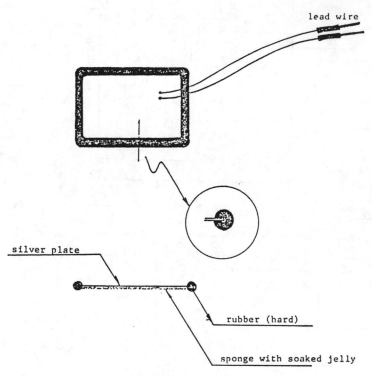

lead wire

silver plate

rubber (hard)

sponge with soaked jelly

Figure G-5. Electrode design.

used to calculate the muscular force and determine the superficial muscles needed for a particular movement. For the considered arm positions and external load, the anatomical angle data are inputed together with the length of the lower arms, the starting and ending point of each muscle, and the external load at the elbow joint. In the special movement cnsidere in the test, it was found that the Biceps Brachii is the main active muscle throughout the slow movement, and that the calculated muscular force has a simple linear relationship to load at each angular position.

FEEDBACK CONTROL

In order to induce a prescribed involuntary slow movement of the loaded forearm, a closed loop feedback control system was developed. The system incorporated the following feedback signals:

1. The processed electrical response of the muscle (PER).
2. The angular movement (position) of the elbow joint.
3. The angular velocity of the elbow joint.

Details of the system can be found in Ref. [10].

EXPERIMENTAL PROCEDURE

The experimental procedure can be summarized as follows:

1. Input Data of (a) positioning the joints, (b) external load, (c) anatomical parameters, (d) selecting only the superficial muscles, and (e) range of motion.
2. Calculate the number of the superficial muscles needed for a particular action as well as the force pattern needed for each muscle.
3. Average the PER-stimulating voltage curve in isometric condition.
4. Generate the stimulating pattern using microcomputer.
5. Start the closed loop control experiment.
6. Monitor PER to observe the actual muscular activities.
7. Record the elbow movement and compare to the prescribed movements.

EXPERIMENTAL RESULTS

Figure 6 shows a typical result with closed loop control using PER, elbow angular movement and angular velocity for feedback. It can be seen that the prescribed movement (curve B) can be induced involuntarily with very good accuracy.

CONCLUSION

It appears from this study that the processed electrical response feedback is the most important element for control of slow movement by electrical stimulation. Control by feedback of angular movement and velocity information alone does not appear to give satisfactory patterns of movement and is not repeatable. When all three states were used for the feedback, the most important control gain was that for the processed electrical response signal (4.27). The other gains for angle and angular velocity were relatively low (0.71 and 0.66, respectively). These gains confirm that the primary control mechanism is the processed electrical

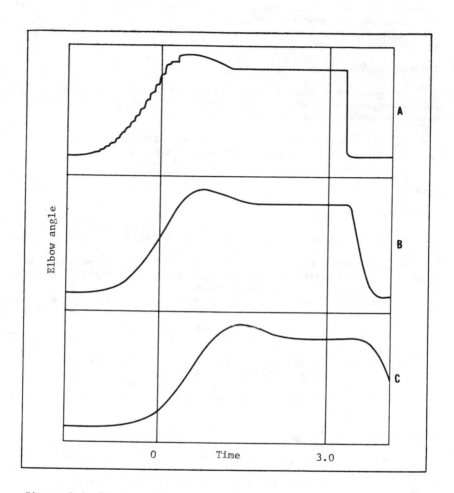

Figure G-6. The motion induced by the system faithfully matches the request. An extended forearm is flexed 90° while pulling a load; the action takes place 2 cm a second. The top figure (a) is the computer-generated stimulated pattern, (b) is a chart of desired movement. As shown in (c) —the resulting involunatry movement with PER and feedback control—the arm motion closely matches the requested pattern.

response feedback. It can therefore be concluded that slow, involuntary movement may be achieved with only appropriately processed electrical response feedback. Angle and angular velocity feedback can be used to insure that no major deviation from the intended movement occurs.

In this feasibility study a situation is chosen where a single muscle controls the movement of the elbow joint. Though this proved to be possible, it would be preferable to use at least two muscles to control a movement having a single degree-of-freedom, such as the elbow joint, in order to use muscular force more effectively. For practical clinical usage, it would be necessary to utilize a multichannel, computer-controlled system for inducing predetermined functional electrical stimulation. Using the musculoskeletal model for calculating the muscle forces and the processed electrical response as a feedback signal for the control of the stimulation shows excellent promise for repeatedly inducing predetermined involuntary movements and forces by functional electrical stimulation.

REFERENCES

1. Arvikar, R. and Seireg, A., "Evaluation of Upper Extremity Joint Forces During Exercise," Advances in Bioengineering, ASME, pp. 71-73, 1980.

2. Bigland-Ritchie, B. and Jones, D. A., et al., "Excitation Frequency and Muscle Fatigue: Electrical Responses During Human Voluntary and Stimulated Contractions," Arch. Phys. Med. Rehabil., Vol. 64, pp. 414-427, 1979.

3. Gracinin, F. and Trokoczy, A, "Optimal Stimulus Parameters for Minimum Pain in the Chronic Stimulation of Innervated Muscle," Arch. Phys. Med. Rehabil., Vol. 56, pp. 243-249, 1975.

4. Liberson, W. T., Holmquest, H. J., Scot, D., and Dow, M., Arch. Phys. Med. Rehabil., 42, pp. 101, 105, 1961.

5. Moreno-Aranda, J. L., and Seireg, A., "Electrical Parameters for Over-the-Skin Stimulation," J. Biomech., Vol. 14(9), pp. 579-585, 1981.

6. Moreno-Aranda, J. L., and Seireg, A., "Investigation of Over-the-Skin Electrical Stimulation Parameters for Different Normal Muscle," J. Biomech., Vol. 14(9), pp. 587-593, 1981.

7. Reswick, R. B., "Functional Electrical Stimulation--Neural Prosthesis for the Disabled," BPR, pp. 1-4, 1976.

8. Trokoczy, A., and Bajd, T., et al., "A Dynamic Model of the Ankle Joint Under Functional Electrical Stimulation in Free Movement and Isometric Conditions," J. Biomech., Vol. 9, pp. 509-519, 1976.

9. Vodovnik, L., and Bajd, T., et al., "Functional Electrical Stimulation for Control of Locomotor Systems," CRC Critical Reviews in Bioengineering, pp. 63-131, 1981.

10. Yamamoto, T., "Control of Movement of a Skeletal Joint by Functional Electrical Stimulation via Surface Electrodes," PhD Thesis, University of Wisconsin-Madison, 1985.

11. Yamamoto, T., and Seireg, A., "Closing the Loop: Electrical Muscle Stimulation and Feedback Control for Smooth Limb Motion," SOME, Engr for the Human Body, Vol. 1, No. 3., Oct. 1986.

Index

Femur (*Cont.*):
  (*See also* Leg)
FES (*see* Functional electrical stimulation)
Fibula, 453
Fingers:
  bones of, 399–400
  joint rotations of, 13
  mechanical heads, 530
  pinch action and, 433–445
  pulling with, 432, 438
Flexor muscles:
  of arm, 102, 110, 112, 222, 263
  of foot, 451, 461–471
  of hand, 213, 400–416
  of leg, 198, 208–211
  quasi-static walking and, 202–203
  stooping posture and, 285
Foot, 447–453
  arch support of, 14, 450
  average density of, 66
  bicycling and, 749
  body-fixed axis for, 476
  bones of, 448, 453
  cane support and, 606–614
  center of gravity for, 147, 283, 504
  computer code for, 699–716
  coordinates of, 474
  equilibrium equation for, 151, 162, 489
  extrinsic muscles of, 455–464
  force model of, 148
  geometric model of, 87–88
  intrinsic muscles, 464–472
  joints of, 449, 453–455, 483
  leaning model and, 163–173, 491–501
  ligaments of, 472–475
  modeling of, 131
  moments of inertia of, 62–63, 92
  pes cavus, 508–524
  pressures on, 9
  relative weight of, 52
  static moments of, 73
  stooping posture and, 284
  vertical force on, 78
  walking model for, 501–515
  (*See also* Leg; Lower extremities)
Footwear, 598
Force models, 2, 148
Force platform, 9–10
Forearm:
  average density of, 66
  kinematics of, 109, 121
  moments of inertia of, 62–63, 92
  (*See also* Arm)
Functional electrical stimulation (FES), 799–807

Gait (*see* Walking)
Gastrocnemius muscle, 136, 140, 146, 198, 264, 741
Geometric models, 79–80
Gluteus muscles, 131, 198, 264
  coordinates of, 146
  model for, 134
  squatting position and, 192
  stooping posture and, 286
  walking cycle and, 208–211
Gracilis muscle, 129–131, 197, 264
  coordinates of, 146
  walking cycle and, 208–211
Graphics, computer, 717, 720, 768–786
Grasping, 530
  (*See also* Hand)
Gravity vector, 727
  (*See also* Center of gravity)
Ground-fixed axis, 116
Gyration radius, 552

Hallus valgus, 516
Hamstrings, 208–211, 606
Hanavan model, 80–93
Hand, 397
  arm comprehensive models and, 121
  average density of, 66
  bones of, 398–399
  carrying load in, 212–225
  center of gravity of, 283
  computer code for, 665–683
  coordinates of, 419–425
  equilibrium equations for, 428
  joints of, 399
  model of, 85
  moments of inertia of, 62–63, 92
  muscles of, 399–433
  pinch action and, 433–435, 531
  pulling with, 432, 438
  relative weight of, 52
  robotics and, 530–534
  stooping posture and, 284
  typing and, 620
  (*See also* Arm; Upper extremities)
Handicapped, assistive devices for, 534
Handi-Man, 532
Hazardous environments, 534
Head:
  average density of, 66
  geometric model of, 83–84
  moments of inertia of, 62–63, 92
  muscles of, 241–247, 262
  relative weight of, 52

813

Neck:
    average density of, 66
    muscles of, 241–247, 262
Newtonian mechanics, 23–46

One-legged stance, 13
Opponens pollicis, 417
Orthotic devices, 799
Osteotomy, 635–647
Overbite, 377

Pain, upper back, 620
Parallel–axis theorem, 552
Paraplegics, 534
Paticle dynamics, 29–30, 34, 41
Pathology:
    of foot, 508–524, 799
    musculoskeletal model and, 587–650
    (See also specific disorders)
Pectoralis muscles, 256, 263, 274, 621,
    624
Pelvis:
    bicycling and, 786
    center of gravity of, 147, 283
    equilibrium equations, 143, 162
    femoral anteversion and, 603
    free body diagram of, 281
    gait analysis and, 9
    leaning position and, 155
    lumbar curve displacement and, 355
    modeling of, 129, 131
    muscles of, 131, 477–482
    stooping posture and, 284
    vertebral column and, 232
Pendulum method, 52–53, 57, 71
Percussion, 2
Peronei muscles, 139, 198, 264
    coordinates of, 146
    foot and, 461, 464
    model of, 141
    stooping posture and, 286
    walking cycle and, 208–211
Pes cavus, 508–524
Phalanges:
    of foot, 447, 453
    of hand, 399
Photographic analysis:
    body mass centers, 11
    volumetric analysis and, 59
Physical handicaps, assistive devices for,
    534
Picture generation, 720, 771, 774–776
Pinch posture, 427, 433–435, 531

Pin joint, 38–39
Plantar ligaments, 472
Polar moments, 551
Polygraph measurements, 75
Post-operative evaluation, 647
Posture, 3
    categories of analysis, 8
    center of gravity for, 77
    control levels for, 15–16, 99
    literature review, 8–15
    moment of inertia for, 77
    oscillation of, 73
    (See also specific postures)
Power, amplification of, 532
Pressure distribution, on foot, 491–517
Pronation, 109
Prosthetic devices, 532, 534
Psoas muscles, 251–253, 262
Pterygoid muscle, 368
Pulling, hand model and, 220, 432

Quadraplegics, 532
Quadratus muscles, 251, 262, 272, 285,
    355, 470
Quadriceps muscles, 131, 136, 198, 208–
    211
Quasi-static walking, 192, 200–206
Quick-release method, 66–71

Radius bone, 100, 114, 399–419
Rectus femoris muscle, 195, 198
Reflexes, 1
Remote manipulators, 534
Respiration, 57, 73
Rhomboideus muscle, 254–255, 263, 271,
    285
Rigid bodies, 30, 36–45
    coordinate transformation matrices and,
        541
    foot as, 453
    moments of inertia and, 551
    spinal column and, 231
Robotics:
    hazardous environments and, 534, 536
    human power amplification, 527
    mechanical hands, 527
    walking machines, 530–534
Rotatores muscles, 247, 262
    coordinates of, 269, 272
    lumbar curve and, 355
    stooping posture and, 285

Tibia:
    foot and, 453
    force model for, 148
    knee motion and, 129
    osteotomy of, 635–647
Tibialis muscle, 131, 198, 264
    coordinates of, 146
    foot and, 455–460
    leaning positions and, 168
    model for, 134–135
    stooping posture and, 286
    walking cycle and, 208–211
Toes, 453
Torso, model of, 83
Tractors, 527
Transducers, 492, 494, 501
Trapezius muscle, 251–254, 263
    coordinates of, 271, 274–275
    model for, 625
    stooping posture and, 285
Triceps brachii muscle, 105–110, 275, 285
Trunk:
    average density of, 66
    moments of inertia of, 62–63, 92
    relative weight of, 52
    (See also Upper extremities)
Typing, 614, 620, 626

Ulna, 100
    center of gravity of, 283
    coordinates of, 110, 114
    hand muscles, and 401–419
    stooping posture and, 284
Underwater environments, 11, 534, 536
Upper extremities:
    computer code for, 665–683
    hand muscle forces and, 440
    muscles of, 263
    typing and, 614, 620, 626
    (See also specific body segments)

Valgus deformities, 587–598, 635, 638
Varus, 587–598, 635, 638
Vastus muscles, 198
Vertebrae, 232–240
    acceleration disc loads, 319–335

coordinates of, 261
disc fusion, 354
free body diagram of, 278–279
joint reactions, 354
(See also Spinal column)
Vibration analysis, 699
da Vinci, Leonardo, 2–4
    on foot, 447
    on hands, 397
    on muscle models, 3, 99
    on posture, 3
    on spine, 231
    on teeth, 367
Vitruvius, 48
Volumetric measurements, 48, 59, 61, 65
Volvar interosseous, 419

Walking:
    analytical categories of, 8
    cane support and, 606–614, 631–635
    computer code for, 660
    control levels of, 15–16
    femoral anteversion and, 603
    foot model for, 491, 501–515
    footwear and, 598
    hip reaction forces, 14
    historical analysis of, 9
    joint deformities and, 587–598
    quasi-static walking, 192, 200–206
    robotics and, 527–528
    tibial osteotomy and, 635–647
    (See also Leg; specific muscles)
Walking dragline, 527
Weber brothers, 4–5, 9
Weight lifting:
    computer code for, 697
    stooping posture and, 284–289, 341,
        347–349
    (See also Load sharing)
Work, Newtonian mechanics and, 35
Wrapping, muscular, 300–304, 489
Wrist, 114
    bones of, 399
    horizontal pull on, 220
    stooping and, 309
    vertical force at, 213
    (See also Hand)